HANDBOOK OF PLASTICS TESTING AND FAILURE ANALYSIS

THE WILEY BICENTENNIAL—KNOWLEDGE FOR GENERATIONS

$\mathcal{E}$ach generation has its unique needs and aspirations. When Charles Wiley first opened his small printing shop in lower Manhattan in 1807, it was a generation of boundless potential searching for an identity. And we were there, helping to define a new American literary tradition. Over half a century later, in the midst of the Second Industrial Revolution, it was a generation focused on building the future. Once again, we were there, supplying the critical scientific, technical, and engineering knowledge that helped frame the world. Throughout the 20th Century, and into the new millennium, nations began to reach out beyond their own borders and a new international community was born. Wiley was there, expanding its operations around the world to enable a global exchange of ideas, opinions, and know-how.

For 200 years, Wiley has been an integral part of each generation's journey, enabling the flow of information and understanding necessary to meet their needs and fulfill their aspirations. Today, bold new technologies are changing the way we live and learn. Wiley will be there, providing you the must-have knowledge you need to imagine new worlds, new possibilities, and new opportunities.

Generations come and go, but you can always count on Wiley to provide you the knowledge you need, when and where you need it!

WILLIAM J. PESCE
PRESIDENT AND CHIEF EXECUTIVE OFFICER

PETER BOOTH WILEY
CHAIRMAN OF THE BOARD

HANDBOOK OF PLASTICS TESTING AND FAILURE ANALYSIS

THIRD EDITION

VISHU SHAH
Consultek
Brea, California

BICENTENNIAL
1807
WILEY
2007
BICENTENNIAL

WILEY-INTERSCIENCE
A JOHN WILEY & SONS, INC., PUBLICATION

Published by John Wiley & Sons, Inc., Hoboken, New Jersey
Published simultaneously in Canada

For general information on our other products and services or for technical support, please contact our Customer Care Department within the United States at (800) 762-2974, outside the United States at (317) 572-3993 or fax (317) 572-4002.

Wiley also publishes its books in a variety of electronic formats. Some content that appears in print may not be available in electronic formats. For more information about Wiley products, visit our web site at www.wiley.com.

Wiley Bicentennial Logo: Richard J. Pacifico

Library of Congress Cataloging-in-Publication Data:
Shah, Vishu, 1951–
    Handbook of plastics testing and failure analysis / Vishu Shah.—3rd ed.
        p. cm.
    Rev. ed. of: Handbook of plastics testing technology. c1998.
    Includes bibliographical references and index.
    ISBN 978-0-471-67189-3
    1. Plastics—Testing—Handbooks, manuals, etc.   I. Shah, Vishu, 1951– Handbook of plastics testing technology.   II. Title.
    TA455.P5S457 2007
    620.1'9230287—dc22
                                                                        2006046129

Printed in the United States of America

10 9 8 7 6 5 4 3

CONTENTS

FOREWORD

The Society of Plastics Engineers is pleased to sponsor and endorse the third edition of *Handbook of Plastics Testing and Failure Analysis* by Vishu Shah. This volume offers a varied compilation of the most common tests used in the plastics industry. The text is an excellent source and reference guide for engineers, chemists and students in plastics testing technology. The author's writing style and knowledge of the subject matter have resulted in an enjoyable presentation allowing the reader to gain meaningful insights into testing and corresponding procedures.

SPE, through its Technical Volumes Committee, has long sponsored books on various aspects of plastics. Its involvement has ranged from identification of needed volumes and recruitment of authors to peer review and approval and publication of new books.

Technical competence pervades all SPE activities, not only in the publication of books, but also in other areas such as sponsorship of technical conferences and educational programs. In addition, the Society publishes periodicals including *Plastics Engineering, Polymer Engineering and Science, The Journal of Injection Molding Technology, Journal of Vinyl & Additive Technology,* and *Polymer Composites,* as well as conference proceedings and other publications, all of which are subject to rigorous techincal review procedures.

The resource of some 36,000 practicing plastics engineers, scientists, and technologists has made SPE the largest organization of its type worldwide. Further information is available from the Society of Plastics Engineers, 14 Fairfield Drive, Brookfield, Connecticut 06804.

PREFACE TO THE 3rd EDITION

During the period that elapsed between the second edition and the date of writing of this third edition, the author shifted his career from a full-time molder to a full-time Consultant/Expert witness/Educator. The opportunity existed for analyzing numerous plastic-part-related premature failures. Failure analysis and testing go together. In order to analyze the failure, it is often necessary to conduct tests. In this third edition, therefore, the decision was made to expand the current chapter on failure analysis substantially and alter the title of the book to *Handbook of Testing and Failure Analysis* to reflect the change appropriately.

Existing books and literature on the subject of failure analysis are too complex, too detailed, sometimes difficult to understand, and more suitable to the persons well-versed in polymer chemistry and physics. This book attempts to simplify a rather difficult subject of failure analysis by focusing on four major types of failures and key reasons behind failure of plastic parts. A step-by-step procedure starting from very basic and simple visual analysis to highly advanced analytical tests is presented. A simple flow chart is included to help with the investigation. To assist with the understanding of the subject matter, several actual case studies are included. This simple approach to analyzing failures is not intended primarily for a specialist but for those who wish to acquire basic knowledge and understanding of the failure mechanism. The author's aim is not to replace excellent books that are in existence on this subject but to supplement and pave the road for more detailed and sophisticated failure analysis techniques in existence today.

All other chapters in the book have been updated with the latest information, diagrams, and photographs of the test equipment. The Appendix section has been updated. Appendix I, which listed the properties of the most common plastics, elastomers, and rubbers in the early edition, has been replaced with information about four major electronic databases for plastic materials. In order to increase the versatility of the book, numerous color photographs depicting photoelastic analysis and color theory along with various animations have been added on to the compact disk that is included with the book. More importantly, a virtual tour of a prominent Plastics Testing Laboratory is included to give the reader an opportunity to visit the laboratory from his or her desk and learn and understand how the tests are conducted and data are collected.

The author wishes to express thanks to all the users of previous editions for their constructive comments and helpful suggestions for changes and improvements for the next edition and to those who helped with this revision. In particular, he wishes to thank Jim Beauregard and Jim Galipeau of Plastics Testing Laboratory for their invaluable contribution, agreeing to the novel concept of virtual laboratory tour and making it possible. The author also wishes to thank Gerard Nelson of Ceast USA, GE Plastics, Kishor Mehta of Plascon Associates, Bayer Corporation, Dr. Alex Redner of Strainoptics, Inc., Paul Gramann of The Madison Group, Jim Rancourt of Polymer Solutions, and Steve Ferry of Micobac Laboratories. Many thanks to Steve Tuszynski of Algoryx for his contribution and to all other companies for providing numerous illustrations and diagrams. The book by Myer Ezrin, *Plastics Failure Guide, Causes and Prevention*, has been an important and valuable source of theoretical and practical information, and the author highly recommends his book for more detailed and in-depth discussion of the subject.

Once again, I would like to thank my family for their encouragement and constant support.

VISHU SHAH

Consultek, LLC,
Brea, CA

PREFACE TO THE SECOND EDITION

Since the publication of the first edition, little has changed as far as the basic concepts and methods of plastics testing. What has changed is the manner in which the data is collected and analyzed. Since the advent of computers and digital instruments, data collection and subsequent analysis and interpretation have become much simpler and faster.

This revised edition attempts to update the book in line with the latest developments in the field of testing, data acquisition, and analysis. The photographs depicting the commercially available testing equipment have been replaced with newer versions. A new chapter covering uniform global testing standards has been added. This chapter also includes current information about computerized material selection, which allows the user to compare various test data and material ranking based on the test data with utmost speed and ease. the entire section on impact properties has been rewritten to include an expanded discussion of instrumented impact testing. The chapters on electrical weathering properties and material characteristics have been revised. Owing to significant changes and developments in flammability testing this chapter has also been updated. The chapter on failure analysis is expanded significantly to further satisfy the need of someone trying to determine a failure mechanism. The discussion on SQC/SPC in the chapter on quality control has been expanded along with the current trend toward "supplier certification." The chapter on nondestructive testing has also been rewritten to include many other NDT techniques and the latest developments. The Appendix has been expanded to include plastics education degree programs and organizations. The list of test equipment suppliers has been updated and now includes appropriate web site addresses. The specification section includes ISO test method designations and ASTM/DOD cross references.

The author wishes to thank all those who helped to make this second edition possible for their constructive and candid comments, support, and guidance. In particular he wishes to thank professor Steven Driscoll (University of Massachusetts Lowell) and Professor Robert Speirs (Ferris State University) for their suggestions and guidance. Special thanks to R. Bruce Cassel of Perkin-Elmer and Kurt Scott of Atlas Electric Devices Company for reviewing and improving the manuscript and to Steve Caldarola of SGS–U.S. Testing who assisted in updating the chapter on flammability. I wish to thank Peter Grady of Ceast U.S.A., John Dechristofaro of Dynisco–Kayeness, Brookfield Engineering, G.E. Plastics, Society of Plastics Industry, Instron Corporation, Underwriters Laboratories, Strain Optics, D.A.T.A. Publishing, Newport Scientific and many others for providing technical assistance and photographs for reproduction. Many thanks to James Galipeau and Mr. James Beauregard of Plastics Testing Laboratories for providing material on advances in plastics testing and for reviewing the entire manuscript.

Last, but not least, I thank my wife Charlene, my son Neerav, and my daughter Beejal for their understanding and patience during many evenings and weekends when I was wrapped up in the preparation of this revised edition.

VISHU SHAH

Performance Engineered Products, Inc.
Pomona, California

PREFACE TO THE FIRST EDITION

The desire to compile this book was initiated mainly because of the virtual non-existence of a comprehensive work on testing of plastic materials. The majority of the literature concerning the testing of plastics is scattered in the form of sales and technical brochures, private organizations' internal test procedures, or a very brief and oversimplified explanation of the test procedures in plastics literature. The main objective of the present book is to provide a general purpose practical text on the subject with the main emphasis on the significance of the test or *why* and not so much on *how* without being extremely technical.

Over the years ASTM (American Society for Testing and Materials) has done an excellent job in providing the industry with standard testing procedures. However, the test procedures discussed in ASTM books lack the theoretical aspects of testing. The full emphasis is not on *significance* of testing but on *procedures* of testing. The ASTM books are also deficient in showing the diagrams and photographs of actual, commercial testing equipment. In this book I have tried to bridge the gap between the oversimplified and less explained tests described in ASTM books and the highly technical and less practical books in existence today.

This handbook is not intended primarily for specialists and experts in the area of plastics testing but for the neophyte desiring to acquire a basic knowledge of the testing of plastics. It is for this reason that detailed discussions and excessive technical jargon have been avoided. The text is aimed at anyone involved in manufacturing, testing, studying, or developing plastics. It is my intention to appeal to a broad segment of people involved in the plastic industry.

In Chapter 1 the basic concepts of testing are discussed along with the purpose of specifications and standards. Also discussed is the basic specification format and classification system. The subsequent chapters deal with the testing of five basic properties: mechanical, thermal, electrical, weathering, and optical properties of plastics. The chapter on mechanical properties discusses in detail the basic stress–strain behavior of the plastic materials so that a clear understanding of testing procedures is obtained. Chapter 7 on material characterization is intended to present a general overview of the latest in characterization techniques in existence today. A brief explanation of the polymer combustion process along with various testing procedures are discussed in Chapter 8. An attempt is made to briefly explain the importance of conditioning procedures. A table summarizing the most common conditioning procedures should be valuable. Several tests that are difficult to incorporate into a specific category were placed in the chapter on miscellaneous tests. End-product testing, an area generally neglected by the majority of processors of plastic products, is discussed along with some useful suggestions on common end-product tests.

Chapter 13 on identification analysis should be important to everyone involved in plastics and particularly useful to plastic converters and reprocessors. The flowchart summarizes the entire identification technique. Since there are so many different tests in existence on the testing of foam plastics, only a brief explanation of each test is given. The chapter on failure analysis is a compilation of methods commonly used by material suppliers. A step-by-step procedure for analyzing product failure should prove valuable to anyone ivolved in failure analysis. Quality control, although not part of the testing, is included in order to explain quality control as it relates to plastics. The section on visual standard, mold control, and workmanship standard is a good example. In this increasing world of product liability, the chapter on product liability and testing should be of value to everyone.

In order to increase the versatility of this book and meet the goal of providing a ready reference on the subject of testing, a large appendix section is given. One will find very useful data: names and addresses of equipment manufacturers, a glossary, names and addresses of trade publications, information on independent testing laboratories, and a guide to plastics specifications. Many useful charts and tables are included in the appendix. Throughout the book, wherever possible, numerous diagrams, sketches, and actual photographs of equipment are given.

A handbook of this magnitude must make inevitable compromises. Depending on the need of the individual user, there is bound to be a varying degree of excess and shortage. In spite of every effort made to minimize mistakes and other short-comings in this book, some may still exist. For the sake of future refinement and improvements, all constructive comments will be welcomed and greatly appreciated.

VISHU H. SHAH

Pomona, California
October 1983

1

BASIC CONCEPTS AND ADVANCEMENT IN TESTING TECHNOLOGY

1.1. BASIC CONCEPTS

Not too long ago, the concept of testing was merely an afterthought of the procurement process. But now, with the advent of science and technology, the concept of testing is an integral part of research and development, product design, and manufacturing. The question that is often asked is "why test?" The answer is simple. Times have changed. The manner in which we do things today is different. The emphasis is on automation, high production, and cost reduction. There is a growing demand for intricately shaped, high-tolerance parts. Consumer awareness, a subject totally ignored by the manufactures once upon a time, is now a major area of concern. Along with these requirements, our priorities have also changed. When designing a machine or a product, the first priority in most cases is safety and health. Manufacturers and suppliers are now required to meet a variety of standards and specifications. Obviously, relying merely on past experience and quality of workmanship is simply not enough. The following are some of the major reasons for testing:

1. To prove design concepts
2. To provide a basis for reliability
3. Safety
4. Protection against product liability suits
5. Quality control
6. To meet standards and specifications

Handbook of Plastics Testing and Failure Analysis, Third Edition, by Vishu Shah
Copyright © 2007 by John Wiley & Sons, Inc.

7. To verify the manufacturing process
8. To evaluate competitors' products
9. To establish a history for new materials

In the last two decades, just about every manufacturer has turned to plastics to achieve cost reduction, automation, and high yield. The lack of history along with the explosive growth and diversity of polymeric materials has forced the plastics industry into placing extra emphasis on testing and on developing a wide variety of testing procedures. Through the painstaking efforts of various standards organizations, material suppliers, and mainly the numerous committees of the American Society for Testing and Materials (ASTM), over 10,000 different test methods have been developed.

The need to develop standard test methods specifically designed for plastic materials originated for two main reasons. Initially, the properties of plastic materials were determined by duplicating the test methods developed for testing metals and other similar materials. The Izod impact test, for example, was derived from the manual for testing metals. Because of the drastically different nature of plastic materials, the test methods often had to be modified. As a result, a large number of nonstandard tests were written by various parties. As many as eight to ten distinct and separate test methods were written to determine the same property. Such practice created total chaos among developers of the raw materials, suppliers, design engineers, and ultimate end-users. It became increasingly difficult to keep up with various test methods or to comprehend the real meaning of reported test values. The standardization of test methods acceptable to everyone solved the problem of communication between developers, designers, and end-users, allowing them to speak a common language when comparing the test data and results.

In spite of the standardization of various test methods, we still face the problem of comprehension and interpretation of test data by an average person in the plastics industry. This is due to the complex nature of the test procedures and the number of tests and testing organizations. The key in overcoming this problem is to develop a thorough understanding of what the various tests mean and the significance of the result to the application being considered (1). Unfortunately, the plastics industry has placed more emphasis on *how* and not enough on *why*, which obviously is more important from the standpoint of comprehension of the test results and understanding the true meaning of the values. The lack of understanding of the real meaning of heat deflection temperature, which is often interpreted as the temperature at which a plastic material will sustain static or dynamic load for a long period, is one such classic example of misinterpretation. In the chapters to follow, we concentrate on the significance, interpretation, and limitations of physical property data and test procedures. Finally, a word of caution: it is extremely important to understand that the majority of physical property tests are subject to rather large errors. As a general rule, the error of testing should be considered ±5 percent. Some tests are more precise than others. Such testing errors occur from three major areas: (1) the basic test itself, (2) the operators conducting the tests, and (3) variations in the test specimens. While evaluating the test data and making decisions based on test data, one must consider the error factor to make certain that a valid difference in the test data exists (2).

1.2. SPECIFICATION AND STANDARDS

A specification is a detailed description of requirements, dimensions, materials, and so on. A standard is something established for use as a rule or a basis of comparison in measuring or judging capacity, quantity, content, extent, value, and quality.

A specification for a plastic material involves defining particular requirements in terms of density, tensile strength, thermal conductivity, and other related properties. The specification also relates standard test methods to be used to determine such properties. Thus, standard methods of test and evaluation commonly provide the bases of measurement required in the specification for needed or desired properties (3).

As discussed earlier, the ultimate purpose of a standard is to develop a common language, so that there can be no confusion or communication problems among developers, designers, fabricators, end-users, and other concerned parties. The benefits of standards are innumerable. Standardization has provided the industry with such benefits as improved efficiency, mass production, superior quality goods through uniformity, and new challenges. Standardization has opened the door to international trade, technical exchanges, and establishment of common markets. One can only imagine the confusion the industry would suffer without the specific definition of fundamental units of distance, mass, and time and without the standards of weights and measures fixed by the government (4).

Standards originate from a variety of sources. The majority of standards originate from industry. The industry standards are generally established by voluntary organizations that make every effort to see that the standards are freely adopted and represent a general agreement. Some of the most common voluntary standards organizations are the American Society for Testing and Materials, the National Sanitation Foundation, the Underwriters Laboratories, the National Electrical Manufacturers Association, and the Society of Automotive Engineers. Quite often, the industry standards do not provide adequate information or are not suitable for certain applications, in which case private companies are forced to develop their own standards. These company standards are generally adapted from modified industry standards.

The federal government is yet another major source of standardization activities. The standards and specifications related to plastics are developed by the U.S. Department of Defense and the General Services Administration under the common heading of Military Standards and Federal Standards, respectively.

After the World War II, there was a tremendous increase in international trade. The International Standards Organization (ISO) was established for the sole purpose of international standardization. ISO consists of the national standards bodies of over 90 countries from around the world. The standardization work of ISO is conducted by technical committees established by agreement of five or more countries. ISO's Technical Committee 61 on plastics is among the most productive of all ISO committees.

1.3. PURPOSE OF SPECIFICATIONS

There are many reasons for writing specifications, but the major reason is to help the purchasing department purchase equipment, materials, and products on an equal basis. The specifications, generally written by the engineering department, allow the purchasing agent to meet his requirements and ensure that the material received at different times is within the specified limits (5). The specifications is intended to ensure batch-to-batch uniformity, as well as remove confusion between the purchaser and supplier—we all know that more often than not what is provided by the supplier is not what is expected by the purchaser.

1.4. BASIC SPECIFICATION FORMAT

Many guidelines and directives have been formulated for writing specifications. The specifications for materials should include the following:

1. A descriptive title and designation
2. A brief but all-inclusive statement of scope
3. Applicable documents
4. A classification system
5. Definitions of related terms
6. Materials and manufacturing requirements
7. Physical (property) requirements
8. A sampling procedure
9. Specimen preparation and conditioning requirements
10. Reference to, or descriptions of, acceptable test methods for determining conformance
11. Inspection requirements
12. Instructions for retest and rejection
13. Packaging and marking requirements

1.4.1 Classification System

Since plastic materials are seldom supplied without the addition of certain additives and fillers, a classification system must be used to avoid confusion. For example, the specification for acetal materials covers three main types of acetal resins: homopolymer, copolymer, and terpolyerm. The resin types are subdivided into classes according to the grade descriptions. The group 1, class 1 represents general-purpose homopolymer acetal resin, and group 2, class 3 represents impact modified copolymer acetal resin. Table 1-1 lists the detailed requirements for acetal materials.

TABLE 1-1. POM Polyoxymethylene (Acetal) Materials, Detail Requirements,[h] Natural Color Only

Group	Description	Class	Description	Grade	Description[a]	Flow Rate, ISO 1133,[b] (g/10min)	Melting Point, ISO 3146/ Method C2,[c] (°C, min)	Density, ISO 1183 (g/cm³)	Tensile Strength, ISO 527,[d] (MPa, min)	Flexural Modulus, ISO 178,[e] (MPa, min)	Izod Impact Resistance, ISO 180/1A,[f] (KJ/m², min)	Deflection Temperature, ISO 75/ Method A,[g] 1.82 MPa (°C, min)
1	Homopolymer	1	General purpose and high flow	1		<8	170	1.39–1.44	65	2400	8.5	100
				2		8–16	170	1.39–1.44	65	2700	5.0	100
				3		16–28	170	1.39–1.44	65	2700	4.5	105
				4		28–55	170	1.39–1.44	65	2700	4.0	105
				0	Other							
		2	Fast cycle	1		11–18	170	1.39–1.44	65	2400	4.5	100
				2		19–33	170	1.39–1.44	65	2400	4.0	100
				0	Other							
		3	UV stabilized	1		<8	170	1.39–1.44	65	2400	8.5	100
				2		8–16	170	1.39–1.44	65	2700	5.0	100
				0	Other							
		4	Impact modified	1		<3	170	1.31–1.37	35	1100	70	70
				2		8–15	170	1.36–1.42	45	1900	9.0	80
				0	Other							
		5	Recycle—general purpose and high flow	1	(Same properties as Group 1 Class 1 Grade 1)							
				2	(Same properties as Group 1 Class 1 Grade 2)							
				3	(Same properties as Group 1 Class 1 Grade 3)							
				4	(Same properties as Group 1 Class 1 Grade 4)							
		6	Recycle—fast cycle	1	(Same properties as Group 1 Class 2 Grade 1)							
				2	(Same properties as Group 1 Class 2 Grade 2)							
		7	Recycle—UV-stabilized	1	(Same properties as Group 1 Class 3 Grade 1)							
				2	(Same properties as Group 1 Class 3 Grade 2)							
		8	Recycle—impact modified	1	(Same properties as Group 1 Class 4 Grade 1)							
				2	(Same properties as Group 1 Class 4 Grade 2)							
		0	Other	0								

TABLE 1-1. *Continued*

Group	Description	Class	Description	Grade	Description[a]	Flow Rate, ISO 1133,[b] (g/10min)	Melting Point, ISO 3146/ Method C2,[c] (°C, min)	Density, ISO 1183 (g/cm³)	Tensile Strength, ISO 527,[d] (MPa, min)	Flexural Modulus, ISO 178,[e] (MPa, min)	Izod Impact Resistance, ISO 180/1A,[f] (KJ/m², min)	Deflection Temperature, ISO 75/ Method A,[g] 1.82MPa (°C, min)
2	Copolymer	1	General purpose and high flow	1		<4	160	1.38–1.43	58	2300	5.5	90
				2		4–7	160	1.38–1.43	58	2300	4.5	90
				3		7–11	160	1.38–1.43	58	2300	4.0	95
				4		11–16	160	1.38–1.43	58	2300	4.0	95
				5		16–35	160	1.38–1.43	58	2300	3.5	95
				6		35–60	160	1.38–1.43	58	2300	3.0	95
				7		60+	160	1.38–1.43	58	2300	2.5	95
				0	Other							
		2	UV-stabilized	1		<4	160	1.38–1.43	58	2300	5.5	90
				2		4–7	160	1.38–1.43	58	2300	4.5	90
				3		7–11	160	1.38–1.43	58	2300	4.0	95
				4		11–16	160	1.38–1.43	58	2300	4.0	95
				5		16–35	160	1.38–1.43	58	2300	3.5	95
				6		35–60	160	1.38–1.43	58	2300	3.0	95
				7		60+	160	1.38–1.43	58	2300	2.5	95
				0	Other							
		3	Impact modified	1		11–26	155	1.34–1.40	40	1500	6.0	70
				2		11–26	155					
				3		4–11	155					
				4		5–11	155					
				5			155					
				0	Other							
		4	Flexural modified	1		<20	155	1.26–1.32	20	800	12.0	50
				0	Other							
		5	Recycle— general purpose and high flow	1	(Same properties as Group 2 Class 1 Grade 1)							
				2	(Same properties as Group 2 Class 1 Grade 2)							
				3	(Same properties as Group 2 Class 1 Grade 3)							
				4	(Same properties as Group 2 Class 1 Grade 4)							
				5	(Same properties as Group 2 Class 1 Grade 5)							

					MFR^b	T_m^c	Density	Tensile[d]	Flexural[e]	Izod[f]	Deflection[g]
		6	Recycle—UV-stabilized	6	(Same properties as Group 2 Class 1 Grade 6)						
				7	(Same properties as Group 2 Class 1 Grade 7)						
				1	(Same properties as Group 2 Class 2 Grade 1)						
				2	(Same properties as Group 2 Class 2 Grade 2)						
				3	(Same properties as Group 2 Class 2 Grade 3)						
				4	(Same properties as Group 2 Class 2 Grade 4)						
				5	(Same properties as Group 2 Class 2 Grade 5)						
				6	(Same properties as Group 2 Class 2 Grade 6)						
				7	(Same properties as Group 2 Class 2 Grade 7)						
		7	Recycle—impact-modified	1	(Same properties as Group 2 Class 3 Grade 1)						
				2	(Same properties as Group 2 Class 3 Grade 2)						
				3	(Same properties as Group 2 Class 3 Grade 3)						
				4	(Same properties as Group 2 Class 3 Grade 4)						
				5	(Same properties as Group 2 Class 3 Grade 5)						
		8	Recycle—flexural modified	1	(Same properties as Group 2 Class 4 Grade 1)						
				0	Other						
		0	Other	0	Other						
3	Terpolymer	1	High melt strength	1		<2	160	1.38–1.43	56	2250	3.5	90
				0	Other						
		2	Recycle—high melt strenght	1	(Same properties as Group 3 Class 1 Grade 1)						
				0	Other						
		0	Other	0	Other						
0	Other	0	Other	0	Other						

[a] No descriptions are listed unless needed to describe a special grade under the class. All other grades are listed by requirements.

[b] Flow rate: 190/2.16 (T/M).

[c] Melting point rate 10°C/min. Tm second melting curve.

[d] Tensile strength shall be determined using a Type I tensile specimen as described in ISO 527. Crosshead speed shall be 50mm/min ± 10%.

[e] Flexural modulus shall be determined by specimen 80 ± 2mm × 10 ± 0.2mm × 4 ± 0.2mm as described in ISO 178.

[f] Izod shall be determined by specimen 80 ± 2mm × 10 ± 0.2mm × 4 ± 0.2mm (Method A).

[g] Deflection temperature shall be determined by specimen 120 ± 2mm × 10 ± 0.2mm × 4 ± 0.2mm.

[h] Data on 4-mm test specimens are limited and the minimum values may be changed in a later revision after a statistical data base of sufficient size is generated.

1.4.2. Requirements

Physical requirements should include specific quantitative information according to the classification.

1.4.3. Sampling and Conditioning

In order to reproduce the test results time after time, sampling and conditioning procedures must be religiously followed. The procedures for sampling and conditioning must be clearly specified.

1.4.4. Test Methods

The specifications, without adequate information regarding test methods, are useless. The standard and generally accepted test methods must be specified. If no standard test methods exist, a detailed description and requirements of the test should be an integral part of the specification.

1.5. ADVANCEMENTS IN TESTING TECHNOLOGY

Over the past decade plastics testing has changed dramatically. Significant advances in materials and the increasingly demanding nature of plastics applications have combined with global demand for uniformity to produce a requirement for data that goes beyond basic material comparison. Data today are used for design purposes and complex models for prediction of a material's end-use compatibility. As a result, the testing technology that was adequate to provide data in the past has, in some cases, become obsolete and major advances in testing equipment sophistication have been developed to create a whole new environment in the plastics laboratory.

The traditional laboratory of the past generated material data using methods that were developed from metals standards and other industries. While adequate for comparing one material to another for basic similarities and differences the information that was generated did not reflect the variables that are specific to polymers—that is, the affect of temperature on properties and the important role that polymer structure plays on its overall performance in the end-use application. With the complexity of the applications for polymers increasing, the need for more sophisticated testing techniques has also increased. State-of-the-art polymer testing laboratories today utilize test equipment that is fully instrumented and capable of collecting data with higher accuracy that not only includes the basic properties but also the more complicated effect of external variables on those properties.

A good example of changes in testing sophistication is the universal tester. In the past, this relatively simple apparatus that generates tensile, compressive, and flexural data (among others) was used largely to generate room-temperature single-point data, such as tensile strength, in North American units of pounds per square inch (psi) by methods from the American Society for Testing and Materials (ASTM). Today, not only has the sophistication of the equipment advanced but it is being used to generate detailed technical data, such as nonambient stress–strain data for inclusion in computerized multi-point databases such as ISO 11403 Acquisition and Presentation of Comparable Multi-Point Data. Another use is the measurement of V-notch shear (losepescu shear) using strain gages that accurately measure strain for precise modulus determination on highly reinforced composites. Most universal testers today utilize computerized data acquisition with Windows-based software that performs nearly all required engineering calculations and provides multiple options for modulus curve fits and energy calculations. Load cells and extensometers (strain measurement device) are self-identifying and calibrating, crosshead speeds are digitally set by computer and real-time graphs of stress–strain curves are generated as evaluations occur. Other relatively new accessories that are used with the universal tester are biaxial extensometers for accurate measurement of axial and transverse strain for Poisson's ratio determinations and bondable strain gauges for accurate evaluation of stiff composites. In the past, grip separation and other analog measurement methods were the primary method of measuring strain; today, strain measurement is performed with strain gages and strain gauge extensometers along with laser extensometers for more accurate determinations.

It is important to note that as data collection accuracy has increased, so have the requirements for accuracy of test equipment and data collection methods set forth in test methods. For example ASTM D638 and ISO 527 (International Standards Organization), both tensile property methods, require the use of extensometers to measure strain for modulus determinations. Grip separation and other methods of measuring strain have been determined to be inadequate for the measurement accuracy required at the low strains necessary for elastic modulus calculations.

Another area that has seen significant advancement in technology is impact testing. Historically, impact testing was limited to Izod impact, which measures the energy remaining in a test specimen under a machine in notch (essentially a notch sensitivity test), and noninstrumented drop dart impact tests such as mean failure height, based on qualitative events (pass or fail). While those two methods are useful for comparison purposes, information relative to the impact event itself is very limited. Today, impact equipment such as multi-axial instrumented impact is available, which not only provides data such as force in newtons required to break a sample, but also complete load-time energy curves, which provide visual data on the impact event itself. Energy values can be chosen for any point on the load/time (or load/deflection) curve. The data along with the curves provide enough information to determine failure modes such as ductile or brittle failure and also initial fracture points.

While the technology associated with single-point tests, such as those described above, has improved their usefulness for evaluation and comparison of polymers, single-point tests themselves are being replaced with more sophisticated multi-point tests that are associated with the unique problems related to designing with polymers. As mentioned earlier, most of the early single point tests for polymers were based on methods that were taken from the metals industry. These methods assumed that the materials would not change much with changes in temperature, which in the case of metals was accurate. For polymers, temperature plays a significant role in how they will perform when subjected to different environments of stress or exposure to chemicals or both. With polymers being used in more "high-end" applications, such as automotive applications, multi-point data is becoming more in demand as design engineers are required to simulate mechanical property responses of polymers when exposed to temperature. Mechanical property data generated with equipment such as dynamic mechanical analyzers (DMA) provide information on the response of mechanical properties (typically dynamic modulus) to temperature and frequency. Other multi-point tests commonly used are outlined in ISO 11403-1, Plastics-Acquisition and Presentation of Comparable Multi-Point Data, and include stress–strain curves at different temperatures, Isochronous stress–strain curves developed from tensile creep data, and charpy impact at multiple temperatures.

The illustrations described above are primarily for mechanical properties, but the increase in testing technology has occurred in almost all other aspects of material testing including thermal and analytical analysis. Examples include the measurement of coefficient of thermal expansion and Fourier transform IR analysis (FTIR).

Coefficient of linear thermal expansion (CLTE), which measures the change in length of a specimen with change in temperature, was previously measured with analog dilatometers and temperature baths that limited the measurement range from –30°C to 30°C according to ASTM D696. With the advent of composites and high temperature polymers the current use temperature can often vary from –40°C to 250°C. Today's technology employs a thermal mechanical analyzer (TMA). The TMA is controlled by a computer and the temperature versus expansion is precisely monitored over a much broader temperature ranges. ASTM D696 has been revised to reference ASTM E228 for higher temperatures, reflecting the increased temperature control required to measure CLTE at elevated conditions.

Fourier transform IR analysis, which measures a materials absorption and transmission of infra red light, has been advancement primarily in the way the spectra, which act as a materials fingerprint, are compared. In the past, the analytical chemist would utilize books of known references and compare scans of the material being analyzed to the references for identifications, often a very time-consuming task. Today, with computerized data acquisition and libraries consisting of thousand of references, the computer performs the peak matching, which makes identification less time-consuming, more consistent, and provides complex spectra subtractions for identification of contaminants.

In addition to technological advances, the testing industry has witnessed a major increase in the acceptance and use of globally recognized testing standards with new emphasis on characterizations performed by ISO and IEC (International Electrochemical Committee) test methods. These procedures are truly internationally agreed upon methods adopted under a one nation one vote system, and provide the common testing language appropriate to an industry dominated by global markets. CAMPUS (Computer-Aided Materials Pre-selection by Uniform Standards), which uses ISO and IEC data, provides material properties in a standardized format so that engineers, designers and plastics purchasers can now choose materials in an apples to apples comparison as never before.

The technical advances referred to in the preceding paragraphs are just a few examples of the many changes that have occurred in polymer testing in the past decade. In order to keep abreast of changes that occur, the polymer testing professional must continually update his/her knowledge of the industry, the information required by the industry, and the technology available to provide data which is both meaningful and accurate. Handbooks, such as the second edition of *The Handbook of Plastics Testing Technology*, provide a way to keep up with changes that are occurring. In addition, participation in standards organizations, such as ASTM and ISO provide first hand knowledge of changes that are taking place in the industry, as well as revisions that are being made to test methods. As applications for polymers become more and more advanced, the information that is generated by the polymer testing laboratory will have to keep pace through technological improvement.

1.6. NEW DEVELOPMENTS AND TRENDS IN TESING TECHNOLOGY

Automation, speed, and instrumentation are the key new developments in testing technology since the new millennium. Increasing number of test equipment manufacturers are engaging themselves in developing instruments that require less human intervention and operate at faster speeds. Automation is accomplished by the use robotics. This type of automated system has been proven to be cost effective for performing tensile, flexure, and impact tests on large quantities of samples while freeing up personnel for more demanding tasks. In combination with a six-axis industrial robot and a bar code scanner, a system can be configured to feed multiple testing machines with a single robot. A system developed by a leading manufacturer (6) uses state-of-the-art web technologies (web cam, e-mail, sms) ensuring a constant process control and remote diagnostic of the system. The data generated by the instrument are analyzed and seamlessly integrated to other office automation programs. Figure 1-1 shows one such highly automated test setup.

Figure 1-1. Automated test setup. (Courtesy of CEAST USA.)

Automation has also led to highly accurate and reproducible specimen preparation. A microprocessor-controlled notching instrument for impact tests can notch up to 50 specimens all in one operation. The notcher uses a linear cutting motion that reduces friction and induced stresses generated by traditional rotary notchers. Figure 1-2 shows an automated notcher.

More and more companies demanding fast trunaround of test results. Equipment manufacturers have met this challenge by developing faster instruments such as a twin-bore capillary rheometer as shown in Figure 1-3. Twin barrels allow the Bagley correction in one run, and they also double the output effectively combining two rheometers into one unit. This means that testing time is considerably reduced; more pressure–velocity increments can be applied to one change of material, and the test can run automatically while the operator performs other tasks. The use of the internet to deliver data to customers is increasing. The customer receives the data within minutes after the tests are completed. Such web-based material data management system (7) has the ability to securely store, display, and output diverse material properties for a variety of materials in complex formats, including CAE material model parameters. Material suppliers now have the ability to store and publish or selectively distribute their data instantly across the globe.

Use of instrumented testing to study impact properties of plastics materials is also on the rise, and industry has witnessed a significant expansion in the application of instrumented testing. Instrumented testers are capable of generating and providing much more meaningful data as compared to conventional instruments. Instrumented instruments are better suited for research and developments as well as simulating real-life conditions. Whereas the noninstrumented tests generally measure the energy necessary to break the specimen, instrumented impact tests provide curves of high-speed stress–strain data that distinguish ductile from brittle failure and crack-initiation from crack-propagation energy. The latter give a more nuanced picture of the "toughness" of a specimen (8). With instrumented impact, the falling dart's tip or the pendulum's hammer is fitted with a load cell. The

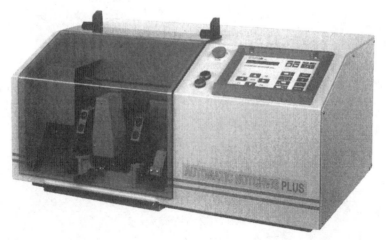

Figure 1-2. Automated notcher. (Courtesy of CEAST USA.)

Figure 1-3. Twin-bore capillary rheometer. (Courtesy of CEAST USA.)

force–time data during the actual impact are stored by a high-speed data-acquisition system. These data can be used to generate curves showing force, energy, velocity, and deformation versus time. By analyzing these curves, one can learn (a) the force, energy, and deformation necessary to initiate a crack and then to cause total failure, (b) the rate sensitivity of a material to impact loading, and (c) the temperature of a material's transition from ductile to brittle failure mode. The advent of piezoelectric sensors for instrumented impact testers is said to provide greatly increased sensitivity, allowing for testing of very light films, foams, and most other materials used in packaging. Developments in fracture mechanics analysis applied to polymers have now been incorporated into ISO 17281, so it is now possible to study intrinsic impact properties of plastic materials. The two properties, critical stress intensity factor (K1C) and the energy release rate (G1C), can be isolated as inherent fracture characteristics independent of the actual geometry and dimensions of the finished product; as such, they can be utilized in CAD tools to determine fracture performance. One such instrumented impact tester is shown in Figure 1-4.

While budget cutbacks and reduced staffing levels have affected nearly every segment of the plastics industry, large research centers operated by major manufacturers have been particularly under scrutiny. As a result, outsourcing of analytical and testing services to independent laboratories has been increasing. According to the latest estimate, nearly $10 billion in analytical services is currently being outsourced, and growth of more than 20 percent per year is

Figure 1-4. Instrumented impact tester. (Courtesy of CEAST USA.)

anticipated. Neolytica™ (9) is a first-of-its kind, single-point-of-contact outsourcing and contracting analytical testing service. This global, web-enabled company allows customers to better leverage capital, equipment, and personnel by managing multiple projects through multiple labs in its vast network of accredited laboratories. By using outside laboratory services, a company can reduce large investment in underutilized lab instrumentation, reduce expensive overhead, and keep in-house laboratory staff focused on higher-value development and problem-solving work.

The use of plastics in increasingly demanding applications has created a need for developing tests that simulate real-life applications. The traditional method of comparing properties of plastics and rating one material against another is no longer adequate, since the results do not reflect the behavior of the product in real-life situations. Laboratories are focusing on testing products under special circumstances and conditions. New regulations have spurred auto manufacturers and suppliers to test and characterize the energy-absorbing properties of the materials used in instrument panels and other components that the drivers and passengers may come in contact with in the event of an accident. High-rate impact testers simulate the dynamic nature of a car crash much better than traditional impact tests. To simulate the conditions found in the interior of an automobile, a outdoor weathering box has been developed by a leading material testing company. This under-glass exposure cabinet was primarily designed to accommodate nonstandard specimen sizes such as complete automotive assemblies as well as standard 10 ×

Figure 1-5. Outdoor weathering box. (Courtesy of Atlas Material Testing, Inc.)

15-mm samples. Various types of windshield or side window glass can be installed for evaluating the effects of different types of glass on automotive interior components. Color and gloss measurements and visual inspections are performed at intervals according to applicable test standards. Figure 1-5 illustrates one such box capable of housing a variety of components, including headphones, remote control devices, CD covers, and window tinting films.

A new type of test methodology for characterizing engineering plastics has been developed. These techniques simulate the extrusion and molding process to show what a polymer would undergo in terms of shear, temperature, pressure, and residence-time deformation. The behavior of a compound can be accurately predicted prior to processing. An online-type rheometer continuously measures the viscosity of the polymer from the die on a real-time basis, and the data is used to make screw speed adjustment to keep the viscosity consistent (10).

REFERENCES

1. Lamond, L., "Right Perspective on Test Data Matches Material to Application Criteria," *Plastics Design and Processing* (Mar 1976), pp. 6–10.

2. "A Guide to Standard Physical Test for Plastics," *Dupont Tech. Rept. No. TR 91.*

3. Schmitz, J. V. (Ed.), *Testing of Polymers*, Interscience, New York, 1965, p. 3.

4. *Ibid*, p. 4.

5. *Ibid*, p. 22.

6. Zwick USA, 1620 Cobb International Blvd., Kennesaw, GA 30152.

7. Datapoint Labs., 95 Brown Road., Ithaca, NY 14850.

8. Sherman, L. M., "Impact: Which Test to Use? Which Instrument to Buy?" Plastics Technology online, www.plasticstechnology.com/articles/200110fal.html.

9. Neolytica. 3606 W. Liberty Road., Ann Arbor, MI 48103.

10. Stewart, R., "Plastics Testing," *Plastics Engineering* (June 2003), pp. 20–28.

2

MECHANICAL PROPERTIES

2.1. INTRODUCTION

The mechanical properties, among all the properties of plastic materials, are often the most important properties because virtually all service conditions and the majority of end-use applications involve some degree of mechanical loading. Nevertheless, these properties are the least understood by most design engineers. The material selection for a variety of applications is quite often based on mechanical properties such as tensile strength, modulus, elongation, and impact strength. These values are normally derived from the technical literature provided by material suppliers. More often than not, too much emphasis is placed on comparing the published values of different types and grades of plastics and not enough on determining the true meaning of the mechanical properties and their relation to end-use requirements. In practical applications, plastics are seldom, if ever, subjected to a single, steady deformation without the presence of other adverse factors such as environment and temperature. Since the published values of the mechanical properties of plastics are generated from tests conducted in a laboratory under standard test conditions, the danger of selecting and specifying a material from these values is obvious. A thorough understanding of mechanical properties, tests employed to determine such properties, and the effect of adverse conditions on mechanical properties over a long period is extremely important.

The basic understanding of stress–strain behavior of plastic materials is of utmost importance to design engineers. One such typical stress–strain (load-deformation) diagram is illustrated in Figure 2-1. For a better understanding of the stress–strain curve, it is necessary to define a few basic terms that are associated with the stress–strain diagram.

Handbook of Plastics Testing and Failure Analysis, Third Edition, by Vishu Shah
Copyright © 2007 by John Wiley & Sons, Inc.

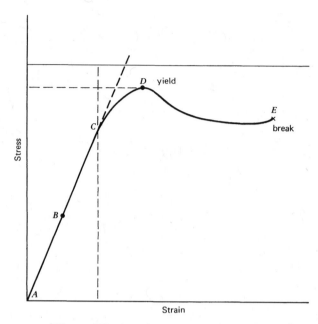

Figure 2-1. A typical stress–strain curve.

Stress. The force applied to produce deformation in a unit area of a test specimen. Stress is a ratio of applied load to the original cross-sectional area expressed in $lb/in.^2$.

Strain. The ratio of the elongation to the gauge length of the test specimen, or simply stated, change in length per unit of the original length ($\Delta l/l$). It is expressed as a dimensionless ratio.

Elongation. The increase in the length of a test specimen produced by a tensile load.

Yield Point. The first point on the stress–strain curve at which an increase in strain occurs without the increase in stress.

Yield Strength. The stress at which a material exhibits a specified limiting deviation from the proportionality of stress to strain. Unless otherwise specified, this stress will be at the yield point.

Proportional Limit. The greatest stress at which a material is capable of sustaining the applied load without any deviation from proportionality of stress to strain (Hooke's Law). This is expressed in $lb/in.^2$.

Modulus of Elasticity. The ratio of stress to corresponding strain below the proportional limit of a material. It is expressed in F/A, usually $lb/in.^2$ This is also known as Young's modulus. A modulus is a measure of material's stiffness.

Ultimate Strength. The maximum unit stress a material will withstand when subjected to an applied load in compression, tension, or shear. This is expressed in $lb/in.^2$.

Secant Modulus. The ratio of the total stress to corresponding strain at any specific point on the stress–strain curve. It is also expressed in F/A or $lb/in.^2$.

The stress–strain diagram illustrated in Figure 2-1 is typical of that obtained in tension for a constant rate of loading. However, the curves obtained from other loading conditions, such as compression or shear, are quite similar in appearance.

The initial portion of the stress–strain curve between points A and C is linear and it follows Hooke's law, which states that for an elastic material the stress is proportional to the strain. The point C at which the actual curve deviates from the straight line is called the proportional limit, meaning that only up to this point is stress proportional to strain. The behavior of plastic material below the proportional limit is elastic in nature and therefore the deformations are recoverable. The deformations up to point B in Figure 2-1 are relatively small and have been associated with the bending and stretching of the interatomic bonds between atoms of plastic molecules as shown in Figure 2-2a. This type of deformation is instantaneous and recoverable. There is no permanent displacement of the molecules relative to each other. The deformation that occurs beyond point C in Figure 2-1 is similar to a straightening out of a coiled portion of the molecular chains (Figure 2-2b). There is no intermolecular slippage and the deformations may be recoverable ultimately, but not instantaneously. The extensions that occur beyond the yield point or the elastic limit of the material are not recoverable (Figure 2-2c). There deformations occur because of the actual displacement of the molecules with respect to each other. The displaced molecules cannot slip back to their original positions and, therefore, a permanent deformation or set occurs. These three types of deformations, as shown in Figure 2-2, do not occur separately but are superimposed on each other. The bonding and the stretching of the interatomic bonds are almost instantaneous. However, the molecular uncoiling is relatively slow. Molecular slippage effects are the slowest of all three deformations (1).

These deformations can be further explained by using a mechanical model that duplicates the behavior of plastics under various external conditions. One such spring and dashpot mechanical model, known as the Maxwell model, is illustrated in Figure 2-3. The spring is perfectly elastic or Hookean and accounts for the

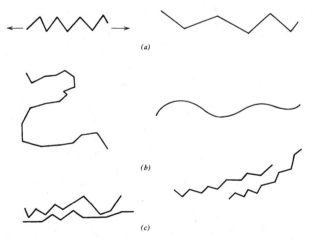

Figure 2-2. Extension types: (a) Bond bending, (b) uncoiling, (c) slippage.

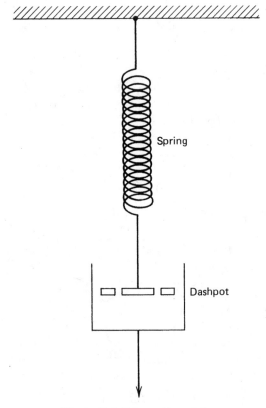

Figure 2-3. Maxwell model.

normal elastic behavior. The spring extensions are analogous to the deformations that occur because of the bending and stretching of interatomic bonds. If a non-linear spring is substituted for a linear one, the deformations are similar to the ones occurring because of uncoiling of portions of molecular chains. The dashpot is filled with a viscous fluid that must leak through holes in the plunger disc before the disc can move. Extensions in the dashpot are not recoverable and correspond to the permanent set. They represent the result of intermolecular slippage.

Many other interesting correlations can be made with the mechanical model, such as the effect of temperature on the mechanical properties of polymers. For example, at higher temperatures, the viscosity of the fluid in the dashpot decreases and the plunger movement is relatively smoother, resulting in greater extensions. Conversely, at lower temperatures, the liquid becomes more viscous and failures occur before appreciable extensions, similar to a brittle fracture. Other correlations involve models with different leakage rates and rate of loading (2). There are many other variations of mechanical models that are discussed in detail by Rodriguez (3), Baer (4), and Williams (5).

The polymeric materials can be broadly classified in terms of their relative softness, brittleness, hardness, and toughness. The tensile stress–strain diagrams serve as a basis for such a classification (6). The area under the stress–strain curve is considered as the toughness of the polymeric material. Figure 2-4*a*

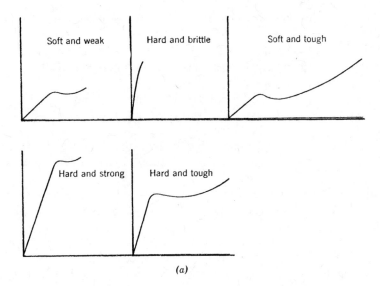

Soft and weak   Hard and brittle   Soft and tough

Hard and strong   Hard and tough

(a)

Toughness Relates to Area Under Curve

The ability of a thermoplastic composite to absorb energy is a function of strength and ductility, which tend to be inversely related. Total absorbable energy is proportional to the area within the lines drawn to the appropriate point on the "curve" from the axes. Material A is rubber-like and is just as tough (equal area) as material C which is metal-like. Most plastics fall between the extremes (material B).

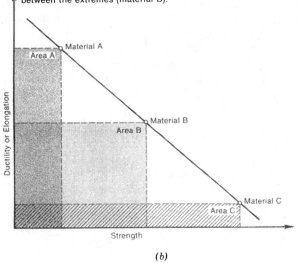

(b)

Figure 2-4. (a) Types of stress–strain curves. (Reprinted with permission of Wiley-Interscience.) (b) Relation between ductility and strength.

illustrates typical tensile stress–strain curves for several types of polymeric materials.

A soft and weak material is characterized by low modulus, low yield stress, and a moderate elongation at break point. Polytetrafluoroethylene (PTFE) is a good example of one such type of plastic material.

A soft but tough material shows low modulus and low yield stress, but very high elongation and high stress at break. Polyethylene is a classic example of these types of plastics.

A hard and brittle material is characterized by high modulus and low elongation. It may or may not yield before breaking. One such type of polymer is general purpose phenolic.

A hard and strong material has high modulus, high yield stress, usually high ultimate strength, and low elongation. Acetal is a good example of this class of materials.

A hard and tough material is characterized by high modulus, high yield stress, high elongation at break, and high ultimate strength. Polycarbonate is considered a hard and tough material. Figure 2-4*b* illustrates the relation between ductility and strength.

Table 2-1 lists the characteristic features of stress–strain curves as they relate to the polymer properties (7). In some applications it is important for a designer to know the stress–strain behavior of a particular plastic material in both tension and compression. At relatively lower strains, the tensile and compressive stress–strain curves are almost identical. Therefore, at low strain, compressive modulus is equal to tensile modulus. However, at a higher strain level, the compressive stress is significantly higher than the corresponding tensile stress. This effect is illustrated in Figure 2-5.

Stress–strain tests are considered short-term tests, which means that the mechanical loading is applied within a relatively short period of time. This limits the usefulness of stress–strain tests in the actual design of a plastic part. Stress–strain tests fail to take into account the dependence of rigidity and strength of plastics on time. This serious limitation can be overcome with the use of creep and stress relaxation data while designing a part.

Stress relaxation is the application of a fixed deformation to a specimen and the measurement of the load required to maintain that deformation as a function of time. Creep is the application of a fixed load to a specimen and measurement of the resulting deformation as a function of time (8). Creep and stress relaxation

TABLE 2-1. Characteristic Features of Stress–Strain Curve as It Relates to Polymer Properties

Description of Polymer	Modulus	Yield Stress	Ultimate Strength	Elongation at Break
Soft, weak	Low	Low	Low	Moderate
Soft, tough	Low	Low	Yield stress	High
Hard, brittle	High	None	Moderate	Low
Hard, strong	High	High	High	Moderate
Hard, tough	High	High	High	High

From Billmeyer, F., *Textbook of Polymer Science*. Reprinted with permission of John Wiley & Sons, Inc.

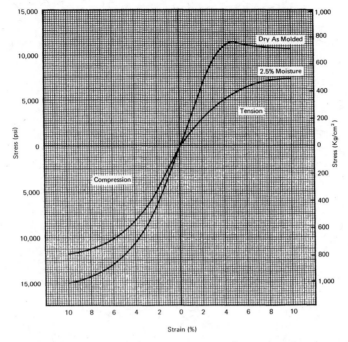

Figure 2-5. Stress–strain curve in tension and compression. (Courtesy of DuPont Company.)

effects are shown schematically in Figure 2-6. Creep and cold flow behavior of plastics are illustrated in Figure 2-7. As soon as the stress is applied, the elastic deformation takes place and continues until the load is removed. A substantial portion of the elastic recovery is not immediate and the material continues to return to its original size or length. The material is considered to have achieved permanent set if the recovery is not complete. The magnitude of permanent set is dependent on the stress applied, the length of time, and the temperature of the material (9). This type of behavior of a polymer in creep can also be represented by a mechanical model that combines a Maxwell and a Voigt element in series (10,11).

2.2. TENSILE TESTS (ASTM D 638, ISO 527-1)

Tensile elongation and tensile modulus measurements are among the most important indications of strength in a material and are the most widely specified properties of plastic materials. Tensile test, in a broad sense, is a measurement of the ability of a material to withstand forces that tend to pull it apart and to determine to what extent the material stretches before breaking. Tensile modulus, an indication of the relative stiffness of a material, can be determined from a stress–strain diagram. Different types of plastic materials are often compared on the basis of tensile strength, elongation, and tensile modulus data. Many plastics are very sensitive to the rate of straining and environmental conditions. Therefore, the data

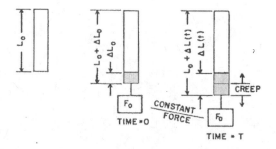

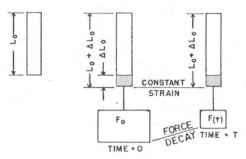

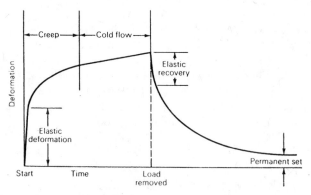

Figure 2-6. Diagram illustrating creep and stress relaxation. (Reprinted with permission of Van Nostrand Reinhold Company.)

Figure 2-7. Diagram illustrating creep and cold flow. (Reprinted with permission of McGraw-Hill Company.)

obtained by this method cannot be considered valid for applications involving load-time scales or environments widely different from this method. The tensile property data are more useful in preferential selection of a particular type of plastic from a large group of plastic materials and such data are of limited use in actual design of the product. This is because the test does not take into account the time-dependent behavior of plastic materials.

2.2.1. Apparatus

The tensile testing machine of a constant-rate-of-crosshead movement is used. It has a fixed or essentially stationary member carrying one grip, and a movable member carrying a second grip. Self-aligning grips employed for holding the test specimen between the fixed member and the movable member prevent alignment problems. A controlled-velocity drive mechanism is used. Some of the commercially available machines use a closed-loop servo-controlled drive mechanism to provide a high degree of speed accuracy. A load-indicating mechanism capable of indicating total tensile load with an accuracy of (1 percent of the indicated value or better is used. Lately, the inclination is toward using digital-type load indicators which are easier to read than the analog-type indicators. An extension indicator, commonly known as the extensometer, is used to determine the distance between two designated points located within the gauge length of the test specimen as the specimen is stretched. Figure 2-8 shows a commercially available tensile testing machine. The advent of new microprocessor technology has virtually eliminated time-consuming manual calculations. Stress, elongation, modulus, energy, and statistical calculations are performed automatically and presented on a visual display or hard copy printout at the end of the test.

2.2.2. Test Specimens and Conditioning

Test specimens for tensile tests are prepared many different ways. Most often, they are either injection molded or compression molded. The specimens may also be

Figure 2-8. Tensile testing machine. (Courtesy of Tinius Olsen Corporation.)

prepared by machining operations from materials in sheet, plate, slab, or similar form. Test specimen dimensions vary considerably depending upon the requirements and are described in detail in the ASTM book of standards. Figure 2-9 shows ASTM D 638 Type I tensile test specimen most commonly used for testing rigid and semirigid plastics.

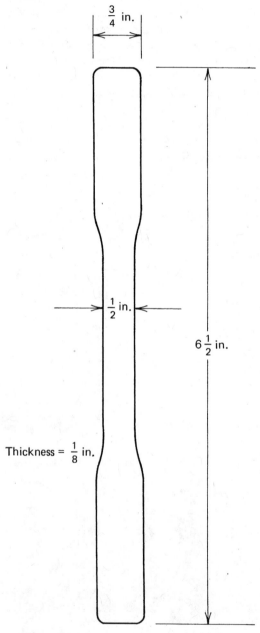

Figure 2-9. Tensile test specimen (Type I).

The specimens are conditioned using standard conditioning procedures. Since the tensile properties of some plastics change rapidly with small changes in temperature, it is recommended that tests be conducted in the standard laboratory atmosphere of $23 \pm 2°C$ and 50 ± 5 percent relative humidity. The Procedure A of ASTM methods D 618 (as explained in Chapter 11) is recommended for this test.

2.2.3. Test Procedures

A. Tensile Strength

The speed of testing is the relative rate of motion of the grips or test fixtures during the test. There are basically five different testing speeds specified in the ASTM D 638 Standard. The most frequently employed speed of testing is 0.2 in./min. Whenever possible, the speed indicated by the specification for the material being tested should be used. If a test speed is not given, appropriate speed that causes rupture between 30 sec and 5 min should be chosen. The test specimen is positioned vertically in the grips of the testing machine. The grips are tightened evenly and firmly to prevent any slippage. The speed of testing is set at the proper rate and the machine is started. As the specimen elongates, the resistance of the specimen increases and is detected by a load cell. This load value (force) is recorded by the instrument. Some machines also record the maximum (peak) load obtained by the specimen, which can be recalled after the completion of the test. The elongation of the specimen is continued until a rupture of the specimen is observed. Load value at break is also recorded. The tensile strength at yield and at break (ultimate tensile strength) are calculated.

$$\text{Tensile strength} = \frac{\text{Force (load) (1b)}}{\text{Cross-section area (sq. in.)}}$$

$$\text{Tensile strength at yield (psi)} = \frac{\text{Maximun load recorded (1b)}}{\text{Cross-section area (sq. in.)}}$$

$$\text{Tensile strength at break (psi)} = \frac{\text{Load recorded at break (1b)}}{\text{Cross-section area (sq. in.)}}$$

B. Tensile Modulus and Elongation

Tensile modulus and elongation values are derived from a stress–strain curve. An extensometer is attached to the test specimen as shown in Figure 2-10a. The extensometer is a strain gauge type of device that magnifies the actual stretch of the specimen considerably. Reliability of the strain measurement is affected by the traditional contact extensometers due to the actual physical contact with the test specimen. The problem magnifies with test specimen that are highly elastic in nature or with brittle, fragile, or very light weight test specimen. Contact extensometers also limit the ability to measure properties over large strain ranges up to a break point and testing specimen at elevated temperatures within a confined chamber. In recent years, many test equipment manufacturers have developed noncontact measurement systems based on optical, video and laser devices to overcome problems associated with contact extensometers. Figure 2-10b shows

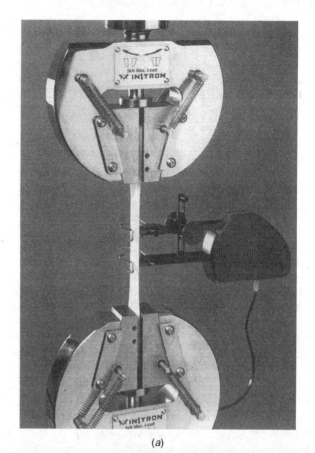

(a)

(b)

Figure 2-10. (*a*) Diagram illustrating an extensometer (strain gauge) attached to the test specimen. (Courtesy of Instron Corporation.) (*b*) Non-contact optical extensometer (Courtesy Zwick USA).

optical extensometer. The simultaneous stress–strain curve is plotted on graph paper. The following stepwise procedure is generally used to carry out the calculations (refer to Figure 2-11).

1. Mark off the units of stress in lb/in.^{2} on the x axis of the chart. This is done by dividing the force by cross-sectional area of the specimen.
2. Mark off the units of strain in in./in. on the y axis. These values are obtained by dividing the chart value by the magnification selected.
3. Carefully draw a tangent KL to the initial straight line portion of the stress–strain curve.
4. Select any two convenient points on the tangent. (Points P and L are selected in this case.)
5. Draw a straight line PQ and LM connecting points P and L with the y axis of the chart.
6. Stress value at L = 8000 psi, corresponding strain value at M = 0.08 in./in. Stress value at P = 3200 psi, corresponding strain value at Q = 0.04.

$$\text{Tensile modulus} = \frac{\text{Difference in stress}}{\text{Difference in corresponding strain}}$$

or

$$\text{Tensile modulus} = \frac{8000 - 3200}{0.08 - 0.04} = \frac{4800}{0.04} = 120,000\,\text{psi}$$

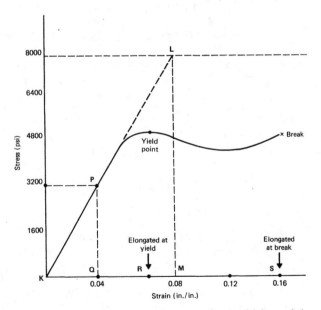

Figure 2-11. Diagram illustrating stress–strain curve from which modulus and elongation values are derived.

7. Elongation at yield

$$\text{Strain} = \frac{\text{Change in length (elongation)}}{\text{Original length (gauge length)}}$$

$$\varepsilon = \frac{\Delta l}{l}$$

or

$$\Delta l = \varepsilon \times l$$
$$\text{Elongation at yield} = 0.06 \times 2 = 0.12\,\text{in.}$$
$$\text{Percent elongation at yield} = 0.12 \times 100 = 12 \text{ percent}$$

8. Elongation at break

$$\text{Elongation} = 0.16 \times = 0.32\,\text{in.}$$
$$\text{Percent elongation at break} = 0.32 \times 100 = 32 \text{ percent}$$

For accuracy, modulus values should not be determined from the results of one stress–strain curve. Several tests should be made on the material and average tensile modulus should be calculated.

2.2.4. Factors Affecting the Test Results

A. *Specimen Preparation and Specimen Size*
Molecular orientation has a significant effect on tensile strength values. A load applied parallel to the direction of molecular orientation may yield higher values than the load applied perpendicular to the orientation. The opposite is true for elongation. The process employed to prepare the specimens also has a significant effect. For example, injection molded specimens generally yield higher tensile strength values than compression molded specimens. Machining usually lowers the tensile and elongation values because of small irregularities introduced into the machined specimen. Another important factor affecting the test results is the location and size of the gate on the molded specimens. This is especially true in the case of glass-fiber-reinforced specimens (12,13). A large gate located on top of the tensile bar will orient the fibers parallel to the applied load, yielding higher tensile strength. A gate located on one side of the tensile bar will disperse the fiber in a random fashion. This effect is shown in Figure 2-12. Tensile properties should only be compared for equivalent sample sizes and geometry.

B. *Rate of Straining*
As the strain rate is increased, the tensile strength and modulus increases. However, the elongation is inversely proportional to the strain rate. The effect of the cross-head speed on the modulus is shown in Figure 2-13.

C. *Temperature*
As discussed earlier in this chapter, the tensile properties of some plastics change rapidly with small changes in temperature. Tensile strength and modulus are decreased while elongation is increased as the temperature increases. Figure 2-14

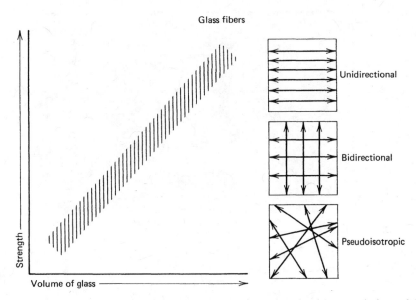

Figure 2-12. The effect of fiberglass orientation. (Reprinted with permission of Van Nostrand Reinhold Company.)

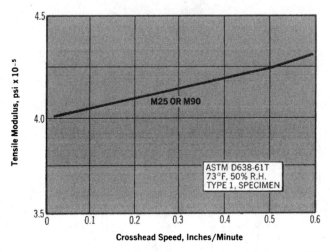

Figure 2-13. The effect of strain rate on modulus. (Courtesy of Ticona.)

illustrates the effect of temperature on the tensile strength of the material. Figure 2-15 is a commercially available environmental test chamber to study the effect of temperature on tensile properties.

2.3. FLEXURAL PROPERTIES (ASTM D 790, ISO 178)

The stress–strain behavior of polymers in flexure is of interest to a designer as well as a polymer manufacturer. Flexural strength is the ability of the material to

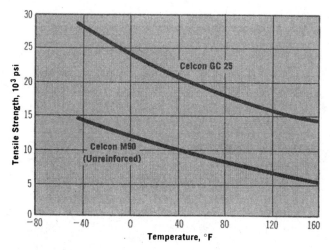

Figure 2-14. The effect of temperature on tensile strength. (Courtesy of Ticona.)

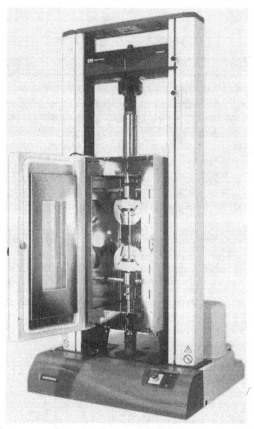

Figure 2-15. Environmental test chamber to study tensile properties at different temperatures. (Courtesy of Instron Corporation.)

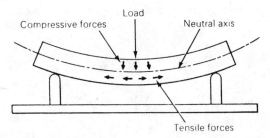

Figure 2-16. Forces involved in bending a simple beam. (Reprinted by permission of McGraw-Hill Company.)

withstand bending forces applied perpendicular to its longitudinal axis. The stresses induced by the flexural load are a combination of compressive and tensile stresses. This effect is illustrated in Figure 2-16. Flexural properties are reported are reported and calculated in terms of the maximum stress and strain that occur at the outside surface of the test bar. Many polymers do not break under flexure even after a large deflection that makes determination of the ultimate flexural strength impractical for many polymers. In such cases, the common practice is to report flexural yield strength when the maximum strain in the outer fiber of the specimen has reached 5 percent. For polymeric materials that break easily under flexural load, the specimen is deflected until a rupture occurs in the outer fibers.

There are several advantages of flexural strength tests over tensile tests (14). If a material is used in the form of a beam and if the service failure occurs in bending, then a flexural test is more relevant for design or specification purposes than a tensile test, which may give a strength value very different from the calculated strength of the outer fiber in the bent beam. The flexural specimen is comparatively easy to prepare without residual strain. The specimen alignment is also more difficult in tensile tests. Also, the tight clamping of the test specimens creates stress concentration points. One other advantage of the flexural test is that at small strains, the actual deformations are sufficiently large to be measured accurately.

There are two basic methods that cover the determination of flexural properties of plastics. Method 1 is a three-point loading system utilizing center loading on a simple supported beam. A bar of rectangular cross section rests on two supports and is loaded by means of a loading nose midway between the supports. The maximum axial fiber stresses occur on a line under the loading nose. A closeup of a specimen in the testing apparatus is shown in Figure 2-17. This method is especially useful in determining flexural properties for quality control and specification purposes.

Method 2 is a four-point loading system utilizing two load points equally spaced from their adjacent support points, with a distance between load points of one-third of the support span. In this method, the test bar rests on two supports and is loaded at two points (by means of two loading noses), each an equal distance from the adjacent support point. This arrangement is shown schematically in Figure 2-18. Method 2 is very useful in testing material that do not fail at the point of maximum stress under a three-point loading system. The maximum axial fiber stress occurs over the area between the loading noses.

Figure 2-17. Close-up of a specimen shown in flexural testing apparatus. (Courtesy of DuPont Company.)

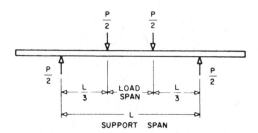

Figure 2-18. Schematic of specimen arrangement for flexural testing (Method II). (Reprinted with permission of ASTM.)

Either method can be used with the two procedures. Procedure A is designed principally for materials that break at comparatively small deflections. Procedure B is designed particularly for those materials that undergo large deflections during testing. The basic difference between the two procedures is the strain rate—Procedure A is 0.01 in./in./min and Procedure B is 0.10 in./in./min.

2.3.1. Apparatus

Quite often, the machine used for tensile testing is also used for flexural testing. The upper or lower portion of the movable crosshead can be used for flexural testing. The dual-purpose load cell that indicates the load applied in tension as well as compression facilitates testing of the specimen in either tension or compression. One such universal testing machine is shown in Figure 2-19. The machine used for this purpose should operate at a constant rate of crosshead motion over the entire range and the error in the load-measuring system should not exceed (1 percent of the maximum load expected to be measured).

The loading nose and support must have cylindrical surfaces. The radius of the nose and the nose support should be at least 1/8 in. to avoid excessive indentation

Figure 2-19. Universal testing machine for testing of the specimen in either tension or compression. (Courtesy of Instron Corporation.)

or failure due to stress concentration directly under the loading nose. A strain-gauge type of mechanism called a deflectometer or compressometer is used to measure deflection in the specimen.

2.3.2. Test Specimens and Conditioning

The specimens used for flexural testing are bars of rectangular cross section and are cut from sheets, plates, or molded shapes. The common practice is to mold the specimens to the desired finished dimensions. The specimens are conditioned in accordance with Procedure A of ASTM methods D618 as explained in Chapter 11 of this book. The specimens of size $1/8 \times 1/2 \times 4$ in. are the most commonly used.

2.3.3. Test Procedures and Calculations

The test is initiated by applying the load to the specimen at the specified crosshead rate. The deflection is measured either by a gauge under the specimen in contact with it in the center of the support span or by measurement of the motion of the loading nose relative to the supports. A load–deflection curve is plotted if the determination of flexural modulus value is desired.

The maximum fiber stress is related to the load and sample dimensions and is calculated using the following equation:

$$\text{Method 1} \quad S = \frac{3PL}{2bd^2}$$

where S = stress (psi); P = load (lb); L = length of span (in.); b = width of specimen (in.); d = thickness of specimen (in.).

Flexural strength is equal to the maximum stress in the outer fibers at the moment of break. This value can be calculated by using the above stress equation by letting load value P equal the load at the moment of break.

For materials that do not break at outer fiber strains up to 5 percent, the flexural yield strength is calculated using the same equation. The load value P in this case is the maximum load at which there is no longer an increase in load with an increase in deflection.

The maximum strain in the outer fibers, which also occurs at midspan, is calculated using the following equation (Method 1).

$$r = 6Dd/L^2$$

where r = strain (in./in.); D = deflection (in.); L = length of span (in.); d = thickness of specimen (in.).

The equations for calculating maximum fiber stress and maximum fiber strain are slightly different in Method 2.

2.3.4. Modulus of Elasticity (Flexural Modulus)

The flexural modulus is a measure of the stiffness during the first or initial part of the bending process. This value of the flexural modulus is, in many cases, equal to the tensile modulus.

The flexural modulus is represented by the slope of the initial straight-line portion of the stress–strain curve and is calculated by dividing the change in stress by the corresponding change in strain. The procedure to calculate flexural modulus is similar to the one described previously for tensile modulus calculations.

2.3.5. Factors Affecting the Test Results

A. Specimen Preparation
The molecular orientation in the specimen has a significant effect on the test results. For example, the specimen with a high degree of molecular orientation perpendicular to the applied load will show higher flexural values than the specimen with orientation parallel to the applied load. The injection-molded specimen usually shows a higher flexural value than a compression-molded specimen.

B. Temperature
The flexural strength and modulus values are inversely proportional with temperature. At higher testing temperatures, flexural strength and modulus values are significantly lower. Figure 2-20 shows the effect of temperature on flexural modulus.

C. Test Conditions
The strain rate, which depends upon testing speed, specimen thickness, and the distance between supports (span), can affect the results. At a given span, the flexural strength increases as the specimen thickness is increased. The modulus of a material generally increases with the increasing strain rate (15).

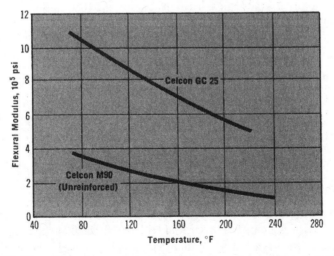

Figure 2-20. The effect of temperature on flexural modulus. (Courtesy of Ticona.)

2.4. COMPRESSIVE PROPERTIES (ASTM D 695, ISO 75-1 AND 75-2)

Compressive properties describe the behavior of a material when it is subjected to a compressive load at a relatively low and uniform rate of loading. In spite of numerous applications of plastic products that are subjected to compressive loads, the compressive strength of plastics has limited design value. In practical applications, the compressive loads are not always applied instantaneously. Therefore, the standard test results that fail to take into account the dependence of rigidity and strength of plastics on time cannot be used as a basis for designing a part. The results of impact, creep, and fatigue tests must be considered while designing such a part. Compression tests provide a standard method of obtaining data for research and development, quality control, acceptance or rejection under specifications, and special purposes. Compressive properties include modulus of elasticity, yield stress, deformation beyond yield point, compressive strength, compressive strain, and slenderness ratio. However, compressive strength and compressive modulus are the only two values most widely specified in design guides.

In the case of a polymer that fails in compression by a shattering fracture, the compressive strength has a definite value. For those polymers that do not fail by a shattering fracture, the compressive strength is an arbitrary one depending upon the degree of distortion that is regarded as indicating complete failure of material. Some material suppliers report compressive stress at 1 or 10 percent deformation of its original height. Some polymers may also continue to deform in compression until a flat disk is produced. The compressive stress continues to rise without any well-defined fracture occurring. Compressive strength has no real meaning in such cases.

2.4.1. Apparatus

The universal testing machine used for tensile and flexural testing can also be used for testing compressive strength of various materials. The machine requirement has been described in detail in the section on tensile testing. A deflectometer or a compressometer is used to measure any change in distance between two fixed points on the test specimen at any time during the test. Figure 2-21 shows a typical setup for compression testing.

2.4.2. Test Specimens and Conditioning

Recommended specimens for this test are either rectangular blocks measuring $1/2 \times 1/2 \times 1$ in. or cylinders $1/2$ in. in diameter and 1 in. long. Specimens may be prepared by machining or molding. The test specimens are conditioned in accordance with Procedure A of ASTM methods D618 as discussed in Chapter 11.

2.4.3. Procedure

The specimen is placed between the surfaces of the compression tool, making sure that the ends of the specimen are parallel with the surface of the compression tool. The test is commenced by lowering the movable crosshead at a specified speed over the specimen. The maximum load carried by the specimen during the test is

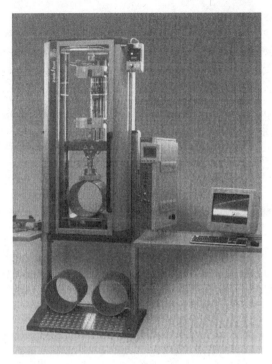

Figure 2-21. A typical test setup for compression testing. (Courtesy of Zwick USA.)

recorded. The stress–strain data are also recorded either by recording load at corresponding compressive strain or by plotting a complete load–deformation curve with an automatic recording device. Compressive strength is calculated by dividing the maximum compressive load carried by the specimen during the test by the original minimum cross-sectional area of the specimen. The result is expressed in lb/in.^{2} Modulus of elasticity or compressive modulus, like tensile and flexural modulus, is also represented by the slope of the initial straight-line portion of the stress–strain curve and is calculated by dividing the change in stress by the corresponding change in strain. The complete procedure to calculate compressive modulus is described in Section 2.2 on tensile properties.

2.5. CREEP PROPERTIES

Today, plastics are used in applications that demand high performance and extreme reliability. Many components, conventionally made of metals, are now made of plastics. The pressure is put on the design engineer to design the plastic products more efficiently. An increasing number of designers have now recognized the importance of thoroughly understanding the behavior of plastics under long-term load and varying temperatures. Such behavior is described in terms of creep properties.

When a plastic material is subjected to a constant load, it deforms quickly to a strain roughly predicted by its stress–strain modulus, and then continues to deform slowly with time indefinitely or until rupture or yielding causes failure (16). This phenomenon of deformation under load with time is called creep. All plastics creep to a certain extent. The degree of creep depends upon several factors, such as type of plastic, amount of load, temperature, and time.

As explained previously in this chapter, the short-term stress–strain data is of little practical value in actual designing the part, since such data does not take into account the effect of long-term loading on plastics. Creep behavior varies considerably among types of plastics; however, under proper stress and temperature conditions, all plastics will exhibit a characteristic type of creep behavior. One such generalized creep curve is shown in Figure 2-22. The total creep curve is divided into four continuous stages. The first stage (OP) represents the instantaneous elastic deformation. This initial strain is the sum of the elastic and plastic strain. The first stage is followed by the second stage (PQ) in which strain occurs rapidly but at a decreasing rate. This stage, where creep rate decreases with time, is sometimes referred to as creep or primary creep. The straight portion of the curve (QR) is characterized by a constant rate of creep. This process is called "cold flow." The final stage (RS) is marked by increase in creep rate until the creep fracture occurs (17–19).

If the applied load is released before the creep rupture occurs, an immediate elastic recovery, substantially equal to elastic deformation followed by a period of slow recovery is observed. The material in most cases does not recover to the original shape and a permanent set remains. The magnitude of the permanent set depends upon length of time, amount of stress applied, and temperature (20).

The creep values are obtained by applying constant load to the test specimen in tension, compression, or flexure and measuring the deformation as a function of

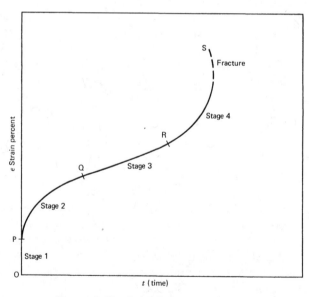

Figure 2-22. Generalized creep curve.

time. The values are most commonly referred to as tensile creep, compressive creep, and flexural creep.

2.5.1. Tensile Creep

Tensile creep measurements are made by applying the constant load to a tensile test specimen and measuring its extension as a function of time. The extension measurement can be carried out several different ways. The simplest way is to make two gauge marks on the tensile specimen and measure the distance between the marks at specified time intervals. The percent creep strain is determined by dividing the extension by initial gauge length and multiplying by 100. The percent creep strain is plotted against time to obtain a tensile creep curve. One such curve is illustrated in Figure 2-23. The tensile stress values are also determined at specified time intervals to facilitate plotting a stress–rupture curve. The more accurate measurements require the use of a strain gauge, which is capable of measuring and amplifying small changes in length with time and directly plotting them on a chart paper. Figure 2-24 illustrates a typical setup for tensile creep testing. The test is also carried out at different stress levels and temperatures to study their effect on tensile creep properties.

2.5.2. Flexural Creep

Flexural creep measurements are also made by applying a constant load to the standard flexural test specimen and measuring its deflection as a function of time. A typical test setup for measuring creep in flexure is shown in Figure 2-25. The deflection of the specimen at midspan is measured using a dial indicator gauge. The electrical resistance gauges may also be used in place of a dial indicator. The deflections of the specimen are measured at a predetermined time interval. The percent flexural creep strain is calculated using the following formula:

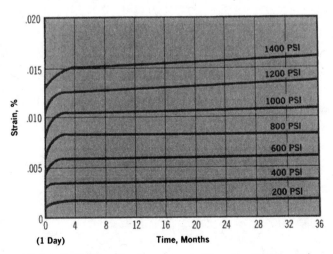

Figure 2-23. Tensile creep curve. (Courtesy of Ticona.)

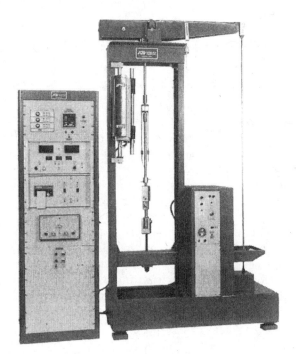

Figure 2-24. Typical test setup for tensile creep testing. (Courtesy of Applied Test Systems Inc.)

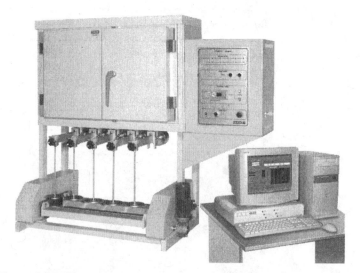

Figure 2-25. Flexural creep testing. (Courtesy of CEAST U.S.A. Inc.)

$$r = \frac{6Dd}{L^2} \times 100$$

where r = maximum percent creep strain (in./in.); D = maximum deflection at midspan (in.); d = depth (in.); L = span (in.).

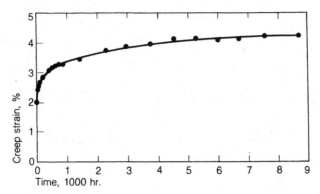

Figure 2-26. Percent creep strain versus time. (Reprinted with permission of McGraw-Hill Company.)

The percent creep strain is plotted against time to obtain a flexural creep curve. An example of one such curve is shown in Figure 2-26. The test is carried out at various stress levels and temperatures and similar flexural creep curves are plotted. If necessary, the maximum fiber stress for each specimen in lb/in.^{2} can also be calculated as follows:

$$S = \frac{3PL}{2bd^2}$$

where S = stress (psi); P = load (lbs.); L = span (in.); b = width (in.); d = depth (in.).

2.5.3. Interpretation and Applications of Creep Data

One of the serious limitations of the earlier creep curves such as the one illustrated in Figure 2-26 was the lack of simplicity of the single-point stress–strain properties such as tensile modulus and flexural strength, especially when one wanted to measure creep as a function of temperature and stress level. Furthermore, the creep data presented in terms of strain were not convenient to use in design or for the purpose of comparing materials. It was obvious that creep curves had to be presented in a more meaningful and convenient way such that they are readily usable. Creep strain curves were easily converted to creep modulus (apparent modulus) by simply dividing the initial applied stress by the creep strain at any time.

$$\text{Creep (Apparent) modulus at time } t = \frac{\text{Initial applied stress}}{\text{Creep strain}}$$

Figure 2-27 shows the same data as shown in Figure 2-26 plotted as creep modulus versus time on Cartesian coordinates. To further simplify the use of the creep curves, the same curve was replotted using logarithmic coordinates as shown in Figure 2-28. These linear plots of creep data are extremely useful in extrapolating the curves to the desired life of the plastic part. Several extrapolation methods are used to extend the range of creep data (21).

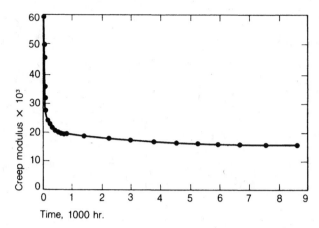

Figure 2-27. Creep modulus versus time on cartesian coordinates. (Reprinted with permission of McGraw-Hill Company.)

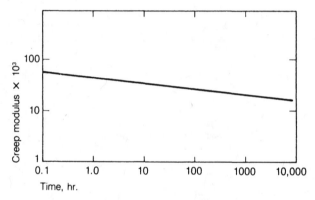

Figure 2-28. Creep modulus versus time on logarithmic coordinates. (Reprinted with permission of McGraw-Hill Company.)

2.5.4. Isochronous Stress–Strain Curves

Quite often, while designing a part, it is necessary to compare various plastic materials. The basic creep curves, such as creep strain versus time or apparent modulus versus time are not completely satisfactory. The comparison is extremely difficult to make, especially when different stress levels are used for different materials.

The method preferred by most design engineers is the use of isochronous (equal time) stress–strain curves. The stress versus corresponding strain is plotted at a specific time of loading pertinent to particular application.

Suppose a designer is asked to design a shelf that is required to withstand a continuous load for 1000 hr. If after 1000 hr of continuous loading, the deformation is not to exceed 2 percent, what should be the maximum allowable stress? To solve this problem, the designer first needs to erect an ordinate on the basic creep strain versus time curve at 1000 hr, as shown in Figure 2-29. From this curve, the designer can determine the strain value at different stress levels. These values allow the designer to plot another curve of stress versus strain at 1000 hr. This isochronous stress–strain curve is shown in Figure 2-30. The maximum allowable stress can be determined by simply erecting an ordinate from 2 percent strain value and reading the corresponding stress, which in this particular case is 3500 psi. A wide variety of materials can easily be compared by studying an isochronous stress–strain curve such as that shown in Figure 2-31 (22).

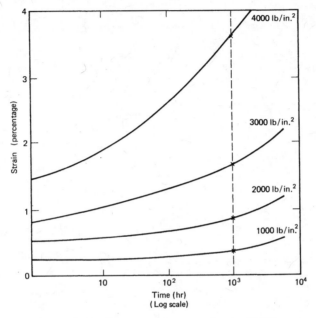

Figure 2-29. Creep strain versus time at 1000 hr.

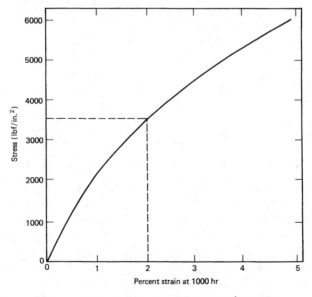

Figure 2-30. Isochronous stress–strain curve.

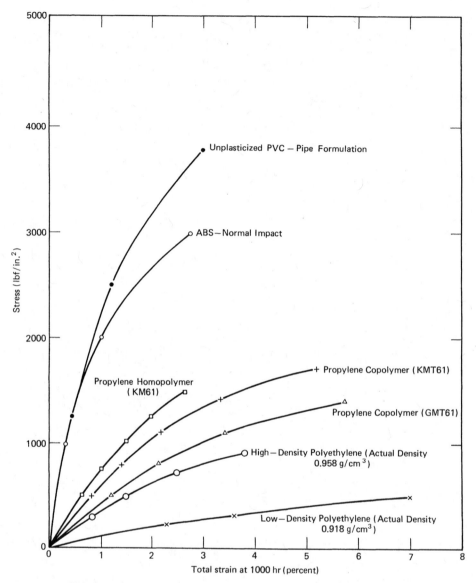

Figure 2-31. Isochronous stress–strain curve for various materials.

2.5.5. Effects of Stress and Temperature on Creep Modulus

The creep modulus is directly affected by the increase in the level of stress and temperature. With the exception of extremely low strains around 1 percent or less, the creep modulus decreases as the amount of stress is increased. This effect is illustrated in Figure 2-32. In a very similar manner, as the temperature is increased, the creep modulus significantly decreases. Figure 2-33 shows the creep modulus versus time plotted at different temperatures. As one would expect, the combined effect of increasing stress level and temperature on creep modulus is much more severe and should not be overlooked.

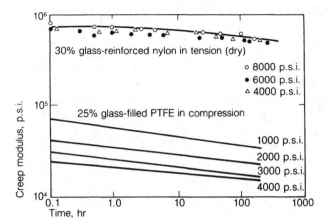

Figure 2-32. Effect of stress on creep modulus. (Reprinted with permission of McGraw-Hill Company.)

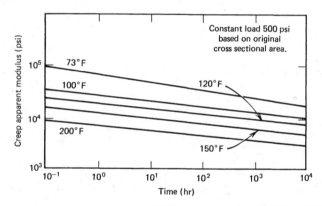

Figure 2-33. Creep modulus versus time at different temperatures. (Reprinted with permission of McGraw-Hill Company.)

2.5.6. Basic Procedures for Developing and Applying Creep Modulus Data

In recent years, much emphasis has been placed on standardizing a method to develop and report creep data that would report creep that would be universally acceptable and understood. A systematic accumulation of data on creep properties of commercially available plastics has been developed with the cooperation of major material suppliers and the results have been published in tabular form, as shown in Table 2-2.

In spite of the wide use and acceptance of creep data, it is evident from the table that a standard creep test method agreed upon by everyone in the plastics industry has not yet been developed. All three basic methods—tension, flexure, and compression—are used by material suppliers to develop creep data. There are also other variables, such as sample fabricating methods, type of specimens, and specimen dimension. These variables have a significant effect on creep properties. A

TABLE 2-2. Creep Property Data of a Commercially Available Plastic Material

Material					Test Specimen[13]				
Type	Supplier	Trade Name and Grade Designation	Description	ASTM, Military, or Other Specification Classification	Molding method	Type or Shape	Dimensions	Type of Load	Strain Measure
ABS	GE	Cycolac DFA-R	Injection molding, medium-impact, high gloss	Fed. Spec. L-P-1183 Type 2	IM	RB	4	F2	3
		Cycolac GSE	Extrusion, high-impact, high modulus	Fed. Spec. L-P-1183 Type 2, 5, and 6	IM	RB	4	F2	3
		Cycolac LS	Extrusion, high-impact, thermoforming	Fed. Spec. L-P-1183	IM	RB	4	F2	3
		Cycolac DH	Injection molding, heat-resistant		IM	RB	4	F2	3
		Cycolac X-37	Injection molding, heat-resistant (D TUL = 230°F),medium-impact		IM	RB	4	F2	3
		Cycolac KJB	Injection molding, medium-impact, flame-retarded (94-VO at 0.060-in. thickness)		IM	RB	4	F2	3
		Cycolac FBK	Structural foam molding, flame-retarded (94-VO at) 0.240-in. thickness		IM Solid Sp. Gr. 1.11	RB	4	F2	3
					IM foam Sp. Gr. 1.02	RB	4	F2	3
		Cycovin KAB	ABS/PVC alloy, extrusion, high-impact, flame-retarded (94-VI at 0.058-in. thickness)		IM	RB	4	F2	3

Creep Test Conditions[13]			Creep Test Data[1,2,3]									Time at Rupture or Onset of Yielding at initial Applied Stress in Air (hr)
			Creep (Apparent) Modulus[3] (thousand psi)									
			Calculated from total creep strain[2] or deflection (before rupture and onset of yielding) at the following test times:									
Special Specimen Conditioning	Test Temp. (°F)	Initial Applied Stress (psi)	1 hr	10 hr	30 hr	100 hr	300 hr	1000 hr	At Latest Test Point	Time at Latest Test Point (hr)	
	73	1,000	345	323	313	278	244	215			
		2,000	345	323	308	274	243	211			
	120	1,000	303	222	187	154	125	94			
		2,000	290	211	167	132	108	90			
	140	1,000	137	93	78	63	53	44			
	73	1,000	294	286	278	256	238	208			
		2,000	294	286	278	256	233	204			
	120	1,000	238	182	159	133	116	100			
		2,000	227	172	149	124	106	90			
	140	1,000	179	122	100	83	71	60			
	160	1,000	105	59	49	41	36	32			
	73	1,000	238	217	204	200	189	159			
		2,000	224	211	200	189	179	155			
	120	1,000	169	135	118	98	83	71			
		2,000	159	121	103	86	74	63			
	140	1,000	133	87	72	61	51	43			
	73	1,000	351	333	303	263	227	200			
	100	1,000	256	208	185	163	147	132			
	120	1,000	238	185	161	139	121	105			
	140	1,000	192	132	111	93	81	70			
	73	1,000	364	339	328	299	267	230			
	100	1,000	304	256	235	208	187	165			
	120	1,000	270	220	200	180	160	143			
	140	1,000	238	179	156	135	119	104			
	180	500	167	80	69	59	50	43			
	73	1,000	244	233	222	204	182	156			
	140	1,000	93	56	46	36	33	28			
	73	1,000	280	260	230	220	200	160			
	73	1,000	230	220	200	190	160	130			
	73	1,000	280	250	235	206	185	160			
	120	1,000	133	86	71	62	54	45			

TABLE 2-2. *Continued*

Type	Supplier	Material: Trade Name and Grade Designation	Material: Description	Material: ASTM, Military, or Other Specification Classification	Test Specimen[13]: Molding method	Test Specimen[13]: Type or Shape	Test Specimen[13]: Dimensions	Test Specimen[13]: Type of Load	Test Specimen[13]: Strain Measure
	Uniroyal	Kralastic W	Pipe extrusion	Fed. Spec. L-P-1183 Types 2, 5, and 6	CM	T1	1	T	1
						RS	2	T	1
		Kralastic MH	Easy flow, injection molding	Fed. Spec. L-P-1183 Types 1 and 2	CM	T1	1	T	1
		Kralastic MV	Sheet extrusion, injection molding	Fed. Spec. L-P-1183 Types 2, 5, and 6	CM	T1	1	T	1
		Kralastic K-2938	Heat-resistant, injection molding	Fed. Spec. L-P-1183 Types 1, 2, 3, and 4	CM	T1	1	T	1
		Karalastic SRS	Sheet extrusion	Fed. Spec. L-P-1183 Types 1, 2, 5, and 6	CM	T1	1	T	1
	Monsanto	Lustran 261	Sheet extrusion	Fed. Spec. L-P-1183 Type 1	ES	T1	1	T	1
		Lustran 461	Sheet extrusion	Fed. Spec. L-P-1183 Type 2	ES	T1	1	T	1
		Lustran 761	Sheet extrusion	Fed. Spec. L-P-1183 Type 5	ES	T1	1	T	1
	Schulman	Polyman 511	Alloy, high temperature, injection molding, flame-retarded		IM	RB	4	F4	3
	LNP	Thermocomp AF-1004	20% glass fiber-reinforced, injection molding		IM	RB	4	F4	3
		Thermocomp AF 1008	40% glass fiber-reinforced, general-purpose, injection molding		IM	RB	4	F4	3

CREEP PROPERTIES** **51**

Creep Test Conditions[13]			Creep Test Data[1,2,3]								Time at Rupture or Onset of Yielding at initial Applied Stress in Air (hr)
			Creep (Apparent) Modulus[3] (thousand psi)								
			Calculated from total creep strain[2] or deflection (before rupture and onset of yielding) at the following test times:						At Latest Test Point	Time at Latest Test Point (hr)	
Special Specimen Conditioning	Test Temp. (°F)	Initial Applied Stress (psi)	1hr	10hr	30hr	100hr	300hr	1000hr			
	73	2,000	230	220	211	200			194	167	
		2,500	223	208	195	180			172	167	
		3,000	211	185	165	130			128	111	
	73	1,500	230	224	217	211	200	188	140	50,000	
	73	1,500	284	278	273	263	254	240			
		2,000	284	268	260	245	230	211			
		2,500					198	211			
		3,000	254	224	203	170	133		125	390	
		3,500	233	165					146	17	
		4,000	167								
	73	3000	216	196	181	161	140		132	83	
		2,500	202	172	152						
		3,000	181	136							
	140	500[5]	161	132	109	77	49				
		1,000[5]	171	136	114	76	44				
		1,500[5]	168	127	99						
	73	2,000	330	323	308	290	270	244	167	580	
		3,000	328	305	288	265	244	217	188	42	
		4,000	308	278	256	230	197				
		5,000	279	237	201						
	160	500	263	250	227	200	147	104			
		1,000	263	233	189	147	118	84			
		1,500	263	203	170	135	105	75			
	73	2,000	282	263	247	233	213	192	125	350	
		2,500	278	253	236	216	192	156	146	3	
		3,000	265	234	214	176	132				
		3,500	206								
	73	2,000	476	465	465	465	465	465			
	73	2,000	571	513	488	417	377	328			
	73	2,000	328	282	250	222	1290	155			
	73	1,500	360	355	350	340	320	305	283	2,200	
		2,000	335	330	321	318	315	300	280	2,200	
	150	1,000	250	230	219	165	138				
	75	2,000		810	800	790	780	780			
		5,000		800	790	780	775	770			
	75	5,000[5]	1,720	1,680	1,650	1,636	1,600				
		10,000[5]	1,760	1,710	1,690	1,670	1,650				

[a] Reprinted by permission of *Modern Plastics Encyclopedia 1980–1981*, McGraw-Hill.
[b] Explanation of test specimen and creep test conditions, codes:

Molding methods

   CA = Compression molded per ASTM D1928
   CB = Comp. molded; annealed 2 hr 140°C; quenched 0°C
   CC = Comp. molded; annealed 2 hr 140°C; slow cooled
   CD = Comp. molded; annealed 2 hr 140°C; cooled 40°C./hr
   CE = Comp. molded and free sintered 700°F; cooled 170°F./hr
   CF = Comp. molded and free sintered 675°F; cooled 175°F./hr
   CG = Comp. molded and free sintered 720°F; cooled 180°F./hr
   CH = Comp. molded and free sintered 720°F; cooled 300°F./hr
   CI = Comp. molded and free sintered 716°F; cooled 54°F./hr
   CL = Calendered sheet
   CM = Compression molded
   CS = Cast sheet
   DS = Direct blend. screw injection molded
   ES = Extruded sheet
   IM = Injection molded
   IZ = Inj. molded; mold temp. 265°F
   M = Machined
   SF = Structural Foam, injection molded
   TM = Transfer molded

Type or shape

   CR = Cylindrical rod
   DC = Die C per ASTM D412
   D1 = Modified Die C with 3.4-in. reduced section
   HC = Hollow cylinder—see Note 17
   I2 = Tensile bar per ISO R527
   RB = Rectangular bar
   RC = Rectangular column
   RS = Rectangular strip
   T = Tensile bar
   TB = Tensile bar—see Note 14
   T1 = Type 1 tensile bar per ASTM D638
   T2 = Type 2 tensile bar per ASTM D638
   T3 = Type 3 tensile bar per ASTM D638

Dimensions, in. (overall for bars and columns, and reduced section for dog bones)

   $1 = \frac{1}{2} \times \frac{1}{8}$
   $2 = 5 \times 1 \times 0.090$
   $3 = 4 \times 1 \times 0.026$
   $4 = 5 \times \frac{1}{2} \times \frac{1}{8}$
   $5 = \frac{1}{4} \times \frac{1}{16}$
   $6 = \frac{1}{2} \times \frac{1}{16}$
   $7 = \frac{1}{2}$ diam. $\times 1$ high
   $8 = \frac{1}{4} \times \frac{1}{8}$
   $9 = \frac{1}{4} \times \frac{1}{12}$
   $10 = 4 \times \frac{1}{8} \times 0.218$
   $11 = 5 \times \frac{1}{2} \times \frac{1}{16}$
   $12 = 0.24 \times 0.075$
   $13 = 5 \times \frac{1}{32} \times 0.392$
   $14 = 5 \times \frac{1}{2} \times \frac{1}{2}$
   $15 = 16 \times 1 \times \frac{1}{8}$ or $\frac{1}{4}$
   $16 = 8 \times \frac{1}{2} \times \frac{1}{8}$
   $17 = 0.190 \times 0.190 \times 2$ high
   $18 = \frac{5}{8}$ID $\times$ F$\frac{3}{4}$OD $\times 2.5$ high
   $19 = 0.024 \times 0.035$
   $20 = 3\frac{1}{2} \times \frac{1}{4} \times \frac{1}{8}$
   $21 = 120 \times 20 \times 6$ mm. ($4.72 \times 0.79 \times 0.25$ in.)
   $22 = 0.39 \times 0.16$
   $23 = 4 \times 0.500 \times 0.175$
   $24 = 6.5 \times 0.5 \times 0.25$
   $25 = 5 \times \frac{1}{2} \times \frac{1}{4}$

Strain measurement

   1 = Strain in reduced seciton
   1A = Strain in reduced section—see Note 15
   1B = Strain in reduced section—see Note 16
   1C = Strain in reduced section—corrected for shrinkage
   2 = Grip separation
   3 = Deflection at center of beam
   4 = Reduction in height
   5 = Extensometer—see Note 17

Type of load

   C = Compression
   C1 = Compression—see Note 11
   F = Flexure—simple beam bending, load at center
   F2 = Flexure—simple beam bending, load at center, 2-in. span
   F3 = Flexure—simple beam bending, load at center, 3-in. span
   F4 = Flexure—simple beam bending, load at center, 4-in. span
   F6 = Flexure—uniform beam bending moment, 6-in. span
   F7 = Flexure—4 point loading, constant outside fiber stress
   T = Tension

From *Modern Plastics Encyclopedia*. 1980–81. Reprinted by Permission of McGraw-Hill Book Company.

designer using the data from such creep tables must be aware of such variables before comparing one material with the other.

The following procedure is recommended for the use of creep modulus data (23):

1. Select the design life of the part.

2. Consult or plot, from data available in the creep chart, the creep modulus curve of the material of interest for the temperature at which the part will be used, extrapolating where necessary. Like creep rupture, creep modulus varies greatly with temperature. In addition it is subject to another variable—stress level. For each material grade, creep modulus data are tabulated in the creep chart, first by test temperature and second by applied (test) stress. For very rigid plastics, such as thermosets, pilled thermoplastics, and amorphous thermoplastics at room temperature and below, the creep modulus curves show minor variation with applied stress. However, for the more flexible and ductile plastics, the creep modulus curves will vary significantly and systematically with stress level. The higher the stress, the lower the creep modulus. This is a consequence of viscoelasticity.

3. To cope with the effect of stress level the designer has two alternatives. If the design problem is such that the stress level is predetermined, the designer should select the creep modulus curve whose stress level is the closest. If the stress level is not known beforehand, the designer should choose a creep modulus curve at a conservative stress level and check the choice after calculating a stress level.

4. Read from the selected creep modulus curve the modulus value corresponding to the design life selected in step 1. This is the design modulus.

5. Apply a safety factor to the design modulus to calculate a working modulus. This is rarely necessary in metal design, but is in plastics design to correct for any uncertainties arising from extrapolations or other compromises that may have been made. Safety factors of 0.5–0.75 are typical.

6. Substitute the calculated working modulus in the part design equation. For example, to calculate the width of a simple rectangular beam required to limit the maximum deflection to specified value when the span and depth of the beam are fixed, the working modulus would be substituted for E in the following design equation:

$$b = \frac{P}{E\Delta}\frac{L^3}{4d^3}$$

where b = width of beam; d = depth of beam; L = span; P = load; Δ = maximum deflection allowed; E = modulus.

2.6. STRESS RELAXATION

Stress relaxation is defined as a gradual decrease in stress with time, under a constant deformation (strain). This characteristic behavior of the polymers is studied by applying a fixed amount of deformation to a specimen and measuring the load required to maintain it as a function of time. This is in contrast to creep measure-

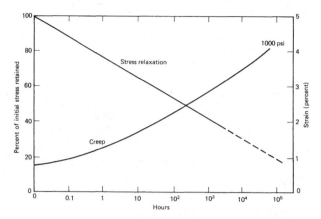

Figure 2-34. Creep and stress relaxation.

ment, where a fixed amount of load is applied to a specimen and resulting deformation is measured as a function of time. This phenomenon of creep and stress relaxation is further clarified schematically in Figure 2-34.

Stress relaxation behavior of the polymers has been overlooked by many design engineers and researchers, partly because the creep data are much easier to obtain and is readily available. However, many practical applications dictate the use of stress relaxation data. For example, extremely low stress relaxation is desired in the case of a threaded bottle closure, which may be under constant strain for a long period. If the plastic material used in the closures shows an excessive decrease in stress under this constant deformation, the closures will eventually fail. Similar problems can be encountered with metal inserts in molded plastics and belleville or multiple cantilever springs use in cameras, appliances, and business machines.

Stress relaxation measurements can be carried out using a tensile testing machine such as that described earlier in this chapter. However, the use of such a machine is not always practical because the stress relaxation test ties up the machine for a long period of time. The equipment for a stress relaxation test must be capable of measuring very small elongation accurately, even when applied at high speeds. Many sophisticated pieces of equipment now employ a strain gauge or a differential transformer along with a chart recorder capable of plotting stress as a function of time. A typical stress–time curve is schematically plotted in Figure 2-35. At the beginning of the experiment, the strain is applied to the specimen at a constant rate to achieve the desired elongation. Once the specimen reaches the desired elongation, the strain is held constant for a predetermined amount of time. The stress decay, which occurs because of stress relaxation, is observed as a function of time. If a chart recorder is not available, the stress values at different time intervals are recorded and the results are plotted to obtain a stress versus time curve. One such stress relaxation curve plotted at various levels of constant strain is shown in Figure 2-36. The stress relaxation experiment is often carried out at various levels of temperature and strain.

The stress data obtained from the stress relaxation experiment can be converted to a more meaningful apparent modulus data by simply dividing stress at a particular time by the applied strain. The curve may be replotted to represent apparent

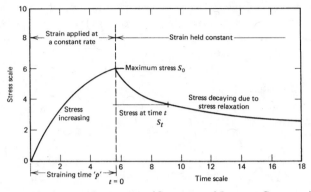

Figure 2-35. Stress–time curve. (Courtesy of Instron Corporation.)

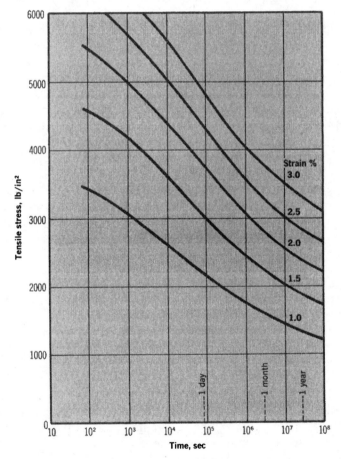

Figure 2-36. Stress relaxation curve plotted at various levels of constant strain. (Courtesy of Ticona.)

modulus as a function of time. The use of logarithmic coordinates further simplifies the stress relaxation data by allowing us to use standard extrapolation methods such as the one used in creep experiments.

2.7. IMPACT PROPERTIES

2.7.1. Introduction

The impact properties of the polymeric materials are directly related to the overall toughness of the material. Toughness is defined as the ability of the polymer to absorb applied energy. The area under the stress–strain curve is directly proportional to the toughness of a material. Impact energy is a measure of toughness. The higher the impact energy of a material, the higher the toughness and vice versa. Impact resistance is the ability of a material to resist breaking under a shock loading or the ability to resist the fracture under stress applied at high speed.

The theory behind toughness and brittleness of the polymers is very complex and therefore difficult to understand. The molecular flexibility plays an important role in determining the relative brittleness or toughness of the material. For example, in stiff polymers like polystyrene and acrylics, the molecular segments are unable to disentangle and respond to the rapid application of mechanical stress and the impact produces brittle failure. In contrast, flexible polymers such as plasticized vinyls have high-impact behavior due to the ability of the large segments of molecules to disentangle and respond rapidly to mechanical stress (24).

Impact properties of the polymers are often modified simply by adding an impact modifier such as butadiene rubber or certain acrylic polymers. The addition of a plasticizer also improves the impact behavior at the cost of rigidity. Material such as nylon, which has relatively fair impact energy, can be oriented by aligning the polymer chains to improve the impact energy substantially. Another way to improve the impact energy is to use fibrous fillers that appear to act as stress transfer agents (25).

Most polymers, when subjected to the impact loading, seem to fracture in a characteristic fashion. The crack is initiated on a polymer surface due to the impact loading. The energy to initiate such a crack is called the crack initiation energy. If the load exceeds the crack initiation energy, the crack continues to propagate. A complete failure occurs when the load has exceeded the crack propagation energy. Thus, both crack initiation and crack propagation contribute to the measured impact energy. There are basically four types of failures encountered due to the impact load (26).

> *Brittle Fracture.* In this type of failure the part fractures extensively without yielding. A catastrophic mechanical failure such as the one in the case of general-purpose polystyrene is observed.
> *Slight Cracking.* The part shows evidence of slight cracking and yielding without losing its shape or integrity.
> *Yielding.* The part actually yields showing obvious deformation and stress whitening but no cracking takes place.

Ductile Failure. This type of failure is characterized by a definite yielding of material along with cracking. Polycarbonate is considered a ductile material.

The distinction between the four types of failures is not very clear and some overlapping is quite possible. Figure 2-37*a* illustrates common failure modes.

Impact behavior is one of the most widely specified mechanical properties of the polymeric materials. However, it is also one of the least understood properties. Predicting the impact resistance of plastics still remains one of the most troublesome areas of product design. One of the problems with some earlier Izod and Charpy impact tests was that the tests were adopted by the plastic industry from metallurgists. The principles of impact mechanisms as applied to metals do not seem to work satisfactorily with plastics because of the plastics' complex structure.

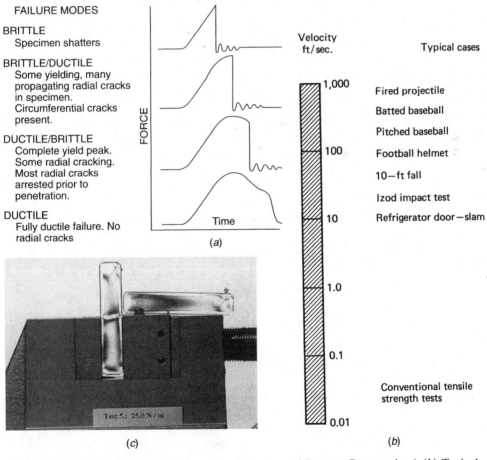

Figure 2-37. (*a*) Common failure modes. (Courtesy of Instron Corporation.) (*b*) Typical velocities of some impact blows. (Courtesy of Dow Chemical Company.) (*c*) Specimen clamping. (Courtesy CEAST U.S.A. Inc.) A full color version of this figure can be viewed on DVD included with this book.

2.7.2 Factors Affecting the Impact Strength

A. *Rate of Loading*
The speed at which the specimen or part is struck with an object has a significant effect on the behavior of the polymer under impact loading. At low rates of impact, relatively stiff materials can still have good impact strength. However at high rates of impact, even rubbery materials may exhibit brittle failure (27). All materials seem to have a critical velocity above which they behave as glassy, brittle materials. Figure 2-37b illustrates some typical velocities encountered during testing and real-life situations. Let us say that we are required to design a football helmet. It is obvious that we cannot use the results from the Izod impact test directly to select our material. The velocity at which the Izod impact test is conducted is approximately 10 times lower than the one encountered in the end-use. A more realistic drop impact test at a velocity closer to the actual use condition must be used.

B. *Notch Sensitivity*
A notch in a test specimen or a sharp corner in a fabricated part drastically lowers the impact energy. A notch creates a localized stress concentration and hence the part failure under impact loading. All plastics are notch-sensitive. The rate of sensitivity varies with the type of plastics. Both notch depth and notch radius have an effect on the impact behavior of materials. For example, a larger radius of curvature at the base of the notch will have a lower stress concentration and, therefore, a higher impact energy of the base material. Thus it is obvious from the above discussion that while designing a plastic part, one should avoid notches, sharp corners, and other factors that act as stress concentrators.

C. *Temperature*
The impact behavior of plastic materials is strongly dependent upon the temperature. At lower temperatures the impact resistance is reduced drastically. The reduction in impact is even more dramatic near the glass transition temperature. Conversely, at higher test temperatures, the impact energy is significantly improved.

D. *Orientation*
The manner in which the polymer molecules are oriented in a part will have a major effect on the impact behavior of the polymer. Molecular orientation introduced into drawn films and fibers may give extra strength and toughness over the isotropic material (28). However, such directional orientation of polymer molecules can be very fatal in a molded part since the impact stresses are usually multiaxial. The impact strength is always higher in the direction of flow.

E. *Processing Conditions and Types*
Processing conditions play a key role in determining the impact behavior of a material. Inadequate processing conditions can cause the material to lose its inherent toughness. Voids that act as stress concentrators are created by poor processing conditions. High processing temperatures can also cause thermal degradation and, therefore, reduced impact behavior. Improper processing conditions also create a

weak weld line that almost always reduces overall impact energy. The compression-molded specimen usually shows a lower impact resistance than the injection-molded specimens.

F. Degree of Crystallinity, Molecular Weight
Increasing the percentage of crystallinity decreases the impact resistance and increases the probability of the brittle failure. A reduction in the average molecular weight tends to reduce the impact behavior and vice versa (29).

G. Method of Loading
The manner in which the part is struck with the impact loading device significantly affects the impact results. A pendulum type of impact loading will produce a different result from the one produced by falling-weight or high-speed ball-impact loading.

H. Specimen Clamping
Excessive clamping force can pre-stress the specimen, particularly behind the notch. Such stress tends to reduce Izod test results (Figure 2-37c).

2.7.3. Types of Impact Tests

In the last two decades, a tremendous amount of time and money has been spent on the research and development of various types of impact tests by organizations throughout the world. Attempts have been made to develop different sizes and shapes of specimens as well as impact testers. The specimens have been subjected to a variety of impact loads including tensile, compression, bending, and torsion impacts. Impact load has been applied using everything from a hammer, punches, and pendulums to falling balls and bullets. Unfortunately, very little correlation exists, if any, between the types of tests developed so far. Numerous technical papers and articles have been written on the subject of the advantages of one method over the other. To this date, no industry-wide consensus exists regarding an ideal impact test method. In this chapter, an attempt is made to discuss as many types of impact tests as possible, along with the respective advantages and limitations of each test. Impact testing is divided into three major classes and subdivided into several classes as follows.

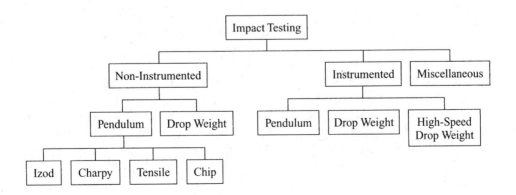

A. Pendulum Impact Tests

Izod–Charpy Impact Test (ASTM D-256, ASTM D4812 ISO 179). The objective of the Izod–Charpy impact test is to measure the relative susceptibility of a standard test specimen to the pendulum-type impact load. The results are expressed in terms of kinetic energy consumed by the pendulum in order to break the specimen. The energy required to break a standard specimen is actually the sum of energies needed to deform it, to initiate its fracture, and to propagate the fracture across it, and the energy expended in tossing the broken ends of the specimen. This is called the "toss factor." The energy lost through the friction and vibration of the apparatus is minimal for all practical purposes and usually neglected.

The specimen used in Izod test must be notched. The reason for notching the specimen is to provide a stress concentration area that promotes a brittle rather than a ductile failure. A plastic deformation is prevented by such type of notch in the specimen. The impact values are seriously affected because of the notch sensitivity of certain types of plastic materials. This effect was discussed in detail in Section 2.7.2.B.

The Izod test requires a specimen to be clamped vertically as a cantilever beam. The specimen is struck by a swing of a pendulum released from a fixed distance from the specimen clamp. A similar setup is used for the Charpy test except for the positioning of the specimen. In the Charpy method, the specimen is supported horizontally as a simple beam and fractured by a blow delivered in the middle by the pendulum. The obvious advantage of the Charpy test over the Izod test is that the specimen does not have to be clamped and, therefore, it is free of variations in clamping pressures. Charpy test is now preferred for testing polypropylene.

Apparatus and Test Specimens. The testing machine consists of a heavy base with a vise for clamping the specimen in place during the test. In most cases, the vise is designed so that the specimen can be clamped vertically for the Izod test or positioned horizontally for the Charpy test without making any changes. A pendulum-type hammer with an antifriction bearing is used. Additional weights may be attached to the hammer for breaking tougher specimens. The pendulum is connected to a pointer and a dial mechanism that indicates the excess energy remaining in a pendulum after breaking the specimen. The dial is calibrated to read the impact values directly in in.-lb or ft-lb. A hardened steel striking nose is attached to the pendulum. The Izod and Charpy tests use different types of striking noses. A detailed list of requirements is discussed in the ASTM Standards Book. One of the newly developed impact testers calculates and digitally displays energy using pulses generated by an optical encoder mounted on the shaft pendulum. Figure 2-38 illustrates a typical pendulum-type impact testing machine. The test specimens can be prepared either by molding or cutting them from a sheet. Izod test specimens are 2 1/2 × 1/2 × 1/8 in. The most common specimen thickness is 1/8 in. but 1/4 in. is preferred since they are less susceptible to bending and crushing. A notch is cut into a specimen very carefully by a milling machine or a lathe. The recommended notch depth is 0.100 in. Figure 2-39 illustrates a commercially available notching machine.

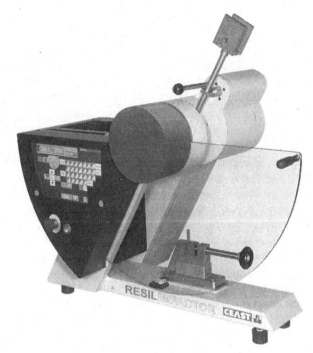

Figure 2-38. Pendulum impact tester. (Courtesy of CEAST U.S.A. Inc.)

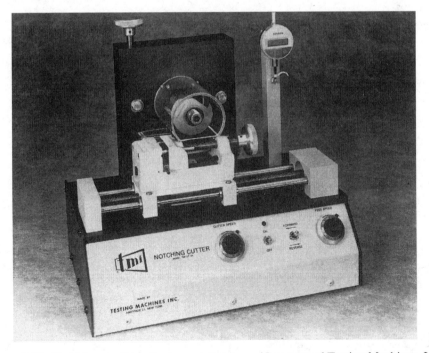

Figure 2-39. Notching machine for impact test bars. (Courtesy of Testing Machines, Inc.)

Test Procedures

Izod Test. The test specimen is clamped into position so that the notched end of the specimen is facing the striking edge of the pendulum. A properly positioned test specimen is shown in Figure 2-40. The pendulum hammer is released, allowed to strike the specimen, and swing through. If the specimen does not break, more weights are attached to the hammer and the test is repeated until failure is observed. The impact values are read directly in in. -lbf or ft-lbf from the scale. The impact strength is calculated by dividing the impact values obtained from the scale by the thickness of the specimen. For example, if a reading of 2 ft-lbf is obtained using a 1/8-in. thick specimen, the impact value would be 16 ft-lbf/in. of notch. The impact values are always calculated on the basis of 1-in. thick specimens even though much thinner specimens are usually used. The reversed notch impact resistance is obtained by reversing the position of a notched specimen in the vise. In this case, the notch is subjected to compressive rather than tensile stresses during impact. As discussed earlier in this chapter, the energy required to break a specimen is the sum of the energies needed to deform it, initiate and propagate the fracture, and toss the broken end (toss factor).

Notching of the test specimen drastically reduces the energy loss due to the deformation and can generally be neglected. Tough plastic materials that have an Izod impact higher than 0.5 ft-lb/in. of notch seem to expend very little energy in tossing the broken end of the specimen. For relatively brittle material, having an Izod impact less than 0.5 ft-lbf/in. of notch, the energy loss due to the toss factor represents a major portion of the total energy loss and cannot be overlooked. A method to determine such energy loss is devised (30).

Charpy Impact Test. This test is conducted in a very similar manner to the Izod impact strength test. The only difference is the positioning of the specimen. Figure 2-41 illustrates one such setup. In this test the specimen is mounted horizontally and supported unclamped at both ends. Only the specimens that break com-

Figure 2-40. Diagram illustrating izod impact test specimen properly positioned in text fixture. (Courtesy of CEAST U.S.A. Inc.)

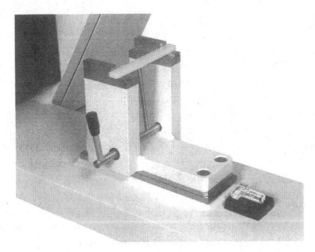

Figure 2-41. Charpy test setup. (Courtesy of CEAST U.S.A. Inc.)

pletely are considered acceptable. The Charpy impact strength is calculated by dividing the indicator reading by the thickness of the specimen. The results are reported in ft-lbf/in. of notch for notched specimens and ft-lbf/in. for unnotched specimens.

Effect of Test Variables and Limitations

Notch. A slight variation in the radius and depth of a notch affects the impact strength results. Many other variables such as the cutter speed, sharpness of the cutting tooth, feed rate, type of plastic, and quality of the notch cutting equipment all seem to have a significant effect on the results. Such variations are difficult to control and nonuniformity between the lots is quite common. Certain heat-sensitive polymers are also affected by the high cutter speed that seems to contribute to the thermal degradation. The notch in the specimen tends to create a stress concentration area that produces unrealistically low impact values in crystalline plastics. Recommendations include, even time interval between notching and testing and also waiting at least 40 hours after notching and before testing.

Specimen Thickness. Although the impact values reported are on the basis of 1-in.-thick specimens, the actual thickness of the specimen used in the test influences the test results. This is especially true in case of polycarbonate.

Specimen Preparation. Injection-molded specimens seem to yield higher impact strength values than compression-molded specimens. This is due to the molecular orientation caused by the injection-molding process. The location of the gate also has a significant effect on the test results, particularly in the case of fiber-reinforced specimens.

Temperature. Impact values increase with the increasing temperature and vice versa.

Fillers and Other Additives. Fillers and reinforcements have a pronounced effect on the test results. For example, unreinforced polycarbonate has an

Izod impact value of 15 ft-lbf/in. of notch, but reinforcing it with 30 percent glass fibers reduces the impact to a mere 3.7 ft-lbf/in. of notch. In contrast, reinforcing polystyrene with 30 percent glass fibers more than doubles the Izod impact. Fillers and pigments generally lower the impact energy.

Limitations. The results obtained from the Izod or Charpy test cannot be directly applied to part design because these tests do not measure the true energy required to break the specimen. The notched Izod impact test measures only the notch sensitivity of the different polymers and not the toughness.

Chip Impact Test (ASTM D4508). The chip impact test was originally developed for measuring the effect of surface microcracking caused by the weathering on impact strength retention. The material toughness is measured by this test as opposed to the material's notch sensitivity, as measured by the notched Izod impact test. The test also allows a user to determine the orientation, flow effects, and weld line strength, properties that are difficult to assess with the conventional impact techniques (31).

The chip impact test is somewhat similar to the Izod impact test. The specimen can be tested using standard Izod pendulum tester. The chip impact test requires the use of a pendulum hammer type of device and a specimen holding fixture. The test specimens are usually 1 in. long × 1/2 in. wide and 0.065 in. thick. The specimens can be prepared either by injection or compression molding or by simply cutting them from a sheet.

The test is carried out by mounting the test specimen as shown in Figure 2-42. The pendulum hammer is released, allowed to strike the specimen, and swing

Figure 2-42. Chip impact test setup.

through. The retained toughness is proportional to the energy absorbed during impact, which is measured by the angle of travel of the pendulum after impact. This is schematically shown in Figure 2-43. The value is expressed in in.-lbf/in.^{2} or ft-lbf/in.^{2}

If the test were used exclusively to study the effect of weathering on a polymer, it would be required that the sample be struck with a pendulum hammer on the weather exposed side. The chip impact test is also useful in measuring the relative toughness of a rather large and a complex-shaped part that is difficult to hold in a conventional fixture. A small chip can be cut from such a part and subjected to the chip impact test.

Tensile Impact Test (ASTM D 1822). The tensile impact strength test was developed to overcome the deficiencies of flexural (Izod and Charpy) impact tests. The test variables, such as notch sensitivity, toss factor, and specimen thickness, are eliminated in the tensile impact test. Unlike Izod–Charpy-type pendulum impact tests, which are limited to thick specimens only, the tensile impact test allows the user to determine the impact strength of very thin and flexible specimens. Many other characteristics of polymeric materials, such as the anisotropy and the orientation effect, can be studied through the use of the tensile impact test.

The early development work on tensile impact testers included the specimen-in-base type of setup (32). This method was excluded by the ASTM Committee D-20 due to the inherent problem of toss factor correction and repeatability. The present acceptable tensile impact test consists of a specimen-in-head type of set-up. In this case, the specimen is mounted in the pendulum and attains full kinetic energy at the point of impact, eliminating the need for a toss factor correction. The energy to break by impact in tension is determined by kinetic energy extracted from the pendulum in the process of breaking the specimen. The test setup requires mounting one end of the specimen in the pendulum with the other end to be gripped by a crosshead member, which travels with the pendulum until the instant of impact. The pendulum is affected only by the tensile force exerted by the specimen through the pendulum's center of percussion.

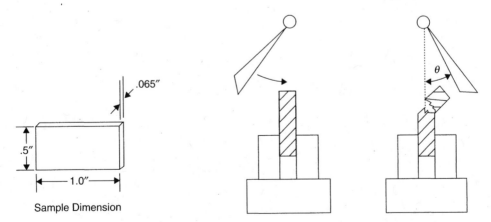

Figure 2-43. Diagram illustrating the principle of chip impact test.

As long as the machine base is rigid enough to prevent the vibrational energy losses, the bounce of the crosshead in the opposite direction can be easily calculated. The tensile impact test is far more meaningful than the Izod–Charpy-type tests since it introduces strain rate as an important test variable. Many researchers have demonstrated that the tensile impact test results correlate better with the actual field failures and are easier to analyze than the Izod impact test results (33–35). However, this is still a uniaxial test and most impact events are multiaxial in real-life situations.

Apparatus and Test Specimens. The tensile impact testing machine consists of a rigid massive base with a suspending frame. The pendulum is specially designed to hold the dumbbell-shaped specimen so that the specimen is not under stress until the moment of impact. Figure 2-44 schematically illustrates a specimen-in-head tensile impact machine. Illustrated in Figure 2-45 is a commercially available tensile impact tester. The specimens are prepared by molding, machining, or die-cutting to the desired shape from a sheet. Two basic specimen geometries are used so that the effect of elongation, the rate of extension, or both can be observed. Type S (short) specimens usually exhibit low extension while the Type L (long) specimen extension is comparatively high. Type S specimens provide a greater occurrence of brittle failures. Type L specimens provide a greater differentiation between materials. Mold dimensions of both types of specimens are shown in Figure 2-46.

Test Procedure. The thickness and width of the test specimen are measured with a standard flat anvil and not ball anvil micrometer. The specimen is clamped to the crosshead while the crosshead is out of the pendulum. The whole assembly is placed in the elevated pendulum. The other end of the specimen is bolted to the pendulum. The pendulum is released and the crosshead is allowed to strike and anvil. The tensile impact energy is measured and recorded from the scale reading. A corrected tensile impact energy to break is calculated as follows:

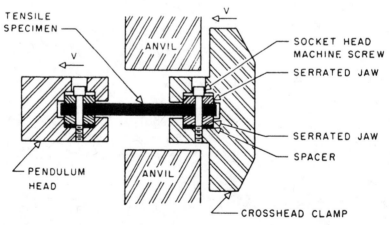

Figure 2-44. Schematic of specimen-in-head tensile-impact machine. (Reprinted with permission of ASTM.)

Figure 2-45. Tensile impact tester. (Courtesy of CEAST U.S.A. Inc.)

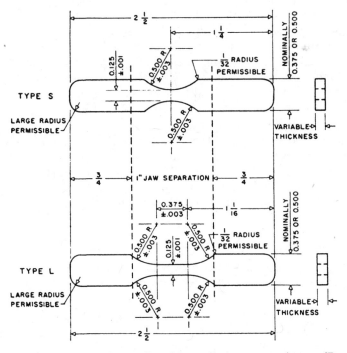

Figure 2-46. Mold dimensions of types S and L tensile-impact specimens. (Reprinted with permission of ASTM.)

$$X = E - Y + e$$

where X = corrected impact energy to break (ft-lbf); E = scale reading of energy of break (ft-lbf); Y = friction and windage correction (ft-lbf); e = bounce correction factor (ft-lbf).

A bounce correction factor curve for each type of crosshead and pendulum employed must be developed.

B. High-Rate Tension Test

The high-rate tension test was developed to overcome the difficulties involved in the evaluation of impact properties using pendulum-type impact testers. Earlier work was done by using conventional tensile testing equipment. At high speeds, such as 20 in./min, a good correlation with the falling-weight test results was obtained (36). The area under stress–strain curve is proportional to the energy required to break the material. This area is also directly proportional to the impact energy of the material if the curve is generated at a high enough speed. The relationship of tensile strength and elongation with the impact energy of the material is also evaluated by the high-rate tension test. Tensile strength seems to increase with the increase in speed of testing, while elongation decreases with the increasing speed. Therefore, the high-rate tension test must be conducted at a speed suitable to the type of polymer in order to obtain some meaningful results.

Ultrahigh-rate tensions testers have also been developed to study the impact behavior of the ductile polymers. These high-speed, hydraulically operated testing machines are capable of providing linear displacement rates over 30,000 in./min (37). High-rate tension tests are usually conducted using a variety of test specimens. Test bars used for the standard tensile test and the tensile impact test are most commonly used. Sample geometries resembling the actual part are also used to study the mechanical behavior at high speeds.

Many other important impact design parameters such as yield stress, energy to yield, initial modulus, and deformation at break can be measured with high-speed tension tests (38). In spite of the capability of high-rate tension tests to provide stress at strain rates that simulate actual service, the tests have not been popular because the delivered stress is uniaxial. Normally, the real-life impact stress is multiaxial.

C. Falling-Weight Impact Test

The falling-weight impact test, also known as the drop impact test or the variable-height impact test, employs a falling weight. This falling weight may be a tup with a conical nose, a ball, or a ball-end dart. The energy required to fail the specimen is measured by dropping a known weight from a known height onto a test specimen. The impact is normally expressed in ft-lb and is calculated by multiplying the weight of the projectile by the drop height.

The biggest advantage of the falling-weight impact test over the pendulum impact test or high-rate tension test is its ability to duplicate the multidirectional impact stresses that a part would be subjected to in actual service. The other obvious advantage is the flexibility to use specimens of different sizes and shapes, including an actual part. Unlike the Izod impact test, which measures the notch sensitivity of the material and not the material toughness, falling-weight impact tests introduce polyaxial stresses into the specimen and measure the toughness. The variations in the test results due to the fillers and reinforcements, clamping pressure, and material orientation are virtually eliminated in the falling-weight impact test. This type of test is also very suitable for determining the impact resistance of plastics films, sheets, and laminated materials.

Three basic ASTM tests are commonly used, depending upon the application:

ASTM D 5420—Impact resistance of flat rigid plastic specimen by means of a falling weight

ASTM D 1709—Impact resistance of plastic film by the free-falling dart method

ASTM D 2444—Test for impact resistance of thermoplastic pipe and fittings by means of a tup

Drop Impact Test. This falling-weight impact test is primarily designed to determine the relative ranking of materials according to the energy required to break flat rigid plastics specimens under various conditions of impact of a striker impacted by a falling weight. A free-falling weight or a tup is used to determine the impact resistance of the material.

Many different versions of test equipment exist today. They all basically operate on the same principle. Figure 2-47 illustrates one such typical commercially available testing machine. It consists of a cast aluminum base, a slotted vertical guide tube, a round-nose striker and striker holder, an 8-lb weight, a die and die support, and a sample platform. The sample platform is used to position a sheet of desired thickness for impact testing. The die is removable from the base in order that the actual parts of complex shapes can be placed onto the base and impact tested.

The test is carried out by raising the weight to a desired height manually or automatically with the use of a motor-driven mechanism and allowing it to fall freely onto the other side of the striker. The striker transfers the impact energy to the flat test specimen positioned on a cylindrical die or a part lying on the base of the machine. The kinetic energy possessed by the falling weight at the instant of impact is equal to the energy used to raise the weight to the height of drop and is the potential energy possessed by the weight as it is released. Since the potential energy is expressed as the product of weight and height, the guide tube can be marked with a linear scale representing the impact range of the instrument in in.-lb. Thus, the toughness or the impact resistance of a specimen or a part can be read directly off the calibrated scale in in.-lb. The energy loss due to the friction in the tube or due to the momentary acceleration of the punch is negligible.

An alternate method for achieving the same result utilizes an instrument that employs a free-falling dart dropped from a specified height onto a test specimen. The dart with a hemispherical head is constructed of smooth, polished aluminum or stainless steel. An electromagnetic, air-operated, or other mechanical release mechanism with a centering device is used for releasing the dart. The dart is also fitted with a shaft long enough to accommodate removable incremental weights. A two-piece annular specimen clamp is used to hold the specimen. One such commercially available piece of equipment is illustrated in Figure 2-48.

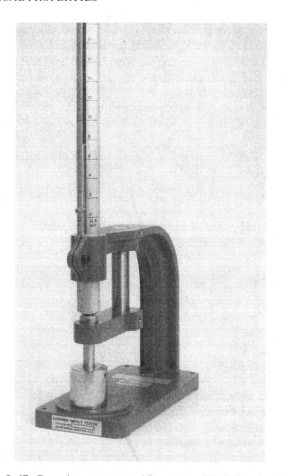

Figure 2-47. Drop impact tester. (Courtesy of Byk-Gardner USA.)

A number of different techniques are used to determine the impact value of a specimen. One of the most common is called the Bruceton Staircase Method, in which testing is concentrated near the mean to reduce the number of specimens required. An alternate method, known as the ultimate nonfailure level (UNF), requires testing the specimen in successive groups of 10. One missile weight is employed for each group and the weight is varied in uniform increments from group to group. These methods are discussed in detail in the ASTM Book of Standards (39).

The impact testing machine, such as the one illustrated in Figure 2-49, is specifically designed for testing plastic pipe and fittings. It employs a much heavier tup with three different sized radii on the tup nose. The drop tube is long enough to provide at least a 10-ft free fall.

Test Variables and Limitations

Specimen Thickness. The drop impact resistance of a specimen, although not linear, is directly proportional to the specimen thickness. As the thickness increase, the energy required to fracture the specimen also increases (40).

Figure 2-48. Falling dart impact tester. (Courtesy of CEAST U.S.A. Inc.)

Brittle polymers such as polypropylene show a smaller increase in the impact strength compared to the ductile polymers like ABS and impact polystyrene (41).

Specimen Slippage. If extreme care were not taken in clamping the specimen, some slippage is bound to occur upon impact, causing distorted results. This is particularly true in the case of films which seem to have a greater tendency toward slippage. Unclamped specimens seem to exhibit somewhat greater impact resistance.

Specimen Quality. The surface quality of the test specimen significantly affects the test results. For example, a notch in the test specimen will alter the impact values (42).

Stress Concentration. When impact testing a fabricated part, care must be taken not to impact the area of extreme stress concentration, such as a bend on a 90° elbow fitting. This usually produces erroneous results because the degree of stress concentration appears to vary with the processing conditions. An area free of stress concentration should be chosen for more reliable results.

Limitations. So far, we have discussed many advantages of falling-weight impact tests over the conventional pendulum impact tests and high-rate tension tests. However, there are serious limitations with the falling-weight tests that cannot be overlooked. One of the biggest disadvantages of this type of test is the large number of samples required to establish and energy level to fail the sample. Since there is no way of determining how many trials will be required

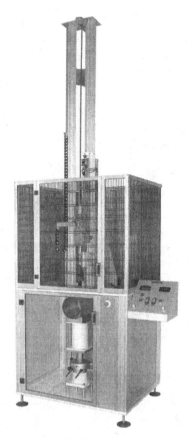

Figure 2-49. Impact tester specifically designed for impact testing pipe and fittings. (Courtesy of CEAST U.S.A. Inc.)

actually to fail the sample, a statistical approach must be used. Another serious limitation of a falling-weight test is the problem of isolating the rate of impact. Although a 2-lb tup dropped from 6 ft produces the same 12 ft-lb impact energy as a 6 lb tup dropped from 2 ft, the effect is not the same. Impact resistance is directly related to the impact and, therefore, a specimen that can survive the impact from a 6-lb tup could fail from the higher velocity impact of the lighter 2-lb tup dropped from a greater height. Many commercially available drop impact testers do not provide flexibility of varying the rate of impact. Furthermore, the velocity impacts produced by such testers are much smaller than the velocity impacts encountered by products in actual use (43). The basic design of the existing drop impact equipment is not suitable for testing large parts such as structural foam cabinets and housings. In case of the falling-weight impact tests, what constitutes a failure is very subjective.

D. Instrumented Impact Testing
One of the biggest drawbacks of the conventional impact test methods is that it provides only one value—the total impact energy—and nothing else. The conven-

tional tests cannot provide additional information on the ductility, dynamic toughness, fracture, and yield loads or the behavior of the specimen during the entire impact event.

This effectively limits the application of noninstrumented impact test methods to quality control and material ranking. Instrumented impact testers are generally suited for research and development as well as advance quality control.

Instrumented impact testers measure force continuously while the specimen is penetrated. The resulting data can be used to determine type of failure and maximum load, in addition to the amount of energy required to fracture the specimen. One of the most common type of failure occurring from ductile to brittle transition at low temperatures can only be observed by studying the load–energy–time curve. The fracture mode of plastics is sensitive to the changes in temperature and can change abruptly at or near the material's transition temperature. Manufacturers of plastics automotive components routinely test materials at low temperatures (−20 to −40°F) to ensure that they will not become brittle in cold weather. By studying the shape of the load–time or load–deflection curve, the type of failure can be analyzed, and important information about its performance in service can be gathered. Figure 2-50 illustrates the changes in behavior of a plastic with temperature. The shape of the curve at the left indicates brittle failure at low temperatures. The same material at a higher temperature undergoes ductile failure (44). The new piezoelectric-equipped strikers offer increased sensitivity, opening the doors for testing a whole new range of materials. Applications involving lightweight products such as foam containers for eggs and ultrathin films used in the packaging industry can now be meaningfully tested.

All standard impact testers can be instrumented to provide a complete load and energy history of the specimen. Such a system monitors and precisely records the entire impact event, starting from the acceleration (from rest) to the initial impact and plastic bending to fracture initiation and propagation to the complete failure. The instrumentation is done by mounting the strain gauges or load cell onto

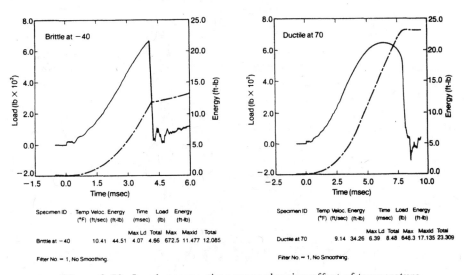

Specimen ID	Temp (°F)	Veloc. (ft/sec)	Energy (ft-lb)	Time (msec)	Load (lb)	Energy (ft-lb)
				Max Ld	Total Max	Maxld Total
Brittle at −40	10.41	44.51	4.07	4.66	672.5	11.477 12.085

Filter No. = 1, No Smoothing.

Specimen ID	Temp (°F)	Veloc. (ft/sec)	Energy (ft-lb)	Time (msec)	Load (lb)	Energy (ft-lb)
				Max Ld	Total Max	Maxld Total
Ductile at 70	9.14	34.26	6.39	8.48	648.3	17.135 23.309

Filter No. = 1, No Smoothing.

Figure 2-50. Load–energy–time curve showing effect of temperature.

striking bit in the case of a pendulum impact tester or onto the tup in the case of a drop impact tester. During the test, a fiber-optic device triggers an oscilloscope just before striking the specimen. The output of the strain gauge is recorded by the oscilloscope, depicting the variation of the load applied to the specimen throughout the entire fracturing process. A complete load–time history of the entire specimen is obtained. The apparent total energy absorbed by the specimen can be calculated and plotted against time. The specimen displacement can be calculated by the double integration of the load–time curve and the load–displacement curve can be plotted (45). With the advent of microprocessor technology, some manufacturers are now capable of offering a unit that automatically calculates the sample displacement and provides a load–displacement curve, eliminating the need for calculations. Many other useful data such as the impact rate, force and displacement at yield, and break, yield, and failure energies, as well as modulus, are calculated and printed out. A commercially available instrumented impact tester is shown in Figure 2-51.

Interpreting Impact Data. Modern instrumented impact test machines generate a complete record of (1) force applied to the specimen versus time, (2) displacement of the impactor (from the point of specimen contact) versus time, (3) velocity

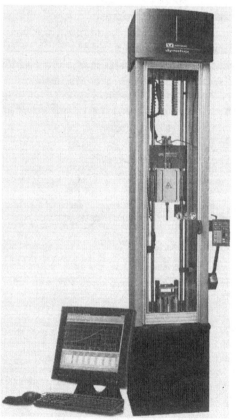

Figure 2-51. Instrumented impact tester. (Courtesy of Instron Corporation.)

of the impactor versus time, and (4) energy absorbed by the specimen versus time.

When plotted, (Figure 2-52*a*), many details of the impact event become clearly visible in the data. Most systems also analyze the data and provide tabulated reports of critical values. Results of instrumented impact tests are interpreted differently depending on the type of material being tested and the failure criteria applied.

For homogeneous materials, four values are critical: (1) maximum load, (2) energy to maximum load, (3) total absorbed energy, and (4) deflection to maximum load.

Maximum (yield) load is simply the highest point on the load–time curve before failure. Often the point of maximum load corresponds to the onset of material damage or complete failure. However, in some cases where the plastics material is reinforced with filler such as carbon fier, the peak load may be higher than the maximum load. This higher value is not useful to the design engineer due the extensive damage already incurred.

Energy to maximum (yield) load is the energy absorbed by the specimen up to the point of maximum load. When maximum load corresponds to failure, the energy to maximum load is the amount of energy the specimen can absorb before failing.

Total energy is the amount of energy the specimen absorbed during the complete test. This number may be the same as energy to maximum load when the specimen abruptly fails at the maximum load point.

Deflection to maximum (yield) load is the distance the impactor traveled from the point of impact to the point of maximum load.

Figure 2-52*b* shows typical test data.

E. High-Speed Impact Tests (ASTM D3763, ISO 6603-2)

An ever-increasing demand for engineering plastics and the need for sophisticated and meaningful impact test methods for characterizing these materials have forced the industry into developing new high-speed impact tests. These tests not only provide the basic information regarding the toughness of the polymeric materials but also yield other important data of interest, such as the load–deflection curve and the total energy absorption. The high-speed impact test overcomes the basic limitations of conventional impact-testing methods, as discussed previously. The rate of impact can be varied from 30 to 570,000 in./min.

High-speed impact testing has gained considerable popularity in recent years because of its ability to simulate actual impact failures at high speeds. For example, conventional impact testing methods are useless in testing to meet advanced automotive crash standards that require the impact simulation at 28 mph (30,000 in./min). High-speed impact testers are able to meet the challenge. As discussed earlier in this chapter, almost all polymers are strain-rate sensitive. Two polymers, when impact tested at one strain rate, may show similar impact values. The same two polymers tested at a high strain rate show a completely different set of values.

Most versatile high-speed impact testing machinery is capable of testing everything from the thin film which may require as low an impact rate as 30 in./min to the plastic automotive bumper which may require an impact rate up to 30,000 in./min. The specimen or product can be tested under a controlled environment of tempera-

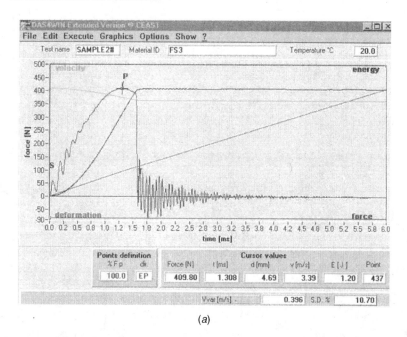

(a)

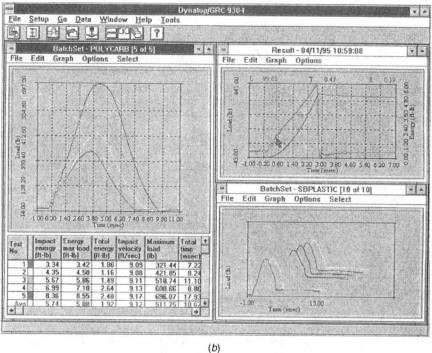

(b)

Figure 2-52. (*a*) Load–energy–time curve showing effect of impact on a specimen. (*b*) A typical test data obtained with instrumented impact tester. (Courtesy of Instron Corporation.)

ture and humidity. The equipment basically consists of a tup attached to a motor wound spring or pneumatically powered actuator along with a plunger displacement measuring system. The force is detected with a fast responding quartz load cell mounted directly on the actuator. The velocity can be set digitally from 30 to 30,000 in./min. Some type of clamp assembly to hold the specimen in place is used. The equipment can also be fitted with an environmental chamber for specialized testing. The tester is equipped with a CRT and x–y plotter that automatically displays load versus displacement data. A built-in microprocessor provides more useful information, such as modulus, yield, and failure energies.

High-speed impact testers have been proven very useful in material evaluation. At low strain rates, some polymers fail in a ductile manner. The same polymers appear to show brittle failure at high strain rates. The point at which this ductile to brittle transition takes place is of particular importance. A high-rate impact test can provide such information in a graphical form. Tests can also be carried out at different temperatures to find ductile–brittle transition points at various temperatures (46). Other useful applications of the high-speed impact tester include the process quality control, design evaluation, and assembly evaluation.

F. Miscellaneous Impact Tests

Depending upon the end-use requirement, many different types of impact tests have been devised to stimulate actual impact conditions. Underwriters Laboratories requires a fully assembled, working electronic cabinet to withstand the 5-ft-lb impact of a swinging ball at the most vulnerable spot to check the damage to the electronics. Small units susceptible to being knocked off a desk must pass a 3-ft drop while the units are plugged into the electrical outlets (47).

The drop impact resistance of a blow-molded thermoplastic container is a standard ASTM D 2463 test. The drop impact resistance is determined by dropping conditioned blow-molded containers filled with water from a platform onto a prescribed surface. A commercially available bottle drop apparatus is shown in Figure 2-53. One other type impact test, called the air cannon impact test (ACIT), is used to determine the toughness of rigid plastic exterior building components. The air cannon impact tester consists of a compressed air gun to propel spherical plastic projectiles at a test specimen. Interchangeable barrels allow projectiles of varying sizes and weights to be used. Projectile velocity is controlled by varying air pressure. Polypropylene and polyethylene molded balls of different sizes are used as projectiles to simulate the effect of different sized hailstones. The weight and the velocity of the projectile are used to calculate the impact force absorbed by the test specimen. This test more realistically simulates end-use environmental conditions than conventional impact tests and provides important information for evaluating product and establishing product quality (48). To determine the damage tolerance of a particular panel, a limited-energy or rebound test may be performed (Boeing, NASA, and McDonnell Douglas each have specifications for this type of test). These tests are only possible on drop-weight machines, and involve dropping a known weight from a known height onto the specimen. The falling weight bounces off the panel and onto a catcher mechanism. The onset of damage can be readily identified on the load–time trace. This procedure is often combined with a compression or tension test of the specimen after the impact to determine the effect of the damage on performance.

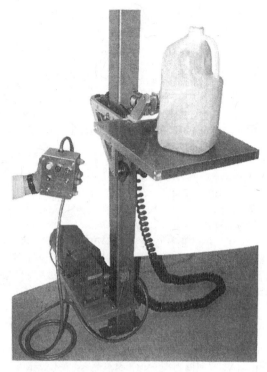

Figure 2-53. Bottle drop impact tester.

2.8. SHEAR STRENGTH (ASTM D 732)

Shear strength of plastic materials is defined as the ability to withstand the maximum load required to shear the specimen so that the moving portion completely clears the stationary portion. The shear strength test is carried out by forcing a standardized punch at a specified rate through a sheet of plastic until the two portions of the specimen completely separate. The shear strength is determined by dividing the force required to shear the specimen by the area of the sheared edge and is expressed as $lb/in.^2$.

2.8.1. Test Specimen and Apparatus

The specimen for the shear strength test can be either molded or cut from a sheet in the form of a 2-in. diameter disc or 2-in.^2 plate. The thickness of the specimen may vary from 0.050 to 0.500 in. A hole with 7/16 in. diameter is drilled through the specimen at its center. A universal testing machine with a constant rate of crosshead movement, such as the one used for tensile, compressive, and flexural strength testing, can be used. The machine requirement has been described in detail in the section on tensile testing. A specially designed shear tool of the punch type is used for shear testing. One such fixture with a sample already mounted in place is shown in Figure 2-54.

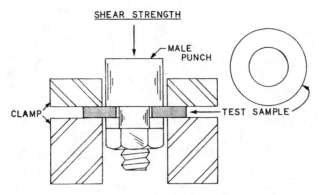

Figure 2-54. Shear strength test setup. (Reprinted with permission of Van Nostrand Reihnold Company.)

2.8.2. Test Procedures

The test is carried out by properly positioning the sample in the shear fixture and lowering the punch onto the specimen until a complete shear-type failure is observed. Shear strength is calculated as follows:

$$\text{Shear strength (psi)} = \frac{\text{Force required to shear the specimen}}{\text{Area of sheared edge}}$$

$$(\text{Area of sheared edge} = \text{Circumference of punch}$$

$$\times \text{Thickness of specimen})$$

2.8.3. Significance and Limitations

Shear strength data is of great importance to a designer of film and sheet products that tend to be subjected to such shear loads. Most large molded and extruded products are usually not subjected to shear loads. The shear strength data as reported in the material suppliers literature should be used with extreme care when designing a product because these shear strength values can be considerably higher than normal. This method allows a user to test a very thin specimen that may stretch excessively and prevent a true shear-type failure, giving high values. For design purposes, shear strength is generally considered to be essentially equal to one-half the tensile strength.

2.9. ABRASION

2.9.1. Introduction

Abrasion resistance of polymeric materials is a complex subject. Many theories have been developed to support the claim that abrasion is closely related to frictional force, load, and true area of contact. An increase in any one of the three generally results in greater wear or abrasion. The hardness of the polymeric materials has a significant effect on abrasion characteristics. For example, a harder material with considerable asperities on the surface will undoubtedly cut through the surface of a softer material to an appreciable depth, creating grooves and scratches. The theory is further complicated by the fact that the abrasion process also creates oxidation on the surface from the buildup of localized high temperatures (49,50). The resistance to abrasion is also affected by other factors, such as the properties of the polymeric material, resiliency, and the type and amount of added fillers or additives.

This all makes abrasion a difficult mechanical property to define as well as to measure adequately. Resistance to abrasion is defined as the ability of a material to withstand mechanical action (such as rubbing, scraping, or erosion) that tends progressively to remove material from its suface.

Resistance to abrasion is significantly affected by such factors as test conditions, type of abradant, and development and dissipation of heat during the test cycle. Many different types of abrasion-measuring equipment have been developed. However, the correlation between the test results obtained from different machines as well as the correlation between test results and actual abrasion-related wear in real life remains very poor. The tests do, however, provide relative ranking of materials in a certain order when performed under a specified set of conditions.

2.9.2. Abrasion Resistance Tests

The material's ability to resist abrasion is most often measured by its loss in weight when abraded with an abraser. The most widely accepted abraser in the industry is called the Taber abraser. A variety of wheels with varying degree of abrasiveness is available. The grade of "calibrase" wheel designated CS-17 with 1000-g load seems to produce satisfactory results with almost all plastics. For softer materials, less abrasive wheels with a smaller load on the wheels may be used. The test specimen is usually a 4-in.-diameter disc or a 4-in.^{2} plate having both surfaces substantially plane and parallel. A 1/2-in.-diameter hole is drilled in the center. Specimens are conditioned employing standard conditioning practices prior to testing. To commence testing, the test specimen is placed on a revolving turntable. Suitable abrading wheels are placed on the specimen under certain set dead-weight loads. The turntable is started and an automatic counter records the number of revolutions. Most tests are carried out to at least 5000 revolutions. The specimens are weighed to the nearest milligram. The test results are reported as weight loss in milligrams per 1000 cycles. The grade of abrasive wheel along with amount of load at which the test was carried out is always reported along with results.

Test methods such as ASTM D 1044 (resistance of transparent plastic materials to abrasion) are also developed for estimating the resistance of transparent plastic materials to one kind of abrasion by measurement of its optical affects. The test is carried out in similar manner to that described above, except that 100 cycles with a 500-g load is normally used. A photoelectric photometer is used to measure the light scattered by abraded track. The percentage of the transmitted light that is diffused by the abraded specimens is reported as a test result. Test method ISO 9352 describes resistance to wear by abrasive wheels.

Another method to study the resistance of plastic material to abrasion is by measuring the volume loss when a flat specimen is subjected to abrasion with loose abrasive or bonded abrasive on cloth or paper. This method is designated as ASTM D 1242. Figure 2-55 illustrates a commercially available abrasion tester.

2.10. FATIGUE RESISTANCE

2.10.1. Introduction

The behavior of materials subjected to repeated cyclic loading in terms of flexing, stretching, compressing, or twisting is generally described as fatigue. Such repeated cyclic loading eventually constitutes a mechanical deterioration and progressive fracture that leads to complete failure. Fatigue life is defined as the number of cycles of deformation required to bring about the failure of the test specimen under a given set of oscillating conditions (51).

The failures that occur from repeated application of stress or strain are well below the apparent ultimate strength of the material. Fatigue data are generally reported as the number of cycles to fail at a given maximum stress level. The fatigue endurance curve, which represents stress versus number of cycles to failure, also known as the *S–N* curve, is generated by testing a multitude of specimens under cyclic stress, each one at different stress levels. At high stress levels, materials tend to fail at relatively low numbers of cycles. At low stresses, the materials can be

Figure 2-55. Abrasion tester. (Courtesy of Taber Industries.)

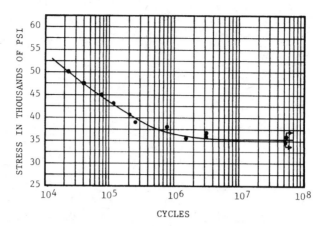

Figure 2-56. Fatigue endurance (*S–N*) curve.

stressed cyclically for an indefinite number of times and the failure point is virtually impossible to establish. This limiting stress below which material will never fail is called the fatigue endurance limit. As shown in Figure 2-56, the *S–N* curve becomes asymptotic to a constant stress line at a certain stress level. The fatigue endurance limit can also be defined as the stress at which the *S–N* curve becomes asymptotic to the horizontal (constant stress) line. For most polymers, the fatigue endurance limit is between 25 and 30 percent of the static tensile strength (52). The fatigue resistance data are of practical importance in the design of gears, tubing, hinges, parts on vibrating machinery, and pressure vessels under cyclic pressures.

Three basic types of tests have been developed to study the fatigue behavior of plastic materials:

1. Flexural fatigue test
2. Tension/compression fatigue test
3. Rotating beam

2.10.2. Flexural Fatigue Test (ASTM D 671)

The ability of a material to resist deterioration from cyclic stress is measured in this test by using a fixed cantilever-type testing machine capable of producing a constant amplitude of force on the test specimen each cycle. The main feature of a fatigue testing machine is an unbalanced, variable eccentric, mounted on a shaft that is rotated at a constant speed by a motor. This unbalanced movement of an eccentric produces alternating force. The specimen is held as a cantilever beam in a vice at one end and bent by a concentrated load applied through a yoke fastened to the opposite end. A counter is used to record the number of cycles along with a cutoff switch to stop the machine when the specimen fails. A typical commercially available fatigue testing machine is illustrated in Figure 2-57. The test specimen of two different geometries are used, as is shown in Figure 5-58. If machined specimens are used, care must be taken to eliminate all scratches and tool marks from the specimens. Molded specimens must be stress-relieved before using.

Figure 2-57. Flexural fatigue tester. (Courtesy of CEAST U.S.A. Inc.)

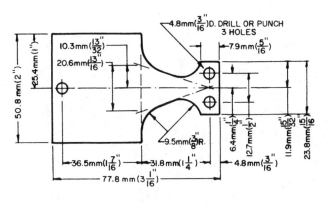

TYPE A

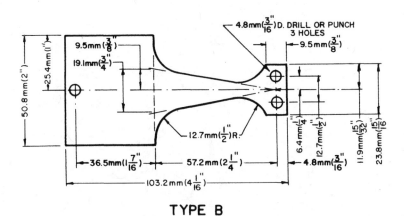

TYPE B

Figure 2-58. Dimensions of constant force fatigue specimen. (Reprinted with permission of ASTM.)

The test is carried out by first determining the complementary mass and effective mass of the test specimen. The load required to produce the desired stress is calculated from these values. The number of cycles required to produce failure is determined. The test is repeated at varying stress levels. A curve of stress versus cycles to failure ($S-N$ diagram) is plotted from the test results.

2.10.3. Tensile Fatigue Test

Unlike the flexural fatigue test, which uses the constant deflection (strain) principle, the tensile fatigue test is conducted under constant load (stress) conditions. The testing machine most commonly used is shown in Figure 2-59. Here, the cyclic load is applied to one end of the test specimen through deflection-calibrated lever which is driven by a variable throw crank. Static and dynamic loads are applied to opposite ends of the specimen, making it possible to maintain constant load on the specimen regardless of dimensional change caused by specimen fatigue.

Figure 2-59. Tensile fatigue testing machine. (Courtesy of Fatigue Dynamics, Inc.)

2.10.4. Rotating Beam Test

The rotating beam fatigue test is generally not used to test plastics materials due to hysteresis phenomena. When used, the test is conducted at a very low testing speed to avoid premature failure due to heat buildup. The test is conducted by mounting both ends of the dumbbell specimen in the testing machine. The specimen is rotated between two spindles, and stress in the form of tension and compression is applied. The specimen is subjected to the number of cycles of stress specified or until fracture occurs.

2.10.5. Factors Affecting the Test Results and Limitation of Fatigue Tests

1. The data obtained from fatigue tests can be directly applied in designing a specific part only when all variables such as size and shape of the part, method of specimen preparation, loading, ambient and part temperature, and frequency of stressing are identical to those used during testing. In practice, such identical conditions never occur and, therefore, it is very important to conduct the trial on an actual part with the end-use conditions simulated as closely as possible. Another reason for not relying heavily on fatigue test data is the possibility of the presence of a notch, a scratch, or voids in the fabricated part that can cause localized stress concentration and lower the overall fatigue resistance considerably.

2. The correlation of test results obtained from different types of machines is very poor. This is due to the differences in specimen preparation techniques.

3. The thickness of the sample greatly influences the test results.

4. The test results are seriously affected by changes in testing temperature, test frequency, and rate of heat transfer. This is particularly true in the case of plastics having appreciable damping.

5. The fatigue life of a polymer is generally reduced by an increase in temperature (53), although this is not always true for elastoplastic materials.

6. Constant deflection or strain testers have a disadvantage in that once a large crack develops, the stress level drops below the fatigue endurance limit and the specimen does not fail for quite some time. However, this test allows one to observe crack propagation in the specimen due to slow rate of failure. In contrast, in the constant stress tester, once the crack develops, the amplitude of deformation increases and failure occurs very rapidly (54).

2.11. HARDNESS TESTS

2.11.1. Introduction

Hardness is defined as the resistance of a material to deformation, particularly permanent deformation, indentation, or scratching. Hardness is purely a relative term and should not be confused with wear and abrasion resistance of plastic materials. Polystyrene, for example, has a high Rockwell hardness value but a poor abrasion resistance. Hardness tests can differentiate the relative hardness of different grades of a particular plastic. However, it is not valid to compare the hardness of various plastics entirely on the basis of one type of test, since elastic recovery along with hardness is involved. The test is further complicated by the phenomenon of creep. Many tests have been devised to measure hardness. Because plastic materials vary considerably with respect to hardness, one type of hardness test does not cover the entire range of hardness properties encountered. Two of the most commonly used tests for plastics are the Rockwell and the Durometer hardness tests. The Rockwell test is used for relatively hard plastics such as acetals, nylons, acrylics, and polystyrene. For softer materials such as flexible PVC, thermoplastic rubbers, and polyethylene, Durometer hardness is measured. The typical hardness values of some common plastic materials are listed for comparison in Table 2-3. A comparison of various types of hardness scales is also given in Appendix I.

2.11.2. Rockwell Hardness (ASTM D 785)

The Rockwell hardness test measures the net increase in depth impression as the load on an indenter is increased from a fixed minor load to a major load and then returned to a minor load. The hardness numbers derived are just numbers without units. Rockwell hardness numbers are always quoted with a scale symbol representing the indenter size, load, and dial scale used. The hardness scales in order of increasing hardness are R, L, M, E, and K scales. The higher the number in each scale, the harder the material. There is a slight overlap of hardness scales and, therefore, it is quite possible to obtain two different dial readings on different scales for the same material. For a specific type of material, correlation in the overlapping regions is possible. However, owing to differences in elasticity, creep, and shear characteristics between different plastics, a general correlation is not possible.

TABLE 2-3. Typical Hardness Values of Some Common Plastic Materials

	Hardness		
	Rockwell		Durometer
Plastic Material	M	R	Shore D
ABS		75–115	
Acetal	94	120	
Acrylic	85–105		
Cellulosics		30–125	
PPO	80	120	
Nylon		108–120	
Polycarbonate	72	118	
H.D.P.E.			60–70
L.D.P.E.			40–50
Polypropylene			75–85
Polystyrene (G.P.)	68–70		
PVC (rigid)		115	
Polysulfone	70	120	

Figure 2-60. Rockwell hardness tester. (Courtesy of Instron Corporation.)

A. Test Apparatus and Specimen

Rockwell hardness is determined with an apparatus called the Rockwell hardness tester. Figure 2-60 illustrates a typical Rockwell hardness tester. A standard specimen of 1/4-in. minimum thickness is used. The specimen can either be molded or cut from a sheet. However, the test specimen must be free from sink marks, burrs, or other protrusions. The specimen also must have parallel, flat surfaces.

B. Test Procedures

The specimen is placed on the anvil of the apparatus and minor load is applied by lowering the steel ball onto the surface of the specimen. The minor load indents the specimen slightly and assures good contact. The dial is adjusted to zero under minor load and the major load is applied within 10 sec by releasing the trip level. After 15 sec the major load is removed and the specimen is allowed to recover for an additional 15 sec. Rockwell hardness is read directly off the dial with the minor load still applied. Figure 2-61 schematically illustrates the operating principle behind the Rockwell hardness tester.

2.11.3. Durometer Hardness (ASTM D 2240, ISO 868)

The Durometer hardness test is used for measuring the relative hardness of soft materials. The test method is based on the penetration of a specified indentor forced into the material under specified conditions.

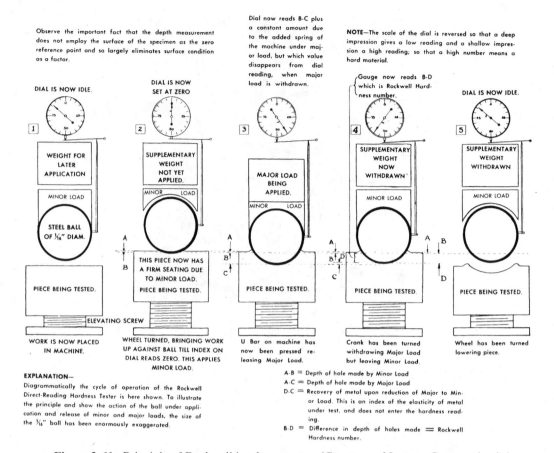

Figure 2-61. Principle of Rockwell hardness tester. (Courtesy of Instron Corporation.)

The Durometer hardness tester consists of a pressure foot, an indentor, and an indicating device. The indentor is spring loaded and the point of the indentor protrudes through the hole in the base. The test specimens are at least 1/4 in. thick and can either be molded or cut from a sheet. Several thin specimens may be piled to form a 1/4 in.-thick specimen but one-piece specimens are preferred. The poor contact between the thin specimens may cause results to vary considerably.

The test is carried out by first placing a specimen on a hard, flat surface. The pressure foot of the instrument is pressed onto the specimen, making sure that it is parallel to the surface of the specimen. The durometer hardness is read within 1 sec after the pressure foot is in firm contact with the specimen.

Two types of durometers are most commonly used—Type A and Type D. The basic difference between the two types is the shape and dimension of the indentor. The hardness numbers derived from either scale are just numbers without any units. The Type A durometer is used with relatively soft material and Type D is used with slightly harder material. A commercially available durometer hardness measuring instrument is shown in Figure 2-62.

Figure 2-62. Durometer hardness tester. (Courtesy of Instron Corporation.)

2.11.4. Barcol Hardness (ASTM D 2583)

The Barcol hardness test was devised mainly for measuring hardness of both rein-forced and nonreinforced rigid plastics. The tester is a portable instrument that can be carried around to measure hardness of fabricated parts as well as the test specimens.

Barcol hardness tests consist of an indentor with a sharp, conical tip and indicat-ing device in the form of a dial with 100 divisions. Each division represents a depth of 0.0003-in. penetration. The test specimens are required to be of 1/16-in. minimum thickness.

The test is carried out by placing the indentor onto the specimen and applying uniform pressure against the instrument. The pressure is applied until the dial indication reaches maximum. The depth of penetration is automatically converted to a hardness reading in absolute Barcol numbers. When measuring the Barcol hardness of the reinforced plastic material, the variation in hardness reading caused by the difference in hardness between resin and filler materials should be taken into account.

Generally, a larger sample size is recommended for reinforced plastic specimens than that used for nonreinforced plastics. Figure 2-63 illustrates a commercially available Barcol hardness tester.

D. Anisotropy
Plastic materials with anisotropic characteristics may cause indentation hardness to vary with the direction of testing.

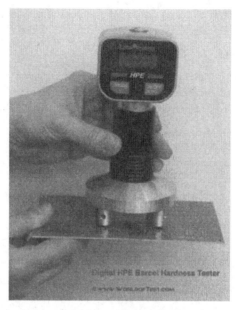

Figure 2-63. Barcol hardness tester. (Courtesy of Qualitest USA)

REFERENCES

1. Kinney, G. F., *Engineering Properties and Application*, John Wiley & Sons, New York, 1957, pp. 182–184.
2. *Ibid.*, p. 187.
3. Rodriguez, F., *Principles of Polymer Systems*, McGraw-Hill, New York, 1970, Chapter 8.
4. Baer, E., *Engineering Design for Plastics*, Reinhold, New York, 1964, Chapter 4.
5. Williams, J. G., *Stress Analysis of Polymers*, Longman Group Limited, London, 1973, Chapter 3.
6. Rodriguez, Reference 3, pp. 229–230.
7. Billmeyer, F. W., *Textbook of Polymer Science*, Interscience, New York, 1962, pp. 109–111.
8. Baer, Reference 4, p. 187.
9. Milby, R., *Plastics Technology*, McGraw-Hill, New York, 1973, pp. 490–491.
10. Billmeyer, Reference 7, p. 193.
11. Baer, Reference 4, p. 309.
12. Lubin, G., *Handbook of Fiberglass and Advanced Plastics Composites*, Reinhold, New York, 1969, p. 171.
13. LNP Corporation, *Tech. Bulletin*, "Predict Shrinkage and Warpage of Reinforced and Filled Thermoplastics," Malvern, PA (1978).
14. Heap, R. D. and Norman, R. H., *Flexural Testing of Plastics*, The Plastics Institute, London, England, 1969, p. 13.
15. *Ibid.*, p. 13.

16. O'Toole, J. L., "Creep Properties of Plastics," *Modern Plastics Encyclopedia*, McGraw-Hill, New York, 1968–1968, p. 48.

17. Kinney, Reference 1, p. 192.

18. Baer, Reference 4, p. 284.

19. Delatycki, O., "Mechanical Performance and Design in Polymers," *Applied Polymer Symposia*, **17**(134), Interscience, New York, 1971.

20. Kinney, Reference 1, p. 192.

21. O'Toole, Reference 16, p. 412.

22. Delatycki, Reference 19, pp. 143–145.

23. "Design Guide," *Modern Plastics Encyclopedia*, McGraw-Hill, New York, 1979–1980, pp. 480–481.

24. Deanin, R. D., *Polymer Structure, Properties and Applications*, Cahners Publishing Co., Boston, MA, 1972, p. 171.

25. Dubois, J. H. and Levy, S., *Plastics Product Design Engineering Handbook*, Reinhold, New York, 1977, pp. 104–105.

26. Vincent, P. I., *Impact Tests and Service Performance of Thermoplastics*, The Plastics Institute, London, England, 1971, p. 12.

27. Deanin, Reference 24, p. 171.

28. Vincent, Reference 26, p. 33.

29. *Ibid.*, p. 44.

30. Spath, W., *Impact Testing of Materials*, Grodon and Breach Science Publishers, New York, 1961, p. 37.

31. Bergen, R. L., "Tests for Selecting Plastics," *Metal Progress* (Nov. 1966), pp. 107–108.

33. Westover, R. F. and Warner, W. C., "Tensile Impact Test for Plastic Materials," *Res. and Stand.*, **1**(11) (1961), pp. 867–871.

33. Bragaw, C. G., "Tensile Impact," *Mod. Plast.*, **33**(10) (1956), p. 199.

34. Westover, R. F., "The Thirty Years of Plastics Impact Testing," *Plastics Technology*, **4** (1958), pp. 223–228, 348–352.

35. Maxwell, B. and Harrington, J. P., "Effect of Velocity on Tensile Impact Properties of PMMA," *Trans. ASME*, **74** (1952), p. 579.

36. Nielsen, L. E., *Mechanical Properties of Polymers*, Reinhold, New York, 1962, pp. 124–125.

37. Baer, Reference 4, p. 799.

38. *Ibid.*, p. 803.

39. ASTM D 1709 (Part 36); ASTM D 3029 (part 35), Annual Book of ASTM Standards, Philadelphia, PA, 1978.

40. Tryson, G. R., Takemori, M. T., and Yee, A. F., "Puncture Testing of Plastics: Effects of Test Geometry," *ANTEC*, **25** (1979), pp. 638–641.

41. Abolins, V., "Gardner Impact Versus Izod, Which Is Better for Plastics?" *Materials Eng.* (Nov. 1973), p. 52.

42. Goldman, T. D. and Lutz, J. T., "Developing Low Temperature Impact Resistance PVC: A New Testing Approach," *ANTEC*, **25** (1979), pp. 354–357.

43. Starita, J. M., "Impact Testing," *Plast. World* (April 1977), p. 58.

44. McMichael, S. and Fischer, S., "Understanding Materials with Instrumented Impact," Materials Engineering, April 1989, p. 47.

45. Tardif, H. P. and Marquis, H., "Impact Testing with an Instrumented Machine," *Metal Progress* (Feb. 1964), p. 58.

46. Starita, Reference 43, p. 59.

47. "The Perils of Izod—Part 1," *Plast. Design Forum* (May/June 1980), pp. 13–20.

48. Tanzillo, J. D., "Development of an Impact Test for Evaluation of Toughness of Rigid Plastic Building Components," *ANTEC*, **15** (1969), pp. 346–349.

49. Nielsen, Reference 36, p. 228.

50. Marcucci, M. A., *S.P.E.J.*, **14** (Feb. 1958), p. 30.

51. Nielsen, Reference 36, p. 230.

52. Lazan, B. J. and Yorgiadis, A., "Symposium on Plastics," *ASTM Spec. Tech. Pub.*, No. 59, (Feb. 1944), p. 66.

53. Nielsen, Reference 36, p. 230.

54. Nielsen, Reference 36, p. 230.

SUGGESTED READING

Dietz, A. G. and Eirich, F. R., *High Speed Testing*, Vol. I, Interscience, New York, 1960.

Lysaght, V. E., *Indentation Hardness Testing*, Reinhold, New York, 1949.

Alfrey, T., *Mechanical Behavior of High Polymers*, Interscience, New York, 1948.

"Simulated Service Testing in the Plastics Industry," ASTM Spec. Tech. Pub. 375.

Ritchie, P. D., *Physics of Plastics*, Van Nostrand, Princeton, NJ, 1965.

Ward, I. M. and Pinnock, P. R., "The Mechanical Properties of Solid Polymers," *Br. J. Appl. Phys.*, **17**(3) (1966).

Vincent, P. I., "Mechanical Properties of High Polymers: Deformation," in *Physics of High Polymers*, Ritchie, P. D. (Ed.), Iliffe, London, 1965, Chapter 2.

Bueche, F., *Physical Properties of High Polymers*, Wiley, New York, 1960.

Tobolsky, A. V., *Properties and Structure of Polymers*, Wiley, New York, 1960.

Murphy, A., *Properties of Engineering Materials*, 2nd ed., International Textbook Co., Scranton, PA, 1947, p. 402.

Brown, W. E., *Testing of Polymers*, Interscience, New York, 1969.

Mark, H. F., Gaylord, N. G., and Bikales, M., Eds., *Encyclopedia of Polymer Science and Technology*, Interscience, New York, 1964–1970.

Schmitz, J. V., Ed., *Testing of Polymers*, Interscience, New York, 1965.

Kulow, P., *Rubber and Plastics Testing*, Chapman and Hall, London, England, 1963.

Davis, H. E., Troxell, G. E., and Wiskocil, C. T., *The Testing and Inspection of Engineering Materials*, McGraw-Hill, New York, 1955.

Liddicoat, R. T. and Potts, P., *Laboratory Manual of Material Testing*, Macmillan Co., New York, 1952.

Findley, W. N., "Creep Characteristics of Plastics," *ASTM Symposium on Plastics*, Philadelphia, PA, 1944.

Turner, S., "Creep Studies on Plastics," *J. Appl. Polym. Sci. Symp.*, **17** (1971).

Gotham, K. V., "A Formalized Experimental Approach to the Fatigue of Thermoplastics," *Plastics and Polymers*, **37**(130) (1969), p. 309.

Larson, F. R. and Miller, J., "A Time–Temperature Relationship for Rupture and Creep Stresses," *Trans. ASME* (July 1952), pp. 765–775.

Constable, I., Williams, J. G., and Burns, D. J., "Fatigue and Cyclic Thermal Softening of Thermoplastics," *J. Mech. Eng. Sci.*, **12**(1) (1970).

Cessna, L. C., Levens, J. A., and Thomson, J. B., "Flexural Fatigue of Glass-Reinforced Thermoplastics," *Polym. Eng. Sci.*, **9**(5) (1969), pp. 339–349.

Horsley, R. A. and Morris, A. C., "Impact Tests—A Guide to Thermoplastics Performance," *Plastics*, **31**(3) (1966), pp. 1551–1553.

Thomas, J. R. and Hagan, R. S., "Meaningful Testing of Plastic Materials for Major Appliances," *S.P.E.J.*, **22** (1966), PP. 51–56.

Hertzberge, R. W. and Manson, J. A., *Fatigue of Engineering Plastics*, Academic, New York, 1980.

McMichael, S. and Fischer, S., "Understanding Materials with Instrumented Impact," Mechical Engineering, April 1989, p. 47.

Mackin, T. J., "A Comparison of Instrumented Impact Testing and Gardner Impact Testing," Instron Corp., Canton, MA.

3

THERMAL PROPERTIES

3.1. INTRODUCTION

The thermal properties of plastic materials are equally as important as the mechanical properties. Unlike metals, plastics are extremely sensitive to changes in temperature. The mechanical, electrical, or chemical properties of plastics cannot be examined without looking at the temperature at which the values were derived. Crystallinity has a number of important effects upon the thermal properties of a polymer. Its most general effects are the introduction of a sharp melting point and the stiffening of thermal mechanical properties. Amorphous plastics, in contrast, have a gradual softening range (1). Molecular orientation also has a significant effect on thermal properties. Orientation tends to decrease dimensional stability at higher temperatures (2). The molecular weight of the polymer affects the low-temperature flexibility and brittleness. Many other factors, such as intermolecular bonding, cross linking, and copolymerization, all have a considerable effect on thermal properties. Hence, it is clear that the thermal behavior of polymeric materials is rather complex. Therefore, in designing a plastic part or selecting a plastic material from the available thermal property data, one must thoroughly understand the short term as well as the long-term effects of temperature on the properties of that plastic material.

3.2. TESTS FOR ELEVATED TEMPERATURE PERFORMANCE

Designers and material selectors of plastic products constantly face the challenge of selecting a suitable plastic for elevated temperature performance. The difficulty arises because of the varying natures and capabilities of various types and grades of plastics at elevated temperatures. Many factors are considered when selecting a plastic for a high-temperature application. The material must be able to support a design load under operating conditions without objectionable creep or distortion. The material must not degrade or lose necessary additives that will cause a drastic reduction in the physical properties during the expected service life.

All the properties of plastic materials are not affected in a similar manner by elevating temperature. For example, electrical properties of a particular plastic may show only a moderate change at elevate temperatures, while the mechanical properties may be reduced significantly. Also, since the properties of plastic materials vary with temperature in an irregular fashion, they must be looked at as a function of temperature in order to obtain more meaningful information. From the foregoing, it is quite clear that a single maximum-use temperature that will apply to all the important properties in high-temperature applications is simply not possible.

When studying the performance of plastics at elevated temperatures, one of the most important considerations is the dependence of key properties such as modulus, strength, chemical resistance, and environmental resistance on time. Therefore, the short-term heat resistance data alone is not adequate for designing and selecting materials that require long-term heat resistance. For the sake of convenience and simplicity, we divide the elevated temperature effects into two categories:

1. Short-term effects
   a. Heat deflection temperature
   b. Vicat softening temperature
   c. Torsion pendulum
2. Long-term effects
   a. Long-term heat resistance test
   b. UL temperature index
   c. Creep modulus/creep rupture tests

3.2.1. Short-Term Effects

A. Heat Deflection Temperature (HDT) (ASTM D 648, ISO 75-1 & 75-2)
Heat deflection temperature is defined as the temperature at which a standard test bar (5 × 1/2 × 1/4 in.) deflects 0.010 in. under a stated load of either 66 or 264 psi. The heat deflection temperature test, also referred to as the heat distortion temperature test, is commonly used for quality control and for screening and ranking materials for short-term heat resistance. The data obtained by this method cannot be used to predict the behavior of plastics materials at elevated temperature nor can it be used in designing a part or selecting and specifying material. Heat deflection temperature is a single-point measurement and does not indicate long-term heat resistance of plastics materials. Heat distortion temperature, however, does distinguish between those materials that lose their rigidity over a narrow temperature range and those that are able to sustain light loads at high temperatures (3).

Apparatus and Test Specimens. The apparatus for measuring heat deflection temperature consists of an enclosed oil bath fitted with a heating chamber and automatic heating controls that raise the temperature of the heat transfer fluid at a uniform rate. A cooling system is also incorporated to fast cool the heat transfer medium for conducting repeated tests. The specimens are supported on steel supports that are 4 in. apart, with the load applied on top of the specimen vertically and midway between the supports. The contact edges of the support and of the piece by which pressure is applied is rounded to a radius of 1/4 in. A suitable deflection measurement device, such as a dial indicator, is normally used. A mercury thermometer is used for measuring temperature. The unit is capable of applying 66- or 264-psi fiber stress on specimens by means of a dead weight. A commercially available heat deflection measuring device with a closeup of a specimen holder is illustrated in Figure 3-1.

More recently, automatic heat deflection temperature testers have been developed. These testers typically replace conventional temperature and deflection measuring devices with more sophisticated electronic measuring devices containing a digital read-out system and a chart recorder that prints out the results. Such an automatic apparatus eliminates the need for the continuous presence of an operator and thereby minimizes operator-related errors.

The test specimens consist of test bars 5 in. in length and 1/2 in. in depth by any width from 1/8 to 1/2 in. The test bars may be molded or cut from extruded sheet as long as they have smooth, flat surfaces and are free of excessive sink marks or flash. The specimens are conditioned employing standard conditioning procedures.

Test Procedure. The specimen is positioned in the apparatus along with the temperature and deflection measuring devices and the entire assembly is submerged into the oil bath kept at room temperature. The load is applied to a desired value (66- or 264-psi fiber stress). Five minutes after applying the load, the pointer is adjusted to zero and the oil is heated at the rate of $2 \pm 0.2°C/min$. The temperature of the oil at which the bar has deflected 0.010 in. is recorded as the heat deflection temperature at the specified fiber stress.

Test Variables and Limitations

Residual Stress. The specimens consisting of a high degree of "molded-in" residual stresses or a high degree of orientation have a tendency to stress

Figure 3-1. Dtul/Vicat tester. (Courtesy of CEAST U.S.A. Inc.)

relieve as the temperature is increased. The specimen warpage occurs in a downward direction owing to the external loading. The warpage due to stress relaxation and deflection, occurring from the softening of the specimen, combined together, yields a false heat deflection temperature. However, if the specimens are annealed in a controlled oven atmosphere prior to testing, the stresses can be relieved and warpage can practically be eliminated. Heat deflection temperature of the unannealed specimen is usually lower than that for a comparable annealed specimen.

Specimen Thickness. Thicker specimens tend to exhibit a higher heat deflection temperature. This is because of the inherently low thermal conductivity of plastic materials. The thicker specimen requires a longer time to heat through, yielding a higher heat deflection temperature.

Fiber Stress. The higher the fiber stress or loading is, the lower the heat deflection temperature. The difference in heat deflection temperature values resulting from different fiber stress varies depending upon the type of polymer.

Specimen Preparation. Injection-molded specimens tend to have a lower heat deflection temperature than compression-molded specimens. This is because compression-molded specimens are relatively stress-free.

B. Vicat Softening Temperature (ASTM D 1525, ISO 306)

The Vicat softening temperature is the temperature at which a flat-ended needle of 1-mm^2 circular cross section will penetrate a thermoplastic specimen to a depth of 1 mm under a specified load using a selected uniform rate of temperature rise. This test is very similar to the deflection temperature under the load test and its usefulness is limited to quality control, development, and characterization of materials. The data obtained from this test is also useful in comparing the heat-softening qualities of thermoplastic materials. However, the test is not recommended for flexible PVC or other materials with a wide Vicat softening range.

The test apparatus designed for deflection temperature under load test can be used for the Vicat softening temperature test with minor modification. The flat test specimen is molded or cut from a sheet with a minimum thickness and width of 0.12 and 0.50 in., respectively.

The test is carried out by first placing the test specimen on a specimen support and lowering the needle rod so that the needle rests on the surface of the specimen. The temperature of the bath is raised at the rate of 50 or 120°C/hr uniformly. The temperature at which the needle penetrates 1 mm is noted and reported at the Vicat softening temperature. A commercially available test apparatus for measuring the Vicat softening temperature is illustrated in Figure 3-2.

C. Torsion Pendulum Test (ASTM D 2236)

The elevated temperature performance of most plastics is dominated by the temperature of occurrence and the temperature range of the glass transition and, in the case of semicrystalline plastics, by the crystalline melting point. The glass transition has a profound negative effect on all mechanical properties except impact, and on certain thermal properties such as the coefficient of expansion. Therefore, a knowledge of the transitional behavior of plastics is necessary in order to understand their elevated temperature capabilities. This knowledge is best derived from

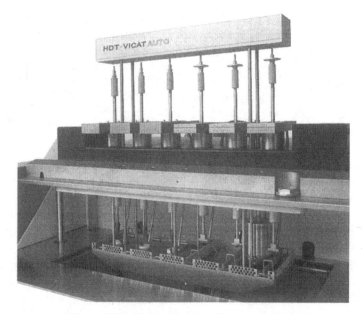

Figure 3-2. Vicat softening point apparatus.

a full temperature modulus plot, especially a dynamic modulus, since the elevated temperature behavior of plastics is too complex and too varied to be described by simple tests, such as Vicat softening temperature and deflection temperature under load (4).

Many different experimental techniques exist today to study the effects of elevated temperature on dynamic mechanical properties of plastics. Perhaps, the most meaningful and simplest technique is the torsion pendulum technique. Using this technique, one can derive a plot of short-time modulus versus temperature up to the beginning of melting and degradation.

The apparatus used for measuring dynamic properties over a wide range of temperatures is quite simple. It consists of a rigidly fixed clamp at one end. The movable clamp is attached to an inertial member, usually a disk or a rod with a known moment of inertia. A differential transformer and strip chart recorder are used to record angular displacement versus time in graphical form as the specimen is oscillating. An insulated chamber is also used, equipped with heating and cooling systems to facilitate testing over a wide range of temperatures. Torsion pendulum test equipment is illustrated in Figure 3-3.

Specimens of rectangular or cylindrical shape and different length and width are used. However, since the thickness of the specimen influences the dynamic results, it is usually specified. To carry out the test, the temperature equilibrium is first established in the chamber. The clamped specimen is put into oscillation and the period and rate of decay of the amplitude of oscillation are measured. The elastic shear modulus is calculated from specimen dimensions, moment of inertia of the movable member, and period of the movement. A quantity known as log decrement is calculated from the rate of amplitude decay. A detailed procedure to calculate these values is given in the literature (5).

Figure 3-3. Torsion pendulum. Test equipment (schematic). (Reprinted with permission of McGraw-Hill Company.)

The log decrement is observed to be an approximate first derivative of the modulus temperature curve and thus has maxima at the temperatures at which the modulus shows a rapid drop (6). This dramatic drop in modulus indicates the effective maximum load-bearing temperature.

Figure 3-4 illustrates typical torsion pendulum test data graphically. The elastic shear modules versus temperature graph shows the effects of glass transition temperature. The modulus is very high at the beginning of the curve and decreases very slowly as the temperature increases. Near the glass transition temperature, the modulus decreases very rapidly in a short temperature span. The damping which is expressed as log decrement is calculated from the rate at which the amplitude of the oscillation decreases. The plot of log decrement versus temperature shows the onset of the transition. This temperature is regarded as the maximum usable load-bearing temperature.

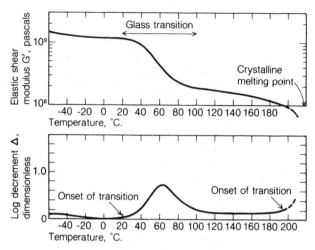

Figure 3-4. Dynamic mechanical properties by torsion pendulum versus temperature of a typical semicrystalline thermoplastic (nylon G, dry) showing effects of glass transition and crystalline melting transition and method of determining their onsets. (Reprinted with permission of McGraw-Hill Company.)

Even though torsion pendulum tests run at elevated temperatures provide very useful information regarding the dynamic mechanical properties of plastics, one must not forget that it is, nevertheless, a short-term test. The data obtained from creep modulus and creep rupture tests conducted at elevated temperatures must be relied on for long-term effects.

3.2.2. Long-Term Effects

The long-term effects of elevated temperature on properties of plastics are extremely important, especially when one considers that the majority of applications involving high heat are long-term applications. During long-term exposure to heat, plastic materials may encounter many physical and chemical changes. A plastic material that shows little or no effect at elevated temperature for a short time may show a drastic reduction in physical properties, a complete loss of rigidity, and severe thermal degradation when exposed to elevated temperature for a long time. Along with time and temperature, many other factors such as ozone, oxygen, sunlight, and pollution combine to accelerate the attack on plastics. At elevated temperatures, many plastics tend to lose important additives such as plasticizers and stabilizers, causing plastics to become brittle or soft and sticky.

Three basic tests have been developed and accepted by the plastics industry. If the application does not require the product to be exposed to elevated temperature for a long period under continuous load, a simple heat-resistance test is adequate. The applications requiring the product to be under continuous significant load must be looked at from creep modulus and creep rupture strength test data. Another widely accepted method of measuring maximum continuous use temperature has been developed by Underwriters Laboratories. The UL temperature index, established for a variety of plastic materials to be used in electrical applications, is the maximum temperature that the material may be subject to without fear of premature thermal degradation.

A. Long-Term Heat-Resistance Test (ASTM D 794)

The long-term heat-resistance test was developed to determine the permanent effect of heat on any property by selection of an appropriate test method and specimen. In ASTM recommended practice, only the procedure for heat exposure is specified and not the test method or specimen.

Any specimen, including sheet, laminate, test bar, or molded part may be used. If a specific property, such as tensile strength loss is to be determined, a standard tensile test bar specimen and procedures must be used for comparison of test results before and after the test. The test requires the use of a mechanical convection oven with a specimen rack of suitable design to allow air circulation around the specimens. The test is carried out by simply placing the specimen in the oven at a desired exposure temperature for a predetermined length of time. The subsequent exposure to temperatures may be increased or decreased in steps of 25°C until a failure is observed. Failure due to heat is defined as a change in appearance, weight, dimension, or other properties that alter plastic material to a degree that it is no longer acceptable for the service in question. Failure may result from blistering, cracking, loss of plasticizer, or other volatile material that may cause embrittlement, shrinkage, or change in desirable electrical or mechanical properties.

Many factors affect the reproducibility of the data. The degree of temperature control in the oven, the type of molding, cure, air velocity over the specimen, period of exposure, and humidity of the oven room are some of these factors. The amount and type of volatiles in the molded part or specimen may also affect the reproducibility.

B. UL Temperature Index

The increased use of plastic materials in electrical applications such as appliances, portable electrically operated tools and equipments, and enclosures has created a renewed interest in the ability of plastics to withstand mechanical abuse and high temperature. A serious personal injury, electric shock, or fire may occur if the product does not perform its intended function (7). Underwriters Laboratories, an independent, not-for-profit organization concerned with consumer safety, has developed a temperature index to assist UL engineers in judging the acceptability of individual plastics in specific applications involving long-term exposure to elevated temperature. The UL temperature index correlates numerically with the temperature rating or maximum temperature in degrees Celsius above which a material may degrade prematurely and therefore be unsafe (8).

Relative Thermal Indices. The relative thermal index of a polymeric material is an indication of the material's ability to retain a particular property (physical, electrical, etc.) when exposed to elevated temperatures for an extended period of time. It is a measure of the material's thermal endurance. For each material, a number of relative thermal indices can be established, each index related to a specific thickness of the material (9).

The relative index of a material is determined by comparing the thermal-aging characteristics of one material of proven field service at a particular temperature level with those of another material with no field service history. A great deal of consideration is given to the properties that are evaluated to determine relative thermal index. For the relative thermal index to be valid, the properties being stressed in the end-product must be included in the thermal-aging program. If, for any reason, the specific property under stress in the end-product is not part of the long-term aging program, the relative thermal index may not be applicable to the use of the material in that particular application.

Relative Thermal Index Based Upon Historical Records. Through experience gained from testing a large volume of complete products and insulating systems over a long period, UL has established relative thermal indices on certain types of plastics. These fundamental temperature indices are applicable to each member of a generic material class. Table 3-1 lists the temperature indices based on past on past field test performance and chemical structure.

Relative Thermal Index Based Upon Long-Term Thermal Aging. The long-term thermal-aging program consists of exposing polymeric materials to beat for a predetermined length of time and observing the effect of thermal degradation. To carry out the testing, an electrically heated mechanical convection oven is preferred; however, with some provisions, a noncirculating static oven may be

TABLE 3-1. Relative Thermal Indices Bases Upon Past Field Test Performance and Chemical Structure

Material	Generic Thermal Index (°C)
Nylon (Type 6, 11, 6/6, and 6/10)[b]	65
Polycarbonate[b]	**80**
Molded phenolic[c,d]	150
Molded melamine[c,d]	130[e]
Molded melamine-phenolic[c,d]	130[e]
Fluorocarbon resins	
(1) Polytetrafluoroethylene	**180**
(2) Polychlorotrifluoroethylene	150
(3) Fluorinated ethylene propylene	150
Silicone rubber (**RTV**)	105
Polyethylene terephthalate film	105
Urea formaldehyde	100
Molded alkyd[c,d]	130
Molded epoxy[c,d]	130
Molded diallyl phthalate[c,d]	130
Molded polyester[c,d] (thermosetting)	130

[a] From UL 746 B. Reprinted with permission from Underwriters Laboratories, Inc.
[b] Includes glass fiber-reinforced materials.
[c] Includes simultaneous heat and high-pressure matched metal die molded compounds only. Excludes low-pressure or low-temperature curing processes such as open-mold (hand lay-up, spray-up, contact bag, filament winding), encapsulation, lamination, and so on.
[d] Includes materials having filler systems of fibrous (other than synthetic organic) types but excludes fiber reinforcement systems using resins that are applied in liquid form.
[e] Compounds having a specific gravity of 1.55 or greater (including those having cellulosic filler material) are acceptable at temperatures not greater than 150°C (302°F).

employed. The specific properties to be evaluated in the thermal-aging program are to be as nearly as possible representative of the properties required in the end-application.

The most common mechanical properties include tensile strength, flexural strength, and Izod impact strength. The electrical properties of concern are dielectric strength, surface or volume resistivity, arc resistance, and arc tracking. The test specimens are standard ASTM test bars, depending upon the type of test. The UL publication "Polymeric Materials—Short-Term Property Evaluations, UL 746A," describes the specimen and test procedures to determine mechanical and electrical properties.

In order to determine the relative thermal index, a control material with a record of good field service at its rated temperature is selected. The control material of the same generic type and the some thickness as the candidate material is preferred. At least four different oven temperatures are selected. The highest temperature is selected so that it will take no more than two months to produce the end-of-life of the material. The next two lower temperatures must produce the anticipated end-of-life of 3 and 6 months, respectively. The lowest temperature selected will take 9–12 months for the anticipated results.

The end-of-life of a material is based upon the assumption that at least a 2:1 factor of safety exists in the applicable physical and electrical property requirements. The end-of-life of a material is the time at each aging temperature, when a property value has decreased 50 percent of its unaged level. A 50 percent loss of property due to thermal degradation is not expected to result in premature, unsafe failure.

Figure 3-5 is a plot of typical time–temperature data obtained from various tests. The insert in Figure 3-5 illustrates the curve obtained as a result of aging a material at four elevated temperatures. The properties of impact strength, tensile strength, and dielectric strength were investigated. At each temperature, the first property to show a reduction of 50 percent of the unaged property value and therefore, failure, is the impact strength. The time-to-failure at each temperature for the impact strength is used to construct an Arrhenius curve B. Curve A represents a plot of a control material having relative thermal index of 100°C. This control material shows a correlation factor of 60,000 hr in this example. The time–temperature plot of the material under investigation crosses the 60,000-hr line at a temperature of 140°C. Therefore, it can be expected to be safe to use at 140°C in the same manner the known material is useful at 100°C.

A relative thermal index of 140°C in this example is applicable to all applications involving all the properties investigated, including impact strength. In applications where impact strength is not a critical property, a higher relative thermal index may be assigned. To avoid underrating the material's relative temperature index, UL publishes the ratings in three categories: applications involving electrical properties only, applications involving both electrical and mechanical properties, and applications involving both electrical and mechanical properties without impact resistance. Underwriters Laboratory publishes such data in its semiannual Plastics Recognized Component Directory. Figure 3-6 shows a page from Plastics Recognized Component Directory. Since the long-term heat-aging resistance of plastics depends on the thickness of the part or test specimen, UL requires that thermal index testing be carried out over a wide range of thicknesses. Finally, it is important

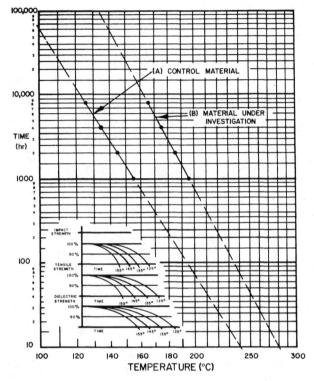

Figure 3-5. Plot typical time–temperature data. (Courtesy of Underwriters Laboratories, Inc.)

QMFZ2 Component - Plastics   Friday, October 24, 2003   E107854

TICONA
FORTRON BUSINESS LINE 86 MORRIS AVE SUMMIT NJ 07901

Material Designation: **1110(a)(b)**

Product Description: Polyphenylene Sulfide (PPS), designated "Fortron" furnished as pellets.

Color	Min. Thick. (mm)	Flame Class	HWI	HAI	RTI Elec	RTI Imp	RTI Str	IEC GWIT	IEC GWFI
NC	0.38	V-0	-	-	130	130	130	-	-
ALL	0.80	V-0	3	4	220	130	130	-	-
	1.5	V-0	2	4	220	130	130	-	-
	3.0	V-0	1	4	220	200	200	-	-

CTI: 4   **IEC CTI: -**   **HVTR: 3**   **D495: 5**   **IEC Ball Pressure** (°C): -

Dielectric Strength (kV/mm): 20   **Volume Resistivity** (10^xohm-cm): 16   **Dimensional Stability**(%): 0
ISO Tensile Strength (MPa): -   **ISO Flexural Strength** (MPa): -   **ISO Heat Deflection** (°C): -
ISO Tensile Impact (kJ/m^2): -   **ISO Izod Impact** (kJ/m^2): -   **ISO Charpy Impact** (kJ/m^2): -

(a)   May be followed by an A, B, D or L to indicate processability

(b)   May be followed by a one digit number 0-9 incl. to indicate molecular weight.

NOTE   These products are also produced and marketed by Ticona GmbH, Postfach 1561, 65444 Kelsterbach, Germany.

Report Date: 12/3/1996   Underwriters Laboratories Inc®

UL94 small-scale test data does not pertain to building materials, furnishings and related contents. UL 94 small-scale test data is intended solely for determining the flammability of plastic materials used in components and parts of end-product devices and appliances, where the acceptability of the combination is determined by ULI.

Figure 3-6. Typical UL yellow card. (Courtesy of Underwriters Laboratories, Inc.)

to understand that the UL temperature index program recognizes that the upper temperature limits of plastics are dependent upon the stresses applied on the end-product in use. Consequently, the temperature index of a particular plastic qualifies it only for those UL applications that UL has specifically approved. UL has developed UL iQ™ for Plastics, a powerful database that allows anyone to search more than 60,000 grades of UL-recognized plastics based on the requirements of specific application. Search can be conducted by product specifications, company name, file name, generic family, grade, and description.

C. Creep Modulus/Creep Rupture Tests

The tests developed to determine creep modulus and creep rupture values of plastic materials are discussed in Chapter 2. Both creep modulus and creep rupture strength decrease significantly as the temperature is increased. Before selecting a material for a load-bearing application at an elevated temperature, one must carefully evaluate the published creep modulus and creep rupture strength test data. This is accomplished by studying bar charts such as the ones shown in Figures 3-7 and 3-8. In one chart, creep modulus of various thermoplastics at three different temperatures is compared. A similar comparison is made using creep rupture strength for the same set of thermoplastic materials. A study of the charts reveals that while most of these thermoplastics have substantial rigidity and strength at room temperature, only a few retain enough of these properties at elevated temperatures to bear significant loads (10).

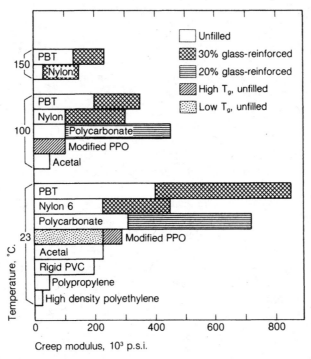

Figure 3-7. Creep modulus versus temperature. (Reprinted with permission of McGraw-Hill Company.)

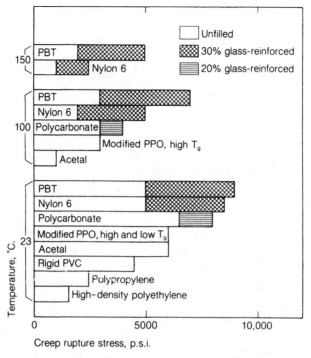

Figure 3-8. Creep rupture strength versus temperature. (Reprinted with permission of McGraw-Hill Company.)

3.3. THERMAL CONDUCTIVITY (ASTM C 177, ISO 8302)

Thermal conductivity is defined as the rate at which heat is transferred by conduction through a unit cross-sectional area of a material when a temperature gradient exists perpendicular to the area. The coefficient of thermal conductivity, sometimes called K factor, is expressed as the quantity of heat that passes through a unit cube of the substance in a given unit of time when the difference in temperature of the two faces is 1°C.

One of the major reasons for the tremendous success of plastics in the last two decades is their low thermal conductivity. Plastics have been readily accepted as materials for cooking utensils and automobile steering wheels. In recent years, cellular plastics, which have the lowest thermal conductivity of all materials, have gained popularity in the field of thermal insulation. The outstanding thermal conductivity of cellular plastics is largely due to the entrapped gases and not to the polymeric material which serves merely as an enclosure for entrapment of gases. As the density of the cellular plastic decreases, the conductivity also decreases up to a minimum value and rises again due to increased convection effects caused by a higher proportion of open cells (11). The quantity of heat flow depends upon the thermal conductivity of the material and upon the distance the heat must flow. This relationship is expressed as (12, 13)

$$Q \sim K/X$$

where Q = quantity of heat flow, K = thermal conductivity, X = the distance the heat must flow. Closed-cell structures provide the lowest thermal conductivity owing to the reduced convection of gas in the cells.

The primary technique for measuring thermal conductivity of insulating materials is the guarded hot plate method. The equipment used for this test is fairly complex and expensive. A schematic assembly of the guarded hot plate is shown in Figure 3-9. The guard heaters are used to prevent the heat flow in all directions except in the axial direction toward the specimens. The test is carried out by placing the specimen between the main heater and the cooling plate (heat sink). The time required to stabilize the input and output temperatures is determined. Thermal conductivity is calculated as follows:

$$K = \frac{Qt}{A(T_1 - T_2)}$$

where K = thermal conductivity (BTU/in./h/ft^2/F); Q = time rate of heat flow (BTU/h); t = thickness of specimen (in.); A = area under test (in^2.); T_1 = temperature of hot surface (°F); T_2 = temperature of cold surface (°F).

Figure 3-10 illustrates a commercially available guarded hot plate equipment. Table 3-2 lists the thermal conductivity of different materials including solid and cellular plastics.

3.4. THERMAL EXPANSION

The coefficient of thermal expansion is defined as the fractional change in length or volume of a material for a unit change in temperature. The coefficient of thermal expansion values for different plastics are of considerable interest to design engineers. Plastics tend to expand and contract anywhere from six to nine times more than materials such as metals. This difference in the coefficient of expansion

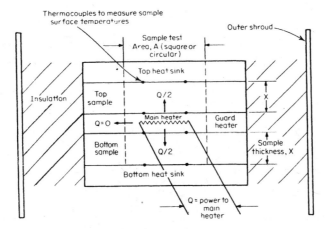

Figure 3-9. Schematic assembly of the guarded hot plate. (Reprinted with permission of McGraw-Hill Company.)

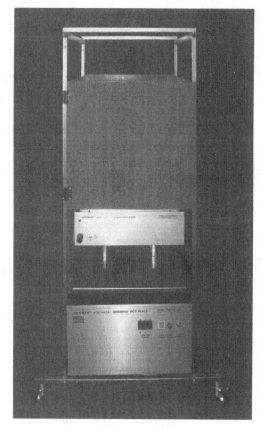

Figure 3-10. Guarded not plate equipment. (Courtesy Anters Corporation.)

develops internal stresses and stress concentrations in the polymer, and premature failures occur. Special expansion joints, which generally require the use of rubber gaskets to overcome the expansion of plastics, are commonly used. The use of a filler such as fiberglass lowers the coefficient of thermal expansion considerably and brings the value closer to that of metal and ceramics.

Two basically similar methods have been developed by ASTM to measure the coefficient of linear thermal expansion (ASTM D 696) and the coefficient of cubical thermal expansion of plastics (ASTM D 864), respectively. The values reported in the trade journals are usually the coefficients of linear thermal expansion.

3.4.1. Coefficient of Linear Thermal Expansion

The test method requires the use of a fused quartz-tube dilatometer, a device for measuring changes in length (dial gauge or LVDT) and liquid bath to control temperature. Figure 3-11 illustrates a schematic configuration of a quartz-tube dilatometer.

The test is commenced by mounting a preconditioned specimen, usually between 2 and 5 in. long, into the dilatometer. The dilatometer, along with the measuring device, is then placed below the liquid level of the bath. The temperature of the

TABLE 3-2. Thermal Conductivity of Different Materials Including Solid and Cellular Plastics

Plastic	Value BTU/(h)(ft²)(°F/in.)
Solid	
Polystyrenes (medium-impact)	0.29–0.87
Polystyrenes (high-impact)	0.29–0.87
Polystyrenes (heat-resistant)	0.29–0.87
Urethanes elastomers	0.49
Polystyrenes (chemical filler)	0.55–0.87
Polyallomer	0.58–1.20
Polypropylene copolymer	0.58–1.20
Polystyrenes (general purpose)	0.70–0.96
Acrylics	0.7–1.7
Polypropylene (unmodified)	0.81
Styrene acrylonitrile copolymer	0.84
Polyvinylidene fluoride (VF2)	0.87
Polypropylene (impact rubber modified)	0.87–1.20
Chlorinated polyether	0.90
PVC flexible (unfilled)	0.99
PVC flexible (filled)	0.99
Styrene butadiene thermoplastic elastomers	1.00
PVC (rigid)	1.10
Nylons (polyamide) type 12 (glass filler)	1.10
Polyphenylene oxides (30 percent glass filler)	1.10
Ethyl cellulose	1.10–2.00
Methylpentene polymer	1.20
Phenolic (fabric and cord filler)	1.20
Phenolic with butadiene–acrylonitrile copolymer (rag filler)	1.20
Acrylics (impact)	1.20–1.50
Acrylics cast	1.20–1.70
Epoxy (low-density)	1.20–1.70
Phenolic with butadiene–acrylonitrile copolymer (asbestos filler)	1.20–1.70
Melamine (phenolic filler)	1.20–2.00
Cellulose acetate	1.20–2.30
Cellulose propionate	1.20–2.30
Cellulose acetate butyrate	1.20–2.30
Phenolic formaldehyde (wood-flour filler)	1.20–2.40
Phenolic formaldehyde (cotton filler)	1.20–2.40
Epoxy encapsulating (glass filler)	1.20–2.90
Epoxy (glass filler)	1.20–2.90
Epoxy encapsulating (mineral filler)	1.20–2.90
Epoxy (mineral filler)	1.20–8.70
Polyaryl sulfone	1.30
Polycarbonate (unfilled)	1.30
ABS (high-impact)	1.30–2.30
ABS (heat-resistant)	1.30–2.30
ABS (medium-impact)	1.30–2.30
ABS (self-extinguishing)	1.30–2.30
Polyphenylene oxides (PPO)	1.33

TABLE 3-2. *Continued*

Plastic	Value BTU/(h)(ft^2)(°F/in.)
Polycarbonate (10 percent glass filler)	1.40
Polycarbonate (up to 40 percent glass filler)	1.40–1.50
Chlorotrifluoroethylene (CTFE)	1.40–1.50
Acrylic multipolymer	1.50
Nylon (polyamide) type 6/10	1.50
Nylon (polyamide) type 6/10 (20–40 percent glass filler)	1.50
Nylon (polyamide) type 12	1.50
Polyphenylene oxides (modified) (Noryl)	1.50
Phenolic with butadiene–acrylonitrile copolymer (wood-flour and flock filler)	1.50–1.70
Diallyl phthalate (synthetic-fiber filler)	1.50–1.70
Diallyl phthalate (glass filler)	1.50–4.40
Acetal homopolymer	1.60
Acetal copolymer	1.60
Cellulose nitrate	1.60
Nylons (polyamide) type 6 (glass filler)	1.60
Ionomers	1.70
Nylons (polyamide) type 6/6	1.70
Nylons (polyamide) type 6	1.70
Polytetrafluorethylene (TFE)	1.70
Flurocarbon (FEP)	1.70
ABS polycarbonate alloy	1.70–2.60
Phenolic formaldehyde (asbestos filler)	1.70–6.40
Melamine (cellulose filler)	1.90–2.50
Nylons (polyamide) type 11	2.00
Polyaryl ether	2.00
Polyesters (thermoplastic)	2.00
Polyesters (18 percent glass filler)	2.00
Polyphenylene sulfides (unfilled)	2.00
Urea formaldehyde	2.00–2.90
Melamine (alpha cellulose filler)	2.00–2.90
Diallyl phthalate (mineral filler)	2.00–7.30
Silicones (glass filler)	2.20
Polyethylenes (low-density)	2.30
Polyethylenes (medium-density)	2.30–2.90
Phenolic (glass filler)	2.40
Nylons (polyamide) type 11 (glass filler)	2.60
Phenolic (mica filler) mineral	2.90–4.00
Melamine (fabric filler phenolic modified)	2.90
Polyester (premixed chopped glass)	2.90–4.60
Melamic (fabric filler)	3.10
Polyethylenes (high-density)	3.20–3.60
Silicones (mineral filler)	3.20–3.80
Polyester alkyd granular putty type (mineral filler)	3.50–7.30
Polyester alkyd granular putty type (glass filler)	4.40–7.30

TABLE 3-2. *Continued*

Plastic	Value BTU/(h)(ft^2)(°F/in.)
Cellular	
ABS	0.58
Cellulose acetate	0.31
Epoxy	0.13–0.36
Phenolics	0.20–0.28
Polyethylene (low-density)	0.26–0.40
Polyethylene (high-density)	0.80–0.92
Polypropylene	0.27–4.2
Polystyrene	0.17–0.26
Polyurethane	0.11–0.52
Silicone	0.36
SAN	0.29–0.32

From *Handbook of Plastics and Elastomers.* Reprinted by permission of McGraw-Hill Book Company.

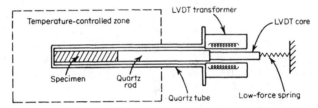

Figure 3-11. Schematic configuration of quartz-tube dilatometer. (Reprinted with permission of McGraw-Hill Company.)

bath is varied as specified. The change in length is recorded. The coefficient of linear thermal expansion is calculated as follows:

$$X = \frac{\Delta L}{L_0 \Delta T}$$

where X = coefficient of linear thermal expansion/°C; ΔL = change in the length of the specimen due to heating or cooling; L_0 = length of the specimen at room temperature; ΔT = temperature difference (°C) over which the change in the length of the specimen is measured.

Table 3-3 compares the thermal expansion values for some common plastics.

3.5. BRITTLENESS TEMPERATURE (ASTM D 746, ISO 974)

At low temperatures, all plastics tend to become rigid and brittle. This happens mainly because, at lower temperatures, the mobility of polymer chains is greatly reduced. In the case of the amorphous polymers, brittle failure occurs at a

TABLE 3-3. Comparison to Thermal Expansion Values for Some Common Materials

Plastic	Coefficient $(10^{-5}\,cm/cm/°F)$
Polyester (low-shrink)	0.366–0.605
Phenolic (glass filler)	0.44–1.14
Phenolic formaldehyde (asbestos filler)	0.44–2.22
Silicones (glass filler)	0.443
Melamine (phenolic filler)	0.555–2.22
Diallyl phthalate (glass filler)	0.555–2.00
Phenolic (fabric and cord filler)	0.555–2.22
Diallyl phthalate (mineral filler)	0.555–2.33
Epoxy (glass filler)	0.610–1.94
Polyester (preformed chopped rovings)	0.66–2.78
Nylons (polyamide) type 6/10 (20–40 percent glass filler)	0.666–1.78
Melamine (glass filler)	0.83–1.10
Nylons (polyamide) type 6/6 (30 percent glass filler)	0.83–1.11
Polyester alkyd granular putty type (glass filler)	0.83–1.39
Polyester woven cloth	0.83–1.66
Phenolic with butadiene acrylonitrile copolymer (wood-flour and flock filler)	0.83–1.94
Phenolic (mica filler)	1.00–1.44
Melamine (fabric filler) (phenolic modified)	1.00–1.55
Styrene acrylonitrile copolymer	1.00–2.50
Polyester (sheet molding compound)	1.11
Polyester (premixed chopped glass filler)	1.11–1.83
Silicones (mineral filler)	1.11–2.22
Melamine (asbestos filler)	1.11–2.50
Polyester alkyd granular putty type (mineral filler)	1.11–2.78
Epoxy (mineral filler)	1.11–2.78
Phenolic with butadiene acrylonitrile copolymer (rag filler)	1.16
Polyphenylene oxides (modified) (20–30 percent glass filler)	1.22
Urea formaldehyde	1.22–2.00
Polyesters (18 percent glass filler)	1.39
Melamine (fabric filler)	1.39–1.55
Styrene acrylonitrile (glass filler)	1.50–2.11
Polypropylene (inert filler)	1.61
ABS (20–40 percent glass filler)	1.61–2.00
Polypropylene (glass filler)	1.61–2.88
Phenolic formaldehyde (wood-flour filler)	1.66–2.50
Epoxy encapsulating (glass filler)	1.66–2.78
Epoxy encapsulating (mineral filler)	1.66–3.33
Polyester alkyd granular putty type (synthetic filler)	1.67–3.06
Nylons (polyamide) type 6 (glass filler)	1.83
Nylons (polyamide) type 11 (glass filler)	1.78
Polystyrenes (medium- and high-impact)	1.88–11.70
Polycarbonate (10 percent glass filler)	1.90
Styrene acrylonitrile copolymer	2.00–2.11
Acetal (20 percent glass filler)	2.00–4.50
Polyphenylene (40 percent glass filler)	2.22
Phenolic with butadiene acrylonitrile copolymer (asbestos filler)	2.22
Polyamides aromatic	2.22–2.78

TABLE 3-3. *Continued*

Plastic	Coefficient $(10^{-5}\,cm/cm/°F)$
Chlorotrifluoroethylene (CTFE)	2.50–3.90
Melamine alpha cellulose	2.22
Melamine (cellulose filler)	2.50
Polyaryl sulfone	2.62
PVC (rigid)	2.77–10.30
Acrylics	2.78–5.00
Polyphenylene oxides (modified) (unfilled)	2.89
Polysulfone	2.89–3.11
Polyphenlene (25 percent asbestos filler)	2.94
Phenoxy	3.00–3.20
Diallyl phthalate (synthetic fiber filler)	3.00–3.33
Polyphenylene sulfides (unfilled)	3.05
Polypropylene (unmodified)	3.22–5.66
Polystyrene (heat and chemical)	3.33
Polyesters (thermoplastic)	3.33
ABS (heat-resistant)	3.33–5.00
Polystyrene (general purpose)	3.33–4.44
Acrylics (impact)	3.33–4.44
Polypropylene (impact rubber modified)	3.33–4.73
Acrylic multipolymer	3.33–5.00
ABS polycarbonate alloy	3.44–4.72
Polyaryl ether	3.61
Polycarbonate (unfilled)	3.64
PVC flexible (unfilled)	3.86–13.90
Nylons (polyamide) type 12 (glass filler)	4.16
Nylons (polyamide) type 6/6	4.44
Chlorinated polyether	4.44
Acetal homopolymer	4.50
ABS (medium-impact)	4.45–5.55
Cellulose nitrate	4.44–6.66
Acetal copolymer	4.72
Polypropylene copolymer	4.44–5.27
Polyallomer	4.60–5.55
Nylons (polyamide) type 6	4.60
Polytetrafluoroethylene (TFE)	4.60–5.80
Cellulose acetate	4.44–10.00
Nylons (polyamide) type 6/10	5.00
ABS (high-impact)	5.27–7.22
Polyethylenes cross-linkable grades	5.55–11.94
Ethyl cellulose	5.55–11.10
Polyethylene (low-density)	5.55–11.10
Urethanes elastomers	5.55–11.10
Fluorocarbon (FEP)	5.60
Nylons (polyamide) type 12	5.78
Polyethylenes (high-density)	6.10–7.22
Cellulose propionate	6.11–9.44
Cellulose acetate butyrate	6.11–9.44
Methylpentene polymer	6.50

TABLE 3-3. *Continued*

Plastic	Coefficient $(10^{-5}\,cm/cm/°F)$
Ionomers	6.66
Polyvinylidene fluoride (VF2)	6.70
Styrene butadiene thermoplastic elastomers	7.22–7.60
Polyethylene (medium-density)	7.77–8.89
Polybutylene	8.32
Nylons (polyamide) type 11	8.33
Ethylene vinyl acetate copolymer	8.88–11.10
Ethylene–ethyl acrylate copolymer	8.88–13.80
Polyterephthalate	4.90–13.00

From *Handbook of Plastic and Elastomers.* Reprinted by permission of McGraw-Hill Book Company.

temperature well below glass transition temperature. The polymer tends to get tougher as it reaches the glass transition temperature. By contrast, in crystalline polymers, the toughness is mainly dependent on the degree of crystallinity, which generally creates molecular inflexibility resulting in only moderate impact strength. The size of the crystalline structure formed also has significant effect on the impact strength of the polymer. The larger the crystalline structure, the lower the impact strength. The polymers exhibiting ductile failure generally show high plastic deformation characterized by material stretching and tearing before fracturing. Brittleness temperature is defined as the temperature at which plastics and elastomers exhibit brittle failure under impact conditions. Yet another way to define brittleness temperature is the temperature at which 50 percent of the specimens tested exhibit brittle failure under specified impact conditions.

3.5.1. Test Apparatus and Procedures

The test apparatus consists of a specimen clamp and a striking member. The specimen clamp is designed so that it holds the specimen firmly as a cantilever beam. A typical specimen clamp is shown in Figure 3-12. The test apparatus most commonly used is the motordriven brittleness temperature tester illustrated in Figure 3-13. The tester has a rotating striking tool that rotates at a constant linear speed of 6.5 ± 0.5 ft/sec. An insulate refrigerant tank with a built-in stirrer to circulate the heat-transfer medium is used. The stirrer maintains the temperature equilibrium. Any type of heat-transfer medium can be used as long as it remains liquid at the test temperature.

The test specimens are usually die punched from a sheet. The specimen dimensions are 1 in. long, 0.25 in. wide, and 0.075 in. thick. Specimen conditioning is carried out in accordance with the standard conditioning procedures.

To perform the test, the specimens are securely mounted in a specimen clamp. The entire assembly is then submerged in the refrigerant. After immersion for a specified time at the test temperature, the striking tool is rotated to deliver a single-impact blow to the specimens. Each specimen is carefully examined for failure. Failure is defined as the division of a specimen into two or more completely

Figure 3-12. Typical specimen clamp for brittleness temperature test. (Reprinted with permission of ASTM.)

Figure 3-13. Motor-driven brittleness temperature tester. (Courtesy of CEAST U.S.A. Inc.)

separated pieces or as any crack in the specimen that is visible to the unaided eye. The temperature is raised by uniform increments of 2 or 5°C per test and the test is repeated. This procedure is followed until both the no-failure and all-failure temperatures are determined.

The percentage of failures at each temperature is calculated by using the number of specimens that failed. Brittleness temperature is calculated as follows:

$$T_b = T_h + \Delta T[(S/100) - (1/2)]$$

where T_b = brittleness temperature (°C); T_h = highest temperature at which failure of all specimens occur (°C); ΔT = temperature increment (°C); S = sum of the percentage of breaks at each temperature.

Brittleness temperature has very little practical value since the data obtained by such a method can only be used in applications in which the conditions of deformation are similar to those specified in the test. The method is, however, useful for specification, quality control, and research and development purpose.

REFERENCES

1. Deanin, R. D., *Polymer Structure Properties and Applications*, Cahners, Boston, MA, 1972, p. 238.
2. *Ibid.*, p. 273.
3. *Symp. on Plastics Testing—Present and Future*, ASTM Publication No. 132, American Society for Testing and Materials, Philadelphia, Pa., 1953, p. 16.
4. "Design Guide," *Modern Plastics Encyclopedia*, McGraw-Hill, New York, 1980–81, p. 49.
5. Nielson, L. E., *Mechanical Properties of Polymers*, Reinhold, New York, 1962, pp. 138–196.
6. Baer, E. *Engineering Design for Plastics*, Reinhold, New York, 1964, p. 185.
7. Miller, R. W., "Considerations in the Evaluation of Plastics in Electrical Equipment," *Plast. Design and Processing* (July 1980), pp. 25–30.
8. Reymers, H., "A New Temperature Index, Who Needs It, What does it Tell," *Modern Plastics* (March 1970), p. 79.
9. *Underwriters Laboratories Publication*, "Polymeric Materials—Long Term Property Evaluations," UL746B.
10. *Modern Plastics Encyclopedia*, Reference 4, p. 492.
11. Ives, G. C., Mead, J. A., and Riley, M. M., *Handbook of Plastics Test Methods*, Iliffe Books, London, England, 1971, p. 335.
12. Deanin, Reference 1, p. 379.
13. Baer, Reference 6, p. 1025.

SUGGESTED READING

Carslaw, H. S. and Jaeger, J. C., *Conduction of Heat in Solids*. Oxford University Press, Oxford, England, 1959.

Jakob, M., *Heat Transfer*. Vols. I and II. Wiley, New York, 1949, 1957.

McAdams, W. H., *Heat Transmission*, McGraw-Hill, New York, 1954.

4

ELECTRIAL PROPERTIES

4.1. INTRODUCTION

The unbeatable combination of such characteristics as ease of fabrication, low cost, light weight, and excellent insulation properties have made plastics one of the most desirable materials for electrical applications. Although the majority of applications involving plastics are insulation-related, plastics can be made to conduct electricity by simply modifying the base material with proper additives such as carbon black.

Until recently, plastics were considered a relatively weaker material in terms of load-bearing properties at elevated temperatures. Therefore, the use of plastics in electrical applications was limited to nonload-bearing, general-purpose applications. The advent of new high-performance engineering materials has altered the entire picture. Plastics are now specified in a majority of applications requiring resistance to extreme temperatures, chemicals, moisture, and stresses. The primary function of plastics in electrical applications has been that of an insulator. This insulator or dielectric separates two field-carrying conductors. Such a function can be served equally well by air or vacuum. However, neither air nor vacuum can provide any mechanical support to the conductors. Plastics not only act as effective insulators but also provide mechanical support for field-carrying conductors. For this very reason, the mechanical properties of plastic materials used as insulators become very important (1). Typical electrical applications of plastic material include plastic-coated wires, terminals, connectors, industrial and household plugs, switches, and printed circuit boards. The following are the typical requirements of an insulator:

Handbook of Plastics Testing and Failure Analysis, Third Edition, by Vishu Shah
Copyright © 2007 by John Wiley & Sons, Inc.

1. An insulator must have a high enough dielectric strength to withstand an electrical field between the conductors.
2. An insulator must possess good arc resistance to prevent damage in case of arcing.
3. An insulator must maintain integrity under a wide variety of environmental hazards such as humidity, temperature, and radiation.
4. Insulating materials must be mechanically strong enough to resist vibration shocks and other mechanical forces.
5. An insulator must have high insulation resistance to prevent leakage of current across the conductors.

The key electrical properties of interest are dielectric strength, dielectric constant, dissipation factor, volume and surface resistivity, and arc resistance.

4.2. DIELECTRIC STRENGTH (ASTM D 149, IEC 243-1)

The dielectric strength of an insulating material is defined as the maximum voltage required to produce a dielectric breakdown. Dielectric strength is expressed in volts per unit of thickness such as V/mL. All insulators allow a small amount of current to leak through or around themselves. Only a perfect insulator, if there is such an insulator in existence, can be completely free from small current leakage. The small leakage generates heat, providing an easier access to more current. The process slowly accelerates with time and the amount of voltage applied until a failure in terms of dielectric breakdown, or what is known as puncture, occurs. Obviously, dielectric strength, which indicates electrical strength of a material as an insulator, is a very important characteristic of an insulating material. The higher the dielectric strength, the better the quality of an insulator. Table 4-1 lists some typical dielectric strength values of common plastics materials. Three basic procedures have been developed to determine dielectric strength of an insulator. Figure 4-1 illustrates the basic setup for a dielectric strength test. A variable transformer and a pair of electrodes are normally employed. Specimens of any desirable thickness prepared from the material to be tested are used. Specimen thickness of 1/16 in. is fairly common. The first procedure is known as the short-time method. In this method, the voltage is increased from zero to breakdown at uniform rate. The rate of rise is generally 100, 500, 1000, or 3000 V/sec until the failure occurs. The failure is made evident by actual rupture or decomposition of the specimen. Sometimes a circuit breaker or other similar device is employed to signal the voltage breakdown. This is not considered a positive indication of voltage breakdown because other factors such as flashover, leakage current, corona current, or equipment magnetizing current can influence such indicating devices.

The second method is known as the slow rate-of-rise method. The test is carried out by applying the initial voltage approximately equal to 50 percent of the breakdown voltage, as determined by the short-time test or as specified. Next, the voltage is increased at a uniform rate until the breakdown occurs.

The step-by-step test method requires applying initial voltage equal to 50 percent of the breakdown voltage, as determined by the short-time test, and then increasing

TABLE 4-1. Dielectric Strength of Various Plastics

Plastics	Dielectric Strength (V/mil)
PFA (fluorocarbon)	2000
CPVC	1200–1500
Rigid PVC	800–1400
Ionomer	1000
Polyester (thermoplastic)	600–750
Polypropylene	650
Polystyrene (high impact)	650
FEP (fluorocarbon)	600
Nylons	350–560
Polstyrene (General Purpose)	500
Acetals	500
PTFE (fluorocarbon)	500
PPO	500
Polyphenylene sulfide	490
Polyethylene	480
Polycarbonate	450
ABS	415
Phenolics	240–340
PVF_2 (fluorocarbon)	260

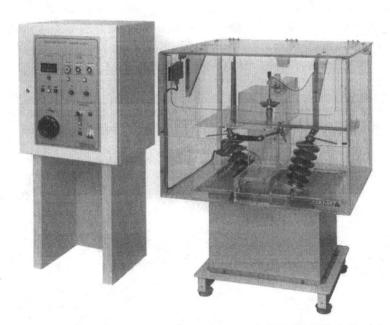

Figure 4-1. Dielectric strength test. (Courtesy CEAST, U.S.A. Inc.)

the voltage in equal increments and holding for specified time periods until the specimen breaks down. In almost all cases, the dielectric strength values obtained by the step-by-step method corresponds better with actual use conditions. However, the service failures are generally at voltage below the rated dielectric strength because of the time factor involved (2). The dielectric strength of an insulating material is calculated as follows:

$$\text{Dielectric strength (V/mil)} = \frac{\text{Breakdown voltage (V)}}{\text{Thickness (mil)}}$$

4.2.1. Factors Affecting the Test Results

A. Specimen Thickness
The dielectric strength of an insulator varies inversely with the fractional power (generally 0.4) of the thickness (3). The thicker specimen requires higher voltage to achieve the same voltage gradient. At higher voltage, a reduction in intermolecular bonds is observed, resulting from thermal expansion created by heat generation. Thicker sections also have internal voids, flaws, moisture, nonuniformity, and greater current leakage, causing early failure of the specimen (4). Thin films with higher dielectric strength value have been used successfully in many critical space-saving applications. Dielectric strength versus thickness curve are readily available from material suppliers.

B. Temperature
Dielectric strength decreases with the increase in the temperature of the specimen. If the design calls for the use of the product at various temperatures, the dielectric strength values at anticipated temperatures should carefully be evaluated. Interestingly enough, the dielectric strength of the material below room temperature is constant and independent of temperature change (5).

C. Humidity
Humidity affects the dielectric strength of the material. Surface moisture as well as moisture absorbed by hygroscopic material affects the results.

D. Electrodes
Dielectric strength of a material is affected by the electrode geometry, the electrode area, and the electrode base material composition. Generally, the breakdown voltage decreases with increasing electrode area.

E. Time
The rate of voltage application significantly alters the test results. As mentioned earlier in this chapter, dielectric strength values obtained using the step-by-step method are lower than those obtained by the short-time test.

F. Mechanical Stress
Mechanical stress tends to reduce the dielectric strength values substantially.

G. Processing

Defects such as poor weld lines, voids, bubbles, and flow lines brought forth by poor processing practices tend to reduce the dielectric strength anywhere from 30 to 60 percent, depending upon the severity of the defect.

4.2.2. Test Limitations and Interpretations

In designing plastic parts for electrical applications, the dielectric strength values must be studied in detail. If the parts are designed based on values derived from published literature without proper attention to the effects of thickness, moisture, temperature, mechanical stress, and actual use conditions, the results could prove disastrous. The actual use conditions of the insulating materials are quite different than the conditions of the dielectric strength test and, therefore, the real emphasis must always be on the performance of the material in actual service.

4.3. DIELECTRIC CONSTANT AND DISSIPATION FACTOR (ASTM D 150, IEC 250)

4.3.1. Dielectric Constant (Permittivity)

The dielectric constant of an insulating material is defined as the ratio of the charge stored in an insulating material placed between two metallic plates to the charge that can be stored when the insulating material is replaced by air (or vacuum). Defined another way, the dielectric constant is the ratio of the capacitance by two metallic plates with an insulator placed between them and the capacitance of the same plates with a vacuum between them.

$$\text{Dielectric constant} = \frac{\text{Capacitance, Material as dielectric}}{\text{Capacitance, Air (or Vacuum) as dielectric}}$$

Simply stated, the dielectric constant indicates the ability of an insulator to store electrical energy.

In many applications, insulating materials are required to perform as capacitors. Such applications are best served by plastic materials having a high dielectric constant. Materials with a high dielectric constant have also helped in reducing the physical size of the capacitors. Furthermore, the thinner the insulating material, the higher the capacitance. Because of this fact, plastic foils are extensively used in applications requiring high capacitance.

One of the main functions of an insulator is to insulate the current-carrying conductors from each other and from the ground. If the insulator is used strictly for this purpose, it is desirable to have the capacitance of the insulating material as small as possible.

For these applications, one is looking for materials with very low dielectric constant. Table 4-2 lists typical dielectric constant values of various plastics. The dielectric constant of air (or vacuum) is 1 at all frequencies. The dielectric constant of plastics varies from 2 to 20.

TABLE 4-2. Dielectric Constant of Various Plastics

Plastics	Dielectric Constant
Melamines	5.2–7.9
Polyvinylidene fluoride (PVF$_2$)	7.5
Cellulose acetate	3.0–7.0
Phenolics	4.0–7.0
Nylons (30 percent glass-filled)	3.5–5.4
Epoxies	4.3–5.1
Polyesters (thermoset)	4.7
Polystyrene (high impact)	2.0–4.0
Acetal copolymer (25 percent glass-filled)	3.9
Nylons	3.5–3.8
Acetals	3.7
Polycarbonate (30 percent glass-filled)	3.48
Polysulfone (30 percent glass-filled)	3.4
Polyester (thermoplastic)	3.2
ABS	3.2
SAN	3.0
Polystyrene (general-purpose)	2.7
Fluorocarbons (PFA, FEP)	2.1

^{a} At 10^6 Hz.

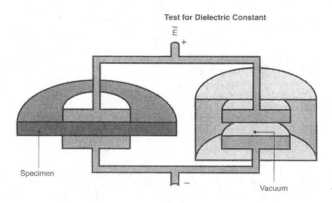

Dielectric constant is the ratio of the system capacitance with the plastic specimen as the dielectric to the capacitance with a vacuum as the dielectric.

Figure 4-2. Schematic of dielectric constant test. (Courtesy of Bayer Corporation.)

The dielectric constant test is fairly simple. The test specimen is placed between the two electrodes, as shown in Figure 4-2, and the capacitance is measured. Next, the test specimen is replaced by air and the capacitance value is again measured. The dielectric constant value is determined from the ratio of the two measurements. Dielectric constant values are affected by factors such as frequency, voltage, temperature, and humidity.

4.3.2. Dissipation Factor

In all electrical applications, it is desirable to keep the electrical losses to a minimum. Electrical losses indicate the inefficiency of an insulator. The dissipation factor is a measure of such electrical inefficiency of the insulating material. The dissipation factor indicates the amount of energy dissipated by the insulating material when the voltage is applied to the circuit (6). The dissipation factor is defined as the ratio of the conductance of a capacitor in which the material is the dielectric to its susceptance or the ratio of its parallel reactance to its parallel resistance. Most plastics have a relatively lower dissipation factor at room temperature. However, at high temperatures, the dissipation factor is quite high, resulting in greater overall inefficiency in the electrical system. The loss factor which is the product of dielectric constant and the dissipation factor, is a frequently used term, which relates to the total loss of power occurring in insulating materials.

4.4. ELECTRICAL RESISTANCE TESTS

As noted, the primary function of an insulator is to insulate current-carrying conductors from each other as well as from ground and to provide mechanical support for components. Naturally, the most desirable characteristic of an insulator is its ability to resist the leakage of the electrical current. The higher the insulation resistance is, the better the insulator. Failure to recognize the importance of insulation resistance values when designing products such as appliances and power tools could lead to fire, electrical shock, and personal injury.

Insulation resistance can be subdivided into:

1. Volume resistance
2. Surface resistance

Volume resistance is defined as the ratio of the direct voltage applied to two electrodes that are in contact with a specimen to that portion of the current between them that is distributed through the volume of the specimen. Simply stated, the volume resistance is the resistance to leakage through the body of the material (7). Volume resistance generally depends upon the material. The term most commonly used by designers is volume resistivity. It is defined as the ratio of the potential gradient parallel to the current in the material to the current density. Thus, the volume resistivity of a material is the electrical resistance between the opposite faces of a unit cube for a given material and at a given temperature (8).

$$\text{Volume Resistivity } (\rho)\,(\Omega\text{-cm}) = \frac{A}{t}(R_v)$$

where A = area; t = thickness of the specimen; R_v = volume resistance (Ω).

High-volume resistivity materials are desirable in applications requiring superior insulating characteristics. Table 4-3 lists typical volume resistivity values of some common plastics.

TABLE 4-3. Volume Resistivity of Various Plastics

Plastics	Volume Resistivity $(\Omega\text{-cm})^a$
Melamine (asbestos filler)	1.2×10^{12}
Urea formaldehyde	10^{12}–10^{13}
Phenolic (asbestos filler)	10^9–10^{13}
Acetal copolymer	10^{14}
Acrylics	10^{14}
Epoxy	10^{14}
Polystyrene	10^{16}
SAN	10^{16}
ABS	5×10^{16}
Polycarbonate	2×10^{16}
PVC (flexible)	10^{11}–10^{15}
Nylons, type 6/6	10^{14}–10^{15}
Acetal homopolymer	10^{15}
PVC (rigid)	10^{15}
Polyethylene	10^{16}
Polypropylene	10^{16}
Thermoplastic polyester	3×10^{16}
Polysulfone	5×10^{16}
PPO	10^{17}
PTFE	10^{18}
FEP	2×10^{18}

^{a}At 50 percent RH and 73°F.

The surface resistance of a material is defined as the ratio of the direct voltage applied to the electrodes to that portion of the current between them that is primarily in a thin layer of moisture or other semiconducting material that may be deposited on the surface. Simply stated, surface resistance is the resistance to leakage along the surface of an insulator. The surface resistance of a material depends upon the quality and cleanliness of the surface of the product. A product with oil or dirt particles on it gives lower surface resistance values.

The test procedures to determine electrical resistance values are rather complex. ASTM D 257 (IEC 93) describes the procedures as well as the complex electrodes and apparatus required to carry out the test in detail.

Temperature and humidity both seem to affect the insulation resistance appreciably. As a rule, the higher the temperature and humidity, the lower the insulation resistance of a material.

4.5. ARC RESISTANCE (ASTM D 495)

Arc resistance is the ability of a plastic material to resist the action of a high-voltage electrical arc, and is usually stated in terms of time required to form material electrically conductive. Failure is characterized by carbonization of the surface, tracking, localized heating to incandescence, or burning. In all applications in which conducting elements are brought into contact, arcing is inevitable. Switches, circuit breakers, and automotive distributor caps are a few good examples of applications where arcing is known to cause failure. Another term that is generally associated with arcing is tracking. Tracking is defined as a phenomenon where a high-voltage source current creates a leakage or fault path across the surface of an insulating material by slowly but steadily forming a carbonized path appearing as a thin, wiry line between the electrodes. Tracking is accelerated by the presence of surface contaminants such as dirt and oil and by the presence of moisture. Resistance to arcing or tracking depends upon the type of plastic materials used. Phenolics tend to carbonize easily and therefore have relatively poor arc resistance. Plastics such as alkyds, melamines, and fluorocarbons are excellent arc-resistance materials. The failures due to arcing are not always because of carbonization or tracking. Many plastics such as acrylics simply do not carbonize. However, they do form ignitable gases that cause the product to fail in a short time. Are resistance of plastics can be improved substantially by the addition of fillers such as glass, mineral, wood flour, asbestos, and other inorganic fillers. Table 4-4 lists the arc resistance of some common plastic materials.

The determination of the arc resistance of plastics using a standard test method has always been a problem because of the numerous ways the test can be conducted. ASTM has developed four basic test methods to satisfy this concern. ASTM D 495, which is high-voltage, low-current, dry arc resistance of solid electrical insulation, has been the most widely used and accepted. This test is only intended for the preliminary screening of materials, for detecting the effect of changes in formulations, and for quality control testing. Since the test is conducted under clean and dry laboratory conditions that are rarely encountered in practice, the prediction of materials from test results is next to impossible. Figure 4-3a and 4-3b illustrates a

TABLE 4-4. Arc Resistance of Various Plastics

Plastic	Arc Resistance (sec)
Polycarbonate (10–40 percent glass-filled)	5–120
Polycarbonate	10–120
Polystyrene (high impact)	20–100
ABS	50–85
Polystyrene (General-purpose)	60–80
Rigid PVC	60–80
Polysulfone	75–190
Urea formaldehyde	80–150
Ionomer	90–140
SAN	100–150
Epoxy	120–150
Acetal (homopolymer)	130
Polyethylene (low-density)	135–160
Polypropylene	135–180
PTFE	>200
Acrylics	No track

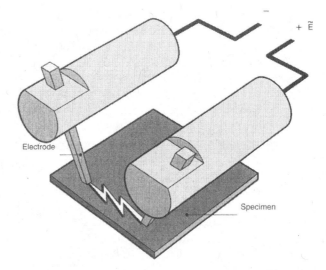

Arc-resistance electrodes intermittently subject the specimen surface to a high-voltage arc until a conductive path is formed.

(a)

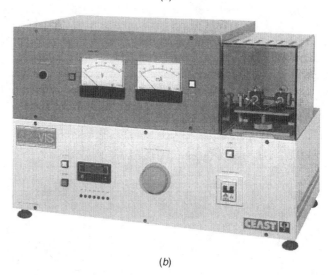

(b)

Figure 4-3. (a) Typical setup for arc resistance test. (Courtesy of Bayer Corporation.) (b) Commercially available arc resistance tester. (Courtesy of CEAST U.S.A. Inc.)

typical setup for an arc resistance test. The voltage is applied intermittently and severity is increased in steps until the failure occurs. Arc resistance is measured in seconds to failure.

ASTM method D 2132 outlines the procedure for determining dust and fog tracking and erosion resistance of electrical insulating materials. The test is carried out in a fog chamber with a standardized dust applied to the sample surface. Failure is characterized by the erosion of the specimen or tracking. ASTM D 2302 also describes the test for differential wet tracking resistance of electrical insulating materials with controlled water-to-metal discharges. The inclined specimen is partially immersed in a water solution of ammonium chloride and a wetting agent. Failure is characterized by tracking. ASTM D 2303 describes the test for liquid-contaminant inclined plane tracking and erosion of insulating materials. In this test, the specimen is inclined at 45°, and the electrolyte is discharged onto the surface at a controlled rate, increasing the voltage at the same time. The failure is marked by erosion and tracking.

4.6. UL REQUIREMENTS

No discussion on electrical properties and testing can be considered complete without including the role of Underwriters Laboratories (UL) in electrical codes, standards, and specifications (9). UL is an independent, non-for-profit organization established with the basic goal of testing for public safety and developing standards for safety. Further discussion regarding UL's organization capabilities is included in Chapter 19.

An increasing number of designers are specifying plastics in a variety of applications. Depending upon the application, plastics material may serve as an insulation, a structural member, or a simple enclosure to house uninsulated electrically live parts. These must perform their intended function; otherwise, fire, electrical shock, or personal injury may result. It is clear that a plastic part used in an electrical application must have adequate properties to minimize the risks related to safety. Underwriters Laboratories engineers have developed a set of requirements that one must meet if UL recognition is desired.

4.6.1. Material Properties

Properties that are considered by UL in the evaluation of polymeric material for suitability in electrical applications are divided into the two broad areas of short- and long-term properties.

A. Short-Term Properties—UL 746 A

UL 746 A covers short-term test procedures. These investigations provide data with respect to the mechanical, electrical flammability, thermal, and other properties of the materials under consideration, and are intended to provide guidance to the material manufacturer, the molder, the end-product manufacturer, and safety engineers. Testing for mechanical properties includes tensile testing, Izod impact, flexural properties, shear properties, bond strength properties of adhesives, and durometer hardness. Electrical properties of interest are dielectric strength, DC resistance or conductance of insulating materials, high-voltage, low-current dry arc resistance of solid electrical insulation, and high voltage arc-tracking-rate index of solid insulating materials. Testing for thermal properties includes deflection temperature of polymeric materials under load, Vicat softening point, and softening point by ring-and-ball apparatus. Ease of ignition tests are also part of short-term property evaluation. Tests such as hot wire ignition, high-current arc ignition, high-voltage arc resistance to ignition, and hot gas ignition are briefly described. Three other short-term tests are dimensional change of polymeric parts, resistance of polymeric materials to chemical reagents, and test for polymer identification. The majority of these test procedures are based on ASTM test methods.

B. Long-Term Properties—UL 746 B

UL 746 B covers the procedures for evaluating long-term properties of polymeric materials. Mechanical, electrical, thermal, flammability, and other properties are considered with the intent to provide guidance for the material manufacturer, molder, and safety engineers.

The behavior of plastic materials under long-term exposure to elevated temperatures is of prime importance to UL engineers. A procedure to determine the relative thermal indices of polymeric materials is discussed in detail in this standard. The relative thermal index of a material is an indication of a material's ability to retain a particular property when exposed to elevated temperatures for an extended period of time. It is a measure of a material's thermal endurance. Other long-term property evaluation tests include environmental exposure, creep, and chemical resistance.

4.6.2. Evaluation of Plastic Materials Used in Electrical Equipments— UL 746 C

UL 746 C provides guidelines for selecting minimum performance levels based on the application. The use of the UL 746 C standard is best understood through a specific example.

Let us assume that we are designing an enclosure of stationary equipment that houses uninsulated electrical live parts. To determine the material requirements we proceed as follows.

First, from UL standard UL 746 C, we need to find the appropriate table that coincides with our requirements. In this case, for stationary equipment, Table 4-5 (taken from UL 746 C and Figure 6-1) is the correct choice. Table 4-5 shows the requirements for enclosure of fixed or stationary equipment. The first consideration is the flammability requirement. Minimum flammability rating Class 5 V material is required for this application. Material classed 5 V should not have any specimens that burn with flaming and/or glowing combustion for more than 60 sec after the fifth flame application. Nor should it have any specimens that drip any particles when tested in accordance with the vertical burning test.

Test considerations in Table 4-5 reflect performance levels obtained on standard specimens or end-product tests that stress the material under actual conditions of use and misuse. We need to carefully study each of the enclosure performance requirements that are clearly spelled out in the standard. In many cases, a mere comparison of performance requirements with published material data is sufficient. If the material complies with the minimum performance levels, no additional testing is required. If it does not, the material may be qualified by an appropriate end-product test.

Similar tables indicating the decision tree and test requirements for portable equipment and fixed equipment have been developed by Underwriters Laboratories. In each case, the procedure for determining the suitability of a plastic material used in electrical equipment is basically the same. If the plastic parts used in this application are required to go through a special process such as plating,

TABLE 4-5. Enclosure Requirement for Fixed or Stationary Equipment

Part 1: Conditions of Use

Start

Material is used to enclose uninsulated live parts or live parts with insulation thickness less than 0.71 mm (0.028 in.)[a,b,c]	Material is used to enclose live parts with insulation thickness of 0.71 mm (0.028 in.) or greater[a,b]

Part 2: Applicable Requirements

Minimum Flammability Rating	5V[i]	5V[i]
Electrical/mechanical prop. per Table 8.1	Yes	Yes
Dielectric withstand per 12.1	Yes	No
Large mass flammability per Section 21	Yes	Yes
Crushing resistance per 23.1	Yes	No
Resistance to impact per Section 24	Yes	Yes
UV resistance per 26.1	Yes[f]	Yes[f]
Water exposure and immersion		
A. Properties per 27.1	Yes[f]	Yes[f]
B. Dimension per 27.2	Yes	No
Abnormal operation per 28.1	Yes	Yes
Severe conditions per 29.1	Yes	Yes[g]
Mold stress—relief distortion per 62.1	Yes	Yes[d]
Input after mold stress—relief distortion per 31.1	Yes	Yes[g]
Conduit connections	Yes[h]	Yes[h]
Strain relief per 32.1	Yes[e]	Yes[e]
Thermal endurance per Sections 33–39	Yes	Yes
Volume resistmty per Section 16	Yes	No

[a]The insulation thickness of the component parts is considered to be equivalent to 0.71 mm (0.028 in.) if the component complies with the requirements for the component.

[b]0.71-mm (0.028-in.) thickness is generally required for internal insulation. However, if insulation or wire insulated with less than 0.71 mm (0.028 in.) insulation thickness (light duty) is provided that is restricted for use in chassis, channels, or other internal areas where contact during user servicing or user operation of the equipment is unlikely, the insulation is considered to be equivalent to 0.71 mm (0.028 in.).

[c]Enamel insulated magnet wire is to be considered as an uninsulated live part.

[d]This test is required only if failure of the material causes a stress on the junction between a lead and a terminal of a component. If the strain–relief test is acceptably performed for components with integral leads either as a separate test or as part of the regular test procedure for the component, it shall be considered that the material does not cause a stress on the junction between the lead and a terminal of the component.

[e]This test is required only if the strain–relief means is mounted to the enclosure or is a polymeric part of the enclosure.

[f]This test is required only if the equipment may be exposed to outdoor weather conditions.

[g]This test is required only if the equipment may be used unattended.

[h]This test is required only if the equipment is permanently connected electrically to the wiring system. The continuity of the conduit system shall be a metal-to-metal contact. If the integrity of the polymeric enclosure is relied upon to provide for bonding between the parts of the conduit system at any location where conduit may be connected, the bonding shall be evaluated by the requirements contained in the Standard for Enclosures for Electrical Equipment, UL 50. If the polymeric enclosure is intended for connection to a rigid conduit system, it shall acceptably perform when tested using the pullout, torque and bending tests as described in the Standard for Industrial Control Equipment, UL 508.

[i]5V = 5VA or enclosure material complies with the 5-in. end-product flame test described in Flammability—127 mm (5 in.) Flame, Section 19.

Courtesy of Underwriters Laboratories, Inc.

adhesive bonding, flame retardant, or metallic coating, some additional testing is required.

4.6.3. Polymeric Materials—Fabricated Parts UL 746 D

This standard covers the requirements for molded or fabricated parts. Material identity-control systems intended to provide tracability of material used for polymeric parts through the handling, molding or fabrication, and shipping operations are described. Guidelines are also provided for the acceptable blending or simple compounding operations that may affect risk of fire, electrical shock, or personal injury.

4.7. EMI/RFI SHIELDING (10)

4.7.1. Introduction

The explosive growth of computers and the telecommunications and electronic devices industry has created an urgent need to understand and develop test and measurement techniques and standards for EMI and RFI shielding. All electronic devices are required to have electromagnetic compatibility. EMI is an acronym for electromagnetic interference. It is a phenomenon, caused by unrestricted electrical and magnetic energy, that escapes from any electrical device and reaches a second unintended device. It is called EMI if the receiving device malfunctions as a result of this pollution. Any electrical signal falling between DC electricity and the gamma ray frequency region of the electromagnetic spectrum can be a source of EMI. For most practical purposes, the majority of EMI problems are limited to that part of the spectrum from 1 kHz to 10 GHz. This portion of the electromagnetic spectrum is known as the radio frequency interface (RFI) band and covers the radio and audio frequencies. EMI of some form is experienced by us every day. Some of the most common sources of EMI are lightning, static electricity, television and radio receivers, electric motors, electric appliances, paging systems and radar transmitters. Some of the most common receptors of EMI are microprocessors, computers, hi-fi equipment, electronic measuring equipment, radio and television receivers, remote control units, aircraft navigation systems, and humans and animals.

The fact that electronic devices are both sources and receptors of EMI creates a twofold problem. Because electromagnetic radiation penetrating the device may cause electronic failure, manufacturers must protect the operational integrity of their product while complying with regulations aimed at reducing electromagnetic radiation emitted into the atmosphere.

The switch from metal to plastics as a housing material for electronic equipment has contributed to the EMI shielding issue. Plastics are insulators to electrical energy, so EMI waves pass freely through unshielded plastic without any impedance or resistance. Metals, being conductors, reflect or absorb electromagnetic energy, and unwanted electrical energy is easily grounded. Shielding, the term used to describe the effective "blocking" of EMI, is provided by a conductive barrier which harmlessly reflects or transmits electrical interference to ground.

EMI/RFI shielding of electronic devices is achieved in many different ways. Two of the most common ways to achieve shielding are to coat the plastic housing with conductive material or to make the housing material itself conductive. The second method of using conductive plastics for shielding effectiveness is relatively new and, for the most part, is in the research and development stage with limited commercial applications. Coating methods on plastic parts are by far the most popular and the most varied shielding techniques. Basically, they all involve coating the finished plastic housing with a layer of conductive material to create a Faraday cage (a concept that an enclosed conductive housing has a zero electrical field), resulting in a EMI shield. The electroless plating method is also gaining popularity as an alternate method to conductive coatings.

4.7.2. Regulations and Standards

In the United States, the Federal Communications Commission (FCC) regulates electromagnetic emission standards. FCC rules apply to both radiated emissions and conducted emissions from the product. The FCC regulates the amount of EMI/RFI radiated into the atmosphere or conducted back onto AC power lines by digital devices. The issue of protecting the device integrity is left up to the manufacturer and end-user. The FCC does not control, nor do they recommend, the type of shielding method used for meeting the requirement.

The regulation divides the digital devices into two classes. Class A applies to equipment sold exclusively for commercial use while Class B applies to equipment sold for use in a residential environment. For each of these classes, the FCC regulation specifies maximum permitted energy levels that can escape a device.

European and Japanese standards for electromagnetic emissions are slightly different than the ones set forth by the FCC. At present, a universal standard for EME does not exist, thus requiring a manufacturer to test a digital device according to each separate standard.

4.7.3. Shielding Effectiveness Measurement

Electromagnetic energy is most effectively measured in terms of attenuation. Attenuation is the amount by which the intensity of an electromagnetic signal is reduced by the introduction of a shielding medium. The attenuation is measured in decibels (dB) and is the ratio of the field strength without the shield to the field strength with the shield. Shielding effectiveness measured in decibels is a logarithmic scale, which means that an attenuation measurement of 50 dB is 10 times more effective than a measurement of 40 dB.

$$\text{Total electromagnetic shielding effectiveness} = 10 \log P_i/P_t \text{ (dB)}$$

where P_i = incident power in watts/m^2; P_t = transmitted power in watts/m^2. Table 4-6 lists shielding attenuation levels required to produce an effective shield.

Four basic techniques for measuring shielding effectiveness have been developed:

**TABLE 4-6. Shielding Attenuation Levels Required to
Produce an Effective Shield**

Shielding Attenuation Levels	
0–10 dB	Very little shielding
10–30 dB	Minimum range for meaningful shielding
30–60 dB	Average shielding
60–90 dB	Good shielding
90–120 dB	Excellent shielding
Over 120 dB	Limit of present "state of the art"

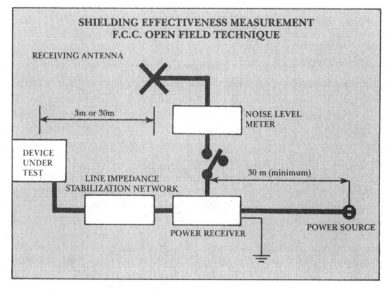

Figure 4-4. Test set up for open field technique. (From "EMI/RFI Shielding Guide."
Reprinted by permission of GE Plastics.)

1. Open-field testing
2. Shielded box
3. Coaxial transmission line
4. Shielded room

A. Open-Field Testing Technique

This technique most closely resembles an actual usage situation. Since the test
results for a metallized test plaque differ from those for a complete commercial
digital device, only tests conducted on the finished product can be used to predict
compliance with agency standards. Figure 4-4 illustrates the test set-up for the
open-field technique. The test is conducted in an open field to allow free space
measurement to be taken for both radiated and conducted emissions. Radiated and
conducted emissions are recorded on a noise level meter and shielding effectiveness
for the device under test is measured.

B. Shielded-Box Technique, Coaxial Transmission Line Technique, and Shielded-Room Technique

Tests conducted using these techniques are mainly designed for a quick comparison between two or more potential shielding mechanisms. The tests are generally done on specimens and not on assembled devices. Furthermore, these tests cannot be relied upon as predictors of agency compliance.

The shielded-box method involves a sealed box with a cut-out, as illustrated in Figure 4-5. The test is conducted simply by recording the EMI signals transmitted from outside the box and comparing them with signals recorded from inside the box. Shielding effectiveness is the ratio of these two signals. The coated plaque is clamped over the aperture, and both transmitted and received emissions are recorded. A pure copper or aluminum plaque is used as a reference measurement. The test method is limited to the measurement of frequencies below 500 MHz.

The coaxial transmission line (ASTM D4935) test method provides a procedure for the electromagnetic shielding effectiveness of a planar material due to plane-wave, far-field EM wave radiation. The test is conducted by mounting the reference test specimens in a specially fabricated specimen holder. The received power (or voltage) is measured at each frequency. Next, the reference specimen is replaced by the load specimen and a similar measurement is made. The shielding effectiveness, which is a ratio of power received with and without a material present for the same incident power, is calculated in decibels, Figure 4-6 illustrates the general test setup and specimens used for this test.

The shielded-room technique is somewhat complex and requires at least four separate Faraday cages, as illustrated in Figure 4-7. The basic principle and test

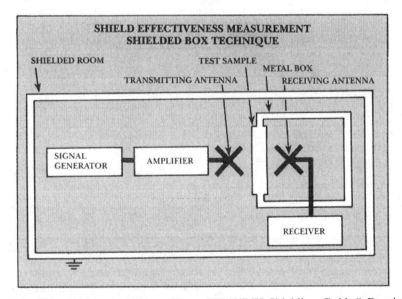

Figure 4-5. Shielded box technique. (From "EMI/RFI Shielding Guide." Reprinted by permission of GE Plastics.)

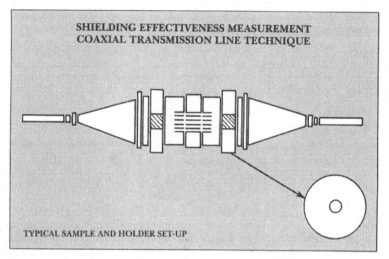

Figure 4-6. Test set-up for coaxial transmission line technique. (From "EMI/RFI Shielding Guide." Reprinted by permission of GE Plastics.)

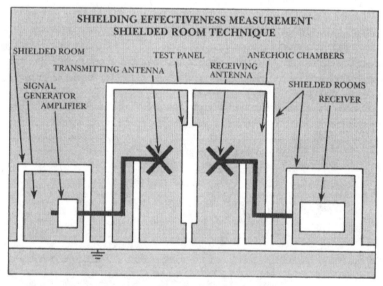

Figure 4-7. Shielding room technique. (From "EMI/RFI Shielding Guide." Reprinted with permission of GE Plastics.)

procedure is very similar to the one adopted for the shielded-box technique; however, the specimen size is much larger for this method. This method offers two distinct advantages over the shielded box, one being the increased frequency range over which the measurements can be made with reasonable accuracy. Second, because the transmitting and recording equipment is outside the measuring chamber, the possibility of interference from this equipment is eliminated.

4.7.4. Conductivity Measurement

In production, conductivity testing becomes an important means for quality control of individual parts as well as for part-to-part testing of assembled enclosures (11). Surface resistivity, volume resistivity, and ohms-per-square measurement are common tests for both conductive coating and conductive plastics and are used as quick and practical quality checks.

A. *Surface Resistivity*

Surface resistivity measures the resistance of a material to moving or distributing a charge over its surface. The lower the surface resistivity value, the easier a material will redistribute an electrical charge over its entire surface area. Surface resistivity is discussed in Section 4.4 and the test procedure is described in ASTM D257.

Surface resistivity meters which allow direct reading in ohms/sq. are most commonly used. One of the meters uses a probe with inner and outer rings. A charge is applied to the inner ring which travels over the surface the material and them is recorded by the outer ring. In addition to these resistivity meters, manufactures often use a digital ohmmeter to measure the point-to-point resistance between two probes in contact with the metallic coating. This method utilizes two spring-loaded probes that are placed at designated positions on the part or between two areas on mating parts. The conductivity between the two points is measured (12). This method is illustrated in Figure 4-8.

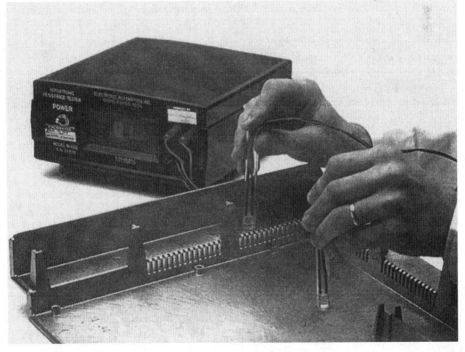

Figure   4-8. Point-to-point   resistance   measurement.   (Courtesy   of   Enthone-EMI Corporation.)

B. Volume Resistivity

Volume resistivity measures the ability of a material to resist moving charge through its volume. The lower the volume resistivity of a material is, the more conductive the material. This test procedure is also described in detail in ASTM D257.

C. Ohms-per-Square Testing

Ohms per square testing involves the placement of a probe on a surface of the part. This probe has two outside blocks of equal geometry and a space of equivalent dimension between the blocks. The probe is simply placed on the part and conductivity is measured. The test is extremely useful in quick checking the coated parts to verify minimum thickness of a conductive film. However, the test is not very effective for determining the overall conductivity of a coating across an actual part.

4.7.5. Adhesion Testing

Proper adhesion between a shielding material and a substrate is an important criterion for selecting a shielding method. If the shield layer begins to separate from the substrate, the loose particles of conductive material may fall onto electronic components for the device and create problems. Loss of adhesion can also result in EMI leakage.

ASTM D3359, Method B is widely used in the industry for adhesion testing and is also specified by Underwriters Laboratories for use with UL Standard 746 C. In this cross-cut tape test, a grid is cut through the coating and into the substrate. A special adhesion tape (Thermacel #99) is used for this procedure. The tape is pressed onto the grid and rubbed on with a pencil eraser. The tape is then pulled off rapidly with a smooth motion. The grid area is inspected using an illuminated magnifier. Figure 4-9 illustrates the classification of adhesion test results. Table 4-7 provides an explanation of each rating.

Adhesion test method ISO 2409 is quite similar, except the cuts are made into the coating using a special knife and results are classified based upon the amount of coating removed from the surface after the cutting process.

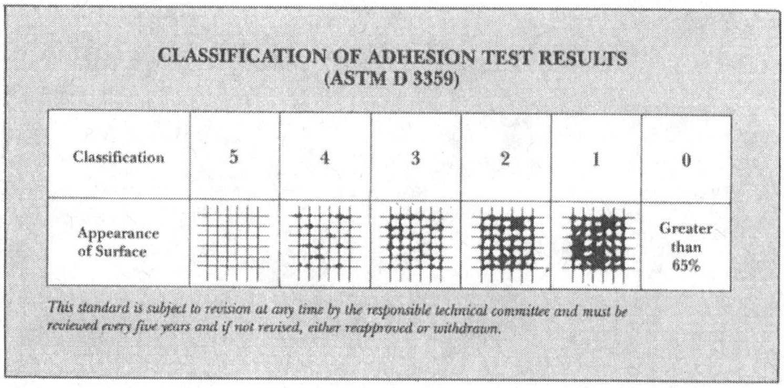

Figure 4-9. Adhesion testing. (From "EMI/RFI Shielding Guide." Reprinted with permission of GE Plastics.)

TABLE 4-7. Explanation of Adhesion Test Ratings

		Adhesion Test Ratings—ASTM D 3359 Method B
UL 746 C Acceptability	Adhesion Rating	Description
Pass	5B	The edges of the cuts are completely smooth; none of the squares of the lattice is detached.
Pass	4B	Small flakes of the coating are detached at intersections; less than 5% of the area is affected.
Fail	3B	Small flakes of the coating are detached along edges and at intersections of cuts. The area affected is 5–15% of the lattice.
Fail	2B	The coating has flaked along the edges and on parts of the squares. The area affected is 15–35% of the lattice.
Fail	1B	The coating has flaked along the edges of cuts in large ribbons and whole squares have detached. The area affected is 35–65% of the lattice.
Fail	0B	Flaking and detachment worse than Grade 1B. The area affected is greater than 65% of the lattice.

REFERENCES

1. Levy, S. and Dubois, H., *Plastics Product Design Engineering Handbook*, Reinhold, New York, 1977, p. 276.
2. Kinney, G. F., *Engineering Properties and Applications of Plastics*, John Wiley & Sons, New York, 1957, P. 222.
3. *Ibid.*, p. 223.
4. Milby, R., *Plastics Technology*, McGraw-Hill, New York, 1973, p. 498.
5. Kinney, Reference 2, p. 223.
6. Harper, C. A., "Short Course in Electrical Properties," *Plast. World* (April 1979), p. 73.
7. Kinney, Reference 2, p. 224.
8. Harper, Reference 6, p. 72.
9. *Underwriters Laboratories Standards for Safety*, UL 746 A-B-C-D, Underwriters Laboratories Inc., Melville, NY, 1997.
10. GE Plastics, "EMI/RFI Shielding Guide." Pittsfield, MA.
11. EMI Shielding, Design Guide, "Performance of Alternative Conductive Coatings," Enthone–OMI Inc., New Haven, CT.
12. *Ibid.*, p. 1.

SUGGESTED READING

Harris, F. K., *Electrical Measurements*, Wiley, New York, 1952.
Birks, J. B., *Modern Dielectric Materials*, Heywood & Company, London, 1960.
Anderson. J. C., *Dielectrics*, Chapman and Hall, London, England, 1964.

Reddish, W., "Chemical Structure and Dielectric Properties of Polymers," *Pure and Appl. Chem.*, **5**(4) (1962).

Von Hipple, A., *Dielectric Materials and Applications*, Wiley, New York, 1954.

Whitehead, S., *Dielectric Breakdown of Solids*, Oxford University Press, Oxford, England, 1951.

Martin, T. and Hauter, J., "Arcing Tests on Plastics," *J. S.P.E.*, **10**(2) (Feb. 1954), p. 13.

5

WEATHERING PROPERTIES

5.1. INTRODUCTION

The increased outdoor use of plastics has created a need for a better understanding of the effect of the environment on plastic materials. The environmental factors have significant detrimental effects on appearance and properties. The severity of the damage depends largely on such factors as the nature of the environment, geographic location, type of polymeric material, and duration of exposure. The effect can be anywhere from a mere loss of color or a slight crazing and cracking to a complete breakdown of the polymer structure. Any attempt to design plastic parts without a clear understanding of the degradation mechanisms induced by the environment would result in a premature failure of the product. The major environmental factors that seriously affect plastics are:

1. Solar radiations—UV, IR, visible X-rays
2. Microorganisms, bacteria, fungus, and mold
3. High humidity
4. Ozone and oxygen
5. Water: vapor, liquid, or solid
6. Thermal energy
7. Pollution: industrial chemicals

The combined effect of the factors mentioned above may be much more severe than the effect of any single factor, and the degradation processes are accelerated many times. Many test results do not include these synergistic effects, which almost always exist in real-life situations.

Handbook of Plastics Testing and Failure Analysis, Third Edition, by Vishu Shah
Copyright © 2007 by John Wiley & Sons, Inc.

5.1.1. UV Radiation

All types of solar radiation have some sort of detrimental effect on plastics. Ultraviolet radiation is the most destructive of all radiations. The energy in ultraviolet radiation is strong enough to break molecular bonds. This activity in the polymer brings about thermal oxidative degradation, which results in embrittlement, discoloration and an overall reduction in physical and electrical properties. Xenon arc lamps, fluorescent lighting, sun lamps, and other artificial sources also emit a similar type of harmful radiation. Other factors in the environment such as heat, humidity, and oxygen accelerate the UV degradation process.

One of the best methods of protecting plastics against UV radiation is to incorporate UV absorbers or UV stabilizers into the plastic materials. The UV absorbers provide preferential absorption to most of the incident UV light and are able to dissipate the absorbed energy harmlessly. Thus, the polymer is protected from harmful radiation at the cost of UV absorbers which are destroyed in the process with time. Several types of organic and inorganic UV absorbers are developed for this purpose. Almost all inorganic pigments absorb UV radiation to a certain extent and provide some degree of protection. Perhaps the most effective pigments are certain types of carbon black that absorb over the entire range of UV and visible radiation and transform the energy into less harmful radiation (1).

Ultraviolet stabilizers, unlike UV absorbers, inhibit the bond rupture by chemical means or dissipate the energy to lower levels that do not attack the bonds. The effectiveness of such additives when incorporated with the polymer can be determined by various test methods which are discussed in this chapter.

5.1.2. Microorganisms

Polymeric materials are generally not vulnerable to microbial attack under normal conditions. However, low-molecular-weight additives such as plasticizers, lubricants, stabilizers, and antioxidants may migrate to the surface of plastic components and encourage the growth of microorganisms. The detrimental effect can readily be seen through the loss of properties, change in aesthetic quality, loss of transmission (optical), and increase in brittleness. The rate of growth depends upon many factors such as heat, light, and humidity. Preservatives, also known as fungicides or biocides, are added to plastic materials to prevent the growth of microorganisms. These additives are highly toxic to lower organisms, but do not affect the higher organisms.

It is necessary to evaluate the effectiveness of various antimicrobial agents both on a laboratory scale and in actual outdoor exposure. Many methods have been devised to perform these tests. The test methods and their serious limitations are also discussed in this chapter.

5.1.3. Oxygen, Moisture, Thermal Energy, and Other Environmental Factors

In addition to UV radiation and microorganisms, many other relevant factors can add to polymer degradation. Even though most polymers react very slowly with oxygen, elevated temperatures and UV radiation can greatly promote the oxidation process. Water is considered relatively harmless; however, it can have at least three kinds of effects leading to early polymer degradation: a chemical effect, that is, the hydrolysis of unstable bonds; a physical effect, that is, destroying the bond between a polymer and filler resulting in chalking; and a photochemical effect. Thermal energy plays and indirect role in polymer degradation by accelerating hydrolyis, oxidation, and photochemical reactions induced by other factors. Many other factors such as ozone and atmospheric contaminants, including dirt, soot, smog, sulfur dioxide, and other industrial chemicals, have a significant effect on polymers (2).

5.2. ACCELERATED WEATHERING TESTS

Most data on the aging of plastics are acquired through accelerated tests and actual outdoor exposure. The latter is a time-consuming method, accelerated tests are often used to expedite screening the samples with various combinations of additive levels and ratios. A variety of light sources are used to simulate the natural sunlight. The artificial light sources include carbon arc lamps, xenon arc lamps, fluorescent sun lamps, and mercury lamps. These light sources, except the fluorescent, are capable of generating a much higher intensity light than natural sunlight. In the same wavelength band, xenon arc lamps can be operated over a wide range from below peak sunlight to twice the sunlight levels. Quite often, a condensation apparatus is used to simulate the deterioration caused by sunlight and water as rain or dew. Modern instruments have direct specimen spray on the front and/or back side of the specimen.

There are three major accelerated weathering tests:

1. Exposure to carbon arc lamps
2. Exposure to xenon arc lamps
3. Exposure to fluorescent UV lamps

The xenon arc, when properly filtered, most closely approximates the wavelength distribution of natural sunlight.

5.2.1. Fluorescent UV Exposure of Plastics (ASTM D4329, ISO 4892-5)

This method simulates the deterioration caused by sunlight and dew by means of artificial ultraviolet light and condensation apparatus (Figures 5-1 and 5-2). Solar radiation ranges from ultraviolet to infrared. Ultraviolet light of wavelengths between 290 and 350 nm is the most efficient portion of terrestrial sunlight that is damaging to plastics. In the natural sunlight spectrum, energy below 400 nm accounts for less than 6 percent of the total radiant energy (3). Since the special fluorescent UV lamps radiate between 280 and 350 nm, they accelerate the degradation process considerably. Figure 5-3 illustrates the spectral energy distribution of sunlight and the fluorescent sunlamp.

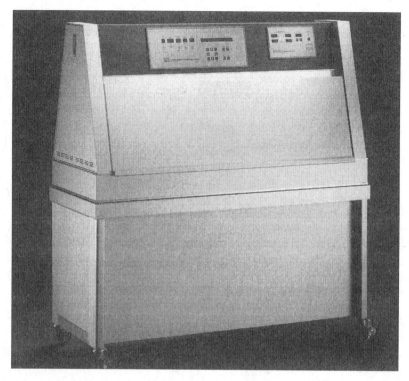

Figure 5-1. Ultraviolet light and condensation apparatus. (Courtesy of Q-Panel Lab Products.)

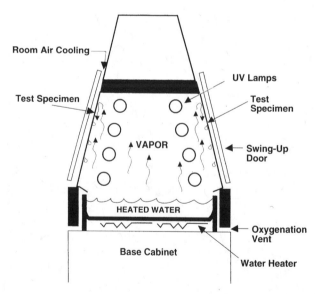

Figure 5-2. Cross section of a UV light and condensation apparatus. (Courtesy of Q-Panel Lab Products.)

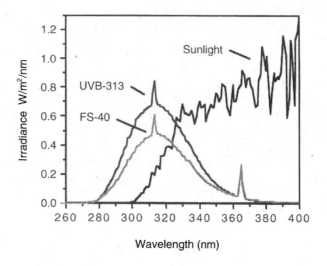

UV-B Lamps Compared to Summer Sunlight

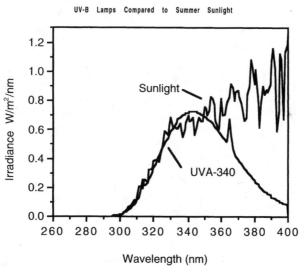

Figure 5-3. Spectral energy distribution of sunlight and fluorescent lamp. (Courtesy of Q-Panel Lab Products.)

In recent years, the UVA-340 lamps have increase in popularity because of the poor results of the conventional FS-40 lamps.

The test apparatus basically consists of a series of UV lamps, a heated water pan, and test specimen racks. The temperature and operating times are independently controlled for both UV and the condensation effect. The test specimens are mounted in specimen racks with the test surfaces facing the lamp. The test conditions are selected based on requirements and programmed into the unit. The specimens are removed for inspection at a predetermined time to examine color loss, crazing, chalking, and cracking.

5.2.2. Filtered Open-Flame Carbon Arc-Type Exposures of Plastics (ASTM D 1499) and Enclosed Carbon Arc Exposures of Plastics (ASTM D 6360)

This method is very useful in determining the resistance of plastic materials when exposed to radiation produced by carbon arc lamps. Two different types of carbon arc lamps used as the source of radiation. The first type is enclosed carbon arc lamp. The second type is known as an open-flame sunshine carbon arc. Figures 5-4 and 5-5 compare both types of light sources with natural sunlight. The enclosed carbon arc apparatus basically consists of either single or twin enclosed carbon arc lamps mounted in a chamber. A bell-shaped borosilicate glass globe surrounds the flame portion of the carbon arc lamp. The globe filters out UV radiation below 275 nm not found in direct sunlight and creates a semi-sealed atmosphere in which the arc burns more efficiently. The globes are normally replaced after 2000 hr of

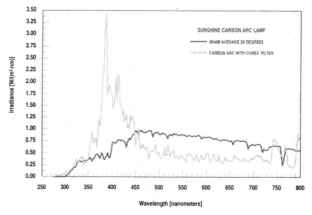

Figure 5-4. Comparison between energy output of sunshine carbon arc lamp and natural sunlight. (Courtesy of Atlas Material Testing Technology, LLC.)

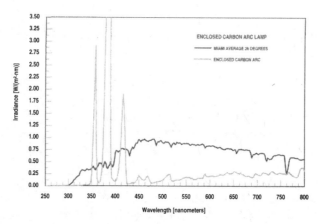

Figure 5-5. Comparison between energy output of enclosed carbon arc lamp and natural sunlight. (Courtesy of Atlas Material Testing Technology, LLC.)

Figure 5-6. Interior of a typical twin enclosed carbon arc apparatus. (Courtesy of Atlas Material Testing Technology, LLC.)

use. Figure 5-6 illustrates the interior of one such type of twin enclosed carbon arc apparatus. The open-flame sunshine carbon arc lamp operates in a free flow of air. This lamp accommodates three upper and three lower electrodes that are consumed during 24–26 hr of continuous operation. Corex glass filters that cut off light below 255 nm surrounds the lamp. The lamp is designed to run with filters in place; however, they can be removed to increase the available UV energy below 300 nm and thereby increase significantly the rate of photodegradation of UV-sensitive materials. However, extreme caution should be exercised while practicing this technique for safety reasons. An interior view of a typical apparatus of this kind is shown in Figure 5-7. The enclosed arc-type apparatus uses an open-ended drum-type cylinder, whereas the sunshine arc type uses a cylindrical framework. Both apparatuses revolve around centrally mounted arc lamps. The provision is also made to expose the specimen to water, which is sprayed through nozzles. The light-on, light-off, and water-spray cycles are independent of each other, and the apparatus can be programmed to operate with virtually any combination. A black panel temperature inside the test chamber can be monitored and controlled by a sensor mounted directly on the revolving specimen rack.

5.2.3. Xenon Arc Exposure of Plastics Intended for Outdoor Applications (ASTM D 2526, G 26, ISO 4892-2)

A water-cooled xenon-arc-type light source is one of the most popular indoor exposure tests since it exhibits a spectral energy distribution of sunlight at the

Figure 5-7. Interior of a typical open-flame carbon arc apparatus. (Courtesy of Atlas Material Testing Technology, LLC.)

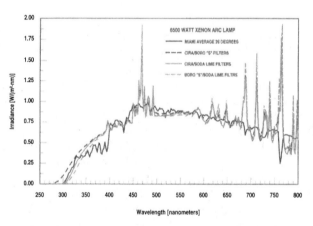

Figure 5-8. Comparison between energy output of natural sunlight and xenon arc lamp with different types of filters. (Courtesy of Atlas Material Testing Technology, LLC.)

surface of the earth (4). The xenon arc lamp consists of a burner tube and a light filter system consisting of interchangeable glass filters used in combination to provide a spectral distribution that approximates natural sunlight exposure conditions, as shown in Figure 5-8 and Figure 5-9. The apparatus has a built-in recirculating system that recirculates distilled or deionized water through the lamp. The water cools the xenon burner and filters out long wavelength infrared energy. For

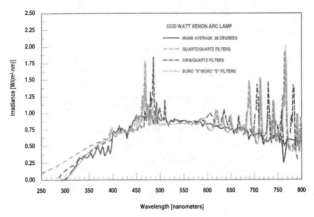

Figure 5-9. Comparison between energy output of natural sunlight and xenon arc lamp with different types of filters. (Courtesy of Atlas Material Testing Technology, LLC.)

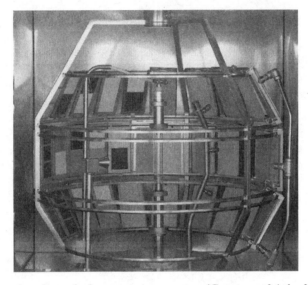

Figure 5-10. Interior of a typical xenon arc apparatus. (Courtesy of Atlas Material Testing Technology, LLC.)

air-cooled lamps, this is accomplished by the use of optical filters. The interior of one such water-cooled, xenon arc apparatus is shown in Figure 5-10.

Five test cycles are listed in Practice D 2565. Cycle 1 is listed because it has been used for plastics by historical convention. Cycle 2 is for general plastics applications. These cycles, along with the three remaining cycles, are described in more detail in the ASTM Standards Manual. In any case, it is highly recommended that a controlled irradiance exposure system be used. This is best accomplished through the use of a continuously controlled monitor that can automatically maintain uniform intensity at preselected wavelengths or wavelength range, when broadband

Figure 5-11. Weathering tester. (Courtesy of Atlas Material Testing Technology, LLC.)

control is being used. Figure 5-11 shows a commercially available weathering tester.

5.2.4. Interpretations and Limitations of Accelerated Weathering Test Results

There has been a severe lack of understanding on the part of users regarding the correlation between the controlled laboratory test and the actual outdoor test and application. The questions often asked are: "How many hours of exposure in a controlled laboratory enclosure is equal to one month of outdoor exposure?" "How do the results obtained from one type of weathering device compare to another type?" There is a general agreement among the researchers, manufacturers, and users that the data from accelerated weathering tests are not easily correlated with the results of natural weathering. However, accurate ranking of the weatherability of most material is possible using improved test methods and sophisticated equipment. The relative weather resistance of unmodified thermoplastics is shown in Table 5-1. ASTM Committee G3 has proposed the following definition for the correlation of exposure results. "For the comparison of the test results to be truly significant, one must satisfy himself that changes in property to be compared are produced by the same or significantly similar chemical reactions throughout the comparison period."

Accelerated weathering tests were devised to study the effect of actual outdoor weather in a relatively short time period. These tests often produce misleading results that are difficult to interpret or correlate with the results of actual outdoor exposure. The reason for such a contradiction is that in many laboratory exposures, the wavelengths of lights are distributed differently than in normal sunlight, possibly producing effects different from those produced by outdoor weathering. All plastics seem to be especially sensitive to wavelengths in the ultraviolet region. If

TABLE 5-1. Relative Weather Resistance of Unmodified Thermoplastics

Polymer	Resistance
Acrylics	High
PTFE and other fluorocarbons	High
Polycarbonate	Medium
Thermoplastic polyester	Medium
Cellulose acetate butyrate	Medium
Nylons	Medium
Polyurethanes	Medium
Polyphenylene oxide	Medium
Polyphenylene sulfide	Medium
Rigid PVC	Medium
Flexible PVC	Low
ABS	Low
Acetal	Low
Polyethylene	Low
Polysulfone	Low
Polystyrene	Low
Polypropylene	Low
Cellulose acetate	Low

the accelerated device has unusually strong emission at the wavelength of sensitivity of a particular polymer, the degree of acceleration is disproportionately high compared to outdoor exposure. The temperature of the exposure device also greatly influences the rate of degradation of a polymer. The higher temperature may cause oxidation and the migration of additives which, in turn, affects the rate of degradation. One of the limitations of accelerated weathering devices is their inability to simulate the adverse effect of most industrial environments and many other factors present in the atmosphere and their synergistic effect on polymers. Some of the newly developed gas-exposure cabinets have partially overcome these limitations. These units are capable of generating ozone, sulfur dioxide, and oxides of nitrogen under controlled conditions of temperature and humidity.

Improved ultraviolet sources and more knowledge of how to simulate natural wetness now make it possible to achieve reliable accelerated weathering results if the following procedures are observed:

1. Include a material of known weather resistance in laboratory tests. If such a material is not available, use another similar product that has a history of field experience in a similar use.

2. Measure or estimate the UV exposure, the temperature of the product during UV exposure, and the time of wetness under service conditions of the product. Try to creat laboratory cycles that duplicate the natural balance of these factors.

3. Do not use abnormal UV wavelengths to accelerate effects unless you are testing small differences in the same material. Evaluating two different materials by this technique can distort results.

4. Natural weathering is invariably a combination of UV and oxidation from wetness. In laboratory tests, UV or water can be excluded to define the problem area. Do not over look this technique.

5. Do not seek to establish a table of how many hours of accelerated testing equals x months of exposure. Natural weathering varies widely, even at one site on one product. Plastic tail-light housings do not experience the same weathering effects as a vinyl roof on the same automobile. Reasonable standards of endurance in laboratory tests can be established for specific products, but one tester, one cycle, and one chart for converting laboratory hours to natural months is an unrealistic objective (5).

Before drawing any final conclusions concerning the ability of a polymer to withstand outdoor environment based on accelerated weathering tests, one must conduct actual outdoor exposure tests for a reasonable length of time.

5.3. OUTDOOR WEATHERING OF PLASTICS (ASTM D 1435, ISO-877)

Since one quarter of all polymers end up in outdoor applications, outdoor weathering tests have become very popular (6).

The test is devised to evaluate the stability of plastic materials exposed outdoors to varied influences that comprise weather exposure conditions that are complex and changeable. The important factors are climate, time of year, and the presence of industrial atmosphere. It is recommended that repeated exposure testing at different seasons and over a period of more than one year be conducted to confirm exposure at any one location. Since weathering is a comparative test, control samples are always utilized and retained at standard conditions of temperature and humidity. The control samples must also be covered with inert wrapping to exclude light exposure during the aging period. However, dark storage does not insure stability (7).

Test sites are selected to represent various conditions under which the plastic product will be used. Arizona is often selected for intense sunlight, wide temperature cycle, and low humidity. Florida, on the other hand, provides high humidity, intense sunlight, and relatively high temperatures.

Exposure test specimens of suitable shape or size are mounted in a holder directly applied to the racks. Racks are positioned at a 45° angle and facing the equator. Many other variations in the position of the racks are also employed, depending upon the requirements.

The specimens are removed from the racks after a specified amount of time and subjected to appearance evaluation, electrical tests, and mechanical tests. The results are compared with the test results from control specimens. Typical aluminum exposure racks are shown in Figure 5-12 and suitably mounted specimens are shown in Figure 5-13. Table 5-2 lists weathering test methods.

5.3.1. Outdoor Accelerated Weathering

To accelerate outdoor weathering, a reliable method for predicting long-term durability in a shorter time frame had to be developed. Outdoor accelerated test methods were developed some 30 years ago by DSET Laboratories. The method employs

Figure 5-12. Typical aluminum exposure racks. (Courtesy of Atlas Material Testing Technology, LLC.)

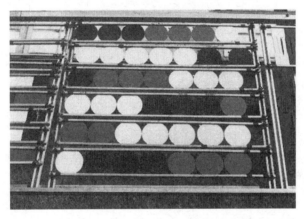

Figure 5-13. Suitably mounted samples. (Courtesy of Atlas Material Testing Technology, LLC.)

Fresnel-reflecting solar concentrators that use 10 flat mirrors to uniformly focus natural sunlight onto specimens mounted in the target plane. High-quality, first-surface mirrors provide an intensity of approximately eight suns with spectral balance of natural sunlight in terms of ultraviolet integrity. The test method provides an excellent spectral match to sunlight, correlating well to subtropical conditions such as southern Florida as well as an arid desert environment such as Arizona.

The test apparatus is a follow-the-sun rack with mirrors positioned as tangents to an imaginary parabolic trough. The axis is oriented in a north–south direction, with the north elevation having the capability for periodic altitude adjustment.

The target board, located at the focal line of the mirrors, lies under a wind tunnel along which cooling air is deflected across the specimens. A nozzle assembly is employed to spray the specimens with deionized water in accordance with established schedules. Nighttime spray cycles can be used to keep specimens moist during the nontracking portion of the test. The entire three-year real-time Florida

TABLE 5-2. Weathering Test Methods

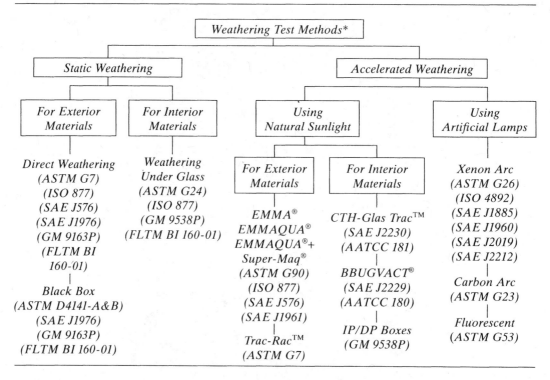

exposure test can be carried out in just six months depending on the program start date. The test is widely used in automotive, agriculture, building, textile, and packaging industries. The testing service is available commercially from Atlas Weathering Services Group, which is listed in the appendix section. Figure 5-14 illustrates outdoor accelerated weathering test apparatus. Table 5-3 lists standards and approvals specifying the EMMAQUA®[1] (equatorial mount with mirrors for acceleration with water) method for accelerated outdoor weathering.

5.4. RESISTANCE OF PLASTIC MATERIALS TO FUNGI (ASTM G 21)

This accelerated laboratory test determines the effect of fungi on plastic materials. The effectiveness of antimicrobial additives is also evaluated by such procedures. The test requires preparing a fungus spore suspension from cultures of various fungi that are known to attack polymers. Many other types of organisms can also be used in the test if necessary. The prepared spore suspension is tested for fungal growth without using plastic specimens as a viability control. The plastic specimens can be of any size or shape, including completely fabricated parts, test bars, or pieces from fabricated parts. Optically clear materials are used to study the effect of fungi on optical reflection or transmission. The specimens are placed onto petri

[1]EMMAQUA® is a registered trademark of Atlas Material Testing Technology, LLC.

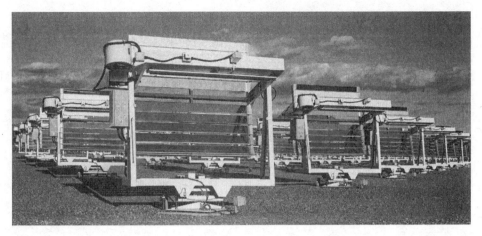

Figure 5-14. Accelerated outdoor weathering testers. (Courtesy of Atlas Material Testing Technology, LLC.)

dishes or any other suitable glass tray covered with nutrient salts agar. The entire surface is then sprayed using a sterilized atomizer with fungus spore suspension. The inoculated test specimens are covered and placed in an incubator maintained at 28–30°C and 85 percent or more relative humidity for 21 days. The fungal growth is visually inspected after 21 days of incubation. To study the effect on physical, optical, or electrical properties, the specimens are washed free of growth and subsequently conditioned, employing standard conditioning procedures. The physical testing is carried out in the usual manner and compared with control.

5.5. RESISTANCE OF PLASTIC MATERIALS TO BACTERIA (ASTM G 22)

This accelerated laboratory test is somewhat similar to the test to determine resistance of plastic material to fungi. The bacterial cell suspension in place of the fungus spore suspension is used to study the effect. The test is conducted in a similar manner by spraying the bacterial cell suspension onto the specimen placed in a petri dish covered with nutrient salts agar. The incubation is carried out for a minimum of 21 days under specified conditions in the incubator. A variation of the standard method is sandwiching the sample between two equal lots of nutrient salts agar and then spraying the bacterial cell suspension on it. This variation provides a more extensive contact between the test bacteria and the specimens.

5.6. LIMITATIONS OF ACCELERATED MICROBIAL GROWTH RESISTANCE TESTING

The data obtained from short-term accelerated testing can be very misleading if the results are not applied correctly and the data is not interpreted properly (8). The limitations of such tests are:

TABLE 5-3. Outdoor Accelerated Test Standards and Approvals

STANDARDS AND APPROVALS SPECIFYING THE EMMAQUA® METHOD						
Standard	Title	EMMA®	EMMA®-UG	EMMAQUA®	EMMAQUA®+	EMMAQUA®-NIW
ASTM D3105	Standard Index of Methods for Testing Elastomeric and Plastomeric Roofing and Waterproofing Materials	■		■		
ASTM D3841	Specification for Glass-Fiber-Reinforced Polyester Plastic Panels			■		
ASTM D4141	Standard Practice for Conducting Accelerated Outdoor Exposure Tests of Coatings			■		
ASTM D4364	Standard Practice for Conducting Accelerated Outdoor Weathering of Plastic Materials Using Concentrated Natural Sunlight	■	■	■		
ASTM D5105	Standard Practice for Performing Accelerated Outdoor Weathering of Pressure-Sensitive Tapes Using Concentrated Natural Sunlight			■		
ASTM E1596	Standard Test Method for Solar Radiation Weathering of Photovoltaic Modules			■		
ASTM C90	Standard Practice for Performing Accelerated Outdoor Weathering of Non-Metallic Materials Using Concentrated Natural Sunlight	■	■	■		
SAE J576	Plastic Materials for Use in Optical Parts such as Lenses and Reflex Reflectors of Motor Vehicle Lighting Devices	■				
SAE J1961	Accelerated Exposure of Automotive Exterior Materials Using a Solar Fresnel-Reflective Apparatus	■		■		
Ford ESB-M16J14-A	Enamel, Thermoset 2-Component Color Coat, Exterior, High-Class			■		
Ford ESB-MSP11-A	Metallized Plastic Parts, Reflectors			■		
ISO 877	Plastics—Methods of Exposure to Direct Weathering, to Weathering Using Class-Filtered Daylight, and to Intensified Weathering by Daylight Using Fresoel Mirrors	■	■	■		
ANSI/NSF 54	Flexible Membrane Liners			■		
JIS Z 2381	Recommended Practice for Weathering Test	■		■		
MIL-T-22085D	Tapes, Pressure-Sensitive Adhesive Preservation and Sealing	■		■		

1. Optimum microbial growing conditions dictate the use of high levels of toxic compounds that would "overkill" in actual commercial applications.
2. Exposure normally is limited to a relatively small number of different species of microorganisms.
3. Exposures are limited in time and conditions.
4. Misinterpretation of data by nonexperts.

It is very important to recognize that some chemicals can remain sufficiently toxic for the short duration of 14–21 days in accelerated testing but may deteriorate or may cause problems on long-term exterior exposures. For example, a fungicide may volatilize in hot weather, be degraded by UV light, or be bleached out by rain and dew.

5.7. OUTDOOR EXPOSURE TEST FOR STUDYING THE RESISTANCE OF PLASTIC MATERIALS TO FUNGI AND BACTERIA AND ITS LIMITATIONS

The outdoor exposure tests to determine the resistance of plastic materials to microbial attack are carried out many different ways (9). The simplest technique is to expose plastic material to an outdoor environment in geographical locations where weather conditions are favorable to microbial growth. The alternate method is called the soil burial method. This method calls for burying the specimens for four weeks and observing the effects of microorganisms of the specimens.

There are several serious limitations of outdoor exposure testing:

1. The chemical composition of the product
2. Angle of exposure to weather
3. Time of year exposures are made
4. Geographic location of exposures

The chemical composition of the product being tested influences the degree of microbial attack. Products having good water resistance and weatherability generally have a greater resistance to microbial attack since there is less plasticizer exudation and therefore, less nutrient available for growth. A product that has the ability to allow microbiocides to partially leach out to the surface is advantageous because the microbial attack is on the surface due to exudation of additives.

The angle of exposure to sunlight and weather conditions in general will influence the degree and duration of microbial attack. Some plastics exude plasticizer or other nutrients more rapidly when exposed at 40° south than on vertical exposures. Time of year and geographic location are also important since microorganisms grow more rapidly in a warm, humid climate than in a cold, dry climate. In conclusion, fungal and bacterial resistance testing must be carried out in two steps: first, a short and accelerated test to screen out unimportant specimens and second, a long-term outdoor exposure test to confirm the results of the laboratory tests.

REFERENCES

1. Mascia, L., *The Role of Additives in Plastics*, Edward Arnold Publishing Company, London, England, 1974, p. 134.
2. Kamal, M. R., "Weatherability of Plastic Materials," *Appl. Poly. Symp.*, No. 4, Interscience Publishers, New York, (1967), p. 2.
3. *Ibid.*, p. 64.

4. *Ibid.*, p. 62.

5. Dregger, D. R., "How Dependable are Accelerated Weathering Tests for Plastics and Finishes?" *Machine Design* (Nov. 1973), pp. 61–67.

6. Winolow, R. M., Matreyek, W., and Trozzolo, A. M., " Polymers Under Weather," *S.P.E.J.* **28**(7), (July 1972), pp. 19–24.

7. Kinmonth, R. A., Saxon, R., and King, R. M., "Sources of Variability in Laboratory Weathering," *Polym. Eng. Sci.* **10**(5), (Sept. 1970), pp. 309–313.

8. Wienert, L. A. and Hillard, M. W., "A Hard Look at Fungal-Resistance Testing," *Plast. Tech.* (Jan. 1977), P. 75.

9. *Ibid.*, p. 77.

SUGGESTED READING

Mullen, A., Kinmonth, R. A., and Searle, N. Z., "Spectral Energy Distributions and Aging Characteristics of Fluorescent Sunlamps and Blacklight," *J. Testing and Evaluation* **3**(1), (Jan. 1975), pp. 15–20.

Reinhart, F. W. and Mutchler, M. K., "Fluorescent Sunlamps in Laboratory Aging Tests for Plastics," *SATM Bull.*, **212**(33), (Feb. 1956), pp. 45–51.

Dunn, J. L. and Heffner, M. H., "Outdoor Weatherability of Rigid PVC," *SPE ANTEC*, **19**(1973), pp. 483–488.

Yaeger, C. C., "The Use of Antimicrobials in Flexible Vinyl Systems," *SPE ANTEC*, **17**(1971), p. 579.

Weinberg, E. L., *Modern Plastics Encyclopedia.* McGraw-Hill, New York, 1976–1977, p. 217.

Borg-Warner, "Weatherability of Cycolac Brand ABS Resins." *Tech. Bull.*, Parkersburg, W. VA, *PB-120A.*

Atlas Electric Devices Co., "Atlas Sun Spots." *Tech. Bull.*, Chicago, published quarterly.

Metzinger, R. and Kinmonth, R. A., "Laboratory Weathering," *Modern Paint and Coatings* (Dec. 1975).

Int. Symp. on the Weathering of Plastics and Rubber, Reprints, Plastics and Rubber Institute, London, England, 1976.

Rosato, D. V., *Environmental Effects on Polymeric Materials.* Interscience, New York, 1967.

Wypych, G., *Handbook of Material Weathering*, Plastics Design Library, Norwich, NY, 1995.

The Effect of UV Light and Weather on Plastics and Elastomers, PDL Handbook Series, Plastics Design Library, Norwich, NY, 1994.

Keller, DM., "Testing to failure of paint on plastics," Advance Coating Technology Conference, Chicago, IL, 1992, pp. 133–144.

Technical and sales literature, "Weathering Services," Atlas Weathering Services Group, Miami, FL.

6

OPTICAL PROPERTIES

6.1. INTRODUCTION

Unique properties, such as excellent clarity and transparency, good impact strength, moldability, and low cost, have made plastics the number-one choice of many design engineers. Successful applications of transparent and translucent plastics include automotive tail-light lenses, safety glasses, window glazing, merchandise display cases, and instrument panels. More recently, plastics have been accepted for more stringent applications such as contact lenses, prisms, low-cost camera lenses, and magnifiers. Plastics are much more resistant to impact than glass and, therefore, in applications such as street-lamp globes and high school windows, glass has been replaced by high-impact vandal-resistant plastic materials such as polycarbonate. However, lack of dimensional stability over a wider range of temperatures and poor scratch resistance have prevented further penetration of plastics in the markets for expensive camera lenses, microscopes, and other precision optics where the use of glass is fairly common.

Almost all plastics, below certain minimum thickness, are translucent. Only a few plastics are transparent. Plastic materials' transparency or translucency depends upon their basic polymer structure. Generally, all amorphous plastics are transparent. Crystallinity increases the density of the polymer, which decreases the speed of light passing through it and this increases the refractive index. When crystals are larger than the wavelength of the visible light, the light passing through many successive crystalline and amorphous areas is scattered, and the clarity of the polymer is decreased. A large single crystal scatters light at wide angles and thus causes haze (1). As a rule, crystalline plastics are translucent. However, the clarity of crystalline plastics can be improved by quenching or random copolymerization (2, 3).

Handbook of Plastics Testing and Failure Analysis, Third Edition, by Vishu Shah
Copyright © 2007 by John Wiley & Sons, Inc.

The primary optical properties are:

Refractive index
Light transmittance and haze
Photoelastic properties
Color
Gloss

6.2. REFRACTIVE INDEX (ASTM D 542, ISO 489)

Refractive index is a fundamental property of transparent materials. Refractive index values are very important to a design engineer involved in designing lenses for cameras, microscopes, and other optical equipment. The refractive index, also known as the index of refraction, is defined as the ratio of the velocity of light in a vacuum (or air) to its velocity in a transparent medium.

$$\text{Index of refraction} = \frac{\text{sin of angle of incidence}}{\text{sin of angle of refraction}}$$

Table 6-1 compares the refractive index values of different materials. Note that the refractive index of plastic materials is very close to the refractive index of glass. Two basic methods are most commonly employed to determine the index of refraction. The first method, known as the refractometric method, requires the use of a refractometer. An alternate method calls for the use of a microscope with a magnification power of at least 200 diameters. The refractometric method is generally

TABLE 6-1. Refractive Index Values for Plastics

Plastic	Value
Polytetrafluoroethylene (PTFE)	1.35
Cellulose acetate butyrate	1.47
Cellulose acetate	1.49
Acetal (homopolymer)	1.48
Acrylics	1.49
Polypropylene	1.49
Polybutylene	1.50
Ionomer	1.51
Low-density polyethylene	1.51
PVC (rigid)	1.52
Nylons (type 66)	1.53
Urea formaldehyde	1.54
High-density polynethylene	1.54
SAN	1.56
Polycarbonate	1.58
Polystyrene	1.60
Polysulfone	1.63
Glass	1.60

preferred over the microscopic method since it is much more accurate. The microscopic method, which is dependent upon the operator's ability to focus, is usually less accurate.

6.2.1. Refractometric Method

The Abbé refractometer is the refractometer most widely used to determine the index of refraction (Figure 6-1). The test also requires a source of white light and a contacting liquid that will not attack the surface of the plastic. The contacting liquid must also have a higher refractive index than the plastic being measured. A test specimen of any size may be used as long as it conveniently fits on the face of the fixed half of the refractometer prism. The surface of the specimen in contact with the prism must be optically flat and polished.

The test is carried out by placing a specimen in contact with the prism using a drop of contacting fluid. Contacting fluid must have a higher refractive index than the sample to be analyzed. Next, looking through the eye piece, the shadowline control knob is adjusted until the shadowline comes into the field of view. The dispersion correction wheel is adjusted to remove all the color from the shadowline. The shadowline should appear black with a sharp boundary, with white to off-white background. Finally the edge of the shadowline is aligned with the center of the crosshairs and the value of the index of refraction is read.

6.2.2. Microscopical Method

In this method, a microscope having a magnifying power of 200 diameters or more is used. A specimen of convenient size, having a fair polish and two parallel

Figure 6-1. Abbe refractometer. (Courtesy of Abbe Engineering Company.)

surfaces, is used. The test is carried out by alternately focusing the microscope on the top and the bottom surface of the specimen and reading the longitudinal displacement of the lens tube accurately. The difference between the two readings is considered the apparent thickness of the specimen. The index of refraction is determined as follows:

$$\text{Refractive index} = \frac{\text{Actual thickness}}{\text{Apparent thickness}}$$

6.3. LUMINOUS TRANSMITTANCE AND HAZE (ASTM D 1003)

Luminous transmittance is defined as the ratio of transmitted light to the incident light. The value is generally reported in percentage of light transmitted. Polymethyl methacrylate, for example, transmits 92 percent of the normal incident light. There is about 4 percent reflection at each polymer–air interface for normal incident light (4).

Haze is the cloudy appearance of an otherwise transparent specimen caused by light scattered from within the specimen or from its surface. Haze is defined as the percentage of transmitted light which in passing through a specimen deviates from the incident beam by forward scattering. It is generally accepted that if the amount of transmitted light is deviated more than 2.5° from the incident beam, the light flux is considered to be haze. Haze is normally caused by surface imperfections, density changes, or inclusions that produce light scattering. Haze is also reported in percentage.

Light transmittance and haze are extremely important from a practical viewpoint. For example, a window glazing material must have high light transmittance characteristics and must be free from haze. In contrast, the housing material for the light fixture must have maximum diffusion and minimum transparency to conceal the bright light source. The housing material must also have high light transmittance (5).

6.3.1. Test Procedure

This procedure employs an integrating sphere hazemeter, as illustrated schematically in Figure 6-2. The test specimen must be large enough to cover the aperture, but small enough to be tangent to the sphere wall. A disc of 2.00 in. in diameter is most commonly used. A commercially available hazemeter is shown in Figure 6-3. The test is conducted by taking four different consecutive readings and measuring the photocell output as follows:

T_1 = specimen and light trap out of position, reflectance standard in position

T_2 = specimen and reflectance standard in position, light trap out of position

T_3 = light trap in position, specimen and reflectance standard out of position

T_4 = specimen and light trap in position, reflectance standard out of position

The quantities represented in each reading are incident light, total light transmitted by specimen, light scattered by instrument, and light scattered by instrument

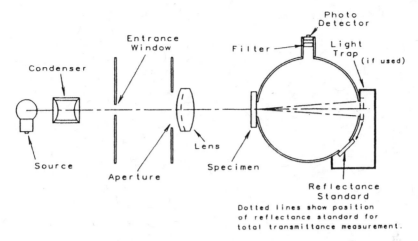

Figure 6-2. Schematic of a hazemeter. (Reprinted with permission of ASTM.)

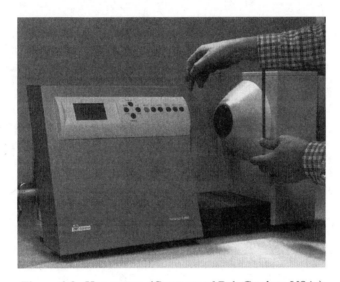

Figure 6-3. Hazemeter. (Courtesy of Byk-Gardner USA.)

and specimen, respectively. Total transmittance T_t and diffuse transmittance T_d are calculated as follows:

$$T_t = \frac{T_2}{T_1}$$
$$T_d = [T_4 - T_3(T_2/T_1)]/T_1$$

The percentage of haze is calculated as follows:

$$\text{Haze percent} = \frac{T_d}{T_t} \times 100$$

6.4. PHOTOELASTIC PROPERTIES

Most plastics, when placed under internally or externally applied stress and viewed through polarized light, exhibit stress-optical properties (6). Light vibrating in one plane within a plastic that is strained travels faster than light vibrating in a plane at right angles. The difference in the two velocities shows up as a birefringence. Stress-optical sensitivity is defined as the ability of some materials to exhibit double refraction (birefringence) of light when placed under stress (7).

Polarized lenses are used to plane polarize the incoherent light. This effect is shown in Figure 6-4. Polarized sunglasses and polarizing lenses that eliminate the glare from shining objects by plane polarizing the light are just a few examples of applications of polarizing mediums. When two such cross-polarizing mediums are used, an optical rotary effect or birefringence is produced (8).

Photoelastic properties of the transparent materials have been used by design engineers for stress analysis and by process engineers for determining residual stress as well as the degree of orientation in molded parts. Many complicated shapes such as bridges, aircraft wings, and gears can be stress analyzed by building a prototype model out of transparent plastic and viewing them through polarized light. Useful information regarding the location of stress concentration, effects of sharp corners, and changes in cross section can be obtained through such analysis. The individual sections of a model can also be physically stressed to observe the effect. A substantial amount of money can be saved in mold modification and the risk potential of the product can be minimized by making a prototype part in a transparent material and analyzing it for stress concentration area. Such analysis also reveals the effects of gate size and location, weld lines, cored holes, molding, annealing, and machining. In the production of lenses and other transparent parts, this technique can be used as a means of controlling quality. This nondestructive technique can also be used to analyze field failures by molding a few parts in transparent material and viewing them through a polarized light to detect stress concentrations.

6.4.1. Stress-Optical Sensitivity Examination

Transparent plastic parts are examined for stress-optical sensitivity by using a relatively simple setup, as shown in Figure 6-5. The object to be examined is placed in

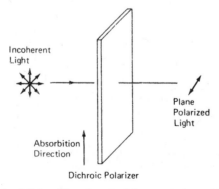

Figure 6-4. Plane polarized light. (Reprinted with permission of Van Nostrand Reinhold Company.)

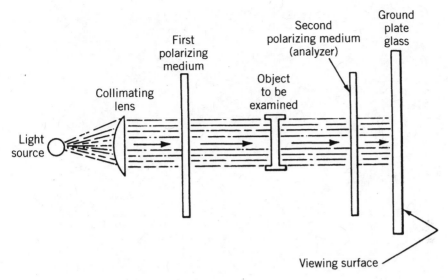

Figure 6-5. Setup for examination of stress-optical sensitivity. (Reprinted with permission of McGraw-Hill Company.)

the center of the polarizing medium. The object is viewed from the opposite side of the light source. The polarized light indicates the number of fringes or rings, which is related to the amount of stress. If white light is used, interesting colored patterns with all colors of the spectrum are seen, but for stress analysis, monochromatic light is preferred because it permits more precise measurements. The stress-optical coefficient of most plastics is on the order of 1000-psi/in. thickness/fringe. This means that a part of 1-in. thickness, illuminated with monochromatic polarized light and viewed through a polarized filter, will show a dark ring for stress of 1000 psi (9).

A typical photoelastic pattern is shown in Figure 6-6. A light box such as the one shown in Figure 6-7 can be constructed by using polarized sheets and a light source. Figure 6-8 illustrates commercially available equipment for the examination.

6.5. COLOR

The ability of plastic materials to color with relative ease and the fact that color can be integrated throughout the entire structure have made plastics one of the most successful materials of this century. To understand color measurement, specification, and tolerances, a proper understanding of color theory is essential.

The color perception process requires three things: a light source, an object, and an observer. The process occurs as follows. A beam of light from the source reaches the surface of the object. A portion of the light is reflected because of the surface interface. Known as the specular reflection or gloss component, this portion is reflected at an angle equal in magnitude to the angle of incidence of the light beam, but opposite in direction. The gloss component contains all wavelengths of the light

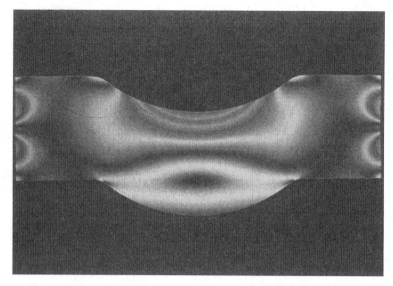

Figure 6-6. Typical photoelastic pattern. (Courtesy of Measurements Group, Inc.) See color insert.

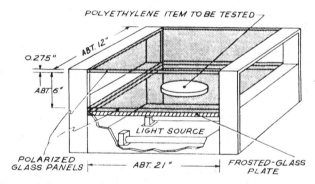

Figure 6-7. Light box for stress-optical sensitivity examination.

source. Therefore, if the light source is white light, the gloss component will also be reflected as white light.

The remainder of the light penetrates the surface of the object, where it is modified through selective absorption, reflection, and scattering by the colorants, polymers, and additives. Selective absorption and reflection by wavelength create color. For example, if an object absorbs all wavelengths other than blue from a white light source, the blue light component will be unmodified and will be reflected or transmitted from the object. The observer will see this reflected or transmitted blue light and say that the object appears blue.

Colors range from dark to light—black being darkest, gray being in the middle, and white being the lightest. These are called neutral colors. This aspect of color is termed "value" or "lightness." Color also has another basic difference: Red differs from blue, green, or yellow. These distinctions are called "hue." Hue is

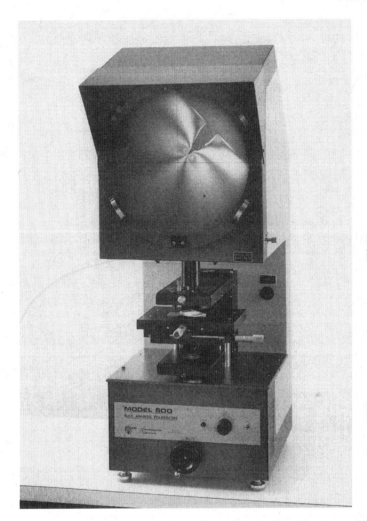

Figure 6-8. Polariscope for two- and three-dimensional photoelastic model analysis. (Courtesy of Measurements Group, Inc.).

defined as the attribute of color perception by means of which an object is judged to be red, yellow, green, blue, purple, or intermediate between some adjacent pair of these. One other dimension called "saturation" or "chroma" is defined as the attribute of color perception that expresses the degree of departure from gray of the same lightness. Thus, the entire color spectrum can be described in terms of value, hue, and chroma. This concept is illustrated in the hue value/chroma chart in Figure 6-9.

The essential difference in the spectrum of light of various hues is the wavelength of the light. The light is dispersed into a spectrum because of the differences in wavelength. The entire visible spectrum ranges from violet light, with the shortest wavelength of about 380 nm, to red light, with the longest wavelength of about 760 nm (10).

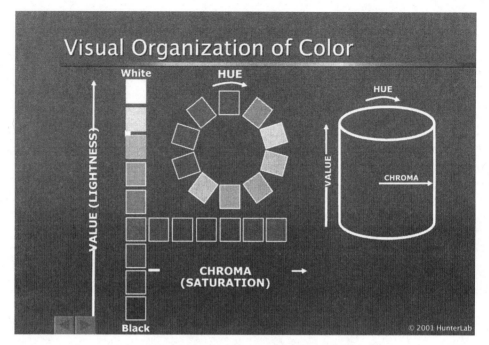

Figure 6-9. Hue value/chroma chart. (Courtesy of Hunter Associates Laboratory, Inc.) See color insert. A full color version of this figure can be viewed on DVD included with this book.

In order to see any object, the object must be illuminated with an illuminant. The type of illuminant, angle of illumination, and angle of viewing all affect the appearance of the object. Therefore, in measuring color, one must consider spectral energy distribution and intensity of the illuminant as it affects the appearance of the object. To standardize the variations among the illuminants, the International Commission on Illuminants, known as CIE (*Commission Internationale de l'Eclairage*) has established standard illuminants. For example, illuminant A represents an incandescent light; illuminant B, noon sunlight; illuminant C, overcast sky daylight. One other factor that must be considered in color measurement is the variations in the color observer. CIE has also established a standard observer. A CIE standard observer is a numerical description of the response to color of the normal human eye (11). CIE spectral tristimulus values are derived by using the CIE standard source, CIE standard observer, and object.

From the above discussion, it is clear that given a CIE standard source, object, and spectral tristimulus values, one can easily measure color. The instrument developed for such color measurement is called a tristimulus colorimeter. A tristimulus colorimeter measures color in terms of three primary colors: red, green and blue, or more properly stated, in terms of three tristimulus values. Many different color scales have been developed to describe the color numerically in terms of lightness and hue. One of the most widely accepted systems is known as the L, a, b tristimulus system. Figure 6-10 illustrates L, a, b color space. The coordinate L is in the vertical direction and corresponds to lightness. A perfect white has an L value of 100 and a perfect black a zero. The variables a_L and b_L identify the hue

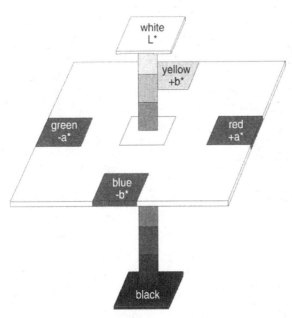

Figure 6-10. L, a, b color space. (Courtesy of Gretag Macbeth.) See color insert. A full color version of this figure can be viewed on DVD included with this book.

and the chroma of the material. A plus value of a_L indicates redness and a minus value greenness. For example, a school bus yellow with the following values describes a color: $L = 70.3$; $a_L = 30.3$; $b_L = 23.7$. This color can be described in common terms as fairly light, as indicated by a high L value, and yellowish red, as indicated by a_L and b_L values (12).

The filter colorimeter described thus far is based on the theory of the standard observer and how the human eye perceives the color. This approach has been described as a "psychophysical" approach to measurement (13). There are some problems associated with the design of filters and the filter material's ability to retain originality with time (14).

In recent years, a new generation of colorimeters based on spectrophotometers have been developed. The spectrophotometric colorimeter does not mimic the human eye. Instead, it makes a spectrophotometric measurement at sixteen 20-nm intervals over the entire range of the visible spectrum. The percentage reflectance value obtained by the spectrophotometer is converted into tristimulus values through the use of a microprocessor. One other useful feature added to these spectrophotometric colorimeters is the ability to select various types of CIE illuminants. Even though the actual light source remains the same, the microprocessor computes the colors that would be seen if the samples were viewed under various illuminants.

For color development work, a colorimeter that provides the tristimulus value and calculates the color difference is simply not enough. A more sophisticated instrument such as a spectrophotometer is generally required. A spectrophoto-meter, when used with a chart recorder or CRT, provides a complete spectral reflectance curve covering the entire range of the visible spectrum from 380 to 700 nm. Additionally, it provides the percentage reflectance value at each 20-nm interval and computes the tristimulus values, chromaticity coordinates, and color difference values.

6.5.1. Instrumented Color Measurement

Two basic types of instruments are used for color measurements depending on what is required. A filter colorimeter or a spectrophotometric colorimeter is employed when the user is only interested in tristimulus values, chromaticity coordinates, and color difference information. Figure 6-11 illustrates a commercially available spectrophotometric calorimeter. Colorimeters are mainly used for production control, quality control, specification, and color-matching requirements.

A spectrophotometer is normally required for color formulation and other color development work. Figure 6-12 illustrates a commercially available spectrophotometer equipped with a monitor for displaying and charting the results. Table 6-2 outlines the differences between colorimeters and spectrophotometers. The operation of this modern spectrophotometer is very simple. The instrument is first calibrated using a calibration standard. Next, the desired CIE standard illuminant is selected. By placing a relatively flat, colored specimen to be measured in a specimen holder and illuminating a light source, one can instantly obtain percentage spectral reflectance values at 16-nm intervals over the visible spectrum. The microprocessor also computes and displays the CIE *Lab* color space and a spectral reflectance curve as a function of wavelength. The color matching is carried out simple by exposing an unknown specimen and a control specimen to the illuminant and then comparing the two overlapping spectral reflectance curves. A typical printout of spectral reflectance values and spectral reflectance curves for two specimens is illustrated in Figure 6-13*a*.

As an added bonus, most available colorimeters and spectrophotometers calculate the yellowness index (ASTM D 1925) and the whiteness index. ASTM D 2244 describes the standard method for instrumental evaluation of color differences of opaque materials in detail.

There have been many new developments in color measurement systems technology in recent years. A major breakthrough is in the area of portable color measurement techniques. Newly developed portable spectrophotometers (Figure 6-13*b*) now make it possible to measure and analyze data on the production floor. Some portable spectrophotometers are also capable of displaying an actual spectral reflectance curve. Bench-top spectrophotometers have been updated to allow more computer control of the lens, UV filter, and specular port. The ability to calibrate the UV component of the color spectrum is an important feature for measuring fluorescent colors accurately. Advances in windows-based software has improved the capabilities for statistical analysis of color measurements, color corrections, and color formulation.

Continuous on-line measurement colorimeters also have been developed. These units are designed to withstand a harsh industrial environment, allowing continuous monitoring of the color of compounded pallets and other extruded or molded products. Color information is received in real-time mode, allowing the user to spot off-color trends and correct the process.

6.5.2. Visual Color Evaluation (ASTM D 1729)

A plastic processor requests a color match from a color supplier. The color supplier, through a rigorous color-matching process using many sophisticated instruments

17
17.11.2

Color
handy-color™
Colorimeter

Catalog 94/11

Technical Data

Geometry:	45° circumferential/ 0° dual beam optics
Measurement Range:	390 - 710 nm
Power Supply:	6 rechargeable AA NiCd batteries approx. 500 measurements per charge
Display:	50 x 50 mm (2 x 2 in), LCD
Languages:	English, German, French, Italian and Spanish
Storage Capacity:	500 samples, 50 standards
Display Options:	Color Scale Pass/ Fail CIELab Plot Lab(h) Plot Color difference in words
Color Scales:	CIELab, CIELCh, Lab(h), Yxy, XYZ
Color Differences:	$\Delta E_{L^*a^*b^*}$, ΔE_{Lab}, ΔE_{CMC} Component deltas
Indices:	Metamerism Yellowness: YI E313, YI D1925 Whiteness: WI E313, CIE Strength (tristimulus) Opacity
Illuminants:	Daylight C, D65 Tungsten A Cool White Fluorescent F2 Narrow Band White Fluorescent F11
CIE-Observer:	2°, 10°
Lamp: Lamp Life:	Pulsed Tungsten approx. 500,000 measurements
Dimensions:	L x W x H 10 x 8.9 x 25 cm 4 x 3.5 x 10 in
Weight:	1.0 kg (2.2 lbs)
Area of View Large area: Small area:	20 mm (0.79 in) 11 mm (0.43 in)

Figure 6-11. Colorimeter. (Courtesy BYK-Gardner USA.)

Figure 6-12. Spectrophotometer. (Courtesy of Gretag Macbeth.)

such as a spectrophotometer or a colorimeter, comes up with a match. When this match is presented to the customer, the customer is totally dissatisfied with the match. A color supplier may have to rematch the color as many as three times before the customer accepts the color match.

This scenario is a common occurrence in plastics and other industries. The reason is very simple. Two objects having the same color, when viewed under one type of illuminant (daylight), appear to match. The same two objects when viewed under different types of illuminants (incandescent) do not match. This phenomenon is known as metamerism. Metamerism is a phenomenon of change in the quality of color match of any pair of colors, as illumination or observer or both are changed.

In order to simplify the visual color evaluation and to minimize the variations brought forth by the variety of lighting conditions, a standard method of visual evaluation of color differences has been established. The test method defines the spectral characteristics of light sources. Three basic types of light sources are used for visual evaluations. The daylight source approximates the color and spectral quality of light from a moderately overcast northern sky. The incandescent source with a color temperature of 2854°K approximates household-type incandescent light. The fluorescent light source is a representative of a production-type cool white fluorescent lamp with a correlated temperature of 4400°K. Figure 6-14 illustrates a commercially available light booth for visual color evaluation. The interior of the light booth is painted with a neutral gray color. The specimens are viewed under specified lighting conditions and compared with the standard. The observation in order of color departure of the specimen from the standard in terms of lightness, saturation, and hue with an indication of the order of prominence is

TABLE 6-2. Colorimeters Versus Spectrophotometers

Colorimeter	Spectrophotometer
An instrument for psychophysical analysis—provides measurements that correlate with human eye-brain perception. Colorimetric data directly read and provided as tristimulus values (XYZ, L, a, b, etc.).	An instrument for physical analysis—provides wavelength-by-wavelength spectral analysis of the reflecting and/or transmitting properties of objects without interpretation by a human. Can indirectly calculate psychophysical (colorimetric) information.
Consists of sensor and simple data processor.	Consists of sensor plus data processor or computer with software.
Has a set observer and illuminant combination, usually 2°/C.	Has many available observer/illuminant combinations that can be used for calculating tristimulus data and metamerism index.
Isolates a broad band of wavelengths using a tristimulus absorption filter.	Isolates a narrow band of wavelengths using a prism, grating, or interference filter.
Is generally rugged and a less complex instrument than a spectrophotometer.	Is a more complex instrument than a colorimeter.
Works well for routine comparisons of similar colors and for adjustment of small color differences under constant conditions. Optimal for quality inspection.	Works well for color formulation, measurement of metamerism, and variable illuminant/observer conditions. Optimal for both quality inspection and research and development.
Example: HunterLab D25-Series instruments are colorimeters.	*Examples:* HunterLab ColorQUESTs, LabScans, MiniScans, and UltraScans are spectrophotometers.

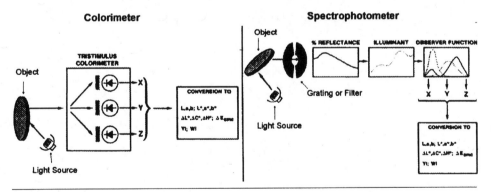

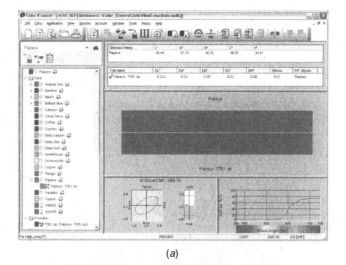

(a)

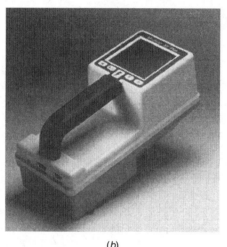

(b)

Figure 6-13. (*a*) A typical spectral reflectance curve. (Courtesy of Gretag Macbeth.) (*b*) A portable spectrophotometer. (Courtesy of Datacolor.)

made. Many color suppliers provide tolerance chips to the buyer so that the degree of color departure from control can easily be assessed.

6.6. SPECULAR GLOSS (ASTM D2457, D523)

Specular gloss is defined as the relative luminous reflectance factor of a specimen at the specular direction. This method has been developed to correlate the visual observations of surface shininess made at roughly corresponding angles. The light beam is directed toward the specimen at a specified angle and the light reflected by the specimen is collected and measured. All specular gloss values are based on a primary reference standard—a highly polished black glass with an assigned

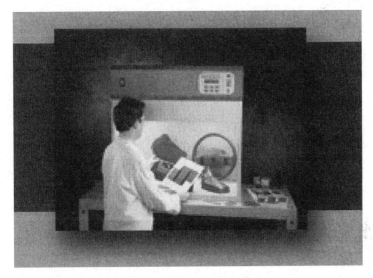

Figure 6-14. Light booth for visual color evaluation. (Courtesy of Gretag Macbeth.)

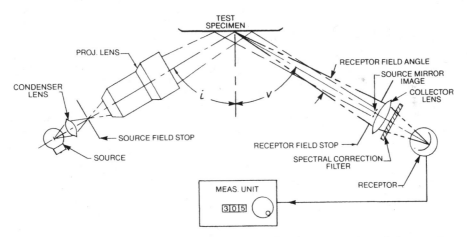

Figure 6-15. Diagram of parallel beam glossmeter showing apertures and source mirror-image position. (Reprinted with permission of ASTM.)

specular gloss value of 100. Three basic angles of incidence—20°, 60°, and 85°—are used for specular gloss measurement of plastic parts. As the angle of incidence increases, the value of gloss of any surface also increases (14).

The glossmeter basically consists of a source-optical assembly, which houses an incandescent light source, a condenser lens, and a projection or source lens. An incident beam, created by this assembly, is directed toward the specimen. A sensitive photodetector collects the reflected light and generates an electrical signal that is amplified to energize an analog or digital panel meter to display the value in gloss units. A schematic diagram of a glossmeter is shown in Figure 6-15.

The operation of a glossmeter is very simple. The instrument is turned on and placed on a black glass primary standard. The control knob is adjusted so that the

Figure 6-16. Glossmeter. (Courtesy Byk-Gardner USA.)

meter indicates the value assigned to the primary standard. Next, the sensor is placed on the specimen surface and the gloss value is read directly from the analog or digital display. The linearity of the instrument is routinely checked by placing the sensor on a white secondary standard, which should read within 1.0 gloss unit of the assigned value of that standard. Figure 6-16 illustrates a commercially available glossmeter.

REFERENCES

1. Deanin, R. D., *Polymer Structure Properties and Applications*, Cahners, Boston, MA, 1972, p. 248.
2. *Ibid.*, p. 248.
3. Byrdson, I. A., *Plastic Materials*, Reinhold, New York, 1970, p. 96.
4. *Ibid.*, p. 241.
5. Ives, C. G., Mead, J. A., and Riley, M. M., *Handbook of Plastics Test Methods*, IlLFFE Books, London, England, 1971, p. 432.
6. Baer, E., *Engineering Design for Plastics*, Reinhold, New York, 1964, p. 604.
7. Milby, R. V., *Plastics Technology*, McGraw-Hill, New York, 1973, p. 548.
8. Levy, S. and Dubois, J. H., *Plastics Product Design Engineering Handbook*, Reinhold, New York, 1977, pp. 309–310.
9. Kinney, G. F., *Engineering Properties and Applications of Plastics*, Wiley, New York, 1957, p. 220.
10. McCamy, C. S., *Color Measurement and Specification*, Macbeth Corporation, Newburgh, NY, Technical Literature, p. 10.
11. Billmeyer, F. W. and Saltzman, M., *Principles of Color Technology*, Interscience, New York, 1966, p. 38.

12. Gardner Laboratories, *Tech. Bull.: Color and Color Related Properties.* Bull. No. 010, Silver Spring, MD.

13. McCamy, Reference 10, p. 9.

14. Gardner Laboratories, *Tech. Bull.: Gloss Measurement.* Bull. No. 073, Silver Spring, MD.

SUGGESTED READING

Webber, T. G. (Ed.), *Coloring of Plastics*, Wiley-Interscience, New York, 1979.

Ahmed, M., *Coloring of Plastics Theory and Practice*, Reinhold, New York, 1979.

Ross, L. and Birley, A. W., "Optical Properties of Polymeric Materials and their Measurements." *J. Phys. D: Appl. Phys.*, **6**(1973), p. 795.

Jenkins, F. A. and White, H. E., *Fundamentals of Optics*, McGraw-Hill, New York, 1957.

Jacobs, D. H., *Fundamentals of Optical Engineering*, McGraw-Hill, New York, 1943.

X-Rite Incorporated, Tech. Bull.: "Understanding Color Communication" Grandville, MI.

Mulholland, B. M., "Introduction to Color Theory," Plastics Engineering, August 97, p. 33.

Technical Bulletins, "How to Choose the Right Refractometer", "Refractive Index to Transparent–Solid Materials," Reichest, Inc., Depew, New York.

7

MATERIAL CHARACTERIZATION TESTS

7.1. INTRODUCTION

The plastics industry has grown phenomenally in the past two decades. Plastic material consumption has quadrupled. Today there are countless numbers of processors consuming in excess of 100 million pounds of material every year. An increasing number of processors are looking into various techniques for characterizing the incoming plastic resin to guard themselves against batch-to-batch variations. Such variations in the properties and the processibility of the polymer have been very costly.

To understand the phenomenon of batch-to-batch variation, one must understand the polymerization process. The polymerization mechanism can often be very complex. Basically, it involves adding a monomer or a mixture of monomers into a reaction kettle along with several other ingredients, such as a catalyst, an initiator, and water, depending on the type of polymerization process. The contents are agitated at a specific speed. The time, temperature, and pressure are carefully controlled. During this process, the monomer is converted into a polymer. The properties of the final product depend on several factors, such as the monomer to water ratio, the presence of water, the degree of agitation, the removal of exothermic heat generated during the polymerization, and the ability of the polymer to dissolve in monomer. This is further complicated by the fact that different polymerization techniques are used to form the same chemical type of polymer. Because it is difficult to manufacture a monodispersed polymer (a polymer in which all molecules have the same size) commercially, we are forced to live with variations in the size and weight of molecules in a polymer. Such variations in the size and the weight of the polymer molecules are also extremely difficult to control. The

Handbook of Plastics Testing and Failure Analysis, Third Edition, by Vishu Shah
Copyright © 2007 by John Wiley & Sons, Inc.

relative proportions of molecules of different weights within a polymer comprise its molecular-weight distribution. Depending on the range of distribution (narrow or broad), the processibility and the properties of a polymer vary significantly. This basic nature of the polymerization process creates a need for material characterization tests to insure against variations in the characteristics of the polymer. Tables 7-1 and 7-2 list various applications using rheometry and thermal analysis techniques.

There are numerous ways of characterizing a polymer. Some are very basic and simple, others are more sophisticated and complex. The five most common and widely accepted tests are:

1. Melt index (melt flow rate) test; rheological tests
2. Viscosity tests
3. Gel permeation chromatography
4. Thermal analysis (TGA, TMA, DSC)
5. Spectroscopy

7.2. MELT INDEX TEST (ASTM D 1238, ISO 1133)

7.2.1. Significance

The melt index, more appropriately known as melt flow rate (MFR), test measures the rate of extrusion of a thermoplastic material through an orifice of specific length and diameter under prescribed conditions of temperature and load. This test is primarily used as a means of measuring the uniformity of the flow rate of the material. The reported melt index values help to distinguish between the different grades of a polymer. A high-molecular-weight material is more resistant to flow than a low-molecular-weight material. However, the data obtained from this test does not necessarily correlate with the processibility of the polymer. This is because plastics materials are seldom manufactured without incorporating additives which affect the processing characteristics of a material, such as stability and flowability. The effect of these additives is not readily observed via the melt index test. The rheological characteristics of polymer melts depend on a number of variables. Since the values of these variables may differ substantially from those in large-scale processes, the test results may not correlate directly with processing behavior.

7.2.2. Test Procedures

The melt index apparatus (Figure 7-1) is preheated to a specified temperature. The material is loaded into the cylinder from the top and a specified weight is placed on a piston. The most commonly used test conditions are shown in Tables 7-3a and 7-3b. The material is allowed to flow through the die. The initial extrudate is discarded because it may contain some air bubbles and contaminants. Depending on the material or its flow rate, cuts for the test are taken at different time intervals. The extrudate is weighed and melt index values are calculated in grams per 10 min.

TABLE 7-1. Material Characterization Through Rheology and Thermal Analysis

Applications + Instruments = Solutions

Applications	Rheometry			Thermal Analysis				
	Torque Rheometer	Rotational Rheometer	Capillary Rheometer	DMA	DSC	TG/DTA	TMA	DES
Bubble stability-blown film	✓							
Compositional analysis	✓	✓				✓		
Compounding	✓		✓					
Creep and stress relaxation	✓	✓		✓			✓	
Curing Studies				✓	✓			✓
Degree of cross linking		✓		✓	✓			
Dielectric properties								
Die swell	✓		✓					
Electro-rheology	✓	✓						
Elongational viscosity		✓	✓					
Gamma, beta transitions				✓			✓	
Gel point		✓	✓		✓	✓	✓	✓
Glass transition				✓	✓		✓	
Haze and gloss of films	✓							
Heat capacities				✓	✓		✓	
Impact resistance		✓		✓				
Lifetime predictions		✓		✓	✓		✓	
Measure degree of dispersion	✓	✓	✓	✓				
Measure rheological properties	✓	✓	✓	✓				
Melt strength	✓	✓	✓	✓				
Melting temperatures	✓	✓	✓	✓	✓	✓	✓	
Molecular weight	✓	✓		✓	✓			
Molecular weight distribution	✓	✓	✓	✓				
Oxidative stability		✓			✓			
Processibility	✓	✓	✓					
Sample preparation	✓							
Shear viscosity	✓	✓	✓					
Shrinkage forces		✓						
Tack & peel of PSA's	✓			✓			✓	
Thermal stability		✓		✓	✓	✓	✓	
Viscoelastic properties		✓	✓	✓			✓	

Courtesy of Haake-USA.

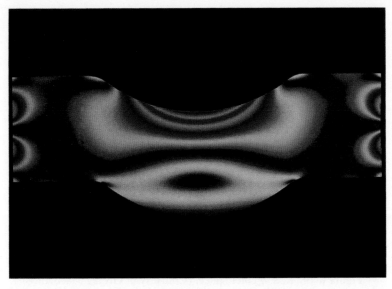

Figure 6-6. Typical photoelastic pattern. (Courtesy of Measurements Group, Inc.)

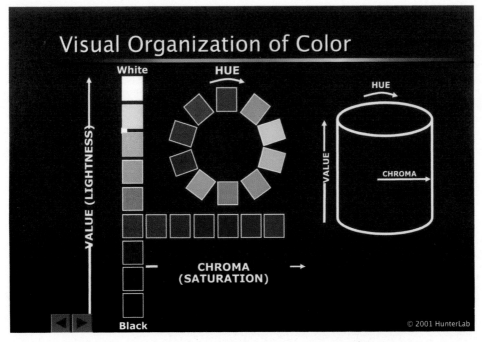

Figure 6-9. Hue value/chroma chart. (Courtesy of Gretag Macbeth.)

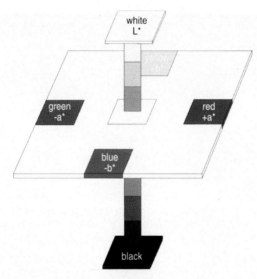

Figure 6-10. *L*, *a*, *b* color space. (Courtesy of Gretag Macbeth.)

TABLE 7-2. Applications of Material Characterization Techniques

Information	Process[a]	Data Obtainable[b]
Composition		
Resins, polymers modifiers	DSC	Types present, plus ratio if copolymer, blend or mixture
	GC	Polymers, oligomers, and residual monomer; amounts present
Additives	DSC	Deduce presence and amount from thermal effects (stabilizers, antioxidants, blowing agents, plasticizers, etc.)
	GC	Detect any organic additive (by MW)
Reinforcement, fillers	TG	Amount (from weight of ash)
Moisture, volatiles	TG	Amounts (from weight loss)
	DSC	Amounts (if sufficient heat absorbed)
Regrind level	DSC	Deduce amount from melting-point shift
	GC	Deduce amount from shift in MW distribution
Processability		
Melting behavior	DSC	Melt point and range (each resin, if blend); melt-energy requirements
Flow characteristics	GC	Deduce from balance of high and low MW polymer in MW distribution
Physical, Mechanical		
Glass transition temperature	DSC	Shown by step-up in energy-absorption as resin heats imarginal for some semicrystalline resins
	TM	Detected by sample's expansion
Crystallinity	DSC	Calculate percentage from heat of fusion during melting: also can find time-temperature conditions for desired percent crystallinity
Tensile, flexibility, impact	GC	Deduce from balance of high and low MW polymer in MW distribution
	DSC	Deduce from overall composition
Cure Characteristics, Thermosets		
Reactive systems	DSC	Shows gel time, cure temperature and cure time
Prepreg	DSC	Measures unreacted resin in B-stage prepreg or after cure (by volatiles)
	GC	As above, but checks by MW analysis of composite residue

[a] DSC, differential scanning colorimeter; GC, gel chromatography; TM, thermomechanical analysis.
[b] These techniques do not specifically identify materials. Materials can be identified by matching the instrument's test carves with those of known polymers or compounds. Otherwise, other analytical procedures must be used.

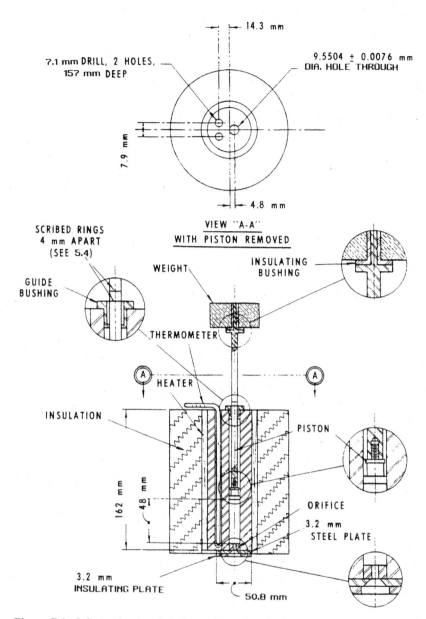

Figure 7-1. Schematic of melt indexer. (Reprinted with permission of ASTM.)

TABLE 7-3a. A Standard Test Conditions

Material	Condition	
Acetals (copolymer and homopolymer)	190/2.16	190/1.05
Acrylics	230/1.2	230/3.8
Acrylonitrile-butadiene-styrene	200/5.0	230/3.8
	220/10	
Acrylonitrile/butadiene/styrene/polycarbonate blends	230/3.8	250/1.2
	265/3.8	265/5.0
Cellulose esters	190/0.325	190/2.16
	190/21.60	210/2.16
Ethylene-chlorotrifluoroethylene copolymer	271.5/2.16	
Ethylene-tetrafluoroethylene copolymer	297/5.0	
Nylon	275/0.325	235/1.0
	235/2.16	235/5.0
	275/5.0	
Perfluoro(ethylene-propylene) copolymer	372/2.16	
Perfluoroalkoxyalkane	372/5.0	
Polychlorotrifluorethylene	265/12.5	
Polyethylene	125/0.325	125/2.16
	2.50/1.2	
	190/0.325	190/2.16
	190/21.60	190/10
	310/12.5	
Polycarbonate	300/1.2	
Polymonochlorotrifluoroethylene	265/21.6	
	265/31.6	
Polypropylene	230/2.16	
Polystyrene	200/5.0	230/1.2
	230/3.8	190/5.0
Polyterephthalate	250/2.16	210/2.16
	285/2.16	
Poly(vinyl acetal)	150/21.6	
Poly(vinylidene fluoride)	230/21.6	
	230/5.0	
Poly(phenylene sulfide)	315/5.0	
Styrene acrylonitrile	220/10	230/10
	230/3.8	
Styrenic thermoplastic elastomer	190/2.16	200/5.0
Thermoplastic elastomer–ether–ester	190/2.16	220/2.16
	230/2.16	240/2.16
		250/2.16
Thermoplastic elastomers (TEO)	230/2.16	
Vinylidene fluoride copolymers	230/21.6	
	230/5.0	
	for $T_m = 100°$ use 120/5.0 or 21.6	

From ASTM D 1238. Reprinted with permission of ASTM.

TABLE 7-3*b*. Standard Test Conditions, Temperature and Load

Condition		Total Load Including Piston (kg)	Approximate Pressure	
Standard Designation	Temperature (°C)		kPa	psi
125/0.325	125	0.325	44.8	6.5
125/2.16	125	2.16	298.2	43.25
150/2.16	150	2.16	298.2	43.25
190/0.325	190	0.325	44.8	6.5
190/2.16	190	2.16	298.2	43.25
190/21.60	190	21.60	2982.2	432.5
200/5.0	200	5.0	689.5	100.0
230/1.2	230	1.2	165.4	24.0
230/3.8	230	3.8	524.0	76.0
265/12.5	265	12.5	1723.7	250.0
275/0.325	275	0.325	44.8	6.5
230/2.16	230	2.16	298.2	43.25
190/1.05	190	1.05	144.7	21.0
190/10.0	190	10.0	1379.0	200.0
300/1.2	300	1.2	165.4	24.0
190/5.0	190	5.0	689.5	100.0
235/1.0	235	1.0	138.2	20.05
235/2.16	235	2.16	298.2	43.25
235/5.0	235	5.0	689.5	100.0
250/2.16	250	2.16	298.2	43.25
310/12.5	310	12.5	1723.7	250.0
210/2.16	210	2.16	298.2	43.25
285/2.16	285	2.16	298.2	43.25
315/5.0	315	5.0	689.5	100.0
372/2.16	372	2.16	298.2	43.25
372/5.0	372	5.0	689.5	100
297/5.0	297	5.0	689.5	100
230/21.6	230	21.6	2982.2	432.5
230/5.0	230	5.0	689.5	100
265/21.6	265	21.6	2982.2	432.5
265/31.6	265	31.6	4361.2	632.5
271.5/2.16	271.5	2.16	298.2	43.25
220/10.0	220	10.0	1379.0	200.0
250/1.2	250	1.2	165.4	24.0
265/3.8	265	3.8	524.0	76.0
265/5.0	265	5.0	689.5	100.0

From ASTM D 138. Reprinted with permission of ASTM.

An alternate method for making the measurement for materials with a high flow rate involves automatic timing of the piston travel by some electrical or mechanical device. The melt index value is calculated by using the following formula:

$$\text{Flow rate} = (426 \times L \times d)/t$$

where L = length of calibrated piston travel (cm); d = density of resin at test temperature (g/cm^3); t = time of piston travel for length L (sec).

A commercial melt indexer is shown in Figure 7-2.

Figure 7-2. Melt indexer. (Courtesy of Dynisco.)

7.2.3. Factors Affecting the Test Results

1. *Preheat Time.* If the cylinder is not preheated for a specified length of time, there is usually some nonuniformity in temperature along the walls of the cylinder even though the temperature indicated on the thermometer is close to the set point. The causes the flow rate to vary considerably. There should be zero thermal gradient along the full length of the test chamber.

2. *Moisture.* Moisture in the material, especially a highly pigmented one, causes bubbles to appear in the extrudate which may not be seen with the naked eye. Frequent weighing of short cuts of the extrudate during the experiments reveals the presence of moisture. The weight of the extrudate is significantly influenced by the presence of the moisture bubbles.

3. *Packing.* The sample resin in the cylinder must be packed properly by pushing the rod with substantial force to allow the air entrapped between the resin pellets to escape. Once the piston is lowered, the cylinder is sealed off, and no air can escape. This causes variation in the test results.

4. *Volume of Sample.* To achieve the same response curve repeatedly, the volume of the sample in the cylinder must be kept constant. Any change in sample volume causes the heat input from the cylinder to the material to vary significantly.

7.2.4. Interpretation of Test Results

The melt index values obtained from the test can be interpreted in several different ways. First, a slight variation in the melt index value should not be interpreted as indicating a suspect material. The material supplier should be consulted to determine the expected reproducibility for a particular grade of plastic material. A significantly different melt index value than the control standard may indicate several different things. The material may be of a different grade with a different flow characteristic. It also means that the average molecular weight or the molecular weight distribution of the material is different than the control standard and may have different properties.

Melt index is an inverse measure of molecular weight. Since flow characteristics are inversely proportional to the molecular weight, a low-molecular-weight polymer will have a high melt index value and vice versa.

7.3. RHEOLOGY

Rheology is defined as the study of flow. Viscosity is the measure of resistance to flow due to internal friction when one layer of fluid is caused to move in relationship to another layer. The greater the friction, the greater the amount of force required to cause this movement, which is called "shear." Shearing occurs whenever the fluid is physically moved or disturbed as in pouring, mixing, and spraying. Highly viscous fluids, therefore, require more force to move then less viscous materials. The velocity gradient is a measure of the speed at which the intermediate layers move with respect to each other. It describes the shearing the liquid experiences and is thus called "shear rate" (1). Shear stress is the stress developing in the polymer melt when the layers in a cross section are gliding along each other. When the ration of shearing stress to the rate of shear is constant, as in the case of water, the fluid is called "Newtonian fluid." In the case of non-Newtonian fluids, the ratio varies with shearing stress. Stated another way, when shear rate is varied, the shear stress does not vary in the same proportion. The viscosity of such fluids will change with the change in shear rate. All polymeric materials exhibit non-Newtonian flow behavior. This information concerning the flow of polymeric material is very useful in identification and characterization of the polymers. Rheological measurements are also helpful in the determination of molecular, elastic, and physical properties of polymers, simulation, determination of large-scale processing conditions, and basic research and development. Instruments developed for measuring the flow properties of polymers are known as rheometers. Many different types of rheometers have been developed. Data analysis and interpretation has been simplified with newly developed Windows-based software.

Until recently, the rheological measurements were confined to test labs for quality control, research, and development purposes. New techniques for on-line rheological measurements have been developed, bringing the analytical instrumentation to the production floor. This is accomplished by employing a closed-loop instrument which diverts a portion of the process stream to a measurement head and then returns it to the process. Within the measurement head, most of the material flows through a bypass channel and a fraction of the stream is pumped across a capillary die, using precision gear pumps. A similar concept, employed by another manufacturer, utilizes a slit die to measure viscosity. An open-loop on-line rheometer, which does not return the diverted melt stream to the process after measurement, splits the diverted melt stream to two capillary dies of different L/D ratios to enable simultaneous determination of viscosity at two different shear rates (2). The following sections describe three major types of rheometers most commonly used in the industry.

7.3.1. Torque Rheometer

Torque rheometry is used extensively for measuring rheological properties and for determining melt flow values, thermal stability of polymers, degradation time, and characterization of different formulations. Chapter 12 discusses torque rheometry in detail.

7.3.2. Rotational Rheometers/Viscometers

Many types of rotational rheometers and viscometers have been developed. The cone and plate, couette (coaxial cylinder), torsional, and disc spindle types are the most common.

The spindle viscometer (ASTMD2393) is one of the simplest instruments to operate. It measures torque required to rotate an immersed element (the spindle) in a fluid. The spindle is driven by a motor through a calibrated spring; the deflection of the spring is measured and mathematically converted into a viscosity value in units of centipoise. By utilizing multiple rotational speeds and interchangeable spindles, a variety of viscosity ranges can be measured.

The sample is preconditioned by placing it in a constant-temperature bath at the specified test temperature. The proper size spindle is allowed to rotate in the sample for up to 2 minutes before recording the reading. Data can be sent to a printer or linked to a PC using a special software package. The newer model viscometers are programmable and have the capability of measuring and displaying viscosity and temperature simultaneously. Figure 7-3a shows a digital programmable viscometer, which runs in stand-alone mode or under PC control.

The cone and plate viscometer is ideal for determining the absolute viscosity of fluids in a small sample volume. The commercially available cone and plate viscometer (e.g., Wells–Brookfield), a precise torque-measuring system, which consists of a calibrated beryllium–copper spring connecting the drive mechanism to a rotating cone, senses the resistance to rotation caused by the presence of sample fluid between the cone and a stationary flat plate. The resistance to the rotation of the cone produces a torque that is proportional to the shear stress in the fluid. The amount of torque is displayed digitally. A commercially available viscometer is illustrated in Figure 7-3b.

7.3.3. Capillary Rheometer

Quite often, the melt index test results simply do not provide sufficient information because the melt index is a single-point test. The prespecified test temperature is not necessarily the actual processing temperature. The test is conducted at a low shear rate, which is significantly lower than the actual processing shear rate. The flow rates are measured at a single shear stress and shear rate performed at one set of temperatures and geometric conditions. Furthermore, since the melt index measurement takes account of the behavior of the polymer at only one point, it is quite

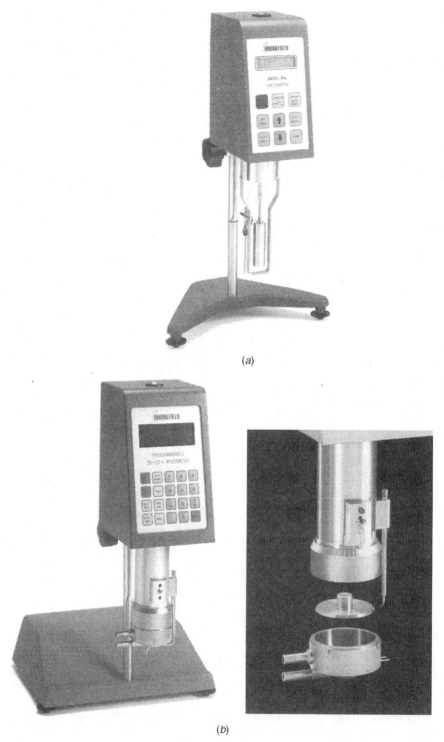

(a)

(b)

Figure 7-3. (*a*) A digital programmable viscometer. (Courtesy of Brookfield Engineering Laboratories.) (*b*) Cone and plate viscometer. (Courtesy of Brookfield Engineering Laboratories.)

possible for two materials with the same melt index values to behave completely differently at shear stresses that are different from the ones used during the melt index measurements. The capillary rheometer measures apparent viscosity or melt index over an entire range of shear stresses and shear rates encountered in compression molding, calendering, extrusion, injection molding, and other polymer melt processing operations. A rheometer is a precision instrument that provides the accuracy and reproducibility necessary for polymer characterization tests.

A quality control test to qualify the incoming material is quick, simple, and accurate. The capillary rheometer can be used to ensure that a material complies with the flow tolerances set for that grade of material. A quality control graph of a polymer flow curve can be generated with upper and lower critical viscosity limits over a wide range of shear rates by running multiple tests on a standard material. Figure 7-4a illustrates one such flow curve. The sample material curve is compared to the control curve. At low shear rates, the sample material curve is well within the tolerance level. However, at high shear rates, where actual processing of the polymer takes place, the sample fails to qualify (3). The graph also shows a clear advantage of the capillary rheometer test over the simple melt index test—at low shear rate, where the melt index test is conducted, the sample material would have qualified, giving false results. The thermal stability of a polymer and resulting processing limitations are of extreme importance to a processor. A capillary rheometer can accurately predict the thermal stability of the polymer. Materials such as PVC can begin to crosslink and exhibit a dramatic increase in viscosity if left in the process too long. This can lead to rejected or failure-prone parts and wasted time and materials. The capillary rheometer allows the residence-time capabilities of materials to be studied over a wide range of processing temperatures and shear rates so the optimal processing temperature can be determined (4). A shear viscosity versus residence time plot for Nylon 66 showing the degradation of the material is illustrated in Figure 7-4b. By generating rheology curves at several different temperatures, the relationship between temperature and viscosity at various shear rates can be determined. This information is valuable to the tool designer using computer-aided flow-simulation programs. When used along with a rheological profile, the temperature sensitivity of the polymer provides the information from which an accurate flow model can be developed and used for mold filling or die-flow design programs (5). The data generated from capillary rheometer measurements can be used to study the processibility of regrind material as well as process optimization. The capillary rheometer consists of an electrically heated cylinder, a pressure ram, temperature controllers, timers, and interchangeable capillaries (Figure 7-5). The plunger can be moved at a constant velocity, which translates to a constant shear rate. The force to move the plunger at this speed is recorded, which determines the shear stress. Alternately, a weight or constant pressure can be applied to the plunger which generates a constant shear stress, and the velocity of the plunger can be determined by cutting and weighing the output. Shear rate can be calculated by knowing the melt density.

The sample material is placed in the barrel of the extrusion assembly, brought to temperature, and forced out through a capillary. The force required to move the plunger at each speed is detected by a load cell. Shear stress, shear rate, and apparent melt viscosity are calculated as follow (6):

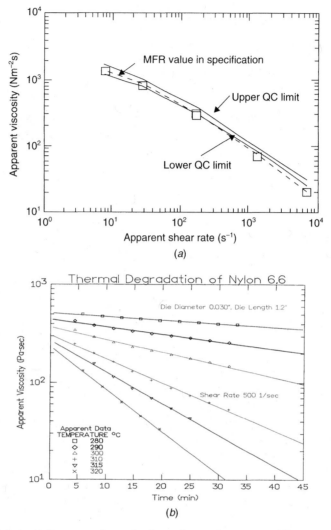

Figure 7-4. (*a*) A quality control graph of a polymer flow curve. (*b*) Apparent viscosity versus resistance time for nylon 66, showing thermal degradation.

$$\tau = \frac{F_r}{2\pi R^2 l}$$

where τ = shear stress (psi); F = load on the ram (lb); r = radius of capillary orifice (in.); R = radius of the barrel (in.); l = length of capillary orifice (in.).

$$\gamma = \frac{4Q}{\pi r^3}$$

where γ = shear rate (sec^{-1}); Q = flow rate (in^3/sec).

$$\text{Apparent melt viscosity (poise)} \; \eta = \frac{\text{Shear stress}}{\text{Shear rate}}$$

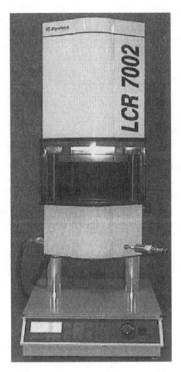

Figure 7-5. Capillary rheometer. (Courtesy of Dynisco.)

Shear stress versus shear rate and apparent viscosity versus apparent shear rate curves are plotted in Figures 7-6 and 7-7.

7.4. VISCOSITY TESTS

Viscosity is defined as the property of resistance to flow exhibited within the body of a material and expressed in terms of a relationship between applied shearing stress and resulting rate of strain in shear. In the case of ideal or Newtonian viscosity, the ratio of shear stress to the shear rate is constant. Plastics typically exhibit non-Newtonian behavior, which means that the ratio varies with the shearing stress. There are two different aspects of viscosity. Dynamic or absolute viscosity, best determined in a rotational type of viscometer with a small gap clearance, is independent of the density or specific gravity of the liquid sample and is measured in poises (P) and centipoises (cP). Kinematic viscosity, usually determined in some form of efflux viscometer equipped with a capillary bore or small orifice that drains by gravity, is strongly dependent on density or specific gravity of the liquid, and is measured in stokes (S) and centistokes (cS). The relationship between the two types of viscosity is

$$\text{stokes} \times \text{specific gravity} = \text{poises}$$

The measurement and control of rheological properties are usually performed with simple devices called "viscometers" or "viscosimeters," which do not measure

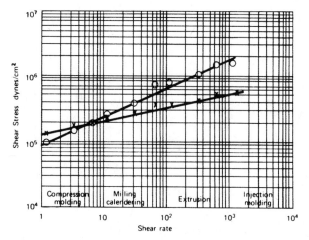

Figure 7-6. Shear stress versus shear rate curve for DVC with two different plasticizers. Note that the two materials exhibit similar characteristics at a shear rate of 5 reciprocal seconds but differ substantially at all others.

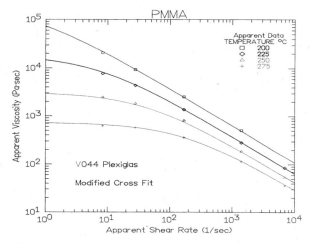

Figure 7-7. Apparent viscosity versus apparent shear rate.

true viscosities of either the dynamic or kinematic type but make relative flow comparisons. Viscosity of plastic materials is measured in three basic ways, employing principles involving liquid deformation due to various forces (7):

1. Downward rate of gravity flow through capillary bores and small orifices
2. Upward speed of a trapped air bubble
3. Torque developed by the liquid drag between moving and stationary surfaces

The first method is basically used for thermoplastic materials. The latter two are more commonly used for thermosetting materials, plastisols, and organosols and are discussed in Section 7.3.2.

7.4.1. Dilute Solution Viscosity of Polymers (ASTM D 2857)

This test method is used to determine dilute solution viscosity for all polymers that dissolve completely without chemical reaction or degradation to form solutions that are stable with time at a temperature between ambient and approximately 150°C. The results of the tests are expressed as relative, inherent, or intrinsic viscosity. Reduced and specific viscosity can also be calculated. Table 7-4 shows recommended test conditions for dilute solution viscosity measurements.

The test apparatus consists of volumetric flasks, transfer pipers, a constant-temperature bath, a timer, a viscometer, and a thermometer. The constant-volume device is recommended for use in which the solution, reduced, or inherent viscosity are to be measured at a single concentration. The second device is called a dilution viscometer and it does not require constant liquid volume for operation. This type is basically used for measuring intrinsic viscosity. Different types of commercially available viscometers are shown in Figure 7-8.

The test sample is prepared by dissolving a specified amount of polymer in the appropriate solvent. The prepared solution and the viscometer are placed in a constant-temperature bath maintained at the test temperature. A suitable amount of solution is transferred into the viscometer using a transfer pipe. Once temperature equilibrium is attained, the liquid level in the viscometer is brought above the upper graduation mark by means of gentle air pressure applied to the arm opposite the capillary. The timer is started exactly when the meniscus passes the upper graduation mark, and stopped exactly when the meniscus passes the lower graduation mark. The test is repeated at least three times for each solution and for each pure solvent. The time for the liquid to pass through the graduation marks is called efflux time. Relative, specific, reduced, inherent, and intrinsic viscosity values are determined in the following manner (8).

A. *Relative Viscosity (Viscosity Ratio)*
This term refers to the ratio of the time it takes for a specified solution of resin in a pure solvent to flow through an orifice to the flow time for an equal quantity of pure solvent through the same orifice.

$$\text{Relative viscosity} = \eta_{\text{rel}} = t/t_0$$

TABLE 7-4. Recommended Test Conditions for Dilute Solution Viscosity Measurements

Polymer	Solvent	Solvent Dissolving Conditions	Test Temperature	Solution Concentration (g/mL)
Polyamide	Formic acid or *m*-cresol	30°C 100°C, 2 hr	30°C	0.0050 ± 0.00002
Polycarbonate	Methylene chloride or *P*-dioxane	30°C 60°C	30°C	0.0040 ± 0.0002
PMMA	Ethylene dichloride	30°C, 24 hr	30°C	0.0020 ± 0.00002
PVC	Cyclohexanone	$85 \pm 10°C$	30°C	0.0020 ± 0.00002

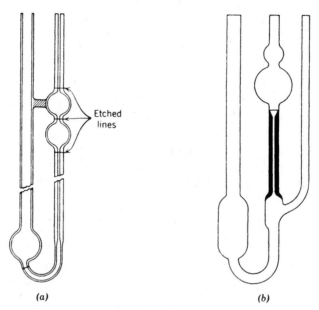

Figure 7-8. Capillary viscometers commonly used for measurement of polymer solution viscosities. (*a*) Ostwald-Fenske, (*b*) Ubbelohde. (Reprinted with permission of Wiley-Interscience.)

where t = efflux time of solution; t_0 = efflux time of pure solvent.
Or

$$\text{Relative viscosity} = \eta_{rel} = \eta/\eta_0$$

where η = viscosity of solution; η_0 = viscosity of solvent.

B. Specific Viscosity
This value is obtained by subtracting 1 from the value obtained for relative viscosity:

$$\eta_{SP} = \eta_{rel} - 1$$

where η_{SP} = specific viscosity; η_{rel} = relative viscosity.

C. Reduced Viscosity (Viscosity Number)
This term is used to describe the ratio of the specific viscosity to the concentration.

$$\eta_{red} = n_{SP}/c$$

where c = solution concentration (g/mL).

D. Inherent Viscosity (Logarithmic Viscosity Number)
The ratio of the natural logarithm of the relative viscosity to the concentration is called inherent viscosity.

$$\eta_{inh} = \frac{\ln \eta_{rel}}{c}$$

where η_{inh} = inherent viscosity; $\ln \eta_{rel}$ = natural logarithm of relative viscosity; c = solution concentration (g/mL).

E. Intrinsic Viscosity (Limiting Viscosity Number)

To determine the intrinsic viscosity of a polymer from dilute solution viscosity data, the reduced and inherent viscosities of solutions of various concentrations are determined at a constant temperature. These values are plotted against the respective concentrations as shown in Figure 7-9. A straight line is drawn through the points and extrapolated to the zero concentration. The intrinsic viscosity is the intercept of the line at zero concentration.

7.4.2. Applications and Limitations of Dilute Solution Viscosity Measurements

The molecular weight of the polymer can be determined by either the absolute or relative method. Absolute methods, such as osmotic pressure and light scattering, are more accurate but are lengthy and complex. In practice, dilute solution viscosity is by far one of the most popular relative methods for characterizing the molecular size of the polymers. Molecular weight from intrinsic viscosity data can be determined as follows (9):

$$\bar{M}_n = \frac{1}{2}(\eta) \times 10^5$$

where $\bar{M}_n$ = number average molecular weight; η = intrinsic viscosity in cyclohexanone at 30°C.

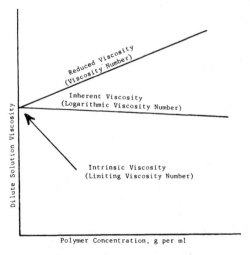

Figure 7-9. Example of plot to determine intrinsic viscosity. (Reprinted with permission of ASTM.)

$$M_w = 0.9 \, (\eta) \times 10^5$$

where $\overline{M}_w$ = weight average molecular weight; η = intrinsic viscosity in cyclohexanone at 30°C.

The determination of dilute solution viscosity provides one item of information to the molecular characterization of polymers. When viscosity data is used in conjunction with other molecular parameters, the properties of polymers, depending on their molecular structure, may be better predicted. Viscosity data alone may be of limited value in predicting the processing behavior of the polymer. However, when used in conjunction with other flow and physical property values, the solution viscosity of polymers may contribute to their characterizations. The viscosity of polymer solutions may be affected drastically by the presence of unknown or known additives in the sample. When dilute solution measurements are used to indicate the absolute physical properties of formulated polymers, many difficulties can arise. Most of these dilute solution measurements are based on 1 percent or less of a polymer that is in complete solution in a pure solvent. This can hardly be called a normal commercial formulation (10). The solution viscosity of a polymer of sufficiently high molecular weight may depend on the rate of shear in viscometer, and the viscosity of a polyelectrolyte (polymer containing ionizable chemical groupings) will depend on the composition and ionic strength of the solvent. Special precautions are required when measuring such polymers. Dilute solution viscosity data are extremely depended upon the purity and type of solvent, the type of viscometer, the temperature, and time. Even the slightest change in one or more of these parameters can seriously affect the result. Satisfactory correlation between solution viscosity and certain other properties is possible from polymers of a single manufacturing process. However, when a polymer is produced by different manufacturing processes, the correlation with other properties of a polymer may be limited.

7.5. GEL PERMEATION CHROMATOGRAPHY

Quite often, traditionally used tests, like melt index or viscosity tests, do not provide enough information about the processibility of the polymer. Such tests only measure an average value and tell us nothing about the distribution that makes up the average. One such example is the average daily temperatures reported for a particular city. These reports can be very misleading because they lack more useful information about extreme high and extreme low temperatures. In a similar manner, the melt index and viscosity tests relate very well to the average molecular weight of the polymer but fail to provide the necessary information about the molecular weight distribution of the polymer. Two batches of resin may have the same melt index, which simply indicates that their viscosity average molecular weights are similar. Their molecular weight distributions, the number of molecules of various molecular weights that make up their averages, can be significantly different. If an excessively high-molecular-weight fraction is present, the material may be hard and brittle. Conversely, if an excessive amount of low-molecular-weight fraction is present, the material may be soft or sticky. This is illustrated in Figure 7-10, which shows MWD curves for three different batches of polyethylene resin.

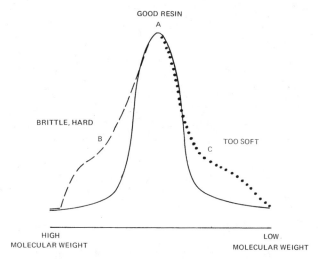

Figure 7-10. Molecular weight distribution curve for material. A represents "good" resin. Material B with excessively high molecular weight is brittle and hard. Material with excessively low molecular weight is soft and sticky. (Courtesy of Waters Corporation.)

When evaluating the nature of the incoming plastics material is important to a processor, he must look for a more reliable technique that will provide the necessary information to allow him to qualify the material, such as the measurement of molecular weight distribution. The molecular weight distribution is the single most fundamental property of the polymer. The molecular weight distribution not only provides basic information regarding the processibility of the polymer, but also gives valuable information for predicting its mechanical properties.

Gel permeation chromatography (GPC) is the method of choice for determining the molecular weight distribution of a polymer. This technique has gained wide acceptance among the plastic material manufacturers and processors because of its relatively low cost, simplicity, and ability to provide accurate, reliable information in a short time. GPC reveals the molecular weight distribution of a polymer compound. It detects not only the resin-based molecules such as polymer, oligomer, and monomer, but most of the additives used in plastic compounds and even low-level impurities. The molecular weight distribution curve is plotted for a well-characterized standard material and the profile of the curve is compared with the test sample. In this manner, batch-to-batch uniformity can be checked quickly as a means of quality assurance.

The separation of polymer molecules by GPC is based upon the differences in their "effective size" in solution. (Effective size is closely related to molecular weight.) Separation is accomplished by injecting the polymer solution into a continuously flowing stream of solvent that passes through highly porous, tiny, rigid gel particles closely packed together in a tube. The pore size of the gel particles may vary from small to very large. As the solution flows through the gel particles, molecules with small effective size (low molecular weights) will penetrate more pores than molecules with large effective sizes and, therefore, take longer to emerge than the large molecules. If the gel covers the right range of molecular sizes, the result will be a size separation with the largest molecules exiting the gel-packed tubes (columns) first.

7.5.1. GPC Instrumentation

Gel permeation chromatography, also known as size exclusion chromatography (sec), basically consists of injecting a polymer in solution into a delivery system that delivers the solution through the columns, detectors, and recorder that records information provided by the detector. This is shown schematically in Figure 7-11.

A. Solvent Delivery System
The delivery pumps produce constant flow rates independent of viscosity differences. The system rapidly pumps the polymer in solution through the system.

B. Injector
An injector accurately places the polymer solution in the flowing solvent stream.

C. Columns
The columns consist of highly porous, tiny, rigid gel particles closely packed together in a tube. The pore size varies from very small to very large. The columns separate the sample and provide the molecular weight distribution.

D. Detectors
Once the polymer is separated into various molecular weight species, they are detected by two basic types of detectors. A universal detector, called a differential refractometer, is used to monitor polymer molecular weight distribution. The UV detector, which provides higher sensitivity than the refractometer, is used for determining low-molecular-weight additives and impurities.

E. Recorder
The separation is recorded immediately to provide a continuous chromatogram.

7.5.2. Test Procedure

A test sample is prepared by dissolving a small amount of polymer in the solvent and filtering the solution to remove the undissolved impurities. The next step is to

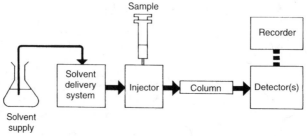

Figure 7-11. Schematic of a GPC system. (Courtesy of Waters Corporation.)

select the proper size columns, connect them, set the sensitivity setting on the detectors, and allow the instrument to equilibrate. A trial analysis is done by injecting the polymer solution into the instrument. The chromatogram is carefully analyzed. If the chromatogram shows all the desired information, the final analysis is carried out. If not, the operating parameters, such as column size, flow rate, and number of columns are optimized. The trial step is repeated before proceeding to the final analysis.

During the final analysis, as the sample flows through the column, the molecules are separated according to size by a simple mechanical effect. Because of their smaller size, the smaller molecules enter into the gel pores more readily and, therefore, take extra time to reach the bottom of the column. The various molecular weight species are separated by the difference in travel time through the column and pass through a detector in descending order of size. The detector measures the concentration of each molecular size and plots the molecular weight distribution of the sample on a strip chart. Commercially available GPC equipment is shown in Figure 7-12.

7.5.3. Interpreting the GPC Curve

A GPC curve can reveal a great deal of information. The regions of the chromatogram can generally be easily and quickly correlated with the molecular weights of the components of the polymer. Molecules always emerge from the instrument and appear on the chromatogram in descending order of molecular size. The polymer is always the first to emerge because its molecules are the largest (Figure 7-13). The intermediate or low-molecular-weight material appears next as sharper peaks in region B of the chromatogram. Last to appear are the smallest molecular weight components—unreacted monomer and low-molecular-weight contaminants such as moisture. In addition, the relative proportions of high and low ends of the polymer can quickly be determined because the height of the curve at any given point is proportional to the concentration of the component emerging at that point.

The broad curve in region A of the chromatogram provides the most useful information regarding the characteristic of the polymer. The distance from the start of the chromatogram to the midpoint of the molecular weight distribution curve is an indication of average molecular weight. This distance, the placement of the curve, and the shape of the molecular weight distribution lead to the prediction of

Figure 7-12. GPC equipment. (Courtesy of Waters Corporation.)

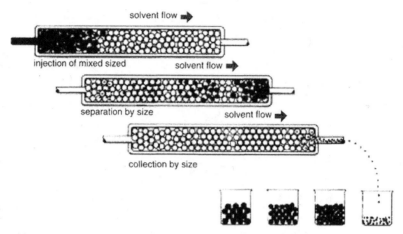

Figure 7-13. A typical chromogram showing emergence of molecules from the point of injection of polymer into a GPC equipment. (Courtesy of Waters Corporation.)

the physical and processing characteristics of the polymer. Any changes in the placement or shape of the curve reflects a change in the behavior of the polymer. An excessive amount of high-molecular-weight fraction present in the polymer is indicated by the skewing of the curve toward the high end. Conversely, an excessive amount of low-molecular-weight fraction is indicated by the skewing of the curve toward the low end. A discrete peak at the high or the low end of the distribution curve shows the change in behavior of the polymer.

The most direct—and sometimes the most informative—use of the GPC curve is in the comparison of different materials by overlaying their chromatograms. Differences in molecular weight distributions, peak shapes, shifts, and tailing are readily observable. Comparisons of additives and other lower-molecular-weight species are straightforward.

Frequently, a master chromatogram representing the acceptable range of GPC profiles is established with samples from "good" batches. All subsequent batches are then chromatographed and compared with the master curve as a rapid quality control method. Figure 7-14 illustrates the process. A number of correlations between the GPC curve and the physical and processing behaviors of a polymer have been developed.

In recent years, significant improvements in solvent delivery systems have been made, allowing the use of expensive solvents such as hexafluoroisopropanol (HFIP) practical. The newer gpc system does not use traditional, mechanically driven pumps with gear-linked pistons. Instead, it incorporates independently driven pistons and dual-pressure transducers with digital signal processing, bringing the software control closer to solvent delivery process. This achieves real-time management and control of the solvent, which allows point of delivery optimization of flow performance and results in superior chromatographic results. The newer software adds to the flexibility in observation, tracking, and reporting of results. The database also provides unlimited information about the structural composition of polymers for advance spectral analysis of copolymers, blends, alloys, and additives.

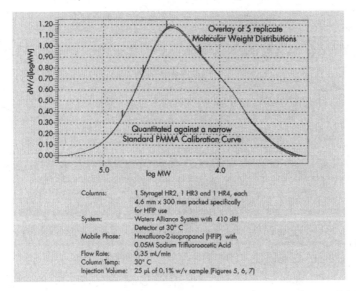

Figure 7-14. A sample material GPC curve is compared with control material GPC curve using a light box. (Courtesy of Waters Corporation.)

7.6. THERMAL ANALYSIS TECHNIQUES

Thermal analysis (TA) consists of a family of analytical techniques in which a property of the sample is monitored against time or temperature while the temperature of the sample is programmed. The properties include weight, dimension, energy take-up, differential temperature, dielectric constant, mechanical modulus, evolved gases, and other, less common attributes. The application of thermal analysis is widespread within the polymer and elastomer industries. Thermal analyzers are used to qualify a material for fitness of use and to troubleshoot processing problems. Thermal analysis consists of three primary techniques that may be used individually or in combination:

1. Differential scanning calorimetry (DSC)
2. Thermogravimetric analysis (TGA)
3. Thermomechanical analysis (TMA)

The maximum benefit of thermal analysis can be gained by using a combination of all three techniques to characterize a polymer. Figure 7-15 shows the three most common techniques and the test methods for each. The following sections describe the more common techniques in terms of analytical capability and standard methodology.

7.6.1. Differential Scanning Calorimetry (ASTM D 3417, ASTM D 3418)

In differential scanning calorimetry (DSC), the most widely used thermal analysis technique, the heat flow rate to the sample (differential power) is measured while

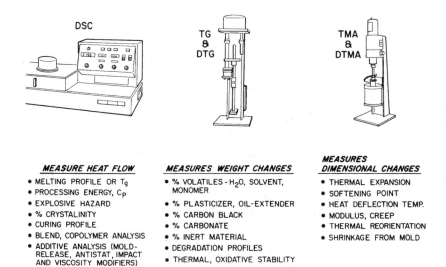

MEASURE HEAT FLOW	MEASURES WEIGHT CHANGES	MEASURES DIMENSIONAL CHANGES
• MELTING PROFILE OR T_g	• % VOLATILES - H_2O, SOLVENT, MONOMER	• THERMAL EXPANSION
• PROCESSING ENERGY, C_p		• SOFTENING POINT
• EXPLOSIVE HAZARD	• % PLASTICIZER, OIL-EXTENDER	• HEAT DEFLECTION TEMP.
• % CRYSTALINITY	• % CARBON BLACK	• MODULUS, CREEP
• CURING PROFILE	• % CARBONATE	• THERMAL REORIENTATION
• BLEND, COPOLYMER ANALYSIS	• % INERT MATERIAL	• SHRINKAGE FROM MOLD
• ADDITIVE ANALYSIS (MOLD-RELEASE, ANTISTAT, IMPACT AND VISCOSITY MODIFIERS)	• DEGRADATION PROFILES	
	• THERMAL, OXIDATIVE STABILITY	

Figure 7-15. Thermal analysis of plastics. (Courtesy of Perkin-Elmer Corporation.)

the temperature of the sample, in a specified atmosphere, is programmed. Because all materials have a finite heat capacity, heating or cooling a sample specimen results in a flow of heat in or out of the sample. The two most common types of commercial DSC measuring cells are shown in Figure 7-16a. The heat flux DSC employs a disk containing sample and reference positions which are heated by a common furnace. The differential heat flow to the sample is proportional to the temperature difference that develops between sample and reference junctions of a thermocouple. The power compensations approach controls a temperature enclosure around the sample and reference individually. Through amplified feedback from platinum resistance thermometers, it records the differential energy flow necessary to maintain the sample on the specified temperature program. With either approach the output is heat flow, normally expressed in milliwatts, watts/gram (normalized), or watts/gram-degree (specific heat units). Figure 7-16b shows a DSC instrument.

The test procedure is simple. A small quantity of sample, usually 5–10 mg, is weighed out into an inert capsule (usually made of aluminum). The encapsulated sample is placed in the DSC sample holder or onto the sample platform of a DSC cell disk. In the attached control module or computer, the operator selects a temperature range and heating rate, or perhaps a more complex temperature program. The test is started. (A hypothetical DSC curve showing both endothermic and exothermic changes in a polymer is shown in Figure 7-17.) Initially, constant energy input is required to heat the sample at a constant rate. This establishes a baseline. At a transition point, the sample requires either more or less energy depending on whether the change is endothermic or exothermic. For example, when the glass transition point is reached, the heat capacity increases. The midpoint is taken as the glass transition temperature (T_g). Adding plasticizers to a formulation lowers T_g. When a polymer reaches the melting point, it requires more energy (endothermic) to melt the crystalline structure. The area of the peak in units of energy is the enthalpy of fusion, the heat of melting. The temperature dependence of the peak

Comparison of Heat Flux and Power Compensation DSC

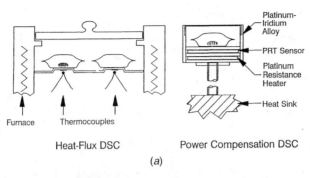

Platinum-Iridium Alloy

PRT Sensor

Platinum Resistance Heater

Heat Sink

Furnace   Thermocouples

Heat-Flux DSC   Power Compensation DSC

(*a*)

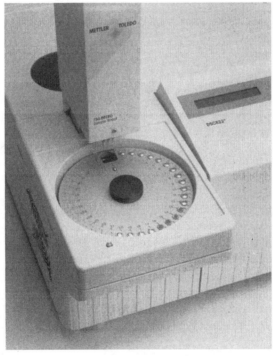

(*b*)

Figure 7-16. (*a*) Two most common types of DSC measuring cells. (Courtesy of Perkin-Elmer Corporation.) (*b*) Differential scanning calorimeter. (Courtesy of Mattler-Toledo Inc.)

and its shape give information about degree of crystallinity, the molecular-weight distribution, degree of branching, copolymer blend ratio, and or processing history. Often quality procedures involve comparing the melting profile to that of a standard "good" material.

When a sample cures, more energy is usually released, and the change is exothermic. The area of the curing peak is proportional to the number of crosslinks that

Thermal Analysis Techniques

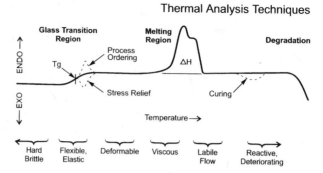

Figure 7-17. A typical DSC thermogram. (Courtesy of Perkin-Elmer Corporation.)

Figure 7-18. Thermogravimetric analysis instrument. (Courtesy of Mattler-Toledo, Inc.)

were formed. This indicates degree of cure. The shape of the curing curve can be analyzed to obtain the reaction kinetic parameters. The use of DSC in foam formulations (11, 12) and epoxy-curing studies (13, 14) is discussed in the literature.

7.6.2. Thermogravimetric Analysis (TGA)

Thermogravimetric analysis (TMA) is a test procedure in which changes in the weight of a specimen is monitored as the specimen is progressively heated. The sample weight is continuously monitored as the temperature is increased either at a constant rate or through a series of steps. The components of a polymer or elastomer formulation volatilize or decompose at different temperatures. This leads to a series of weight-loss steps that allow the components to be quantitatively measured. A typical high-performance apparatus consists of an analytical balance supporting a platinum crucible for the specimen, the crucible situated in an electric furnace (Figure 7-18). Variations in instrumentation include horizontally mounted furnaces and top-loading balances.

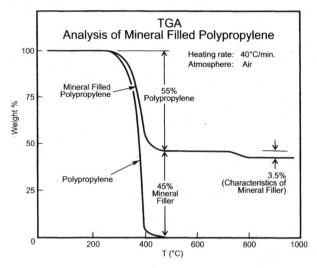

Figure 7-19. A typical TGA thermogram. (Courtesy of Perkin-Elmer Corporation.)

A simple thermal separation can be seen in the determination of mineral filler in polypropylene (Figure 7-19). The sample is heated in an air atmosphere to decompose completely the polypropylene such that the remainder is mineral filler. The amount of volatiles (here, none), polymer, and filler is obtained by analysis of the weight-loss curve. The small weight loss (3.5 percent) occurring over the temperature range of 750–800°C is characteristic of the particular mineral filler used in this sample. An unfilled polypropylene thermogram is also run to compare with the filled formulation. TGA is very useful in characterizing polymers containing different levels of additives by measuring the degree of weight loss. It can also be used to identify the ingredients of blended compounds according to the relative stabilities of individual components. The thermal stability of a polymer can be obtained through a kinetic analysis of the decomposition profile. The use of TGA in determining the track resistance of epoxy compounds is also described (15).

7.6.3. Thermomechanical Analysis (TMA)

When a sample is heated, its dimensions change because of thermal expansion, stress reorientation, and deformation under an imposed stress. Thermomechanical analysis (TMA) measures these properties using a constant force on the sample. Under a no-load or fixed load, it measures the dimensional change in the vertical direction as the sample temperature is controlled. The TMA equipment consists of (Figure 7-20) a probe mechanically connecting three entities: (1) a force transducer to control the force applied to the sample, (2) a position transducer to measure the displacement, and (3) a temperature-controlled sample specimen. Samples to be examined are cut to a defined geometry and then deformed by the TMA probe in a defined manner. Sample deformation includes compression, tension, and bending geometries. Once mounted, the sample is surrounded by a furnace and monitored by a·thermocouple.

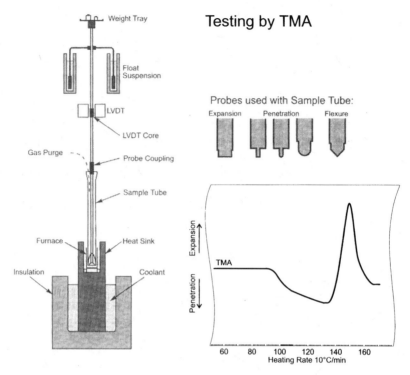

Figure 7-20. Testing by TMA (schematic). (Courtesy of Perkin-Elmer Corporation.)

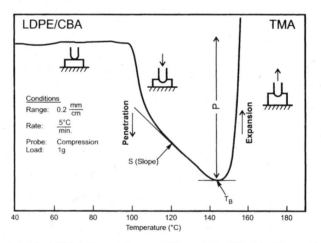

Figure 7-21. A typical TMA thermal curve. (Courtesy of Perkin-Elmer Corporation.)

A typical TMA thermal curve is shown in Figure 7-21. Here a fused silica rod is used as a compression probe. The probe is placed onto the sample, and the thermal expansion or contraction is recorded as a function of temperature. The polymer used in the test contains a chemical blowing agent. Over the first 60°C, the TMA probe is slightly pushed up by the expansion of the polymer solid. As the

Figure 7-22. TMA instrument. (Courtesy of Mettler-Toledo, Inc.)

melting point is reached, the lightly loaded probe penetrates the sample at a rate related to the viscosity. This penetration is later reversed by the expansion of the foam as the blowing agent is released. Polymer characterization through the use of TMA is accomplished by determining the glass transition temperature, the coefficient of expansion, and the elastic modulus. The TMA data also correlate with the Vicat softening point and the heat distortion temperature. Figure 7-22 shows a commercially available TMA instrument.

7.6.4. Dynamic Mechanical Testing

For linear dynamic mechanical testing (DMT), the transducers are identical or functionally similar to those for TMA. However, in addition to the static forces applied to the sample, there is an oscillatory applied force (or applied strain). This results in the ability to measure modulus (which is the ratio of stress, which is normalized force, per unit strain, which is normalized displacement). From each cycle, the DMA calculates both the modulus and the phase, that is, the response delay caused by viscoeleastic changes taking place in the sample. From these fundamental measurements can be calculated the storage and loss modulus, compliance, and viscosity. These parameters can be calculated as a function of applied stress, strain, and frequency as well as temperature. Many tests can be carried out under controlled stress or strain conditions. Most common sample geometries include clamped or knife-edge bending for solids, clamped extension for films and fibers, and parallel-plate compression geometries for polymer melts.

This technique allows the quantitative characterization of viscoeleastic properties. That is, it allows quantifying the ability of a material to store and/or dissipate mechanical energy of a specific duration. These properties correlate with such traditional empirical tests as hardness, strength, yield point, impact, and some viscosity tests. Because of the wide range of operational modes, geometries, and test methods, DMT can only treated here in a brief overview. Figure 7-23 shows a commercially available DMT instrument.

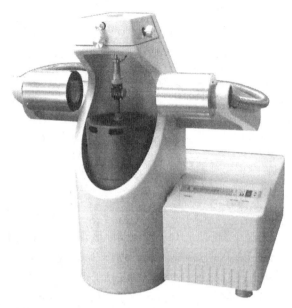

Figure 7-23. DMA instrument. (Courtesy of Mettler-Toledo, Inc.)

7.7. SPECTROSCOPY

Infrared spectroscopy has been one of the most widely used material analysis techniques for over 70 years. An infrared spectrum represents a fingerprint of a sample with absorption peaks that correspond to the frequencies of vibrations between the bonds of the atoms making up the material. Because each different material is a unique combination of atoms, no two compounds produce the exact same infrared spectrum. This fact allows a positive identification of polymeric materials. By studying the size of the peaks in the spectrum, one can also determine the amount of material present.

A newly developed technique called Fourier transform infrared (FT–IR) spectrometry overcomes the limitations encountered with the traditional infrared technique. The original infrared instruments were of dispersive type, which separated the individual frequencies of energy emitted from infrared source using a prism or grating. These instruments measured each frequency individually, making the entire process painfully slow. Modern FT-IR instruments can process as many as 100 samples per day compared to only 2 to 4 samples per day. FT-IR spectroscopy is fast, precise, and simple, requiring a very small amount of sample for a successful amalysis. In infrared spectroscopy, IR radiation is passed through a sample. Some of the infrared radiation is absorbed by the sample and some of it is passed through (transmitted). The resulting spectrum represents molecular absorption and transmission, creating a molecular fingerprint of the sample. Figure 7-24 graphically illustrates this process.

The sample analysis process is quite simple using a modern spectrometer coupled with a powerful computer. Infrared energy is emitted from a glowing black-body source. This beam passes through an interferometer where the "spectral encoding"

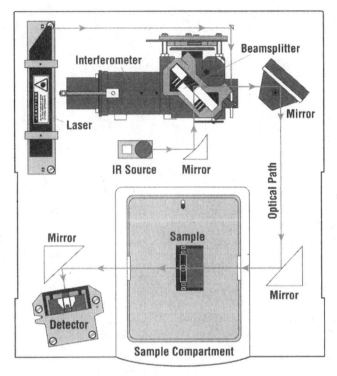

Figure 7-24. Sample spectrometer layout. (Courtesy of Nicolet Instruments.)

takes place. The resulting interferogram signal then exits the interfermeter. Next, the beam enters the sample compartment where it is transmitted though or reflected off the surface of the sample. This is where the specific frequencies of energy, which are uniquely characteristic of the sample, are absorbed. The beam finally passes through a detector and the signal is sent to computer where a mathematical technique called the Fourier transformation takes place. The infrared spectrum is displaced on the CRT for analysis and interpretation (16).

The computer is equipped with a collection of thousands of known polymer and additive FT–IR spectra for easy comparison and identification. A spectral library search can identify a polymer, additive, or contaminant within minutes, making the entire process fast and efficient. Figures 7-25 and 7-26 illustrate the entire FT–IR process and a commercially available FT–IR instrument.

7.8. MATERIAL CHARACTERIZATION TESTS FOR THERMOSETS

Thermoplastic materials are usually purchased in a ready-to-process form. Most thermoplastic materials do not require further treatment in terms of blending and compounding except for the addition of color. The material supplier is generally expected to provide quality resin and only a few characterization tests are conducted to determine the uniformity of the resin, as discussed earlier in this chapter.

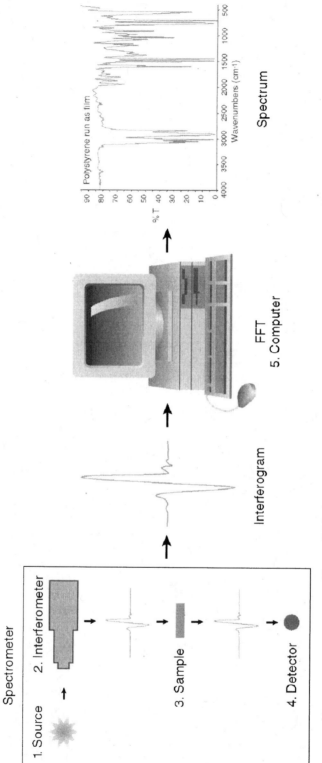

Figure 7-25. Sample analysis process. (Courtesy of Nicolet Instruments.)

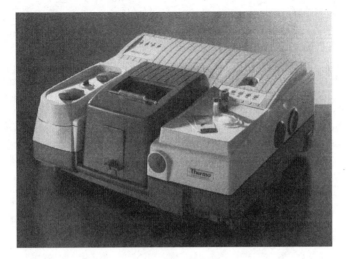

Figure 7-26. Spectrometer. (Courtesy of Nicolet Instruments.)

Thermoset materials, unlike thermoplastics, are purchased in an unpolymerized intermediate chemical stage. The polymerization takes place during processing. To ascertain the quality of the raw materials, material characterization tests are routinely conducted by thermoset processors.

7.8.1. Apparent (Bulk) Density, Bulk Factor, and Pourability of Plastic Materials (ASTM D 1895)

The information regarding the bulk factor of the material is very important to a molder prior to molding for several reasons. The bulk factor of the material directly affects the preform size, the loading space in the cavities, and the design of the loading trays. Any change in the bulk factor can cause serious molding problems, such as porous or incomplete moldings, since the molding charge is measured rather than weighed. The bulk factor is defined as the ratio of the volume of any given quantity of the loose plastic material to the volume of the same quantity of the material after molding or forming. The bulk factor is also equal to the ratio of density after molding or forming to the apparent density of material as received. Bulk factor is a measure of volume change that may be expected in fabrication. Apparent density is defined as the weight per unit volume of a material including voids inherent in the material as tested. Apparent density is a measure of the fluffiness of a material.

A. Determination of Apparent Density

This is a simple test requiring a measuring cup and a funnel of a specified size. The funnel is closed from the small end with the hand or a suitable flat strip. The material is poured into the funnel. After the material is poured, the bottom of the funnel is quickly opened and the material is allowed to fall into a cylindrical measuring cup. The excess material laying on top of the cup is immediately scraped off with a straight edge without shaking the cup. The material in the cup is weighed to the nearest tenth of a gram. Apparent density values are reported as grams per cubic centimeter or pounds per cubic foot. This method is used primarily to determine the apparent density of free-flowing powders and fine granules.

An alternate method requires a funnel with a larger-diameter opening at the bottom so that a coarse, granular material in the form of pellets or dice can easily pass through. The procedure and calculations are carried out in a similar manner.

Yet another variation of the same test is used for measuring the apparent density of coarse flakes, chips, cut fibers, or strands. Such materials cannot be poured through the funnels described in the first two methods. Also, since they are ordinarily very bulky when loosely poured and are usually compressible to a lesser bulk, even by hand, a measure of their density under a small load is appropriate. An apparatus for all three apparent density tests is shown in Figure 7-27.

B. Determining Bulk Factor

The density of molded or formed plastic material is determined as described in Chapter 10, Section 10.3. The calculation is carried out as follows:

$$\text{Bulk factor} = \frac{\text{Average density of molded part}}{\text{Average apparent density of meaterial prior to molding}}$$

C. Pourability

Pourability is defined as a measure of the time required for a standard quantity of material to flow through a funnel of specified dimensions. Pourability characterizes the handling properties of finely divided plastic material. The procedure simply calls for pouring the material through a specified size funnel and measuring the time required for material to completely pass through the funnel.

7.8.2. Flow Tests

The ability of the material to flow is measured by filling a mold with the plastics material under a specified condition of applied temperature and pressure with a controlled charge mass. The flow tests are used as a quality control test and as an acceptance criterion for incoming raw materials.

A. Factors Affecting Flow

Resin Types. All resins flow differently because of basic differences in the structure of the polymers. For example, melamine formaldehyde exhibits longer flow than urea formaldehyde. Phenolics, because of the variety of resin types, enable the molder to select the flow best suited for a particular design.

Type of Fillers. The small particle size of wood flour, mica, and minerals creates less turbulence and less frictional drag during mold filling. The size of the glass fibers, short or long, can adversely affect the flow.

Degree of Resin Advancement. The degree of advancement is generally controlled by the resin manufacturers. Molders can advance resin polymerization with oven or radiant heat or electronic preheating.

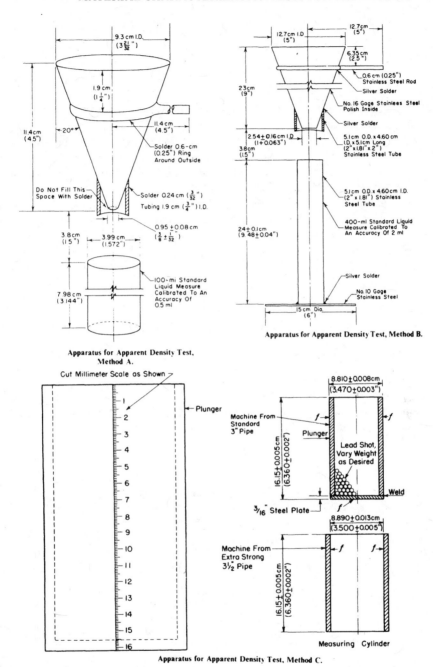

Apparatus for Apparent Density Test, Method A.

Apparatus for Apparent Density Test, Method B.

Apparatus for Apparent Density Test, Method C.

Figure 7-27. Apparatus for apparent density test. (Reprinted with permission of ASTM.)

Storage Time. All resins have a natural tendency to polymerize in storage, causing partial precure which reduces flow. An exception might be a polyester in which catalyst decomposition slows or prevents curing, which increases flow duration (17).

B. Spiral Flow of Low-Pressure Thermosetting Compounds (ASTM D 3123)

The spiral flow of a thermosetting molding compound is a measure of the combined characteristics of fusion under pressure, melt viscosity, and gelation rate under specific conditions. The test requires a transfer molding press, a standard spiral flow mold, and a thermosetting molding compound. The molding temperature, transfer pressure, charge mass, press cure time, and transfer plunger speed are preselected as specified. The preconditioned compound is forced through a sprue into a spiral flow mold. Once the curing is complete, the part is removed and the spiral flow length is read directly from the molded specimen. Compounds are classified as low (1–10), medium (11–22), and high (23–40) plasticity. A typical molded specimen is shown in Figure 7-28.

C. Cup Flow Test (ASTM D 731)

Molding Index of Thermosetting Molding Powder. This test is primarily useful for determining the minimum pressure required to mold a standard cup and the time required to close the mold fully. The preconditioned and preweighed material is loaded into the mold. The mold is closed using sufficient pressure to form a required cup. The pressure is reduced step by step until the mold cannot close. The next higher pressure and time to close the mold is reported as the molding index of the material. Figure 7-29 illustrates a cup mold to produce a molded cup.

Figure 7-28. Spiral flow test specimens.

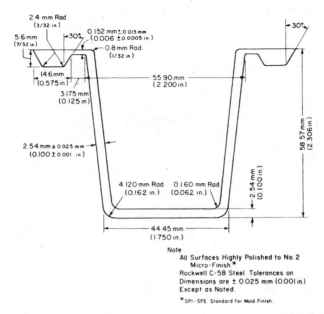

Figure 7-29. Cup mold. (Reprinted with permission of ASTM.)

7.8.3. Viscosity Tests for Thermosets

Viscosity tests are very important in assessing the quality of incoming thermosetting materials. Variation in the viscosity can occur because of chemical changes due to storage conditions, manufacturing variations, and the presence of impurities. Viscosity limits are of particular interest in achieving uniform impregnation and wetting of fiber/fabric reinforcements and flow during molding and curing operations. Viscosity measurements also serve as an indicator of working life in the case of catalyzed resin systems. The increase in viscosity indicates advancing stages of polymerization or solvent evaporation.

Bubble Viscometer
In a bubble viscometer, a liquid streams downward in the ring-shaped zone between the glass wall of a sealed tube and a rising air bubble. The rate at which the bubble rises is a direct measure of the kinematic viscosity. The rate of bubble rise is compared with a set of calibrated bubble tubes containing liquids of known viscosities. Bubble viscometers are shown in Figure 7-30.

7.8.4. Gel Time and Peak Exothermic Temperature of Thermosetting Resins (ASTM D 2471)

Gel time and peak exothermic temperature are two of the most important parameters for a thermosetting material processor. Gel time is the interval of time between introduction of the catalyst and the formation of gel. Such information regarding viscosity change with time of resin–catalyst mixture helps to determine

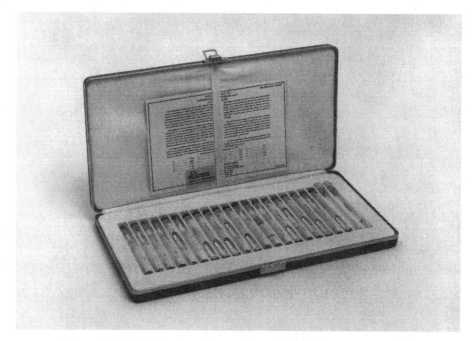

Figure 7-30. Bubble viscometers. (Courtesy of Byk-Gardner USA.)

working-life characteristics of the material. The maximum temperature reached by reacting thermosetting plastic composition is called peak exothermic temperature. Resin producing high exothermic heat is more susceptible to cure shrinkage and craze cracking. This is due to the thermal expansion that occurs as the heat is generated and shrinkage that takes place when the thermosetting three-dimensional network forms. In the laminating process, excessive heat buildup during cure tends to contain weak interlaminar bonds and, hence, relatively poor physical properties. To ensure against storage changes and batch variations, the processor should measure gel characteristics before using the materials.

The test method requires sample containers, a wooden probe, a constant-temperature bath, a stopwatch, and a temperature-measuring device. All the components are placed in the temperature-controlled bath for a specified amount of time and temperature. When all the components have reached the test temperature, they are combined in the recommended ratio. The start of mixing is recorded as starting time. Components are agitated with a stirrer. Simultaneously, the temperature of the reactants are monitored and recorded. Every 15 sec, the center surface of the reacting mass is probed with an applicator stick. When the reacting material no longer adheres to the end of a clean probe, the "gel time" is recorded as the elapsed time from the start of mixing. The time and temperature recording is continued. The highest temperature is recorded as "peak exothermic temperature." "Peak exothermic time" is recorded as the elapsed time from the start of mixing.

Commercially available mechanical gel time meters or gel timers can be used in place of manual stirring to determine gel time. There are various types of gel time meters available in the market. One such version measures a liquid's gel point by

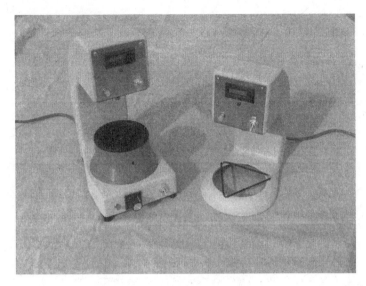

Figure 7-31. Gel time meter. (Courtesy of Shyodu Instrument INC.)

dropping a definite weight into a sample once per minute. Once the "gelation point" is reached, the weight does not sink during the downstroke. At this point, the equipment automatically turns off and records the elapsed time from the start of the experiment. Yet another version of a gel timer (Figure 7-31) consists of a motor-driven stirrer. As the gelation commences, drag soon exceeds torque, and the motor stalls. Gel time can then be directly read off the counter to the nearest 0.1 min. Gel time of resins at elevated temperature can also be determined by using a thermo-statically controlled heated pot in place of a regular pot.

REFERENCES

1. Brookfield Engineering Labs, Inc., "More Solutions to Sticky Problems," Technical Bulletin, Stoughton, MA.
2. Lobo, H., Bonilla, J., and Riley, D., "Plastics Analysis—Improved Characterization of Polymer Behavior and Composition," *Plastics Engineering*, Nov. 96.
3. Delaney, D. and Houlston, S., "The Capillary Rheometer: Applications in Plastics Industry," *Materials World*, Sept. 1996.
4. Dynisco–Kayeness Polymers Test Systems, Technical Bulletin, Sept. 1996.
5. Delaney, Reference 3.
6. Slysh, R. and Guyler, K. E., "Prediction of Diallyl Phthalate Molding Performance from Laboratory Tests," *SPE ANTEC*, **23** (1977), p. 4.
7. Gardner Laboratories, *Tech. Bull.: Rheology*, Silver Spring, MD, Sept. 1976.
8. Parks, R. A., "Re-evaluation of Dilute Solution Viscosity Test Methods Proves Boon to PVC Quality Control," *Plast. Design and Processing* (Aug. 1974), pp. 24–25.
9. McKinney, P. V., *J. Polym. Sci.*, **9** (1965), pp. 583–587.
10. Parks, Reference 8, p. 26.

11. May, W. P., "New Test Methods for Plastisol Foams," *Plast. Tech.* (June 1977), p. 97.

12. Breakey, D. and Cassel B., "What Foam Processors Should Know about Thermal Analysis Techniques," *Plast. Tech.* (Nov. 1979), p. 75.

13. Abofalia, O. R., "Application of DSC to Epoxy Curing Studies," *SPE ANTEC*, **15** (1969), p. 610.

14. Prime, R. B., "DSC of Epoxy Cure Reaction," *SPE ANTEC*, **19** (1973), p. 205.

15. Jaegers, G. and Gedmer, T. J., "The Use of TGA in Determining Track Resistance of Epoxy Compounds," *SPE ANTEC*, **16** (1970), p. 450.

16. Nicolet Instruments Corporation, "Introduction to Fourier Transform Infrared Spectrometry," Technical Bulletin, 1996.

17. Milby, R. V., *Plastics Technology*, McGraw-Hill, New York, 1973, pp. 457–474.

GENERAL REFERENCES

Waters Associates, *Tech. Bull.: Gel Permeation Chromatography.*

Instron, *Tech. Bull.: Rheological Instruments.*

Perkin-Elmer, *Tech. Bull.: Thermal Analysis.*

Nicolet Instrument Corporation, *Tech. Bull.: FT–IR.*

SUGGESTED READING

Sward, G. G. (Ed.), *Paint Testing Manual*, American Society for Testing and Materials, Philadelphia, PA, 1970.

Shida, M. and Cancino, L. V., "Prediction of High Density Polyethylene Resin Processibility from Rheological Measurements," *SPE ANTEC*, **16** (1970), pp. 620–624.

Macosko, C. and Starita, J. M., "Polymer Characterization with a New Rheometer," *SPE ANTEC*, **17** (1971), p. 595–600.

Terry, B. W. and Yang, K., "A New Method for Determining Melt Density as a Function of Pressure and Temperature," *S.P.E. J.*, **20**(6), p. 37 (June 1964).

Merz, E. H. and Colwell, R. E., "A High Shear Rate Capillary Rheometer for Polymer Melts," *ASTM Bull.* No. 232 (Sept. 1958).

Mendelson, R. A., "Melt Viscosity," *Encyclopedia of Polymer Science and Technology*, Vol. 8, John Wiley & Sons, New York, 1970, pp. 587–619.

Bernhardt, E. C., *Processing of Thermoplastic Materials*, Reinhold, New York, 1959.

McKelvey, J. M., *Polymer Processing.* John Wiley & Sons, New York, 1960.

Van Wurzer, J. R., Lyons, J. W., Kim, K. Y., and Colwell, R. E., *Viscosity, and Flow Measurement, A Laboratory Handbook of Rheology.* John Wiley & Sons, New York, 1963.

Hertel, D. L. and Oliver, C. K., "Versatile Capillary Rheometer," *Rubber Age* (May 1975).

Miller, B., "Why Good Resin Makes Bab Parts—and What you Can Do About It," *Plast. World* (Feb. 1976), p. 44.

Ekmanis, J. and Church, S., "Simple Test of Incoming Resins Rates Batch to Batch Quality Level," *Plast. Design and Processing* (March 1977), pp. 30–34.

Willard, P. E., "Determination of Cure of DAP Using DSC," *SPE ANTEC*, **17** (1971), pp. 464–468.

Perkin–Elmer, Thermal Analysis Literature, *TA Application Studies Bulletin*: TAAS-I9, "Characterization of Thermosets"; TAAS-20, "Polymer Testing by TMA"; TAAS-22, "Characterization and Quality Control of Engineering Thermoplastics by Thermal Analysis"; TAAS-25, "Applications of TA in the Electrical and Electronic Industries"; TAAS-26, "Applications of TA in the Automotive Industries"; TAAS-29, "Use of Thermal Analysis Method in Foam Research and Development."

Cassel, B. and Gray, A. P., "Thermal Analysis Simplifies Accelerated Life Testing of Plastics," *Plast. Eng.* (May 1977), pp. 56–58.

Cassel, B. and Breakey, D., "What Foam Processors Should Know About Thermal Analysis Techniques," *Plast. Tech.* (Nov. 1979), pp. 75–78.

Slade, P. E. and Jenkins, L. T. (Eds.), *Techniques and Methods of Polymer Evaluation*, Vol. 1, *Thermal Analysis*, 1966; Vol. 2, *Thermal Characterization Techniques*, 1970, Marcel Dekker, New York.

8

FLAMMABILITY

8.1. INTRODUCTION

Because of its increased use in homes, buildings, appliances, automobiles, aircraft, and many other sectors of our lives, plastic materials have been under considerable pressure to perform satisfactorily in situations involving fire. A good deal of time and money has already been spent on research and in-depth study of the behavior of polymeric materials exposed to fire. Many new tests for preliminary screening of materials and simulating actual fire have been developed. Before getting into a detailed discussion about tests and testing procedures, it is necessary to understand polymers as they relate to flammability.

When a polymeric material is subjected to combustion, it undergoes decomposition, which produces volatile polymer fragments at the polymer surface. The fuel produced in this process diffuses to the flame front, where it is oxidized, producing more heat. This, in turn, causes more material decomposition. A cyclic process is established—solid material is decomposed, producing fuel which burns, giving off more heat, which results in more material decomposition (1). This process is illustrated in Figure 8-1.

To reduce the flammability of a material, this cycle must be attacked in either the vapor phase or at the solid material surface. In the vapor phase, the cycle can be inhibited by adding certain additives to the polymer that disrupt the flame chemistry when vaporized. Bromocompounds and chlorocompounds with antimony oxide operate in this manner and are commonly used in polystyrene or ABS structural foams. Solid-phase inhibition may be achieved by including additives in the polymer that promote the retention of fuel as carbonaceous char and provide a protective insulating layer. This layer prevents further fuel evolution. Such an

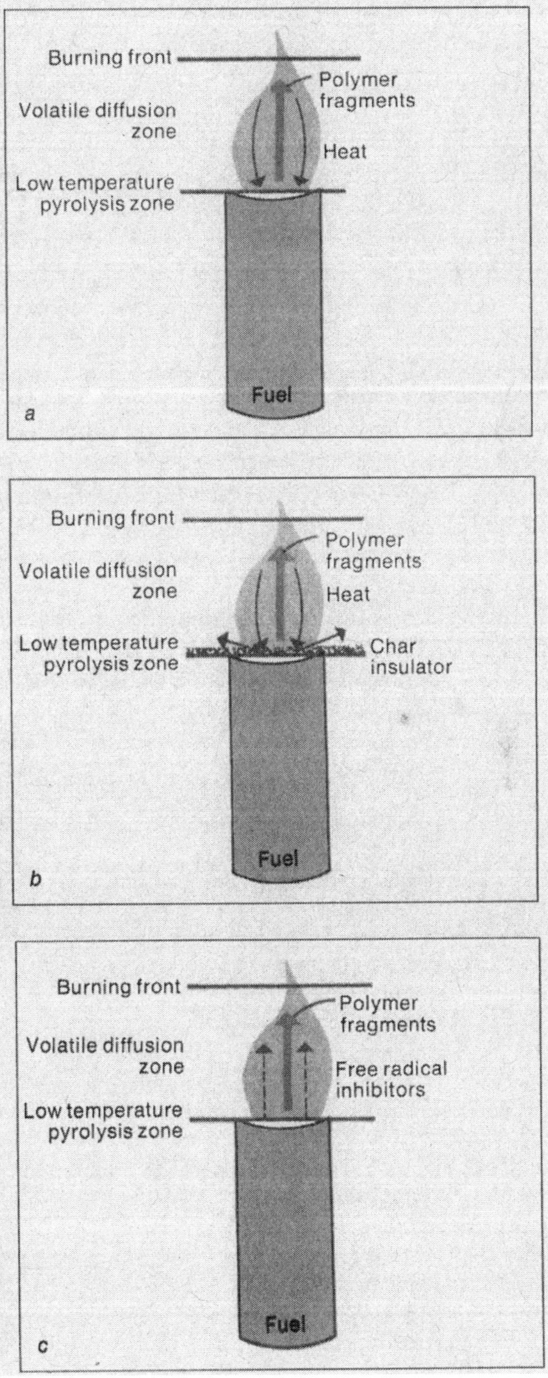

Polymer combustion occurs in a continuous cycle. Heat generated in the flame produces volatile polymer fragments. The polymer decomposes into fuel in the low-temperature pyrolysis zone, and free radical reactions take place in the volatile diffusion, or flame, zone

Solid phase inhibition. Extensive crosslinking at the polymer surface forms char, insulating the polymer from heat and preventing production of new fuel

Vapor phase inhibition. Volatile free-radical inhibitors formed during burning can inhibit branching radical reactions, interrupting the burning cycle

Figure 8-1. Polymer combustion process. (Reprinted with permission of *Chemistry.*)

approach is effective in polycarbonate and polyphenylene oxide-based structural foams. Other solid-phase approaches involve the use of heat sinks, such as hydrated alumina, which absorb heat and release water of hydration when heated, or alter the decomposition chemistry to consume additional heat in the decomposition process (2).

Polymer's inherent flammability can be divided into the basic classes listed in Table 8-1. The first group consists of inherently flame-retardant structures containing either halogen or aromatic groups that confer high thermal stability as well as the ability to form char on burning. The second group of materials is relatively less flame retardant. Many of the second group can be made more flame retardant by incorporating appropriate flame retardants. The third group consists of quite flammable polymers that are difficult to make flame retardant because they decompose readily, forming large quantities of fuel (3).

No discussion on polymer's flammability can be considered complete without discussing the formation of smoke and the generation of toxic gases. Smoke impairs the ability of occupants to escape from a burning structure as well as the ability of firefighters to carry out rescue operations. All polymers, to some extent, produce smoke and toxic gases, although some polymers inherently generate more smoke and toxic gases than others. Many tests have been developed to measure smoke density and toxicity.

The material's ability to burn depends upon fire conditions as well as polymer composition. Actual fire conditions are difficult to simulate and, therefore, we are forced to rely upon small- and large-scale laboratory tests to predict combustibility,

TABLE 8-1. Polymers and Flammability

Inherently Flame Retardant
Polytetrafluoroethylene
Aromatic polyethersulfone
Aromatic polyamides
Aromatic polyimides
Aromatic polyesters
Aromatic polyethers
Polyvinylidene dichloride

Less Flame Retardant
Silicones
Polycarbonates
Polysulfone

Quite Flammable
Polystyrene
Polyacetal
Acrylics
Polyethyleneterphthalate
Polypropylene
Polyethylene
Cellulose
Polyurethane

Reprinted from *Chemistry*, June 1978, p. 23.

smoke density, and toxicity. The flammability of materials is influenced by several factors (4):

Ease of ignition—how rapidly a material ignites

Flame spread—how rapidly fire spreads across a polymer surface

Fire endurance—how rapidly fire penetrates a wall or barrier

Rate of heat release—how much heat is released and how quickly

Ease of extinction—how rapidly the flame chemistry leads to extinction

Smoke evolution

Toxic gas generation

Table 8-2 lists existing flammability tests and other pertinent information. Table 8-3 lists smoke evolution tests and related information.

8.2. FLAMMABILITY TEST (NONRIGID SOLID PLASTICS) (ASTM D 4804)

This test was developed for nonrigid specimens that may distort or sag when tested in a horizontal position using test methods ASTM D 3801 or ASTM D 635. The method covers a small scale laboratory screening procedure for comparing extinguishing characteristics and determining relative rate of burning of plastic materials. The test is primarily useful for screening, comparing, quality control, and meeting specification requirements in terms of flammability of plastics.

Two basic test methods have been developed. Method A requires the use of a laboratory burner, a clamp for holding a test specimen, a test chamber, and a suitable timing device such as a stopwatch. The test specimen of size $2 \times 8 \times$ thickness of the specimen as prepared is used. A gauge mark is drawn across the specimen 5 in. from the bottom. After appropriate conditioning, the specimen is clamped vertically in a holding clamp. The laboratory burner flame is ignited and adjusted to a specified height. The tip of the flame is applied to the end of the specimen for 3 sec and withdrawn. The flaming time before extinguishment is recorded. Once the flaming of the specimen ceases, the specimen is immediately subjected to another 3-sec flame application. The duration of the flaming time and glowing time is recorded in seconds. The report also includes whether any flaming material drips from the specimen and whether the specimen burns up to the gauge mark. Figure 8-2a shows a typical test setup.

Test method B requires the use of a specially designed test fixture to allow the specimen to be held horizontally at a 45° angle. The test specimens are marked by scribing two lines 1.0 in. from one end of the specimen. The free end of the test specimen is subjected to a test flame for 30 sec. The time required for the flame front to reach from the 1.0-in. scribe line to a 3.9 in. scribe line is recorded and the rate of burning is determined. A typical test setup is shown in Figure 8-2b. The application of the test is limited to screening, development work, production control, specification testing, and quality control. The test results cannot be used as a criterion for fire hazard. Many papers have been presented (5, 6) to illustrate the poor correlation between such small-scale tests and large-scale flame tests that are closer to predicting real fire behavior.

TABLE 8-2. Flammability Tests

Sponsoring Organization	Name of Test	Procedure Identification Number	Specimen Size (in.)	Number of Specimens	Angle of Specimen	Ignition Source	Properties Measured
ASTM	Rate of burning nonrigid solid plastics	D4804 Method A	$2 \times 8 \times$ thickness	10	Vertical	Laboratory burner	Flaming and glowing time before extinguishment
		D4804 Method B	$\frac{1}{2} \times 5 \times$ thickness	10	Horizontal	Laboratory burner	Burning rate
ASTM	Rate of burning of self-supporting plastic in horizontal position	D-635	$\frac{1}{2} \times 5 \times$ thickness	10	Horizontal	Bunsen burner	Burning rate, and average time and extent of burning
ASTM	Rate of burning of self-supporting plastic in vertical position	D-3801	$\frac{1}{2} \times 5 \times$ thickness	10	Vertical	Laboratory burner	After flame time and after glow time
ASTM	Incandescence resistance of rigid plastics	D-757	—	—	—	—	Discontinued
ASTM	Ignition properties of plastics Procedure B	D-1729	3 gwt. Sheet size $\frac{3}{4} \times \frac{3}{4} \times t$	14	Horizontal	Hot air ignition furnace	Flash and self-ignition temperatures Visual characteristics
ASTM	Oxygen index flammability test	D-2863	$\frac{1}{4} \times 1/8 \times 5$	10	Vertical	Hydrogen or propane flame in oxygen and nitrogen atmosphere	Oxygen index

ASTM	E-84	Surface burning characteristics of building material	20×300	1	Horizontal	Gas/air mixture	Smoke developed
ASTM	D-3014	Flame height, time of burning, and loss of weight, cellular plastics, vertical position	$10 \times \frac{3}{4} \times \frac{3}{4}$	6	Vertical	Gas burner	Loss of weight and time to extinguishment
ASTM	D-1692	Rate of burning cellular plastics horizontal position	—	—	—	—	Discontinued
ASTM	D-1433	Rate, extent, and time of burning of flexible thin plastic sheeting	3×9	10	45°	Butane burner	Burning rate, extent and time burning, visual characteristics
Federal Test Method Standard	FTMS 406 Method 2023	Flame resistance of difficult to ignite plastics	$5 \times \frac{1}{2} \times \frac{1}{2}$	5	Vertical	Electric coil and spark plugs	Ignition time, burning time, flame travel
UL	UL V-O	Flammability of plastic materials	$\frac{1}{2} \times 5 \times$ thickness	5	Vertical	Laboratory burner	Time to extinguish
UL	UL HB	Flammability of plastic materials	$\frac{1}{2} \times 5 \times$ thickness	3	Horizontal	Laboratory burner	Time to extinguish

TABLE 8-3. Smoke Evolution Tests

Sponsoring Organization	Name of Test	Procedure Identification Number	Specimen Size (in.)	Number of Specimens	Angle of Specimen	Ignition Source	Properties Measured
ASTM	Smoke density from burning of plastics	D-2843	$1 \times 1 \times \frac{1}{4}$	3	Horizontal	Propane burner	Percent light absorption, smoke density
ASTM	Smoke density test	ASTM E662/ NFPA 258	3×3	6	Vertical	Electric furnace	Smoke density
ASTM	Smoke particulates from burning of plastics	ASTM D-4100	$1\frac{1}{2} \times \frac{1}{2} \times \frac{1}{8}$	6	Horizontal	Micro-Bunsen burner	Percent smoke
ASTM	Radiant panel test	E-162	$6 \times 18 \times$ thickness	4 minimum	60°	Acetylene	Flame spread index
OSU	OSU release rate test	—	—	—	Horizontal or vertical	Gas-fired radiant panel of electric globar	Maximum smoke release rate

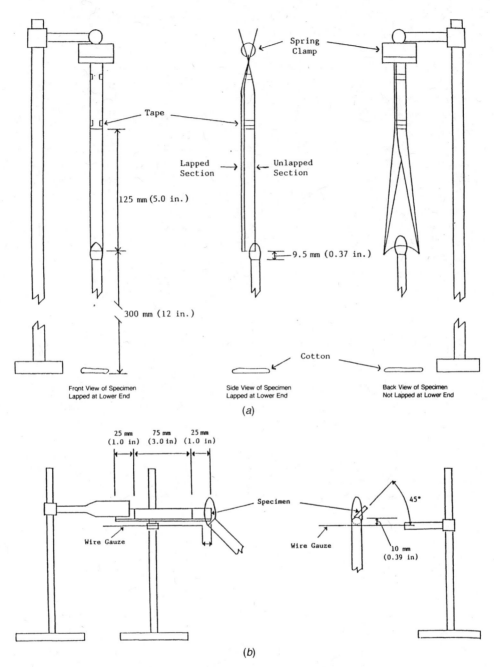

Figure 8-2. (*a*) Flammability test setup, Method A. (Reprinted with permission of ASTM.) (*b*) Flammability test setup, Method B. (Reprinted with permission of ASTM.)

8.3. FLAMMABILITY TEST (SELF-SUPPORTING PLASTICS IN HORIZONTAL POSITION) (D 635)

This method covers the procedures for determining relative rate of burning and/or extent and time of burning of self-supporting plastics in a horizontal position. Self-supporting plastics are defined as those plastics which, when mounted with the clamped end of the specimen 0.4 in. above the horizontal screen, do not sag initially so that the free end of the specimen touches the screen.

This test is very similar to Method A for determining the flammability of non-rigid solid plastics except for the specimen support. Unlike the previous test, the specimen is supported horizontally at one end. The free end is exposed to a specified gas flame for 30 sec. The time and extent and relative rate of burning are measured and reported along with visual observations.

The test is mainly useful for quality control, production control, comparison screens and specification testing. Poor correlation between test results obtained by this method and flammability under actual use conditions exist and therefore cannot be used as a criterion for fire hazard. Figure 8-3 shows a typical test setup for this test.

8.4. FLAMMABILITY TEST (SOLID PLASTICS IN VERTICAL POSITION) (D 3801)

The test determines the comparative burning characteristics of solid plastic materials held in a vertical position. The test method was developed for the plastic materials used for parts in devices and appliances. The results are intended to serve as a preliminary indication of their acceptability with respect to flammability for a particular application. The classification system for determining the comparative burning characteristics is developed for quality assurance and preselection of component materials for products. This is further discussed in detail in Section 8–12 on UL94 flammability testing.

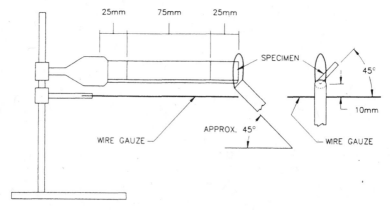

Figure 8-3. Flammability test setup for self-supporting plastics in horizontal position. (Reprinted with permission of ASTM.)

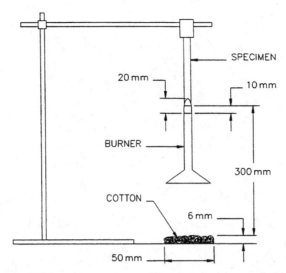

Figure 8-4. Vertical burning test setup. (Reprinted with permission of ASTM.)

The test procedure is quite simple and consists of positioning the test specimen vertically using a clamp and applying flame for two separate 10-sec applications. The afterflame time (i.e., the length of time a material continues to flame after the ignition source has been removed) is recorded after the first flame application. The afterflame and afterglow times are recorded after the second flame application. Information is also recorded about whether flaming material drips and ignites the cotton. Figure 8-4 illustrates a typical test setup for a vertical burning test. Figure 8-5 shows a commercially available test chamber for conducting flammability tests.

8.5. IGNITION PROPERTIES OF PLASTICS

8.5.1. Introduction

Ignition properties of plastics are very important when one considers that without ignition there would be no fire and no need for fire containment. Ignitability is defined as the ease of ignition or the ease of initiating combustion (7). The flash-ignition temperature is the lowest initial temperature of air circulating the specimen at which a sufficient amount of combustible gas is evolved to be ignited by a small external pilot flame. The self-ignition temperature is the lowest initial temperature of air circulating the specimen at which, in the absence of an ignition source, the self-heating properties of the specimen lead to ignition or ignition occurs by itself, as indicated by an explosion, a flame, or a sustained glow.

Two basic test methods have been developed to measure ignitability. The first (ASTM D 1929) determines the temperatures that are necessary to cause sufficient decomposition to generate volatile fuel that will either spontaneously ignite or that can be ignited by a pilot flame. The second approach for measuring ignitability involves the application of a Bunsen burner flame to polymer samples for short

Figure 8-5. Flammability test chamber. (Courtesy of Atlas Material Testing Technology, LLC.)

periods of time and determination of whether sustained ignition occurs after removal of the flame.

8.5.2. Ignition Temperature Determination (ASTM D 1929)

This method covers the laboratory determination of the self-ignition and flash-ignition temperatures of plastics using a hot-air ignition furnace. The equipment used is known as a Setchkin apparatus, as illustrated in Figure 8-6. A schematic cross section view of a hot-air ignition furnace assembly is shown in Figure 8-7. The apparatus consists of a furnace tube and an inner ceramic tube placed inside the furnace tube that is capable of withstanding 750°C temperature. The provision is made to introduce clean, metered, outside air, tangentially near the top of the annular space between the ceramic tubes. Air is heated with an electrical heating unit, circulated in the space between the two tubes, and allowed to enter the inner

Figure 8-6. Setchkin self-ignition cell and console. (Courtesy of Custom Scientific Instruments, Inc.)

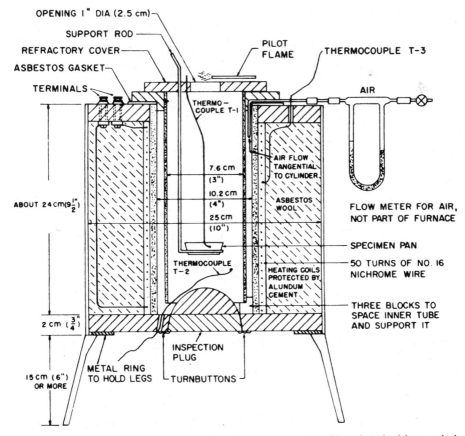

Figure 8-7. Cross section of hot-air ignition furnace assembly. (Reprinted with permission of ASTM.)

furnace tube at the bottom. Copper tubing attached to a gas supply and placed horizontally above the top surface of the divided disc provides a pilot flame. A specimen pan is used to hold the specimen in place inside the furnace. Two thermocouples, one located near the specimen and the other located below the specimen holder, are used to measure specimen and air temperature, respectively. Thermoplastic materials are tested in pellet form normally supplied for molding. Sheet specimens of size $3/4 \times 3/4$ in. are also used. The specimens are conditioned using standard conditioning procedures prior to testing.

Two procedures are used to determine ignition temperatures. Procedure A is quite lengthy and therefore less commonly used. Procedure B accomplishes similar results in a considerably smaller period of time. To commence the test, the air flow rate is set at 5 ft/min at 400°C. Once the constant air temperature is attained, the specimen holder is lowered into the furnace. Immediately after that, the timer is started and the pilot flame ignited. A mild explosion of combustible gases or flash indicates flash ignition. This is usually followed by continuous burning of the specimen. If the specimen ignites before the end of a 5-min period, the test is repeated at a lower temperature setting using a fresh specimen. If ignition has not occurred at the end of 5 min, the temperature is raised, and the test is repeated. The lowest air temperature at which a flash is observed is recorded as the minimum flash-ignition temperature. The self-ignition temperature is determined in a similar manner without the gas pilot flame. The lowest air temperature at which a specimen burns is recorded as the minimum self-ignition temperature. Visual observations such as melting, bubbling, and smoking are also recorded.

This test is useful in comparing the relative ignition characteristics of different materials. The test cannot be regarded as the sole criterion for fire hazard. An alternate method for screening and comparing materials for ignitability has been reported (8).

8.5.3. Ignition Response Test (ASTM D 3713)

This test procedure is primarily used for characterizing the response of a plastic to an ignition source consisting of a small flame of controlled intensity applied to the base of a sample held in a vertical position.

The test is carried out by subjecting test specimens to a standard flame. The flame is applied in increasing 5-sec increments, to a maximum of 60 sec, using a new specimen at each increment. If the endpoint occurs, the time of application is decreased until 10 specimens testing consecutively, at the same duration, pass the test. This duration, along with the sample thickness and letters identifying the mode of response, is reported as ignition response index. Ignition response index (IRI) is defined as the response of a sample of specified thickness or shape to the thermal energy produced by a small flame. It consists of the maximum flame impingement time withstood by the sample without it being totally consumed, or burning or glowing or both, more than 30 sec after removal of ignition source, producing droplets that ignite cotton.

The application of the test is limited to measuring and describing the response of materials, products, or assemblies to heat and flame under controlled conditions and should not be used to describe or appraise fire hazard or fire risk under actual fire conditions.

8.6. OXYGEN INDEX TEST (ASTM D 2863, ISO 4589)

8.6.1. Introduction

Oxygen index is defined as the minimum concentration of oxygen, expressed as volume percent, in a mixture of oxygen and nitrogen that will just support flaming combustion of a material initially at room temperature under specified conditions. The oxygen index test is considered one of the most useful flammability tests because it allows one to precisely rate the materials on a numerical basis and simplifies the selection of plastics in terms of flammability. The oxygen index test overcomes the serious drawbacks of conventional flammability tests. These drawbacks are variation in sample ignition techniques, variation in the description of the endpoint from test to test, and operation of tests under nonequilibrium conditions (9). Table 8-4 compares the oxygen index values of a variety of materials. Note that red oak wood has a higher oxygen index (lower flammability) than polystyrene, but a much lower oxygen index compared to polycarbonate or PVC.

8.6.2. Test Procedures

The test determines the minimum concentration of oxygen in a mixture of oxygen and nitrogen flowing upward in a test column that will just support combustion. This process is carried out under equilibrium conditions of candlelike burning. It is necessary to establish equilibrium between the heat removed by the gases flowing past the specimen and the heat generated from the combustion. The equilibrium can only be established if the specimen is well ignited and given a chance to reach equilibrium when the percent oxygen in the mixture is near limiting or critical value (10).

The equipment used for measuring the oxygen index consists of a heat-resistant glass tube with a brass base. The bottom of the column is filled with glass beads, which allows the entering gas mixture to mix and distribute more evenly. A specimen-holding device to support the specimen and hold it vertically in the center of the column is used. A tube with a small orifice having propane, hydrogen, or other

TABLE 8-4. Oxygen Index Rating of Some Materials

Material	Oxygen Index, Percent
Red oak	24.6
Acetal	16.2
Polyethylene	17.4
Polypropylene	17.4
Polystyrene	18.3
Polycarbonate	27.0
Nylon 6-6	28.0
40 percent glass-filled polycarbonate	30.5
PVC	47.0
Polyvinylidene chloride	60.0
PTFE	95.0

gas flame, suitable for inserting into the open end of the column to ignite the specimen, is used as an ignition source. A timer, flow measurement, and control device are also used. Figure 8-8 illustrates a typical equipment layout. A commercially available oxygen index tester is shown in Figure 8-9.

The test specimen used in the experiment must be dry since the moisture content of some materials alters the oxygen index. Four different types of specimens are specified. They are physically self-supporting plastics, flexible plastics, cellular

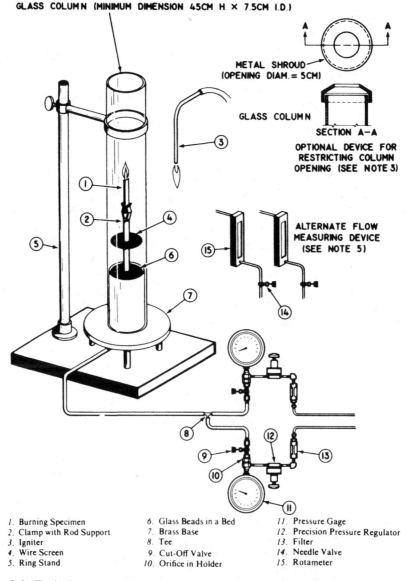

1. Burning Specimen
2. Clamp with Rod Support
3. Igniter
4. Wire Screen
5. Ring Stand
6. Glass Beads in a Bed
7. Brass Base
8. Tee
9. Cut-Off Valve
10. Orifice in Holder
11. Pressure Gage
12. Precision Pressure Regulator
13. Filter
14. Needle Valve
15. Rotameter

Figure 8-8. Typical equipment layout for oxygen index test. (Reprinted with permission of ASTM.)

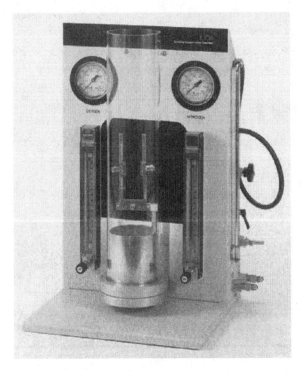

Figure 8-9. Oxygen index tester.

plastics, and plastic film or thin sheet. The dimension of the specimen varies according to the type.

The specimen is clamped vertically in the center of the column. The flow valves are set to introduce the desired concentration of oxygen in the column. The entire top of the specimen is ignited with an ignition flame so that the specimen is well lighted. The specimen is required to burn in accordance with set criteria, which spell out the time of burning or the length of specimen burned. The concentration of oxygen is adjusted to meet the criteria. The test is repeated until the critical concentration of oxygen, which is the lowest oxygen concentration that will meet the specified criteria, is determined. The oxygen index is calculated as follows:

$$\text{Oxygen index percent} = (100 \times O_2)/(O_2 + N_2)$$

where O_2 = volumetric flow of oxygen at the concentration determined; N_2 = volumetric flow of nitrogen.

8.6.3. Factors Affecting the Test Results

1. *Thickness of Specimen.* As the specimen thickness increases, the oxygen index also increases steadily.
2. *Fillers.* Fillers such as glass fibers tend to increase the oxygen index up to a certain percentage loading. In the case of polycarbonate, the oxygen index

peaks at about 25 percent loading. Higher loading beyond this point subsequently decreases the oxygen index.

3. *Flame Retardants.* Flame retardants increase the oxygen index, making polymers more suitable for applications requiring improved flammability.

8.7. SURFACE BURNING CHARACTERISTICS OF MATERIALS

Two major tests have been developed to study the surface burning characteristics of materials.

8.7.1. Surface Flammability of Materials Using a Radiant Heat Energy Source (ASTM E 162)

This test, also known as the radiant panel test, is one of the most widely used laboratory-scale flame-spread tests. Although not recommended for use as a basis of ratings for building code, the test does provide a basis for measuring and comparing the surface flammability of materials when exposed to a prescribed level of radiant heat energy.

The test apparatus consists of a radiant panel with an air and gas supply, a specimen holder, a pilot burner, a stack, thermocouples, a hood with an exhaust blower, a radiation pyrometer, a timer, and an automatic potentiometer recorder. Figure 8-10 is a schematic diagram of a radiant panel test apparatus.

A test specimen of size 6×18 in. is employed. A specimen thickness of 1 in. or less is desirable. Prior to the test, the specimens must be dried for 24 hr at 140°F and then conditioned to equilibrium at 73°F and 50 percent relative humidity. The specimen is mounted in the specimen holder so that it is inclined at 30° to a 12×18 in. vertical radiant panel maintained at 1238°F. A small pilot flame is positioned at the top of the specimen to ignite the developing volatile gases. Once started, the flame front progresses downward. The time for the flame to reach each of the 3-in. marks on the specimen holder is recorded. Observations such as dripping of the burning specimen are also recorded.

The flame-spread index of a specimen is determined by measuring the rate of burning down of the specimen and the heat rise associated with the burning in the stack above the burning specimen.

$$I_s = F_s \times Q$$

where I_s = flame spread index; F_s = flame spread factor; Q = heat evaluation factor.

The reproducibility of this test within and between laboratories is very poor for thermoplastics. A similar test method for surface flammability of flexible cellular materials using a radiant heat energy source (ASTM D 3675) has also been developed.

8.7.2. Surface Burning Characteristics of Building Materials (ASTM E 84)

This method for determining surface-burning characteristics of building materials, commonly known as the "E-84 tunnel test," is applicable to any type of building

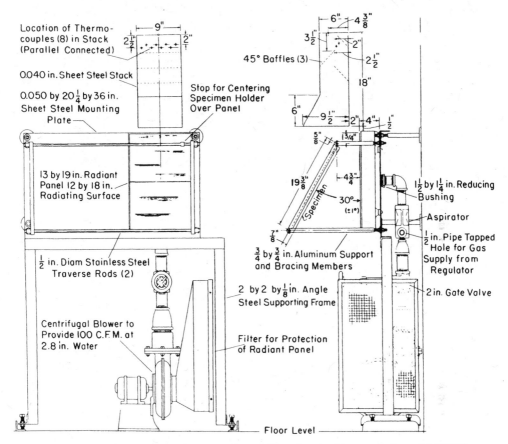

Figure 8-10. Schematic radiant panel test. (Reprinted with permission of ASTM.)

material capable of supporting itself or being supported in a test furnace at a thickness comparable to its recommended use. The purpose of this test is to determine the comparative burning characteristics of the material and smoke density.

The test is carried out by placing a 2-in-wide and 24-ft-long specimen in a fire-test chamber. The specimen may be continuous or sections joined end to end. Figure 8-11 shows a schematic diagram of a typical test furnace. One end of the test chamber is designated as the "fire end," where two gas burners capable of delivering the flames upward against the surface of the test specimen are positioned. The other end is designated as the "vent end" and is fitted with an induced draft system.

The flame-spread rating is determined by exposing the specimen to a fire at one end of a 25-ft tunnel for 10 min. The flame-spread rating is stated as a comparative measurement of the progress of flame over the surface of the tested material on the 0–100 scale where inorganic reinforced cement board is rated at 0 and red oak is rated at 100. A photoelectric cell is used to measure the smoke density.

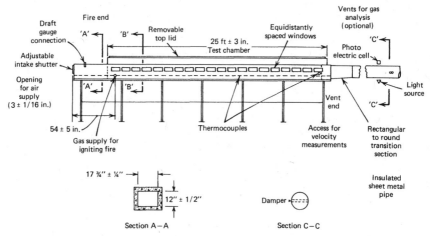

Figure 8-11. Schematic "tunnel test." (Reprinted with permission of ASTM.)

8.8. FLAMMABILITY OF CELLULAR PLASTICS—VERTICAL POSITION (ASTM D 3014)

This method covers a small-scale laboratory screening procedure for comparing the relative extent and time of burning and loss of weight of rigid thermoset cellular plastics. The test is carried out by mounting a specimen vertically on support pins in a vertical chimney with a glass front and igniting it with a Bunsen burner for 10 sec. The flame height, time of burning, and weight percent retained by the specimen are determined. A commercially available test apparatus is shown in Figure 8-12. This test is useful only for comparing relative flammability of cellular plastics and does not give any information regarding the behavior of cellular plastics in actual fire conditions.

8.9. FLAMMABILITY OF CELLULAR PLASTICS—HORIZONTAL POSITION (ASTM D 1692)

This test covers a small-scale laboratory screening procedure for measuring the rate of burning and/or extent and time of burning of rigid or flexible cellular plastics. It was discontinued in 1978 and has not been replaced.

8.10. FLAME RESISTANCE OF DIFFICULT-TO-IGNITE PLASTICS (FEDERAL STD. No 406 METHOD 203)

This federal test method was developed primarily to measure ignition time, burning time, and flame travel of plastics that are difficult to ignite. An apparatus called an ignition tester, illustrated in Figure 8-13, is employed. It consists of an enclosure that houses heating coils, spark plugs, specimen supports, and a flame-travel gauge. The specimen supports hold the specimen in vertical position. Two spark plugs with

Figure 8-12. Apparatus to determine flammability of rigid cellular plastics.

Figure 8-13. Ignition tester.

extended electrodes are spaced 1/8 in. from the surface of the specimen. The function of these spark plugs is to ignite the gases emitted from the specimen preheated by the heating coil. A suitable electric circuit is provided to maintain continuous sparking at the electrodes during the specified time. A shatterproof glass window allows clear viewing of the interior and protects the operator.

The test is started by positioning the specimen ($5 \times 1/2 \times 1/2$ in.) in the specimen support and applying a constant current of 55 amps to the heating coil. The spark plugs are energized at the same time and the timer is started. As soon as the ignition starts, the timer is stopped, and the time required to ignite the specimen is recorded. Ignition is considered as occurring when the flame transfers from the escaping gases to the surface of the specimen and continues there. Heating is discontinued 30 sec after ignition occurs. If ignition does not take place in 600 sec, the test is discontinued. The electrical supply to the spark plugs is also discontinued immediately after ignition occurs and the plugs are moved away from the flame. The maximum distance that the flame travels along the surface of the specimen, measured from the top of the heater coil before extinction, is considered the flame-travel distance. The burning time is the total time that the specimen continues to burn until the cessation of all flaming after the heater coil is turned off.

8.11. SMOKE GENERATION TESTS

There is a growing sense of urgency regarding the need to study and develop realistic smoke generation tests. The smoke generation from burning plastics is of as much concern as the flames from burning structures, because the smoke obscures visibility and seriously impairs the ability of a person to escape a fire hazard. Many small- and large-scale laboratory tests have been developed. However, there is a considerable controversy over the practicality, usefulness, and reliability of these test methods. Some tests required a large number of specimens while others are cumbersome and time-consuming. Some are too expensive, some are misleading, and others simply lack correlatability. An attempt is made in this section to briefly describe many of the current tests, test equipment, procedures, advantages, and limitations. Table 8-3 summarizes smoke generation tests.

8.11.1. Smoke Density Test (ASTM D 2843)

This test measures the loss of light transmission through a collected volume of smoke produced under controlled, standardized conditions. The test employs a $12 \times 12 \times 31$ in. aluminum test chamber with a heat-resistant glass observation door. The chamber is completely sealed except for 1×9 in. openings on four sides of the bottom of the chamber. A specimen holder holds the specimen in a horizontal position. A photoelectric cell and a light source are used to measure light absorption. A test specimen of size $1 \times 1 \times 1/4$ in. is placed in the specimen holder and exposed to a propane–air flame so that the flame is directly under the specimen. The percentage of light absorbed by the photoelectric cell is measured and recorded at 15-sec intervals for 4 min. A visual comparison of smoke density is made by observing the illuminated "EXIT" sign. The light absorption data (light absorption in percent) is plotted versus time on a graph recorder. The total smoke produced is determined by measuring the area under the curve. The percentage of the area under the curve is the smoke density rating. The maximum smoke density is the highest point on the curve. Figure 8-14 illustrates a commercially available smoke density chamber.

8.11.2. Surface Flammability Tests (ASTM E 84)

This test is discussed in detail in Section 8.7.

8.11.3. Specific Optical Density of Smoke Generation (ASTM E 662)

This test measures the specific optical density of smoke generated by a specimen mounted vertically in a smoke density chamber. A photometric system is used to measure varying light transmission as smoke accumulates from exposure of specimens to the flaming and nonflaming conditions.

Figure 8-14. Smoke density chamber. (Courtesy of Testing Machines, Inc.)

The test is conducted by mounting a 3 × 3 in. specimen in a special holder with a specific orientation and fixed area of exposure. Next, the specimen is exposed to a precisely adjusted and regulated radiant energy source for radiant nonflaming tests. For the flaming condition, the specimen is subjected to a multijet impingement by a special flaming jet with precise setting of both air and gas supply. The measurement of the attenuation of a light beam by smoke is made using a light source and a photodetector. This vertical positioning of the photo detector and light source minimizes measurement differences due to smoke stratification that could occur with horizontal photometer path at a fixed height (11). The light transmission measurements are used to calculate specific optical density of the smoke generated during the time period to reach the maximum value. The test results are affected by small variations in sample geometry, positioning of the specimen, surface orientation, specimen thickness, and small differences in conditioning. A commercially available NBS (NIST) smoke density chamber is shown in Figure 8-15.

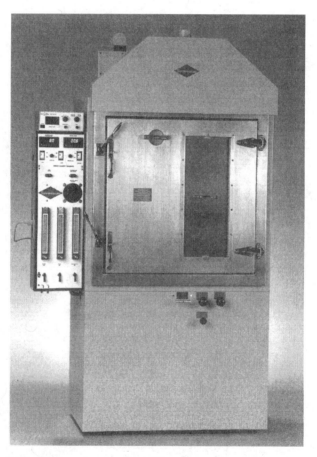

Figure 8-15. NBS smoke chamber. (Courtesy of Newport Scientific, Inc.)

8.11.4. Gravimetric Determination of Smoke Particulates from Burning of Plastic (ASTM D 4100)

This test method was developed for gravimetric determination of smoke particulate matter produced from the combustion or pyrolysis of plastic materials. This smoke test is a gravimetric test as opposed to an optical test in the sense that the smoke evolution is measured gravimetrically by weight of the smoke rather than optically by the light obscuration caused by the particulates (12). The test was introduced to provide the industry with a quick, simple, and repeatable technique for measuring smoke generated from burning plastics.

The test equipment, shown schematically in Figure 8-16, consists of a chimney extending from a cylindrical combustion chamber; a filter assembly, positioned at the top of the chimney, is connected to a high-capacity vacuum source; a micro-Bunsen burner, located at the bottom of the combustion chamber, is mounted at an angle of 10° from horizontal; a specimen holder holds a 1 1/2 × 1/2 × 1/8 in. specimen horizontally. The specimen is exposed to the flame for 30 sec. At the end of 30 sec, the burner is shut off, and the specimen is extinguished. The smoke generated by the burning specimen is drawn up the chimney by the draft caused by vacuum suction and is collected on the surface of filter paper. The weight of smoke deposited is determined by weighing the filter paper. The value reported is percent smoke produced from burning plastics. This test has been withdrawn by ASTM without the replacement.

8.11.5. Radiant Panel Test (ASTM E 162)

A radiant panel apparatus consists of a radiating heat source maintained at 1238°F. A vertically mounted, porous refractory panel acts as a radiating heat source. A

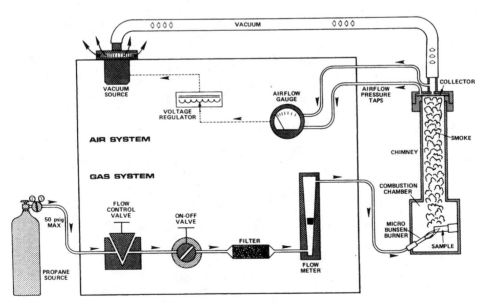

Figure 8-16. Schematic of smoke generation test.

specimen of size 6 × 18 in. is supported in front of the panel. The specimen is ignited from the top by a pilot flame so that the flame travels downward along the underside, exposed to the radiant panel. A glass-fiber filter paper is positioned at the top of the stack. The smoke particles are collected on the surface of the filter by drawing air through the filter. The filter paper is weighted before and after the test and the difference in weight is reported as smoke deposit in milligrams. Figure 8-17 shows commercial test apparatus.

8.11.6. OSU Release Rate Test

The Ohio State University (OSU) release rate apparatus consists of a chamber 35 × 16 × 8 in., with a top section shaped like a pyramid connected to an outlet. The chamber contains an electrically heated radiant panel, a gas-fired radiant panel, or electrically heated elements. This test offers the flexibility of orienting the specimen either vertically or horizontally to simulate wall and floor applications (13). A photocell and a light source positioned above the outlet measures the light absorption.

8.12. UL 94 FLAMMABILITY TESTING

UL 94, developed by Underwriters Laboratories, is one of the most widely used and most frequently cited sets of flammability tests for plastic materials. These tests are for the flammability of plastic materials used for parts in devices and appliances. The results are intended to serve as a preliminary indication of the material's suitability with respect to flammability for a particular application. The UL

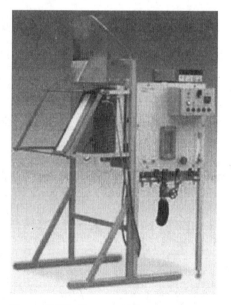

Figure 8-17. Commercial test apparatus.

flammability tests include a standard burning test applied to vertical or horizontal test bars, from which a general flammability rating is derived.

The UL 94 standard for flammability of plastic materials consists of five basic tests for classifying materials in different categories. These five basic tests are:

1. Horizontal burning test for classifying materials (HB)
2. Vertical burning test for classifying materials (V-0, V-1, V-2)
3. Vertical burning test for classifying materials (5V, A or B)
4. Vertical burning test for classifying materials (VTM-O, VTM-1, or VTM-2)
5. Horizontal burning test for foam materials (HBF-1 or -2)

8.12.1. Horizontal Burning Test for Classifying Materials (HB)

Materials are classified HB if they burn over a 3-in. span in a horizontal bar test at a rate of not more than 1.5 in./min for specimens 0.120–0.500 in. thick and not more than 3 in./min for specimens less than 0.120 in. thick, or if the specimens cease to burn before the flame reaches the 4.0-in. reference mark.

The apparatus employed for the test consists of a test chamber, an enclosure or laboratory hood, a laboratory burner, wire gauze, technical-grade methane gas, a ring stand, and a stopwatch. The test is conducted in a humidity- and temperature-controlled room. Test specimens are usually $1/2 \times 5$ in. and should have smooth edges. Before conducting the test, the specimens are marked across the width with two lines, 1.0 and 4.0 in. from one end of the specimen. The specimen is clamped in the ring stand, as shown in Figure 8-18. The burner is ignited to produce a 1-in.-high blue flame. The flame is applied so that the front edge of the specimen, to a depth of approximately 1/4 in., is subjected to the test flame for 30 sec without changing the position of the burner, and is then removed from the burner. If the specimen burns to the 1.0-in. mark before 30 sec, the flame is withdrawn. If the specimen continues to burn after removal of the flame, the time for the flame front

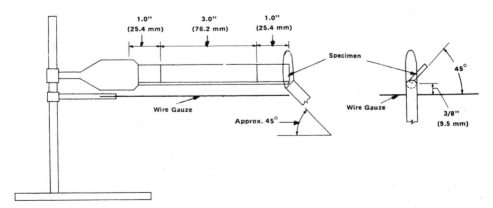

Figure 8-18. Horizontal burning test for HB classification. (Reprinted with permission of Underwriters Laboratories.)

TABLE 8-5. Summary of UL 94 Vertical Burning Test for Classifying Materials V-0, V-1, V-2

Criteria	Classification		
	V-2	V-1	V-0
Number of test specimens	5	5	5
Number of ignitions	2	2	2
Maximum flaming time per specimen per flame application, sec	30	30	10
Total flaming time, five specimens, 2 ignitions, sec	250	250	50
Flaming drips ignite cotton	Yes	No	No
Maximum afterflow time, per specimen, sec	60	60	30
Burn to holding clamp	No	No	No

Reprinted from Plastics Compounding, May/June 1978.

to travel from the mark 1.0-in. from the free end to the mark 4.0 in. from the free end is determined and rate of burning is calculated.

8.12.2. Vertical Burning Test for Classifying Materials (V-0, V-1, V-2)

In this test, the specimens are clamped vertically. The materials are classified V-2, the least stringent classification, or V-1 or V-0, which are the most stringent or highest classifications for this test. Table 8-5 summarizes the requirements for each classification.

The apparatus employed for vertical burning tests are similar to the ones employed in the horizontal burning test, except for the addition of dessiccator, a conditioning oven, and dry absorbent surgical cotton. The test is conducted on a 1/2- × 5-in. specimen. A small 3/4-in.-high blue flame is applied to the bottom of the specimen for 10 sec, withdrawn, and then reapplied for an additional 10 sec; the duration of flaming and glowing is noted as soon as the specimen has extinguished. A layer of cotton is placed beneath the specimen to determine whether dripping material will ignite it during the test period.

8.12.3. Vertical Burning Test for Classifying Materials (5V)

For any material to achieve this somewhat stringent classification, the test specimens must not burn with flaming and/or glowing combustion for more than 60 sec after the fifth flame. Also, the test specimens must not drip.

The test apparatus consists of a test chamber, a laboratory burner, an adjustable ring stand for vertical positioning of the specimens, a gas supply, a mounting block capable of positioning the burner at an angle of 20° from the vertical, a stop watch, a dessiccator, dry absorbent surgical cotton, and a conditioning oven. The test specimens are in two forms: 1/2- × 5-in.-long bars of 6- × 6-in. plaques.

Procedure 1 for the testing of specimens 1/2 × 5 in. involves positioning the test specimen vertically on the ring stand and supporting the burner on the inclined plane of a mounting block so that the burner tube may be positioned 20° from the vertical. This arrangement is shown in Figure 8-19. The burner is ignited at a remote

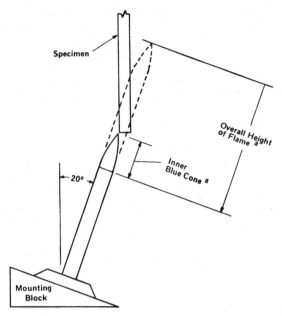

Figure 8-19. Vertical burning test for 5V classification. (Reprinted with permission of Underwriters Laboratories.)

location from the specimen and adjusted so that the overall flame height is 5 in. and the height of the inner blue cone is 1.5 in. The flame is then applied to one of the lower corner of the specimen at a 20° angle from the vertical so that the tip of the blue cone touches the specimen. The flame is applied for 5 sec and removed for 5 sec. This process is repeated four additional times. Duration of flaming plus glowing, the distance the specimen burned, dripping, and deformation of the specimen are observed and recorded.

Procedure 2, which involves the use of test plaques, is similar to Procedure 1, the only difference being the positioning of the test plaques. The five different positions include: (1) plaque vertical with the flame applied to the lower corner of the plaque, (2) plaque vertical with the flame applied to the lower edge, (3) plaque vertical with the flame applied to the center of one side of the plaque, (4) plaque horizontal with the flame applied to the center of the bottom surface of the plaque, and (5) plaque horizontal with the flame directed downward to the top surface of the plaque.

This test is also conducted using an actual molded part. The flame is applied to the most vulnerable areas of the part (14).

8.12.4. Vertical Burning Test for Classifying Materials (VTM-0, VTM-1, or VTM-2)

This particular test is designed for very thin materials which may distort, shrink, or be consumed up to the holding clamp if tested using previously described test conditions. Such materials must possess physical properties that will allow an 8-in.-long by 2-in.-wide specimen to be wrapped longitudinally around a 0.5 in.

TABLE 8-6. Summary of UL 94 Vertical Burning Test for Classifying Materials, VTM-0, VTM-1, VTM-2

Criteria	Classification		
	VTM-0	VTM-1	VTM-2
Number of test specimens	5	5	5
Number of ignitions	2	2	2
Maximum flaming time per specimen per flame application (sec)	10	30	30
Total flaming time, two flame applications, five specimens (sec)	50	250	250
Flaming drips ignite cotton	No	No	Yes
Maximum afterflow time, per specimen (sec)	30	60	60
Burn to 5-in. bench mark	No	No	No

Reprinted from *Plastics Compounding*, May/June 1978.

diameter mandrel. The requirements for three different classifications are summarized in Table 8-6.

The specimens are cut from the sheet to the specified size and conditioned prior to the test. A 5-in. mark is made across the specimen width. The specimen is wrapped around a 1/2-in. diameter mandrel. The test is carried out by subjecting the specimens to a 3/4-in. high blue flame for 3 sec and withdrawing it until flaming of the specimen ceases. The second application is made immediately and repeated. Observations such as the duration of the flame after the first flame application, the duration of flaming after the second application, the duration of flaming plus glowing after the second application, whether specimens burn up to a 5-in. mark, and whether specimens drip flaming particles which ignite the cotton are recorded.

8.12.5. Horizontal Burning Test for Classifying Foam Materials (HBF-1 or -2)

This test is very similar to the horizontal burning test for classifying conventional materials (HB) discussed in Section 8.12.1.

8.12.6. Factors Affecting UL 94 Flammability Testing

The UL 94 tests are considered to be very subjective tests. Identical specimens tested by different operators using the same testing equipment can give different flammability ratings. Such variations are attributed to differences in interpretation of endpoints, differences in observation techniques, and differences in operators.

The application of the flame to the test specimen is also critical. If proper care in observing the procedure is not taken, the specimen may preheat, overheat, underheat, or unevenly heat, giving inconsistent results. Other factors affecting the test results are molding of the test specimens and variations in calibration procedures, equipment, test burners, and testing environments (15).

TABLE 8-7. Standard Cross References

ASTM	FM	IEC	ISO	NFPA	UBC	UL
D 635	—	707	1,210	—	—	94
D 1929	—	—	871	—	—	—
D 2863	—	—	4,589	—	—	—
D 3801	—	707	1,210	—	—	94
D 4804	—	—	9,773	—	—	94
D 4986	—	—	—	—	—	94
D 5048	—	707	10,351	—	—	94
E 84	—	—	—	255	—	723
E 108	—	—	—	256	—	790
E 119	—	—	—	251	—	263
E 603	—	—	—	—	17-5	1,715
	4,880	—	—	—	—	1,040
	4,450	—	—	—	—	1,258

8.13. MEETING FLAMMABILITY REQUIREMENTS

In this day and age of rules and regulations, no design engineer can afford to over-look the flammability requirements put out by government agencies, private consumer protection institutions, and insurance underwriters. Today, no matter what one is designing, whether it is a TV cabinet, an appliance housing, furniture, business machine components, or a building product, it is more than likely that the plastics materials used in these applications will be required to meet specific criteria regarding their ability to withstand ignition and burning. Meeting flammability requirements is simply not enough, one must also consider smoke formation and generation of toxic gases and how they affect the overall flammability picture. Table 8-7 cross-references various tests used throughout the industry.

8.13.1. Agencies Regulating Flammability Standards

A. Government Agencies

National Bureau of Standards (NBS). NBS develops and issues a variety of standards and testing techniques. The NBS smoke test is one of the most widely accepted smoke density tests.

The department of Housing and Urban Development (HUD). This agency sets the standards for products used in the building and construction industry.

Consumer Product Safety Commission. This agency issues standards requiring voluntary as well as mandatory compliance in areas relating to consumer safety.

Department of Transportation (DOT). DOT's main concern is in the area of flammability as it relates to public safety in transportation. Many tests have been developed by DOT to study burning of simulated mass transit interiors.

Federal Aviation Administration (FAA). The FAA regulates the standards and specifications concerning the flammability of materials used in the interior of aircraft. One of the main concerns of the FAA is the formation of smoke and toxic gases from a burning object.

B. Industry Associations

Society of Plastics Industry (SPI). This association, representing the plastics industry, is actively involved in developing realistic flammability tests and standards. SPI works closely with many government and private agencies to develop new standards.

Society of Automotive Engineers (SAE), Manufactured Housing Institute, and Manufacturing Chemists Association. These are also actively involved in developing flammability standards.

C. Private Institutions

Underwriters Laboratories. "Underwriters Laboratories is an independent, not-for-profit third-party testing and certification organization evaluating products, devices and systems, and developing standards in the interest of public safety."

American Society for Testing and Materials (ASTM). ASTM is a scientific and technical organization whose prime function is to develop standards on characteristics and performance of materials, products, systems, and services. Numerous test methods concerning flammability of plastics have been developed by ASTM.

D. Insurance Underwriters

An increasing number of insurance underwriters, such as FM Global, have become interested in the flammability of plastic materials. Many realistic tests have been developed by FM Global Research.

E. Other Agencies

National Fire Protection Association (NFPA), Southern Furniture Manufacturers Association, Building Officials and Code Administrators (BOCA), National Electric Manufacturers Association (NEMA), and Southern Building Code Congress (SBCC).

8.13.2. Steps in Meeting Flammability Requirements

1. The first step in meeting flammability requirements is to carefully define the application in detail. This will help narrow the list of agencies you may have to deal with.

2. Determine the appropriate agency that deals with your application. For example, if the application has something to do with the building industry, you may want to contact one of the building and construction organizations. If the application is a plastic cabinet that houses electrical components, UL is the organization to contact. One good source of general information is the Society of Plastics Industry.

3. Once the application is defined and the governing agency is narrowed down, you may proceed with the designing and material selection. An important thing to remember at this stage is to specify the material to your design and not vice versa. The material selection process can be expedited by consulting published sources, such as the UL Plastics Recognized Component Directory, which lists plastics according to manufacturer and flammability classifications as tested by the relevant UL Standard(s). The *Modern Plastics Encyclopedia* flammability chart is another source of information for preliminary screening of the materials.

4. Once the preliminary decisions have been made on the type of material that will meet the requirement of the application, a resin supplier or a custom compounder can be consulted for specific grade of material.

5. If the code or standard requires testing, an independent laboratory should be consulted.

REFERENCES

1. Nelson, G. L., "Flame Tests for Structural Foam Parts," *Plast. Tech.* (Nov. 1977), p. 88.
2. *Ibid.*, p. 89.
3. Nelson, G. L., "Fire and Polymers," *Chemistry* **51** (June 1978), p. 23.
4. *Ibid.*, p. 26.
5. Hill, B. J., "How Predictive Are Small-Scale Flame Tests?" *SPE ANTEC* **24** (1978), p. 587.
6. Fang, J. B., NBS Technical Note 879, U. S. Department of Commerce, National Bureau of Standards, Washington, D. C., June 1976.
7. Hilado, C. J. and Murphy, R. M., "Screening Materials for Ignitability," *Mod. Plast.* (Oct. 1978), p. 52.
8. *Ibid.*, p. 52.
9. Goldblum, K. B., "Oxygen Index: Key to Precise Flammability Rating," *S.P.E. J.*, **25**(2) (Feb. 1969), p. 50.
10. *Ibid.*, p. 51.
11. Hilado, C. J., Cumming, H. J., and Machdo, A. M., "Screening Materials for Smoke Evolution, *Mod. Plast.* (July 1978), p. 62.
12. Kracklauer, J., Sparkes, C., and Legg, R., "New Smoke Test—Fast, Simple, Repeatable," *Plast. Tech.* (Mar. 1976), pp. 46–49.
13. Hilado, C. J., Reference 7, p. 62.
14. Nelson, G. L., Reference 3, p. 89.
15. Howard, J. M., "Factors Affecting UL 94 Flammability Testing," *SPE ANTEC*, **25** (1979), p. 942.

SUGGESTED READING

Hilado, C. J., *Flammability Test Methods Handbook*, Technomic, West Port, 1973.

Hilado, C. J., *Flammability Handbook for Plastics*, Technomic, West Port, 1969.

Zabetakis, J. M. "Flammability Characteristics of Combustible Gases and Vapors," *U.S. Bureau of Mines Bull.*, **627** (1965).

Bradley, J. N., *Flame and Combustion Phenomena*, Methuen and Co., London, England, 1969.

Lyons, J. W., *The Chemistry and Uses of Fire Retardants*, John Wiley, New York, 1970.

Hilado, C. J., *Fire and Flammability Series*, Technomic, West Port, 1973. Vol. 1, "Flammability of Cellulosic Materials"; Vol. 2, "Smoke and Products of Combustion"; Vol. 3, "Flammability of Consumer Products"; Vol. 4, "Oxygen Index of Materials"; Vol. 5, "Surface Flame Spread"; Vol. 6, "Flame Retardants."

Troirzsch, J., *Plastics Flammability Handbook*, Hanser Gardner, Cincinnati, 1990.

9

CHEMICAL PROPERTIES

9.1. INTRODUCTION

Chemical resistance of plastics is a complex subject. The test results are often misinterpreted by engineers and designers. Material selection is made without a proper understanding of the tests' limitations and how the results are derived. Extremely strong and tough plastic like polycarbonate has limited applications because of its poor chemical resistance. Polypropylene, on the other hand, has poor physical properties but is impervious to most chemicals and solvents. The resistance of plastics to chemicals is best understood through the study of its basic polymer structure. The type of polymer bonds, the degree of crystallinity, branching, the distance between the bonds, and the energy required to break the bonds are the most important factors to consider while studying the chemical resistance of plastic materials (1). For example, highly crystalline structure, lack of branching, and the presence of very strong covalent bonds between carbon and fluorine atoms in the main chain makes polytetrafluoroethylene resistant to almost all chemicals and solvents. Similarly, in the case of polyamides (nylons), the regular symmetrical structure and the molecular flexibility that produces high crystallinity and the presence of greater intermolecular forces help the polymer to be rigid, strong, and resistant to chemicals (2). Polycarbonate is easily attacked by most common solvents due to its intermediate polarity and lack of major intermolecular attraction (3). Excessive molecular inflexibility and low intermolecular attraction combine to make polystyrene rigid but unable to withstand surface attack by surfactants and solvents (4). Another important consideration when studying the chemical resistance of plastics is the effect of additives such as plasticizers, fillers, stabilizers, and colorants.

Chemical resistance tests are conducted using four different methods:

Handbook of Plastics Testing and Failure Analysis, Third Edition, by Vishu Shah
Copyright © 2007 by John Wiley & Sons, Inc.

1. Immersion test
2. Stain-resistance test
3. Solvent stress-cracking resistance
4. Environmental stress-cracking resistance

Plastic materials should not be selected solely on the basis of published chemical resistance data. The type of test conducted, test temperature, media concentration, duration of exposure, type of loading, and additives used in the base polymer must be considered, since each of the above-mentioned factors can have a significant effect on the chemical resistance of plastics. The risk potential of premature failure can be minimized by conducting the test under anticipated end-use conditions and media (5).

9.2. IMMERSION TEST (ASTM D 543, ISO 175)

The method of measuring the resistance of plastics to chemical reagents by simple immersion of processed plastic specimens is a standard procedure used throughout the plastics industry. The method can only be used to compare the relative resistance of various plastics to typical chemical reagents. The test results do not provide a direct indication of suitability of a particular plastic for end-use application in certain chemical environments. The limitation influencing the results, such as duration of immersion, temperature of the test, and concentration of reagents should be considered when studying the test data. For applications involving continuous immersion, the data obtained in short-time tests are useful only in screening out the most unsuitable materials.

The test equipment consists of a precision chemical balance, micrometers, immersion containers, an oven or a constant-temperature bath, and a testing device for measuring physical properties. The dimensions and type of test specimens are dependent upon the form of the material and tests to be performed. At least three test specimens are used for each material being tested and each reagent involved. For studying the weight and dimension change, each specimen is weighed and its thickness is measured. The specimens are totally immersed in a container for seven days in a standard laboratory atmosphere, in such a way that no contact is made with the wall or the bottom of the container. After seven days, the specimens are removed from the container and weighed. The dimensions are remeasured. The procedure remains unchanged for studying the mechanical property changes after immersion of the test bars in reagents. The mechanical properties of nonimmersed and immersed specimens are determined in accordance with standard methods for tests prescribed in the specifications and comparisons are made. Observations such as loss of gloss, swelling, clouding, tackiness, crazing, and bubbling are also reported in the test results.

9.3. STAIN RESISTANCE OF PLASTICS

Plastics have deeply penetrated the household products market in the last two decades. Determination of stain resistance of plastic materials has become increasingly important since such household products come in contact with many types of chemicals and staining reagents everyday. The test developed for determining stain resistance applies only to the incidental contact of plastic materials with miscellaneous staining reagents. Any long-term intimate contact of the reagent with plastics must be dealt with in a different manner. Certain types of additives in plastic materials seem to contribute substantially to the staining process.

The test requires an oven, an applicator, and closed glass containers for low-viscosity liquids. A wide variety of staining reagents are used. The most common ones are found among food, cosmetics, solvents, detergents, pharmaceuticals, beverages, and cleansing agents. Jelly, tea, blood, coffee, bleach, shoe polish, crayons, lipstick, and nail-polish remover are some examples of staining reagents.

A test specimen of any size may be used as long as it has a flat, smooth surface and is large enough to permit the test and visual examination. It is recommended that all thermosetting decorative laminates be wet-rubbed with a grade FF or equivalent grade of pumice to remove the surface gloss and then washed with mild soap. The staining reagent is applied onto the specimen with an applicator, forming a thin coat. In the case of low-viscosity liquids, the specimen is immersed in a liquid-staining reagent kept in a glass container. The container is then closed and the specimens are placed in an oven at $50 \pm 2°C$ for 16 hr.

Excess staining material is removed from the surface after exposure and the specimen is visually observed for residual staining. Depending upon the specific requirement, the residual staining may or may not be acceptable. The color of plastic has a significant bearing on the noticeability of stains and, therefore, one must consider testing end-use color specimens.

9.3.1. Resistance of Plastics to Sulfide Staining (ASTM D 1712)

Many plastic compositions contain salts of lead, copper, and antimony in the form of pigments, stabilizers, fillers, and other additives. When these materials come in contact with external materials containing sulfide, such as hydrogen sulfide, they stain easily. For example, if a lead-stabilized PVC compound is mixed with a tin-stabilized PVC compound that contains sulfide, the staining is quite evident. Industrial fumes and rubber are two other major sulfide-containing external agents.

The test to determine the resistance of plastics to sulfide staining is simple and requires only a freshly prepared solution of hydrogen sulfide and a test specimen of any size or shape. The specimen is partially immersed in a saturated hydrogen sulfide solution for 15 min along with a control specimen with a known tendency to sulfide stain. After 15 min, the specimens are removed and examined for staining. The comparison between control, unexposed, and exposed specimens is made to determine the relative degree of staining.

9.4. SOLVENT STRESS-CRACKING RESISTANCE

One of the most difficult challenges a design engineer faces is selecting the right plastic for the right application. The chemical resistance of plastics is a prime consideration in selecting the proper material. The chemical resistance data published by material suppliers is the most convenient source of information. Such published data is usually derived from a simple immersion test, such as described earlier in this chapter. Most polymers will undergo stress cracking when exposed to certain chemical environments under high stress for a given period of time. Such cracking will occur even though some chemicals have no effect on unstressed parts and, therefore, simple immersion of test specimens is an inadequate measure of chemical resistance of polymers (6). At this point, it is important to understand how solvent stress cracking occurs in a polymer. Initially, the polymer-to-polymer bond is replaced by a polymer–solvent bond by lowering the cohesive bond energies of the surface layers of the affected materials. These new polymer–solvent bonds cannot contribute to the overall strength of the material. If the stresses present exceed the cohesive strength of the weakened polymer, rupture occurs. The type and number of such fractures depend upon the stress pattern present in the material. The solvent penetrates deeper and cracks becomes more extensive with time (7).

The solvent stress-cracking phenomenon occurs in all plastics at varying degrees. However, the presence of stress, internal or external, is essential. The internal or molded-in stresses pose the biggest problem since complete removal of such stresses is practically impossible. The internal stresses can be minimized through proper design, optimizing processing conditions, and annealing the parts after fabrication. When a polymeric material is exposed simultaneously to a chemical and a stress, it can be characterized as exhibiting "critical stress," below which chemical media has no apparent effect. Critical stress is defined as the stress at which the first sign of crazing is observed when a specimen is exposed to a chemical environment. Two different tests have been developed to determine critical stress. One test, often referred to as the calibrated solvent test, employs a tensile testing machine along with a standard tensile test bar. The test is carried out by stressing the tensile bar specimen to a known stress level and immediately exposing it to a chemical environment. This is accomplished by either spraying the chemical onto the specimen of continuously wetting the specimen using a wick. The specimen is exposed to the chemical for 1 min and is examined by any sign of crazing with the naked eye. If no such crazing is evident, the experiment is repeated at a higher level of stress using a fresh specimen each time until crazing is observed. The material is considered safe to use in that particular chemical environment if no crazing is observed at the yield point of the material.

One of the disadvantages of the calibrated solvent test is that it requires a large number of specimens to determine the critical stress level. One other factor is how long the specimen is exposed to chemicals. It is quite possible that the chemical may attack the polymer if exposed for a long time period. Because it is no practical to expend a long time for such visual testing, an accelerated method of testing must be developed. This is generally accomplished by carrying out the test at elevated temperature and high stress. As always, there is no substitute for testing an actual part by simulating the service condition; however, this test does provide some useful information regarding the behavior of the polymer exposed to a chemical environment at different stress levels. The critical stress value established for a particular polymer–solvent combination is very useful in determining the level of molded-in stresses in a part. This is further discussed in Chapter 15.

An alternate method for measuring solvent stress cracking, developed several years ago, has a few advantages over the previous method (8). This method employs a specimen of size $4 \times 1 \times 0.03$ in. strapped to an elliptical jig. The entire assembly, as

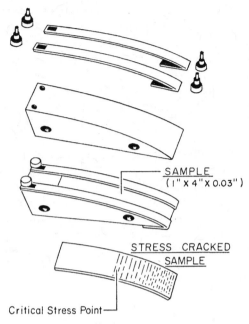

SAMPLE
(I" X 4" X 0.03")

STRESS   CRACKED
SAMPLE

Critical Stress Point

Figure 9-1. Jig for solven stress cracking test. (Reprinted with permission of Wiley Interscience.)

shown in Figure 9-1, is immersed in a reagent. Because of the elliptical design of the jig, the stress at the high end of the jig is extremely low. Conversely, the stress at the low end of the jig is extremely high. The level of stress in the specimen at different points on the jig can be calculated. After 1 min, the specimen is observed for crazing. The point at which the crazing stops is considered the critical stress point. The critical stress value at this point is determined from a previously calculated value. If no crazing is observed after 1 min, the test is continued for several hours. The test may also be carried out at elevated temperatures to accelerate the stress-cracking process. The biggest advantage of this method is that one can look at the stress-cracking process over the entire range of stress values using only one specimen.

9.5. ENVIRONMENTAL STRESS-CRACKING RESISTANCE (ASTM D 1693, ISO 4599)

Environmental stress cracking is the failure in surface-initiated brittle fracture of a polyethylene specimen, or a part under polyaxial stress, in contact with a medium in the absence of which fracture does no occur under the same conditions of stress. Combinations of external and/or internal stresses may be involved, and the sensitizing medium may be gaseous, liquid, semisolid, or solid.

Several conditions must be present for environmental stress cracking to occur. First, the presence of a "stress riser" or a "notch" is a very important factor. The need for some type of stress, "molded-in" or external, is inevitable. Finally, without the presence of an external sensitizing agent environmental stress cracking is impossible (9). Environmental stress cracking should not be confused with other types of stress cracking, such as solvent stress cracking and thermal stress cracking. Environmental stress cracking describes the tendency of polyethylene products to prematurely fail in the presence of detergents, water, sunlight, oil, or other active environments, usually under conditions of relatively high strain. It is a purely physical phenomenon that involves no swelling or similar mechanical weakening of the material. Polyethylene products are most susceptible to such cracking or crazing under load when exposed to certain chemicals and environments. This phenomenon was first recognized in polyethylene-coated wire, which often was lubricated with surface-active materials to facilitate installation in conduits. Under these conditions, polyethylene, which appeared to perform satisfactorily in the laboratory, rapidly developed severe cracks that propagated completely through to the conductor (10). The stress-cracking resistance of polyethylene can be improved by increasing molecular weight, reducing stresses by proper fabrication practices, and incorporating elastomers in the formulation. It is further observed that narrow molecular-weight distributions considerably improve the resistance of a polymer of given density and average molecular weight. Large crystalline structures and molecular orientations appear to aggravate the problem (11).

9.5.1. Test Procedure

The test specimens of size 1.5 × 1 in. are cut very precisely. The rectangular specimen is nicked to a fixed length and depth using a sharp blade mounted in the nicking jig (Figure 9-2). The nicked specimen is then bent through 180° so that the

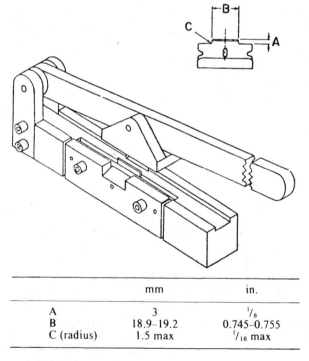

	mm	in.
A	3	$^1/_8$
B	18.9–19.2	0.745–0.755
C (radius)	1.5 max	$^1/_{16}$ max

Figure 9-2. Nicking Jig. (Reprinted with permission of ASTM.)

nick is on the outside of the bend and at a right angle to the line of bend. The samples are mounted onto the holder. The holder is inserted in the test tube. Immediately after that, the test tube is filled with fresh reagent to submerge the samples. The reagent can be a surface-active agent, soap, or any other liquid organic substance. One of the most commonly used reagents is Igepal C0–630, manufactured by Rhone-Poulenc, Cranbury, NJ. The tube is placed in a constant-temperature bath maintained at 50 ± 0.5°C or 100.0 ± 0.5°C, depending upon the conditions selected for the test. Test specimens are removed after a specified time and observed for crazing. Figure 9-3 illustrates a test specimen, specimen holder, and test assembly.

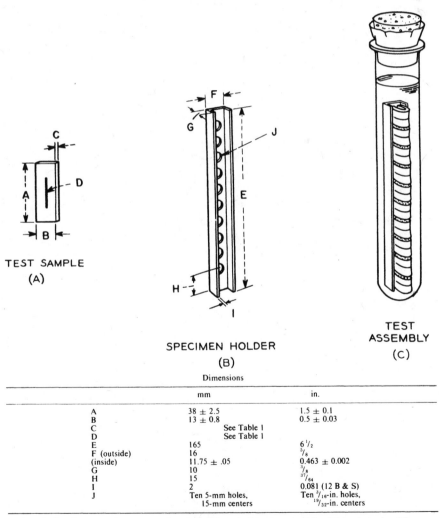

TEST SAMPLE
(A)

SPECIMEN HOLDER
(B)

TEST
ASSEMBLY
(C)

Dimensions		
	mm	in.
A	38 ± 2.5	1.5 ± 0.1
B	13 ± 0.8	0.5 ± 0.03
C	See Table 1	
D	See Table 1	
E	165	$6\frac{1}{2}$
F (outside)	16	$\frac{5}{8}$
(inside)	11.75 ± .05	0.463 ± 0.002
G	10	$\frac{7}{8}$
H	15	$^{37}/_{64}$
I	2	0.081 (12 B & S)
J	Ten 5-mm holes, 15-mm centers	Ten $^3/_{16}$-in. holes, $^{19}/_{32}$-in. centers

Figure 9-3. Test specimen, specimen holder and test assembly. (Reprinted with permission of ASTM.)

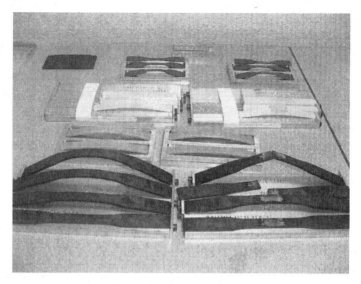

Figure 9-4. Simple fixture.

9.5.2. Alternate Constant Strain Test

This constant strain test method is an inexpensive way to test environmental stress crack resistance of a polymer without investing in expensive equipment. A simple fixture as shown in Figure 9-4 is fabricated. Test samples are molded, machined from bar stock or can also be obtained from material suppliers. By adjusting the fixtures, various amount of strain is applied to the test samples. Test is carried out by applying small amount of stress cracking agent for a predetermined length of time. Following this, the samples are inspected for crazing and cracking. Test bars can also be subjected tensile and impact tests to determine deterioration of mechanical properties. One of the disadvantages of this method is stress decay with time due to stress relaxation.

REFERENCES

1. Richardson, T. A., *Modern Industrial Plastics*, Howard W. Sams and Co., Indianapolis, IN, 1974, p. 112.
2. Deanin, R. D., *Polymer Structure, Properties and Applications.* Cahners, Boston, MA, 1972, P. 455.
3. *Ibid.*, p. 449.
4. *Ibid.*, p. 427.
5. Borg-Warner Corporation, "Chemical Resistance." *Tech.* Bull. Design Tip No. 6, Parkersburg, W. VA.
6. Smith, W. M., *Manufacture of Plastics*, Vol. 1, Reinhold, New York, 1964, p. 443.
7. Baer, E., *Engineering Design for Plastics*, Reinhold, New York, 1964, p. 778.
8. Bergen, R. L., Jr., "Stress Cracking of Rigid Thermoplastics," *SPE ANTEC*, **8** (1962).
9. Baer, Reference 7, p. 772.
10. *Modern Plastics Encyclopedia*, McGraw-Hill, New York, 1967, p. 238.
11. Brydson, J. A., *Plastics Materials*, Reinhold, New York, 1970, p. 117.

10

ANALYTICAL TESTS

10.1. INTRODUCTION

Analytical tests are important to material suppliers and processors. These tests provide basic information that is necessary for characterizing and qualifying the material. Analytical tests such as density and specific gravity tests are used as a means of assuring product uniformity. Very few plastics are sold today without additives and modifiers. These additives and modifiers tend to alter the physical properties of the base material depending upon the amount and type used. Compounders of such additives make specific gravity value part of their product specification.

Unlike metals and ceramics, plastics absorb water. The amount of water absorption depends upon the specific type of plastic. The key properties—mechanical, electrical, and optical—are seriously affected. Water also tends to act as a plasticizer and lowers the softening temperature of the part (1). Plastic materials that absorb a large amount of water normally affect the dimensional stability of the product. The plastic product designer must take into account the water absorption characteristics of the plastic materials to avoid premature failures.

Another important and frequently used test throughout the industry is the moisture analysis test. This simple but effective test provides useful information regarding the processibility of plastic materials. Excessive moisture can cause many processing and visual problems such as splay marks. Bulk density tests and sieve analysis tests also help to predict the material's behavior during mixing, compounding, and processing. Chapter 7 discusses analytical tests often used in material characterization.

Handbook of Plastics Testing and Failure Analysis, Third Edition, by Vishu Shah
Copyright © 2007 by John Wiley & Sons, Inc.

10.2. SPECIFIC GRAVITY (ASTM D 792)

Specific gravity is defined as the ratio of the weight of the given volume of a material to that of an equal volume of water at a stated temperature. The temperature selected for determining the specific gravity of plastic parts is 23°C.

Specific gravity values represent the main advantage of plastics over other materials, namely, light weight. All plastics are sold today on a cost per pound basis and not on a cost per unit volume basis. Such a practice increases the significance of the specific gravity considerably in both purchasing and production control. Two basic methods have been developed to determine specific gravity of plastics depending upon the form of plastic material. Method A is used for a specimen in forms such as sheet, rods, tubes, or molded articles. Method B is developed mainly for material in the form of molding powder, flakes, or pellets.

10.2.1. Method A

This method requires the use of a precision analytical balance equipped with a stationary support for an immersion vessel above or below the balance pan. A corrosion-resistant wire for suspending the specimen and a sinker for lighter specimens with a specific gravity of less than 1.00 are employed. A beaker is used as an immersion vessel. A typical setup for the specific gravity test is shown in Figure 10-1a. The test specimen of any convenient size is weighted in air. Next, the specimen is suspended from a fine wire attached to the balance and immersed completely in distilled water. The weight of a specimen in water (and sinker, if used) is determined. The specific gravity of the specimen is calculated as follows:

$$\text{Specific gravity} = \frac{a}{(a+w)-b}$$

where a = weight of specimen in air; b = weight of specimen (sinker, if used) and wire in water; w = weight of totally immersed sinker (if used) and partially immersed wire.

10.2.2. Method B

This method, which suitable for pellets, flakes, or powder, requires the use of an analytical balance, a pycnometer, a vacuum pump, and a vacuum desiccator. The test is started by first weighing the empty pycnometer. The pycnometer is filled with water and placed in a water bath until temperature equilibrium with the bath is attained. The weight of the pycnometer filled with water is determined. After cleaning and drying the pycnometer, 1–5 g of material is added and the weight of the specimen plus the pycnometer is determined. The pycnometer is filled with water and placed in a vacuum desiccator. The vacuum is applied until all the air has been removed from between the particles of the specimen. Last, the weight of the pycnometer filled with water and the specimen is recorded. The specific gravity is calculated as follows:

(a)

(b)

Figure 10-1. Specific gravity test. (Courtesy of Ohaus Scale Corporation.) (b) Electronic version of the apparatus to measure density and specific gravity. (Courtesy Qualitest USA.)

$$\text{Specific gravity} = \frac{a}{(b + a - m)}$$

where a = weight of the specimen; b = weight of the pycnometer filled with water; m = weight of the pycnometer containing the specimen and filled with water.

If another suitable immersion liquid for the water is substituted, the specific gravity of the immersion liquid must be determined and taken into account in calculating the specific gravity.

Figure 10-1*b* shows electronic version of the apparatus to measure density and specific gravity.

10.3. DENSITY-BY-DENSITY GRADIENT TECHNIQUE (ASTM D 1505, ISO R 1183)

The density of plastic materials is defined as the weight per unit volume and is expressed in grams per cubic centimeter or pounds per cubic foot. The test method, developed to determine the density of plastics very accurately, is based on observing the level to which a test specimen sinks in a liquid column exhibiting a density gradient in comparison with standard specimens of known density. A number of calibrated glass floats of precisely known density are introduced into the density gradient and allowed to sink in the column to a point where the glass floats' density matches that of the solution. A series of such floats of differing densities within the range of the column serves as a means of calibrating the column (2). The float position versus float density is plotted on a chart large enough to be read accurately to ±1 mm to obtain a calibration line. When a specimen of unknown density is introduced into the column, the measurement of its position upon reaching equilibrium, when referred to the calibration line, gives an accurate measurement of its density.

An alternate method of density determination requires numerical calculation. Table 10-1 lists a number of liquid systems recommended for use in density gradient columns. Figure 10-2 illustrates a typical commercially available density gradient column. A number of papers have been presented on this subject (3–6).

TABLE 10-1. Liquid Systems Recommended for Use in Density Gradient Columns

System	Density Range (g/mL)
Methanol–benzyl alcohol	0.80–0.92
Isopropanol–water	0.79–1.00
Isopropanol–diethylene glycol	0.79–1.11
Ethanol–carbon tetrachloride	0.79–1.59
Ethanol–water	0.79–1.00
Toluene–carbon tetrachloride	0.87–1.59
Water–sodium bromide	1.00–1.41
Water–calcium nitrate	1.00–1.60
Zinc chloride–ethanol–water	0.80–1.70
Carbon tetrachloride–1,3-dibromopropane	1.60–1.99
1,3-Dibromopropane–ethylene bromide	1.99–2.18
Ethylene bromide–bromoform	2.18–2.89
Carbon tetrachloride–bromoform	1.60–2.89
Tetrachloroethylene–bromoform	1.55–2.70

Courtesy of Techne, Inc.

Figure 10-2. Density gradient column. (Courtesy of Techne, Inc.)

10.4. BULK (APPARENT) DENSITY TEST (ASTM D 1895)

Apparent density is a measure of the fluffiness of a material. Bulk density is defined as the weight per unit volume of a material, including voids inherent in the material as tested. Bulk density is commonly used for materials such as molding powders. The test method to determine bulk density has been discussed in detail in Chapter 7.

10.5. WATER ABSORPTION (ASTM D 570, ISO 62)

The tendency of plastics to absorb moisture simply cannot be overlooked since even the slightest amount of water can significantly alter some key mechanical, electrical, or optical property. Water absorption characteristics of plastics depend largely upon the basic type and final composition of a material. For example, materials containing only hydrogen and carbon, such as polyethylene and polystyrene, are extremely water resistant, whereas plastics having oxygen or oxy-hydrogen groups are very susceptible to water absorption. Cellulose acetate and nylons are good examples of the preceding type. Materials containing chlorine, bromine, or fluorine are water repellent. Fluorocarbon, such as PTFE, is one such type of water-repellent material (7). Water absorption characteristics of plastic materials are altered by the addition of additives such as fillers, glass fibers, and plasticizers. These additives show a greater affinity to water, especially when they are exposed to the outer surface of the molded article. Some plastics absorb very little water at room temperature, but at higher temperatures they tend to take in a considerable amount of water and lose properties rapidly. Washing machine agitators, plastic dinnerware, irrigation valves, and sprinklers are examples of applications requiring low water absorption. Table 10-2 lists typical water absorption values of some common plastics.

The test to determine the water absorption of plastics is relatively simple. Only two pieces of equipment are required—an analytical balance and an oven capable of maintaining a uniform temperature. The test specimen may be a molded disk

TABLE 10-2. Water Absorption of Common Plastics

Plastic Material	Percent Absorption
ABS	0.20–0.45
Acetal	0.22–0.25
Alkyd	0.50–0.25
Acrylic	0.30–0.40
Cellulose acetate	2.00–7.00
Cellulose acetate butyrate	0.90–2.20
Cellulose propionate	1.20–2.80
CTFE	0.00
Epoxy (unfilled)	0.08–0.15
FEP	0.01
Nylon	
Type 6	1.30–1.90
Type 66	1.50–2.0
Type 610	0.40
Type 612	1.5
Type 11	1.10
Polycarbonate	0.15–0.35
Polyester (thermoplastic)	0.8–0.38
Polyethylene	0.010
PPO (Noryl)	0.06–0.07
Polypropylene	0.010
Polysulfone	0.22
Polystyrene	0.03–0.6
SAN	0.2–0.3
TFE	0.01
Urea formaldehyde (cast)	0.02–1.50
PVC	0.07–0.75

From Milby, R. *Plastics Technology*. Reprinted with permission of McGraw-Hill Book Company.

TABLE 10-3. Immersion Temperature and Periods

Immersion Period	Immersion Temperature	Comments
24 hr	23°C	Average materials
2 hr	23°C	Materials with relatively high rate of absorption
Long term	23°C	Test continued until specimen saturation
2 hr	100°C (boiling water)	Water absorption at elevated temperature
$\frac{1}{2}$ hr	100°C (boiling water)	Materials with relatively high rate of absorption
Cyclic immersion	23°C 100°C	For special applications such as dinnerware and washing machine agitators

or a piece cut from a sheet, rod, or tube. Dimensions vary according to the type of specimen. A special conditioning procedure must be followed before actual testing. The specimens are dried in an oven at a specified temperature for a predetermined time, cooled in a desiccator, and immediately weighed. Table 10-3 shows commonly used immersion temperatures and periods.

The percent increase in weight during immersion is calculated as follows:

$$\text{Increase in weight percent} = \frac{\text{Wet weight} - \text{Conditioned weight}}{\text{Conditioned weight}} \times 100$$

10.6. MOISTURE ANALYSIS

The hygroscopic nature of plastic materials causes processing as well as dimensional stability problems. Materials like ABS and polycarbonate must be dried thoroughly before processing to avoid splay marks on molded parts, loss of impact strength, and loss of other properties. The presence of moisture also tends to produce a weak weld or knit lines, further weakening the molded part. Many processors conduct routine moisture analysis tests on materials prior to processing. Five basic methods have been developed and are most frequently used.

10.6.1. Loss on Drying (LOD) Method

This method has been used to determine the level of moisture in plastic materials for quite some time. The test is carried out by heating the material at a preset temperature and determining the resulting loss of weight by evaporation. The test can be conducted using either a simple laboratory oven and an accurate gram scale or a newly developed sophisticated instruments.

A. LOD Using a Laboratory Oven

The test requires the use of an oven, a weighing pan, and an extremely sensitive scale capable of weighing up to 0.0001 g accurately. The test is carried out by weighing a small quantity of material in the weighing pan and placing it in an oven at a specified temperature and drying to a constant weight. The pan with the material is removed from the oven at the end of the test period and placed in a desiccator for 30 min and allowed to cool. The pan is reweighed to the nearest 0.001 g. The percentage of moisture is calculated as follows:

$$\text{Percent moisture} = \frac{A - C}{A - B} \times 100$$

where A = weight of the pan and material; B = weight of the empty pan; C = weight of the pan and the material after drying.

B. LOD Using an Instrument (ASTM D 6980)

Sophisticated instruments employing LOD principles have been developed to determine moisture content accurately with relative speed and simplicity. The moisture analyzer is preprogrammed with material data such as ideal parameters for sample weight, temperature, minimum testing time, and endpoint setting. The test is carried out by preselecting the material to be tested, pressing the start button, and simply placing the material sample in the pan. The moisture content is displayed in terms of parts per million or as a percentage. Alternatively, data can be downloaded to printers and computers or displayed graphically on a built-in monitor. Figure 10-3 illustrates a commercially available moisture analyzer.

10.6.2. Karl Fisher Method

Karl Fisher titration is a complex but accurate method for moisture determination. The test involves heating a small sample in a pyrolysis oven, transfer of the resulting

Figure 10-3. Moisture analyzer employing LOD principle. (Courtesy of Omnimark Instrument Corporation.)

vapors to the reagent vessel, and electrical columetry of the water in the vapor system. The test method requires the use of toxic reagents and the system must be free of leaks for accurate results. The Karl Fisher method is more suitable for the analytical laboratory atmosphere and not manufacturing environment.

10.6.3. Sensor-Based Technology

The newly developed sensor-based technology claims to overcome many of the problems associated with the LOD or Karl Fisher techniques. The system consists of a cylindrical sample bottle heater, a dry carrier gas flow system, and a relative humidity transducer. The preweighed sample material is placed in a septum bottle, and heated to a preset temperature. The resulting volatiles are carried to an analysis cell, where the relative humidity of the flowing gas is measured. A microprocessor converts the integrated signal and displays the result in micrograms of water, percentage moisture, or PPM moisture. For most products, the entire test can be carried out in 3 to 10 minutes with the results graphically displayed and available for downloading to a computer or printing to an external printer. A commercially available sensor-based moisture analyzer is shown in Figure 10-4.

10.6.4. Microwave Technology

This method uses microwave technology to determine moisture content. The test is conducted by placing the sample of material pellets in a tube and inserting the tube into a heater assembly. Ambient air is pumped through and predried using a molecular sieve-drying tube. A metering valve and a flowmeter regulates the dry air before it enters the heater assembly. The dry air passes through the heated chamber containing the material sample and carries with it the volatiles to the cellulose acetate absorber. The absorber is subjected to a microwave beam with the amount of moisture affecting the output of the microwave energy applied to the

Figure 10-4. Sensor-based moisture analyzer. (Courtesy of Arizona Instrument.)

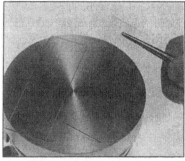

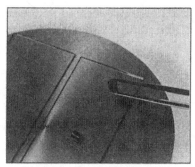

1. Plug in hot plate (be sure surface is clean) and calibrate it to a surface temperature of 550° ± 50°F (288° ± 10°C). Place two glass slides on surface for 1-2 minutes.

2. After two minutes the glass surface temperature should have reached 500-550°F (260-288°C). Use the tweezers to place four or five pellets on one of the glass slides.

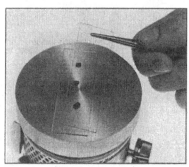

3. Place a second hot slide over the first one to sandwich the pellets between them.

4. Press a tongue depressor on the top of the sandwich until the pellets flatten out to about 1/2 inch (12.7 mm) diameter.

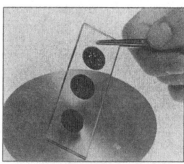

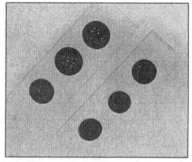

5. Remove sandwich and allow to cool. Amount and size of bubbles indicate percentage of moisture, correlating bubbles with moisture.

6. Here are typical results. Slide at lower right indicates dry material; slide at upper left indicates moisture-laden material. One or two bubbles may be only trapped air.

Figure 10-5. G.E. moisture test. (Courtesy of G.E. Plastics.)

solid state electronics. Only the moisture in the absorber is detected and the result is displayed digitally in terms of percent moisture.

10.6.5. TVI Drying Test

This test was developed by a General Electric engineer and is called the Thomasetti Volatile Indicator (TVI) test. This is a low-cost, fast, simple method for determining the readiness of moisture-sensitive thermoplastic materials for processing. The test, however, does not determine the moisture content of the material, but indicates the absence or presence of moisture in the material. A hot plate capable of maintaining up to $600 \pm 25°F$, glass microscope slides, tweezers, and a wooden tongue depressor are required for the test. Figure 10-5 shows the step-by-step procedure for conducting this test. This test cannot be used to determine the presence or absence of moisture in glass-reinforced thermoplastics.

10.7. SIEVE ANALYSIS (PARTICLE SIZE) TEST (ASTM D 1921)

The particle size and particle size distributions are important because these two characteristics of materials have a great effect on compounding, processing, and bulk handing. Large and fairly uniform particles are easier to handle and process. Fine particles are difficult to handle and difficult to process. Fine particles, when mixed with large particles, tend to cause uneven melting and hence, nonuniform mold filling, orange peel, and other surface problems. In the case of PVC dry blend-

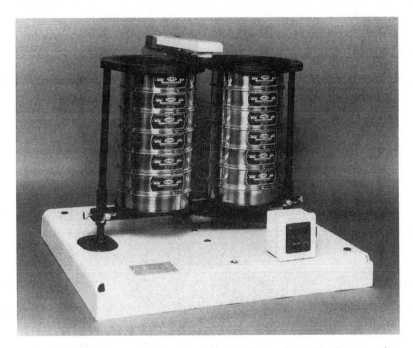

Figure 10-6. Sieve analysis. (Courtesy of Fisher Scientific Company.)

ing operations, fine particles do not allow the plasticizer to be absorbed evenly throughout the batch. Conversely, oversized particles are unable to absorb the plasticizer in sufficient amounts, resulting in poor fusing and creating the possibility of gels in the end-products (8). Particle size and particle distribution of dispersion resins affects the viscosity and stability of plastisols and organosols. Larger particles in plastisol compounds fuse more slowly and therefore yield poor physical properties (9). The large particle size of certain fillers, such as calcium carbonate, tends to increase the wear on the extruder screws and barrel and to reduce the physical properties of the end-products (10).

The test method used to determine the particle size and particle size distribution employs a series of sieves with various opening sizes. The material is simply poured from the top, allowed to pass through a series of sieves, and collected at the bottom. The quantity of material retained on each sieve is determined by weighing the sieves before and after the test. A shaker is employed to facilitate separation of various-sized particles. Figure 10-6 illustrates a typical setup for sieve analysis.

REFERENCES

1. Levy, S. and DuBois, J. H., *Plastics Product Engineering Handbook*, Reinhold, New York, 1977, p. 211.
2. Techne Inc., Princeton, NJ, *Tech. Bull.: Density Gradient Column*, Techne Catalog No. 202.
3. Boyer, R. F., Spencer, R. S., and Wiley, R. M., "Use of Density Gradient Tube in the Study of High Polymers," *J. Polym. Sci.*, **1** (1946), p. 249.
4. Tung, L. H. and Taylor, W. C., "An Improved Method of Preparing Density Gradient Tubes," *J. Polym. Sci.*, **21** (1956), p. 144.
5. Mills, J. M., "A Rapid Method of Construction of Linear Density Gradient Columns," *J. Polym. Sci.*, **21** (1956), p. 585.
6. Wiley, R. E., "Setting up Density Gradient Laboratory," *Plast. Tech.* **8**(3) (1962), p. 31.
7. Milby, R., *Plastics Technology.* McGraw-Hill, New York, 1973, pp. 534–536.
8. Schoengood, A. A., "PVC Primer," *Plast. Eng.* (Dec. 1973), p. 29.
9. *Ibid.*, p. 30.
10. Prust, R. S., "Quality Control in PVC Compounding," *Plast. Compounding* **1**(3) (May–June 1978), p. 25.

11

CONDITIONING PROCEDURES

11.1. CONDITIONING (ASTM D 618, ISO 291)

A true material comparison is possible only when property values are determined by identical test methods under identical conditions (1). Generally speaking, physical and electrical properties of plastics and electrical insulating materials are affected by temperature and humidity. Plastic materials tested above room temperature will yield relatively higher impact strength and lower tensile strength and modulus. High humidity tends to alter the electrical property test results. Obviously, in order to make reliable comparisons of different materials and test results obtained by different laboratories, it is necessary to establish standard conditions of temperature and humidity.

Conditioning is defined as the process of subjecting a material to a stipulated influence or combination of influences for a stipulated period of time (2). Three basic reasons for conditioning specimens are:

1. To bring the material into equilibrium with normal or average room conditions
2. To obtain reproducible results regardless of previous history or exposure
3. To subject the material to abnormal conditions of temperature and humidity in order to predict its service behavior

Standard Laboratory Temperature. Standard laboratory temperature is defined as 23°C (73.4°F) with a standard tolerance of ±2°C (±3.6°F).

Standard Laboratory Atmosphere. Standard laboratory atmosphere is defined as an atmosphere having a temperature of 23°C (73.4°F) and a relative humid-

Handbook of Plastics Testing and Failure Analysis, Third Edition, by Vishu Shah
Copyright © 2007 by John Wiley & Sons, Inc.

TABLE 11-1. Conditioning Procedures

Conditioning Procedure	Specimen Thickness[a] (in.)	Duration (hr)	Temperature (°C)	Humidity (Percent)	Special Requirement	Application
A	X	40	23 ± 2	50 ± 5	None	Majority of tests
	Y	88	23 ± 2	50 ± 5	None	
B	X	48	50 ± 2	—	Coll to room temperature for 5 hr in desiccator over anhydrous calcium chloride	Thermosetting materials
	Y	48	50 ± 2	—	Cool to room temperature for 15 hr in desiccator over anhydrous calcium chloride	
C	—	96 ± 2	35 ± 1	90 ± 2		Studying effect of severe atmospheric moisture
D	—	24 ± $\frac{1}{2}$	23 ± 1	—	Immersion of specimen in distilled water	Electrical and mechanical tests
E	—	48 ± $\frac{1}{2}$	50 ± 1	—	Cool specimen to 23°C by immersing in distilled water for 1 hr	Electrical and mechanical tests
F	—	—	23 ± 1	96 ± 1	Time as specified in applicable material specification	—

[a] X = 0.250 and less than 0.250 in.; Y = 0.250 in. and over.

ity of 50 percent with a standard tolerance of ±2°C (±3.6°F) and ±5 percent, respectively.

11.2. DESIGNATION FOR CONDITIONING

Conditioning of a test specimen is designated as

$$A/B/C$$

where A = a number indicating duration of conditioning (hr); B = a number indicating conditioning temperature (°C); C = a number indicating relative humidity (percent or a word) to indicate immersion in liquid.

A sequence of conditions is denoted by the use of a plus (+) sign between successive conditions.

Example. 40/23/50 indicates conditioning for 40 hr at 23°C at 50 percent RH. 48/50 + Des indicates conditioning for 48 hr at 50°C followed by desiccation.

Table 11-1 summarizes conditioning procedures.

REFERENCES

1. Borg–Warner Tech. Rept.: Measurement, Reporting and Interpretation of Thermoplastic Properties, Parkersburg, W VA, Report No: P-127.
2. Lever, A. E. and Rhys, J. A., *The Properties and Testing of Plastic Materials*, Temple Press, Feltham, England, 1968, p. 7.

SUGGESTED READING

Schmitz, J. V. (Ed.), *Testing of Polymers*, Vol. 1, Interscience, New York, 1965, pp. 41–85.

12

MISCELLANEOUS TESTS

12.1. TORQUE RHEOMETER TEST (ASTM D 2538)

The torque rheometer is one of the most versatile pieces of equipment for research and development, production control, and quality control work on polymeric materials. It is a laboratory tool often used to predict processing behavior and to simulate realistic-use conditions. Some typical applications include determining melt flow values and the stability of polymers and their degradation times at varying shear rates, studying rheological properties and pigment dispersion, characterizing different formulations, and observing the effects of changing ingredients and temperatures.

The torque rheometer is a torque-measuring rheometer based on the dynamometer principle. A sample of material to be tested is placed in the mixing head where it is subjected to shear by means of two rotating blades. The sample material is also subjected to high temperatures. The dynamometer is suspended freely between two bearing blocks, while it drives the rotors of the measuring head. The shear rate, measured by the angular velocity of the rotors, is set according to a tachometer. The measuring head rotors encounter a resistance torque from the test material that causes the dynamometer to rotate in the opposite direction. The reaction torque is balanced out through the lever system against the torque indicator scale, simultaneously recording on a strip chart recorder. An oil dashpot dampens the movement of the lever system. By sliding the weight on the arm, the zero can be suppressed several times, thus increasing the range without influencing its sensitivity setting. The measuring head is either electrically or oil heated. Figure 12-1 illustrates a torque rheometer schematically.

Handbook of Plastics Testing and Failure Analysis, Third Edition, by Vishu Shah
Copyright © 2007 by John Wiley & Sons, Inc.

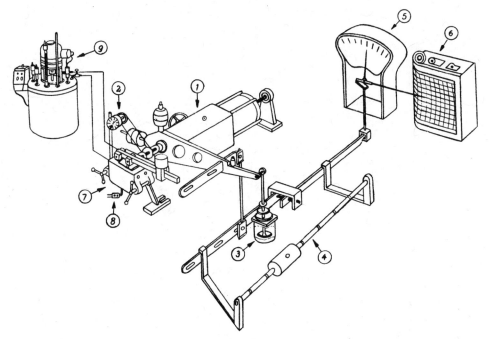

Figure 12-1. Schematic torque rheometer. (Courtesy of C. W. Brabender Instruments, Inc.)

In the newly developed units, the dynamometer is mounted together with the high-precision electronic torque-measuring unit on a stable and distortion-free base plate. The torque of the linear dynamometer is measured directly using a load cell without any intermediate member, thereby increasing the overall reliability of the torque rheometer.

The torque rheometer is used most extensively in the PVC compounding operation. A great deal of useful information can be derived from a simple fusion test. The fusion test is carried out by weighing a predetermined amount of PVC compound and introducing the charge into the preheated mixing head with rotor blades rotating at a specified rpm. The material is allowed to flux and reach fusion point and continue to the point of degradation. The torque is recorded as a function of time on a strip chart recorder. Figures 12-2a and 12-2b illustrates a typical fusion chart showing fusion point, time to flux, maximum torque at fusion, and total stability of the compound. By altering the quantity of the different additives, such as lubricants and stabilizers, one can observe the effect of these variations in terms of fusion torque, fusion time, and total stability. The correlation between the values obtained from the torque rheometer experiment and the actual manufacturing processes, such as extrusion and blow molding can be established. The torque rheometer can also be used to perform capillary flow analysis by simply attaching a small extruder to the dynamometer in place of the mixing head (1). Figure 12-3 illustrates a commercially available torque rheometer.

BRABENDER° Plastogram
PLASTI-CORDER and Mixer Measuring Head
Fusion Behaviour / Version 3.2.6

Test Conditions

Order	: C. W. BRABENDER Instrument	Speed	:	60 rpm
Operator	: AY	Mixer Temp.	:	190 °C
Date	: 1/28/2005 12:43	Start Temp.	:	190 °C
Drive Unit	: DR2051	Meas. Range	:	5000 mg
Mixer	: Roller Type 6 Elect	Damping	:	3
Loading Chute	: Manual + 5 kg	Test Time	:	5 min
Sample	: Rigid PVC	Sample Mass	:	60.0 g
Additive	: None	Code Number	:	Fusion Test A

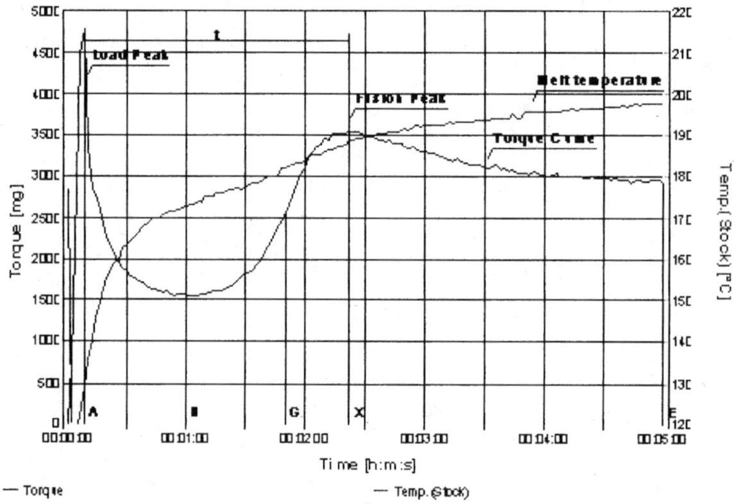

— Torque   — Temp. (Stock)

Name		Time [h:m:s]	Torque [mg]	Stock Temp. [°C]
Loading Peak	A	00:00:08	4791	129
Minimum	B	00:01:00	1560	173
Inflection Point	G	00:01:50	2555	182
Maximum	X	00:02:22	3542	189
End	E	00:05:00	2939	198

Integration / Energy

Loading Peak	to	Minimum	A - B	6.2	kJ
Minimum	to	Maximum	B - X	12.1	kJ
Maximum	to	End	X - E	30.6	kJ
Loading Peak	to	Maximum	A - X	18.4	kJ
Loading Peak	to	End (W)	A - E	48.9	kJ
Specific Energy(W/Sample Mass)				0.8	kJ/g
Gelation Area above B			B - X	4.2	kJ

Results

Fusion Time t		A - X	00:02:14 [h:m:s]
Gelation Speed v			2953 [mg/min]

(a)

Figure 12-2. (*a*) Typical fusion curve. (Courtesy of C. W. Brabender Instruments, Inc.) (*b*) Fusion curve, heat, and shear stability. (Courtesy of C. W. Brabender Instruments, Inc.)

12.2. PLASTICIZER ABSORPTION TESTS

The ability of polyvinyl chloride (PVC) resin to absorb a plasticizer is of considerable interest to a PVC compounder. The amount of plasticizer that can be added to PVC resin depends upon the type of compound formulated. The amount may vary anywhere from 20 PHR in the case of a flexible extrusion compound to 80

BRABENDER Plastogram

PLASTI-CORDER and Mixer Measuring Head
Heat & Shear Stability / Version .32.6

Test Conditions

Order	: C. W. BRABENDER Instrument	Speed	:	60 rpm
Operator	: AY	Mixer Temp.	:	200 °C
Date	: 1/28/2005 13:05	Start Temp.	:	200 °C
Drive Unit	: DDRV752	Meas. Range	:	5000 mg
Mixer	: Roller Type 6 Elect	Damping	:	3
Loading Chute	: Manual + 5 kg	Test Time	:	30 min
Sample	: Rigid PVC	Sample Mass	:	60.0 g
Additive	:	Code Number	:	Heat and Shear Stability

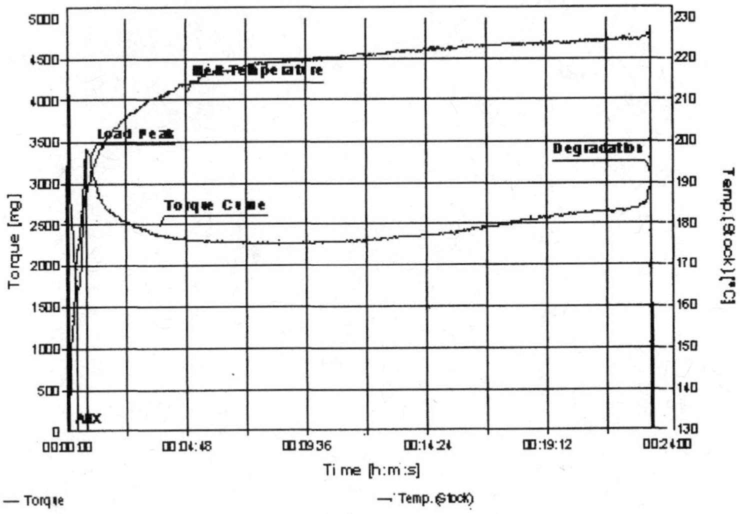

Name		Time [h:m:s]	Torque [mg]	Stock Temp. [°C]
Loading Peak	A	00:00:06	4574	135
Minimum	B	00:00:26	1718	174
Maximum	X	00:00:48	3415	188
Decomposition	D	00:29:56	0	0

Integration / Energy

Loading Peak	to	Minimum	A - B	2.8	kJ
Minimum	to	Maximum	B - X	3.4	kJ
Maximum	to	Decomposition	X - D	203.7	kJ
Loading Peak	to	Maximum	A - X	6.2	kJ
Loading Peak	to	Decomposition (W)	A - D	210.0	kJ
Specific Energy (W/Sample Mass)				3.5	kJ/g

Results

Fusion Time t		A - X	00:00:42 [h:m:s]
Decomposition Time tD		A - D	00:29:50 [h:m:s]

(b)

Figure 12-2. *Continued*

PHR in the case of plastisols. In dry blending, a flexible profile formulation for extrusion, it is imperative that the plasticizer added to the PVC resin gets fully absorbed, yielding a dry compound. A semidry powder blend can cause processing as well as conveying problems. The plasticizer absorption efficiency is related to the rate of heating of the resin-plasticizer mix, the type of resin, the particle size and distribution, the surface-volume ratio, and the type of plasticizer. Additives such as filler and impact modifiers also have an effect on plasticizer absorption.

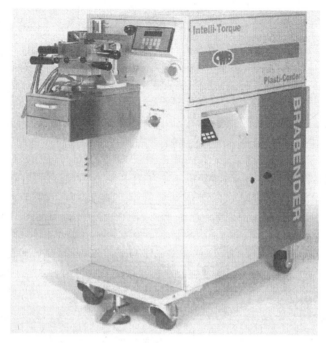

Figure 12-3. Brabender torque rheometer. (Courtesy of C. W. Brabender Instruments, Inc.)

Three basic methods have been developed to study plasticizer absorption characteristics:

1. Plasticizer absorption—Burette method
2. Plasticizer absorption using a torque rheometer
3. Plasticizer absorption under applied centrifugal force

12.2.1. Plasticizer Absorption—Burette Method

This is a quick and simple method for determining the ability of resin to absorb a plasticizer in the standard laboratory atmosphere. A burette with a shortened tip to increase the rate of flow, a titration stand with a glazed tile base, a spatula, and a balance are employed for the test. The burette is filled with commercial grade DOP (di-2 ethylhexylphthalate). The next step is to weigh out 5 ± 0.01 g of resin accurately and place it under the burette on the glazed tile. A small amount of plasticizer is slowly added to the resin and distributed throughout the resin using the spatula. This dropwise addition of plasticizer is carried out until the flow point is reached. The flow point is the point at which the resin plasticizer mixture will flow off the spatula. The entire test is repeated to verify the reproducibility. The calculation is carried out as follows:

$$\text{Plasticizer (PHR)} = \frac{\text{cc Plasticizer} \times 100 \times \text{Density of plasticizer}}{5}$$

12.2.2. Plasticizer Absorption Using a Torque Rheometer (ASTM D 2396)

This method determines the powder-mixing characteristics of the polyvinyl chloride resin. The function of the torque rheometer is described in Section 12.1.

The test requires the use of a torque rheometer equipped with a sigma-style mixer measuring head, as shown in Figure 12-4. The mixer is heated either electrically or by circulating heat-transfer oil through the jacket. To obtain consistency, a standard formulation is established as follows:

Resin	$225 \pm 0.1\,\text{g}$
Clay	$40 \pm 0.1\,\text{g}$
DIDP plasticizer	$124 \pm 0.1\,\text{g}$

Because of lot-to-lot variations in the quality of plasticizer and clay, it is recommended that the laboratory maintain a large enough inventory of these additives to establish control standards. A standard powder-mix curve should also be generated using standard additives and kept on file for comparison purposes.

The following standard test conditions are used:

Temperature	$88 \pm 1°\text{C}$
Mixer speed	$60 \pm 1\,\text{rpm}$

All ingredients except the plasticizer are weighed into the container and mixed thoroughly. The mixer is preheated and allowed to run at a specified speed for

Figure 12-4. Sigma-style mixer for plasticizer absorption test. (Courtesy of C. W. Brabender Instruments, Inc.)

30 min to obtain equilibrium conditions. All dry additives are added to the mixer and allowed to mix for 5 min. Next, the plasticizer is poured quickly into the mixer and mixing is continued for 10 min beyond the dry point. Figure 12-5 illustrates the entire mixing process graphically. The torque value increases abruptly at T_1, as the plasticizer is added and a wet lumpy condition occurs in the mixing bowl. As the plasticizer gets absorbed into the resin and additives, the mix begins to change into a free-flowing powder and torque value starts to drop until the dry point occurs, as indicated by T_2. Powder mixing time is determined by drawing two lines at T_1 and T_2, as shown in Figure 12-5 and subtracting time at T_2 from time at T_1. The test is conducted alternately by using a planetary mixing head in place of a sigma mixing head.

12.2.3. Plasticizer Absorption Under Applied Centrifuge Force (ASTM D 3367 ISO 4608)

This method provides a quantitative measure of plasticizer absorption of PVC resin under standard temperature conditions using a controlled centrifugal force. A small quantity of PVC resin (0.500 ± 0.0050 g) is weighed accurately into the plastic screening tube, which is prepacked with cotton to cover the tube orifice. Then 1 mL of plasticizer (DOP) is added to the screening tube from a pipe. The PVC and

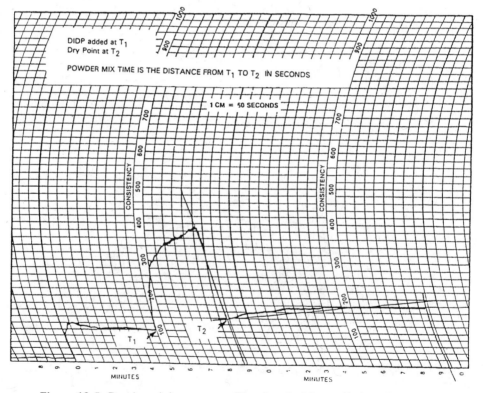

Figure 12-5. Powder mixing process. (Reprinted with permission of ASTM.)

plasticizer are subjected to centrifugal force of 3000 rpm for 40 min. The plasticizer that is not absorbed by the PVC resin is removed by centrifugation through the orifice of the screening tube. The cotton prevents the PVC particles from escaping through the orifice. After centrifuging, the screening tube is weighed and the percentage plasticizer absorption is calculated from the difference in the weight of the resin–plasticizer mix.

12.3. CUP VISCOSITY TEST

As the name implies, cup viscosity tests employ a cup-shaped gravity device that permits the timed flow of a known volume of liquid through an orifice located at the bottom of the cup. Under ideal conditions, this rate of flow would be proportional to the kinematic viscosity that is dependent upon the specific gravity of the draining liquid. However, the conditions in a simple flow cup cannot be considered ideal for true measurements of viscosity. Cup viscosity tests, however imprecise, are practical, easy-to-use instruments for making flow comparisons under strictly comparable conditions (2, 3).

In the plastisol hot-dipping operation, a preheated mold is dipped into the plastisol for a predetermined amount of time. At the end of dwell time, the mold is withdrawn and the part is removed. As the process continues, the viscosity of the plastisol reduces considerably. Such a change in the flow properties of plastisol can have a significant effect on the appearance and thickness of the fused coating. The viscosity of the plastisol is conveniently measured by using a flow cup and any necessary adjustments in the viscosity are made.

A Zahn-type viscosity cup such as the one shown in Figure 12-6 is most commonly used. The test is carried out by simply dipping the cup in plastisol or other liquid to be measured and measuring the time interval in seconds from the moment of withdrawal until the stream of material flowing from the cup orifice breaks. Zahn cups of varying orifice diameters are available for measuring all types of liquids with varying viscosities. Many other types of viscosity cups, such as Shall and Ford viscosity cups, have also been developed (4).

12.4. BURST STRENGTH TEST

Plastics are used in a variety of applications requiring internal stress applied by the transporting fluid. Plastic pipes, fittings, valves, tanks, and containers are some of the typical examples of pressure vessels.

Two basic tests of primary interest are:

1. Quick-burst strength test
2. Long-term burst strength test

12.4.1. Quick-Burst Strength Test (ASTM D 1599)

This method was developed to determine the ability of a plastic pressure vessel to resist rupturing when it is pressurized for a short period of time. Surging is

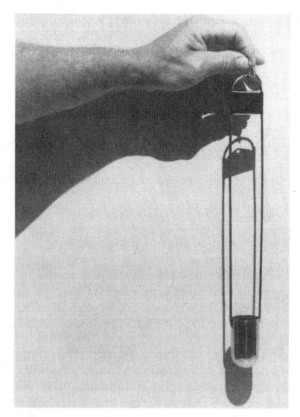

Figure 12-6. Zahn viscosity cup. (Courtesy of Byk-Gardner USA.)

a common phenomena in a fluid-transfer system. Surging is a pressure rise in a pipeline caused by a sudden change in the rate of flow or stoppage of flow in the line.

In such cases, pressure vessels are subjected to very high internal pressures for .a relatively short period. The short-time rupture strength of pressure vessels is determined by continuously increasing internal hydraulic pressure while the vessel is immersed in a controlled-temperature environment until rupture occurs.

A hydraulic burst-strength tester such as the one shown in Figure 12-7 is employed. A pressure intensifier such as the one shown in Figure 12-8 can also be used. This latter device is relatively simple and only requires the use of shop air and water or other suitable fluid. The test is carried out by simply pressurizing the specimen and uniformly increasing the pressure until the failure occurs. ASTM D 1599 requires the time to failure for all specimens to be between 60 and 70 sec. The system must be bled thoroughly, to avoid entrapment or air bubbles, prior to commencing each test. The specimen is considered to have failed when it develops a leak, crack, or rupture. The hoop stress can be calculated as follows:

$$S = \frac{P(D-t)}{2t} \quad \text{or} \quad S = \frac{P(d+t)}{2t}$$

Figure 12-7. Quick burst test apparatus. (Courtesy of Applied Test Systems, Inc.)

Figure 12-8. Pressure intensifier.

where S = hoop stress (psi); P = internal pressure (psi); D = average outside diameter (in.); d = average inside diameter (in.); t = minimum wall thickness (in.).

Hoop stress is defined as the circumferential stress in a material of cylindrical form subjected to internal or external pressure.

12.4.2. Long-Term Burst Strength Test (ASTM D 1598)

The long-term burst strength of plastic pressure vessels is determined by subjecting the pressure vessels to constant internal pressure and observing time-to-failure. This test is a static pressure test, as opposed to the dynamic quick-burst test described earlier.

A hydrostatic pressure tester such as the one shown in Figure 12-9 is employed. It consists of a pressurizing system capable of continuously applying constant internal pressure on the specimen. The apparatus is equipped with a pressure gauge and an individual timing device that is capable of measuring time-to-failure accurately. The specimens are filled with test fluid or gas, and pressure is applied to produce the desired loading. The timers are started immediately after reaching the desired pressure. A constant-temperature system may be employed if so desired. The test must be carried out in a standard laboratory atmosphere since any variation in temperature and humidity can cause results to change drastically.

The specimen failure is marked by continuous loss of pressure, bursting, abnormal ballooning, and leakage. The hoop stress in the specimens is calculated by using the formula described in Section 12.4.1.

Figure 12-9. Hydrostatic pressure tester. (Courtesy of Applied Test Systems, Inc.)

12.4.3. Developing Long-Term Hydrostatic Design Stress Data and Pressure Rating

The Plastic Pipe Institute of the Society of the Plastics Industry has developed a method of obtaining a long-term hydrostatic design stress and pressure rating a thermoplastic pressure pipe. Hydrostatic design stress is defined as the estimated maximum tensile stress in the wall of the pipe in the circumferential orientation caused by internal hydrostatic pressure that can be applied continuously with a high degree of certainty that failure of the pipe will not occur. Pressure rating is the estimated maximum pressure that the medium in the pipe can exert continuously with a high degree of certainty that failure of the pipe will not occur.

Long-term hydrostatic design stress is obtained by essentially extrapolating the stress-time regression line on data obtained in Section 12.4.2. The following is a summary of procedures used to obtain long-term hydrostatic design stress (5).

1. Specimens of plastic pipe are subjected to constant internal water pressure at different levels of pressure and the time to rupture is measured. Stress on each specimen is calculated by means of a formula applicable to plastic pipe in the 1/2- to 48-in. range.

$$S = \frac{Pd}{2t}$$

where S = hoop stress (psi); P = pressure (psi); d = mean diameter (in.); t = average wall thickness (in.).

The specimens are tested for 10,000 hr under specified conditions and a linear plot of hoop stress versus time to rupture on log–log coordinates is generated. One such plot is shown in Figure 12-10.

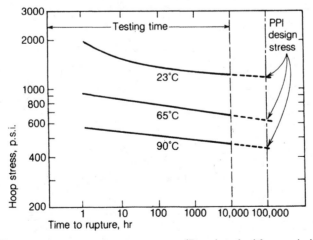

Figure 12-10. Hoop stress versus time to rupture. (Reprinted with permission of McGraw-Hill Company.)

2. The stress rupture data is analyzed by statistical regression to generate a hoop stress versus time equation. This equation is extrapolated mathematically one decade of time to 100,000 hr (approximately 11.4 yr) to obtain a 100,000-hr design stress.

3. Next, the entire design stress scale is divided into continuous increments, each of which is approximately 25 percent larger than the one below it. These increments in psi are 800, 1200, 1600, 2000, 2500, 3200, 4000 psi, and so on. Each material is arbitrarily assigned the threshold value of the increment in which its 100,000-hr design stress falls. The working stresses are calculated from this fundamental stress, called hydrostatic design basis.

4. A safety factor, depending upon temperature and type of service, is applied to the hydrostatic design basis to obtain a working stress. This value is then substituted in the pipe design equation to obtain pressure rating or required wall thickness. Typical safety factors are:

Water service at 23°C: 0.5
Water service at 38°C: 0.4
Natural gas, class location 1: 0.32

Although this method was primarily developed for plastic pressure pipe, it is not limited to plastic pipe and can be applied to obtain pressure ratings of other materials and pressure vessels. ASTM D 2837, ASTM D 2992, and PPI TR3 describe this method in full detail.

12.5. CRUSH TEST

Parts made from plastic materials are often subjected to compressive loads. The compressive strength values for plastic materials obtained by testing standard test specimens (see Section 2.4) are simply not good enough for determining relative crush resistance of molded articles. The ability of a molded article to resist compressive loading depends upon several factors such as molded-in stress, part design, and processing conditions.

A simple test was devised to study load-deflection characteristics of molded and extruded articles under parallel plate loading. A crush tester such as the one illustrated in Figure 12-11 is employed. It consists of a variable speed drive, two parallel plates, one of which is stationary, and a load cell to measure applied force. A dial indicator is used to measure deflection. More sophisticated compression testers such as the one shown in Figure 2-21 can also be used.

The test is carried out by simply placing a specimen between the parallel plates and applying load until failure occurs. The specimen is considered to have failed if it cracks or fractures. Quite often, requirements such as "the part shall not crack or fracture when deflected 10 percent of its original dimension" are placed on the part drawing. The crush test is used as a routine quality control procedure. ASTM D 2412 discusses the external loading properties of plastic pipe by parallel plate loading.

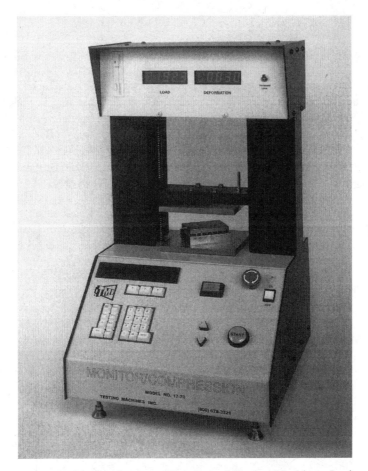

Figure 12-11. Crush tester. (Courtesy of Testing Machines, Inc.)

12.6. ACETONE IMMERSION TEST (ASTM D 2152)

This method was developed to determine the quality of rigid PVC pipe and fittings as indicated by their reaction to immersion in anhydrous acetone. An unfused PVC compound attacked by anhydrous acetone causes the material to swell, flake, or completely disintegrate. A properly fused PVC compound is impervious to anhydrous acetone and only a minor swelling, if any, is observed.

The test is carried out by placing a small specimen cut from the molded or extruded article in reagent grade acetone and observing the effect of acetone immersion after 20 min. The presence of water in acetone reduces its effectiveness and therefore acetone must be dried by shaking it with anhydrous calcium sulfate, which is removed from the acetone by filtering. An acetone immersion test is used as an ongoing quality control test by many PVC pipe and fittings manufacturers.

12.7. ACETIC ACID IMMERSION TEST (ASTM D 1939)

This method evaluates the residual stresses in extruded or molded ABS parts by immersing them in glacial acetic acid and observing the effect of immersion. The presence of excessive residual stresses in ABS parts is indicated by the cracking of the specimen upon immersion in glacial acetic acid. This test is very useful in determining residual stresses in ABS parts that are going to be plated. The plating process requires the parts to be stress-free; therefore, many processors to ABS plastics have adopted this method as a production control test.

The specimen, regardless of the size, is immersed into reagent grade glacial acetic acid for 30 sec. Immediately after the immersion period, the specimen is removed, rinsed in running water, and dried. The specimen is examined for cracking. The same specimen is reimmersed for an additional 90 sec or a new specimen is immersed for 2 min. After rinsing and drying, the specimen is again examined for cracks. The time taken to develop cracks and the degree of cracking indicates the magnitude of residual stress in the specimen.

12.8. END-PRODUCT TESTING

In spite of numerous field failures and ever-increasing product liability problems, processors of plastic products continue to neglect end-product testing. Too much emphasis is placed on testing raw materials and blaming raw material suppliers for providing substandard material, and not enough importance is given to end-product testing. All major and minor raw material suppliers are well-equipped with sophisticated quality control equipment and generally adhere to strict quality standards of their own. Plastics are polymerized under specified conditions in controlled environments. Plastics are not processed under controlled conditions. In the injection-molding process, for example, there are at least six major variables and numerous minor variables that can affect the quality of the molded part. Material temperature, injection pressure, mold temperature, regrind/virgin mix, injection speed, and packing are the six major variables. Too high a melt temperature can cause material to degrade and consequently lose physical properties very rapidly. Even the best quality material cannot save the product from failing if the product is improperly processed. The test data supplied with the raw material are derived by testing certain sized specimens molded under a controlled environment. Unfortunately, plastic parts are not molded under the exact same ideal conditions and they vary in size. Therefore, such data are basically of little or no value to a processor from the standpoint of end-product testing. The quality of the molded product cannot be assessed from raw material data provided by the material supplier without proper consideration of molding variations. One other factor to consider is the molded-in stresses that are usually present in all parts, depending upon the part geometry, mold design, and molding practices. These stresses, unless relieved by annealing, can cause warpage and premature failure.

It is very clear from the foregoing discussion that end-product testing is imperative if one is to control the quality of the product going out in the field. End-product testing offers numerous advantages. First, it protects consumers from premature product failure and resulting possible injuries. The manufacturer is equally protected by not having to worry about product liability suits. Second, it verifies the manufacturing process for any possible mishaps. More importantly, it reduces unnecessary misunderstandings between the custom processor and the buyer of the product. An end-product testing specification generated by a buying party helps the custom processor to understand the product requirement and assures the buyer of receiving a quality product. As we all know, just because the part looks right visually, this does not necessarily mean it has adequate physical properties. Such properties can only be verified by actual testing of the product.

There are many ways to perform end-product testing. The following is a partial list of common end-product test.

Simulated actual use (functional) testing
Impact testing
Torque (shear) test
Crush test
Pull test
Chemical test

Many companies devise their own end-product tests that simulate actual use conditions. This requires designing the proper fixtures and test equipment and training personnel. Energy-to-break test is the best indication of long-term mechanical performance of a product. This can be accomplished by testing the end-product for impact as well as crush resistance. The impact test, however, is preferred over the crush test. The torque test is often used to verify shear strength of threaded components. The test is fairly simple and requires only a torque wrench and suitable fixtures. A simple pull test is employed to determine the force required to pull apart two components. Figures 12-12 and 12-13 illustrate a torque test apparatus and an inexpensive pull tester. Chemical tests such as acetic acid and acetone immersion tests (see Sections 12.6 and 12.7) are useful in verifying processing conditions.

Many end-product tests are self-devised and so is the equipment. Care must be taken in devising such tests to ensure that they are reliable and the test results are reproducible. Whenever possible, commercially available testing equipment should be employed and guidelines and procedures must be established and followed.

12.9. ASH CONTENT (ASTM D 5630, ASTM D 2584, ISO 3451)

This test method was developed to determine the inorganic content of inert substances used as fillers and reinforcement such as minerals and glass fibers in plastics. The test method is based on a loss in weight of a thermoplastic sample that is heated and combusted to oxidize all organic matter. A small sample weighing approximately 2 g in powder or pellet form is necessary for this test. The specimen is accurately weighed and placed into a pre-dried crucible. The crucible is then placed in a muffle furnace at 600°C until the entire polymer has burned off. The crucible is allowed to cool in a desiccator for at least 30 min. The difference between the weight of the specimen before and after the burn-off and degree of weight loss

Figure 12-12. Torque tester. (Courtesy of Mountz, Inc.)

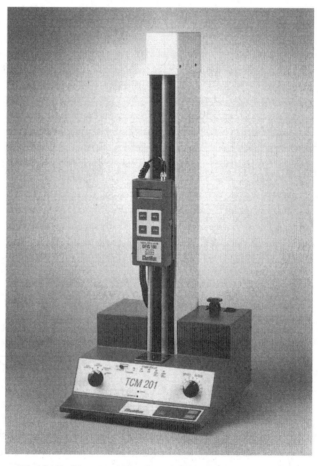

Figure 12-13. Pull tester. (Courtesy of John Chatilon and Sons, Inc.)

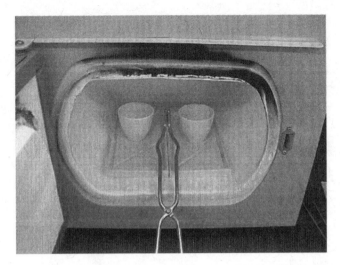

Figure 12-14. Typical ash content test setup. (Courtesy Plastics Technology Laboratory, Inc.)

is calculated to determine percentage of additive in the polymer composition. The residual ash can be viewed under the microscope to determine and identify the basic content such as mineral or fiber. Figure 12-14 shows a typical ash content test setup.

REFERENCES

1. Mentovay, L. W. and Yasenchak, L. P., "Capillary Flow Analysis," *Plast. Design and Processing* (March 1973), p. 18.
2. Technical literature, *Catalog Section C, Rheology*, Gardner Laboratory, Silver Spring, MD, Sept. 1976.
3. Sward, G. G., "Paint Testing Manual," *Am. Soc. for Testing and Materials* Philadelphia, PA, 1972, pp. 181–185.
4. *Ibid.*, pp. 184–185.
5. "Design Guide," *Modern Plastics Encyclopedia*, McGraw-Hill, New York, 1979–1980.

GENERAL REFERENCES

Technical Literature on *Torque Rheometer*, C. W. Brabender Instruments, Inc. South Hackensack, NJ.

Park, R. A., "Characterizing Fluid Plastics by Torque Rheometer," *Plast. Eng.* (Nov. 1976), p. 59.

Allen, E. O. and Willium, R. F., "Prediction of Polymer Processing Characteristics Using C. W. Brabender Plasticorder Torque Rheometer," *SPE ANTEC* (1971), p. 587.

13

IDENTIFICATION ANALYSIS OF PLASTIC MATERIALS

13.1. INTRODUCTION

Plastic products are manufactured using a variety of processing techniques and materials. It is practically impossible to identify a plastic material or product by a visual inspection or a simple mechanical test. There are many reasons that necessitate the identification of plastics. One of the most common reasons is the need to identify plastic materials used in competitive products. Defective products returned from the field are quite often put through rigorous identification analysis. Sometimes it is necessary to identify a finished product at a later date in order to verify the material used during its manufacture. The custom compounders of reprocessed materials may also need to identify already processed material purchased from different sources. Quite often, processors find substantial quantities of plastic material, hot stamp foils, and decals in the warehouse without any labels to identify the particular type. A little knowledge of the identification process can save time and money.

On a rare occasion, the buyers of molded parts may choose to verify the material specified in the product by performing a simple identification analysis. The development of new material is another reason for such analysis.

There are two ways plastic materials can be identified. The first technique is simple, quick, and inexpensive. It requires very few tools and little knowledge of plastic materials. The second approach is to perform a systematic chemical or thermal analysis. The latter technique is very complex, time-consuming, and expensive. The results can only be interpreted by a person well-versed in polymer chemistry. Plastic materials are often copolymerized, blended, and modified with filler or compounded with different additives such as flame retardants, blowing agents,

Handbook of Plastics Testing and Failure Analysis, Third Edition, by Vishu Shah
Copyright © 2007 by John Wiley & Sons, Inc.

lubricants, and stabilizers. In such cases, simple identification techniques will not yield satisfactory results. The only true means of positive identification is a complex chemical or thermal analysis.

The first technique is laid out in a flowchart for easy step-by-step identification by process of elimination. This is shown in the Plastics Identification Chart, which appears on pages 304–305 and is also included as a seperate insert. There are some basic guidelines one must follow in order to simplify the procedure. The first step is to determine whether the material is thermoplastic or thermoset. This distinction is made by simply probing the sample with a soldering iron or a hot rod heated to approximately 500°F. If the sample softens, the material is thermoplastic. If not, it is thermoset. The next step is to conduct a flame test. It is desirable to use a color-less Bunsen burner. A matchstick can also be used in place of a Bunsen burner. However, care must be taken to distinguish between the odor of the mater-ials used in the match and the odor given off by burning plastic materials. Before commenc-ing the burning test, it is advisable to be prepared to write down the following observations:

1. Does the material burn?
2. Color of flame
3. Odor
4. Does the material drip while burning?
5. Nature of smoke and color of smoke
6. The presence of soot in the air
7. Self-extinguishes or continues to burn
8. Speed of burning—fast or slow

To identify the material, compare the actual observations with the ones listed in the flowchart. The accuracy of the test can be greatly improved by performing similar tests on a known sample. While performing the identification tests, one must not overlook safety factors. The drippings from the burning plastic may be very hot and sticky. After extinguishing the flame, inhale the smoke very carefully. Certain plastics like acetals give off a toxic formaldehyde gas that may cause a severe burning sensation in the nose and chest.

The results of the simple identification technique can be further confirmed by the following tests:

1. Melting-point test
2. Solubility test
3. Copper wire test
4. Specific gravity test

13.1.1. Melting Point Determination

Two basic methods are used for melting point determination. For the first method, a Fisher–Johns melting point apparatus as shown in Figure 13-1 is most commonly used. The apparatus consists of a rheostatically controlled heated block, a

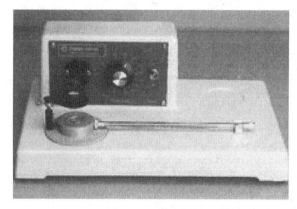

Figure 13-1. Fisher–Johns melting point apparatus. (Courtesy of Fisher Scientific Company.)

thermometer, and a viewing magnifier. A small pellet or a sliver of the plastic material to be tested is placed on the electrically heated block along with a few drops of silicone oil. A cover glass is placed over the material and the heat is gradually increased until the sample material melts or softens enough to deform.

The meniscus formed by the oil is viewed through the magnifier. The temperature at which the meniscus moves is considered the melting point. The expected accuracy of the test is within ±5°F of the published literature value. This method can be used for both crystalline and amorphous plastics. All crystalline plastics have a sharp melting point and the transition is much easier to detect. In contrast, amorphous plastics melt over a wide range and an exact melting point is difficult to determine.

The second method, known as the Kofler method, is used only for semicrystalline polymers. It consists of heating the sample by hot-stage unit mounted under a microscope and viewing it between crossed polarizers. When crystalline material melts, the characteristic double refraction from the crystalline aggregates disappears. The point at which the double refraction or birefringence (typically a rainbow color) completely disappears is taken as the melting point of the polymer. Figure 13-2 shows a commercially available apparatus. The use of a control sample for comparison is particularly helpful in both methods.

13.1.2. Solubility Test

The behavior of plastic materials in various organic solvents often indicates the type of material. The solubility data found in the literature is of a general nature and consequently difficult to use at times. A partial solubility of some plastics in different solvents and a high concentration of additives such as plasticizers further complicate identification by the solubility test. However, a solubility test is very useful in distinguishing between the different types of the same base polymer. For example, cellulose acetate can be distinguished from cellulose acetate butyrate because the acetate is completely soluble in furfuryl alcohol whereas the butyrate is only partially soluble (1). Types of nylons and polystyrenes can be identified

Figure 13-2. Kofler method, melting point determination apparatus. (Courtesy of Ceast U.S.A.)

similarly. The solubility test is best conducted by placing a silver of the sample in a small test tube, adding the solvent, and gently stirring it. Ample time should be allowed before passing judgment regarding solubility of the sample in a particular solvent.

13.1.3. Copper Wire Test

The presence of chlorine such as in polyvinyl chloride can easily be confirmed by simply conducting the copper wire test. The tip of the copper wire should be heated to a red-hot temperature in a flame. A small quantity of material is picked up by drawing the wire across the surface of the sample. The tip of the wire is returned to the flame. A green-colored flame indicates the presence of chlorine in the material. Fluorocarbons can also be identified by detecting the presence of fluorine.

13.1.4. Specific Gravity Test

The increasing used of plasticizers, fillers, reinforcing agents, and other additives makes the identification of plastics by the specific gravity test very difficult. The test is described in detail in Chapter 10.

13.2. ADVANCED METHODS FOR IDENTIFICATION

As discussed in introduction section of this chapter, a complete and positive identification analysis of plastics materials is very complex and time-consuming task, requiring in-depth knowledge of analytical chemistry, experience, and sophisticated equipment. Plastics materials are often blended, copolymerized, and modified with various additives to enhance the performance. These modifications alter the fundamental behavior of the material in terms of burning characteristics, odor, and color of flame, making simple identification techniques unacceptable for positive identification. Furthermore, if the amount of sample available is very small, only advanced methods described in this chapter can be employed. Only a few milligrams of material are necessary for spectroscopy, thermal analysis, microscopy, or chromatography. The following techniques are those most commonly used today to positively identify plastic materials and additives. Table 13-1 lists the advanced identification techniques used for identifying polymers and additives.

The following advanced methods are used for the positive identification of plastics materials.

1. Fourier transform infrared/near infrared (FTIR/NIR)
2. Thermogravimetric analysis (TGA)
3. Differential scanning calorimetry (DSC)
4. Thermomechanical analysis (TMA)
5. Nuclear magnetic resonance (NMR)
6. Chromatography
7. Mass spectrometry

TABLE 13-1. Identification Techniques for Polymer and Additives (2)

Technique	Identification
LC/GPC	Polymer molecular weight distribution. Phenols, phosphites, plasticizers, lubricants
GC	Residual monomers
	Nonpolymeric compounds
	Plasticizers
IR	Polymer type
	Additives
Thermal	Fillers
	Lubricants
	Polymer molecular weight
X-ray	Fillers
	Flame retardants
	Stabilizers
NMR	Polyesters
	Silicones
	Phenols
Wet Chemistry	Lubricants
	Flame retardants
	Catalysts

8. X-ray analysis
9. Microscopy

13.2.1. Fourier Transform Infrared Analysis

Fourier transform infrared (FTIR) analysis is by far the most widely used and accepted technique for plastics identification by the professionals and scientists. The test is conducted simply by passing an infrared beam through a prefabricated sample, whereby some of the infrared radiation is absorbed or transmitted. The resulting spectrum, like a fingerprint unique to a particular polymer, is graphically displayed on a computer screen. Since no two unique molecular structures produce the same infrared spectrum, the resulting spectra can be compared with known material spectra and the material can be positively identified.

Fast identification of polymers using near-infrared spectroscopy (NIR) has gained popularity in recent years. The specimen to be identified is illuminated with a near-infrared wavelength ranging from 800 to 2000 nms. Polymer molecules absorb this radiation in different ways, generating unique spectra allowing one to quickly identify a particular polymer. NIR technology offers plastics recyclers low-cost, high-speed alternative to traditional FTIR methods.

13.2.2. Thermogravimetric Analysis

In thermogravimetric analysis (TGA), changes in the weight of a specimen are recorded as the specimen is progressively heated. The test procedure is quite simple. A typical apparatus consists of an analytical balance, a programmable electrical furnace, and a recorder. TGA is very useful in characterizing polymers containing different levels of additives and fillers by measuring the degree of weight loss. For example, the percentage of glass fiber or mineral filler in a polymer can be determined by completely combusting the polymer in an inert atmosphere such that the remaining residue is only glass or inert filler. TGA is also used to identify the ingredients of blended compounds according to the relative stability of individual components.

13.2.3. Differential Scanning Calorimetry (DSC)

Differential scanning calorimetry (DSC) measures the amount of energy absorbed or released in terms of heat by a sample as it is heated or cooled or held at a constant temperature. DSC is one of the most powerful techniques available for determining melting profiles of polymers including glass transition temperature, melting point, crystallization temperature, and degradation temperature. In the case of a semicrystalline polymer, many important physical properties are dependent upon crystallinity. DSC measurements provide important information concerning percent crystallinity along with rate of crystallization. The presence or absence of anti-oxidant in a polymer along with oxidative stability can be determined with this technique. DSC can be used to determine the ratio of polymer in a blend, block, or random copolymer through measuring the thermal characteristics of the melting region. Many applications requiring quantitative determination of additives such as mold release agents, antistats, UV stabilizers, and impact modifiers can be achieved through the use of DSC technique. By studying a typical DSC thermogram, one can characterize a polymer by studying (a) the behavior of the material all the way from glass transition temperature to degradation temperature and (b) any changes that take place between the two extremes.

13.2.4. Thermomechanical Analysis (TMA)

TMA is designed to measure the physical expansion or contraction of the material, or changes in its modulus or viscosity, as a function of temperature. It is used to identify softening points and characterize viscoelastic performance over a temperature range. Thermomechanical analysis is fairly simple, and it consists of applying a constant force with a probe and measuring dimensional changes in vertical direction as the sample is heated under a no-load or fixed-load condition. TMA is very useful in polymer characterization, to determine properties such as melting point, glass transition temperature, crosslink density, degree of crystallization, and coefficient of expansion accurately.

13.2.5. Nuclear Magnetic Resonance (NMR)

NMR spectroscopy is a very powerful analytical technique for identifying organic molecules and determining their chemical structures. Some of the nuclei in a molecule possess the property of spin, which can be in one of two orientations. If a magnetic field is applied to the molecule, these differences in spin cause the nuclear energy level to split. The molecule is then subjected to an additional weak, oscillating magnetic field. At a precise frequency the nuclear magnets resonate, and it is this that is recorded and amplified (3). NMR provides characterization of compound structure as well as absolute identification of specific ingredients in simple mixtures. The technique can also provide a general characterization by functional group that cannot be obtained by other analytical methods (4). In the plastics-polymer field, carbon-13 NMR is most widely used for identification analysis. Identification of monomeric substances such as plasticizers, stabilizers, and lubricants from their NMR spectra is relatively simple and straightforward.

13.2.6. Chromatography

Chromatography is an analytical technique used to separate the components of a mixture, which flow at different rates along some separating medium. The fixed material over which the mixture passes is called the stationary phase and is usually solid or gel. The moving fluid, often a liquid but sometimes a gas, is called moving phase. The mixture is dissolved in a solvent called eluent and passed through a column or series of columns. The different strengths of the interatomic force between the molecules being separated and the molecules of the stationary phase and the eluent are responsible for the separation of the mixture. The different components may be identified and in some cases may be determined quantitatively (5).

Both gas chromatography (GC) and liquid chromatography (LC) are used for identification analysis. However, gel permeation chromatroghy (GPC) is most used and accepted in the plastics industry.

13.2.7. Mass Spectrometry (MS)

A mass spectrometer is a very useful tool in obtaining detailed structural information about a polymeric compound from very minute quantity of material. The molecular weight and atomic composition can be determined by studying and evaluating the spectra of organic compounds, and additives are positively identified. The combined use of GC (gas chromatography) and MS (mass spectrometry), called GCMS, provides even more powerful tool for identification analysis. In mass spectrometry the substance to be analyzed is heated and placed in a vacuum. The resulting vapor is exposed to a beam of electrons which causes ionization to occur, either of the molecules or their fragments; the ions thus produced are accelerated by an electric impulse and then passed through a magnetic field, where they describe curved paths whose direction depend on the speed and mass-to- charge ratio of the ions. This has the effect of separating the ions according to their mass (electromagnetic separation). Because of their greater kinetic energy, the heavier ions describe a wider arc than the lighter ones and can be identified on this basis. The ions are collected in appropriate devices as they emerge from the magnetic field (6).

13.2.8. X-Ray Fluorescence

X-ray fluorescence spectrometry is primarily used for qualitatively and quantitatively identifying the additives used in a variety of polymers, determining the presence of impurities and contaminants in a compound, and detecting and determining trace elements in polymers and monomers.

Two types of instruments are available for X-ray fluorescence spectrometry: wavelength-dispersive (WDXRF) and energy-dispersive (EDXRF).

13.2.9. Microscopy

Optical microscopy (OM) excels in providing information about surface morphology, identifying contaminants and impurities, and analyzing polyblends and alloys. The technique is used extensively for study of elemental composition in thin films. Optical microscopy is further divided into two classes, SEM (scanning electron microscopy) and TEM (transmission electron microscopy), the latter being capable of providing greater resolution capabilities. The image can be magnified over 100,000 times from the original object. SEM image is produced by scanning a finely focused electron beam over a surface and using the scattered secondary electrons to create highly magnified image of the material surface (7). By contrast, transmitting the electrons through the specially prepared specimen produces a TEM image.

In special cases, more advance techniques such as SPM (scanning probe microscopy) and AFM (atomic force microscopy) are used.

13.3. IDENTIFICATION OF PLASTIC MATERIALS

13.3.1. Thermoplastics

A. ABS

ABS is an amorphous terpolymer with a specific gravity of 1.04. It is soluble in toluene and ethylene dichloride. It burns with a yellow flame and an acrid odor and continues to burn after the removal of the flame source with black smoke, soot, and drippings. Infrared spectroscopy is the best method for the positive identification of ABS. The same technique can also be used to further identify a percentage of each polymer in the terpolymer composition.

B. Acetal

Acetal is available as a homopolymer and a copolymer. Both types burn with a blue flame and give off a toxic formaldehyde odor. They are both crystalline with a specific gravity of 1.41. The homopolymer can be distinguished from the copolymer by a Fisher–Johns melting-point test. The copolymer acetal has a lower melting point than the homopolymer. Both types are soluble in hexafluoroacetone sesquihydrate. However, owing to the extreme toxicity of this solvent, it is rarely used for identification purposes. A differential thermal analysis (DTA) can also be used to distinguish a homopolymer from a copolymer.

C. Acrylic

Acrylic is an amorphous polymer with a specific gravity of 1.18. Acrylic is one of the few transparent plastics used in outdoor applications. It burns with a blue flame with a yellow tip, gives off a fruity odor, and does not produce smoke while it burns. Acrylic is soluble in acetone, benzene, and toluene. Infrared spectroscopy is the best means of identification of this polymer.

D. Cellulose Acetate

Cellulose acetate is an amorphous material with a specific gravity of 1.30. It burns very slowly with a yellow flame, drips, gives off an odor of vinegar, and has black smoke with soot. It is soluble in acetone, furfuryl alcohol, and acetic acid. Cellulose acetate is positively identified by infrared spectroscopy.

E. Cellulose Acetate Butyrate

This polymer burns with a yellow flame with a blue tip, drips, and gives off an odor of rancid butter. The specific gravity is 1.24, and it is soluble in acetone and trichloromethane. Cellulose acetate butyrate is also one of the few weatherable transparent plastics.

F. Cellulose Propionate

The cellulose propionate is also an amorphous material like other celluloids with a specific gravity of 1.20. It burns with a dark yellow flame and an odor of burnt sugar. The material drips while burning and the drippings also burn like other celluloids. Acetone, carbon tetrachloride, and trichloromethane are the most common solvents.

G. Fluorocarbons (FEP, CTFE, PTFE, PVF)

Fluorocarbon plastics do not actually burn when exposed to a flame. They can easily be identified by a copper wire test that indicates the presence of fluorine by a bright green-colored flame. The fluoroplastics have a very high melting point. PTFE has a waxy surface and a specific gravity of 2.15. Fluoroplastics are practically impossible to dissolve in any chemical.

H. Nylons

There are many types of nylons with a specific gravity ranging from 1.04 to 1.17. All types burn with blue flames and yellow tips and give off a burnt wool or hair odor. Nylon self-extinguishes on removal of the flame. Phenol, *m*-cresol, and formic acid are the most common solvents. Different types of nylons can be identified by the Fisher–Johns melting-point test (ASTM D 789). Solubility and specific gravity tests are also used to differentiate between the types of nylons. However, infrared spectroscopy is the best method for positive identification.

I. Polycarbonate

Polycarbonate is one of the toughest transparent thermoplastics. It has a specific gravity of 1.2, burns with a yellow or orange flame with the odor of phenol, and gives off black smoke with soot in the air. Polycarbonate is soluble in methylene dichloride and ethylene dichloride. Infrared spectroscopy is the best method of identification for polycarbonate.

J. Thermoplastic Polyester

Thermoplastic polyester burns with a yellow flame with a blue edge, and as it burns it drips and gives off black smoke with soot. The characteristic odor is one of burning rubber.

K. PVC

This self-extinguishing amorphous plastic has a specific gravity of 1.2–1.7. It burns with a yellow flame and a green tip, and gives off the odor of hydrochloric acid and white smoke. Tetrahydrofuran and methyl ethyl ketone are common solvents for PVC. PVC can easily be identified by the copper wire test which indicates the presence of chlorine by a bright green flame. There are many different types of vinyl polymers such as polyvinyl acetate, polyvinylidene chloride, and ethylene vinyl acetate. The type of vinyl can be positively identified by infrared spectrometric techniques.

L. Polyethylene

Polyethylene is one of the very few crystalline plastics that will float on water. The specific gravity ranges from 0.91 to 0.96. It burns quickly with a blue flame with a yellow tip. Polyethylene drips while it burns and gives off a paraffin odor similar to a burning candle. Polyethylene is impervious to most common solvents. However, it can be dissolved in hot toluene or hot benzene. Infrared spectroscopy is used to confirm the identity of polyethylene.

M. Polypropylene

This polyolefin family polymer has a specific gravity of 0.88 and floats on water. It burns with a blue flame with a yellow tip and gives off an acrid odor similar to diesel fumes. The material is soluble in hot toluene. Polypropylene is positively identified by an infrared spectrometer.

N. Polystyrene

Polystyrene is an amorphous plastic with a specific gravity of 1.09. It burns with a yellow flame and gives off the odor of illuminating gas or marigolds. The material burns and drips, creating a dense black smoke with soot in the air, also dripping while continuing to burn. Polystyrene is soluble in acetone, benzene, toluene, and ether. Polystyrene is often modified with rubber to improve its impact strength. These different types of polystyrenes can be identified by infrared spectroscopy.

O. Polyphenylene Oxide (PPO)

PPO has a specific gravity of 1.06. It burns with a yellow-orange flame without dripping and gives off an odor of phenol or burning gas. It exhibits self-extinguishing characteristics. PPO is soluble in toluene and dichloroethylene. Like other polymers, PPO can be identified by infrared spectroscopy.

P. Polysulfone

This self-extinguishing amorphous thermoplastic has a specific gravity of 1.24. It burns with an orange flame, gives off the pungent odor of sulfur, and produces black smoke with soot in the air while it burns. Polysulfone is soluble in methylene chloride.

Q. Polyurethane (Thermoplastic)

This generally easy-to-ignite plastic burns with a yellow flame with a faint odor of apple. It produces black smoke. Thermoplastic polyurethane is soluble in tetrahydrofuran and dimethyl formamide. The specific gravity of this material is 1.2.

R. Polyphenylene Sulfide (PPS)

This self-extinguishing thermoplastic material burns with a yellow-orange flame and gives off an odor of a faint rotten egg caused by sulfur dioxide off-gassing from the material. PPS burns very slowly and does not drip while burning. It gives off a gray smoke with traces of black. The resulting char is black and glossy. The sample generally deforms slightly with some swelling. PPS is positively identified by Differential Thermal Analysis.

13.3.2. Thermosetting Plastics

A. Diallyl Phthalate (DAP)

The DAP compound's specific gravity ranges from 1.30 to 1.85, depending upon the type of filler. It burns with a yellow flame, produces black smoke, and gives off the faint odor of phenol. It is self-extinguishing in nature and is difficult to ignite.

B. Epoxy

The specific gravity of epoxy compounds ranges from 1.10 to 2.10. It burns with a yellow flame and gives off a black smoke that gives off a pungent amine odor. Different types of epoxies can be characterized by infrared spectroscopy and thermal analysis.

C. Phenol Formaldehyde

Phenol formaldehyde burns with a yellow flame and gives off the odor of phenol. Depending upon the type of filler, it may or may not be self-extinguishing. Phenolic compounds are only available in dark colors. The specific gravity ranges from 1.30 to 1.90. Phenol formaldehyde is soluble in acetone and acetic acid.

D. Urea Formaldehyde
Urea formaldehyde burns with a yellow flame with a greenish blue edge. It gives off a strong odor of formaldehyde and the material usually swells and cracks as it continues to burn.

E. Melamine Formaldehyde
This self-extinguishing thermoset has a specific gravity of 1.47 to 1.8. It burns with a yellow flame with a blue edge and gives off a fishlike odor. The material swells, cracks, and turns white at the edges of a burned section. Melamine formaldehyde is soluble in acetone and cyclohexanone.

F. Polyesters
Polyesters burn with a yellow flame with blue edges and produce black smoke with soot in the air. They continue to turn without dripping after the removal of the ignition source and give off a sour cinnamon odor. The specific gravity ranges from 1.30 to 1.50.

G. Silicones
Silicones burn with a bright yellow flame without odors and continue to burn after the removal of the flame soruce. The specific gravity ranges from 1.05 to 2.82. The positive identificaiton is made by an infrared spectrometer.

13.3.3. Elastomers

A. Styrenic TPE (S-TPE)
Styrenics derive their elastomeric properties from their two-phase chemical structure. The polymeric phase, a styrene that is hard at room temperature, is dispersed in a softer elastomeric phase (8). The elastomeric phase generally consists of butadiene or isoprene or olefin pair such as ethylene-propylene or ethylene-butylene. Styrenic TPEs are one of the most widely used elastomers because of their ability to be processed on conventional equipment and comparable properties to thermoset rubbers. The specific gravity ranges from 0.9 to 1.2. Hardness ranges from 20 to 95 shore A. S-TPEs exhibit good low-temperature performance, good abrasion resistance, good oxidative and weathering resistance, and fair compression set. S-TPEs are used in a variety of applications, including brightly colored handles and grips for sporting goods, kitchen utensils, personal care products, and medical devices.

S-TPEs burn with relative ease and speed, produce yellow color flame with black smoke, and give off rubber odor. Two most common methods for positive identification are PY-GC and FTIR.

B. Polyurethanes (TPUs)
Two types of thermoplastic polyurethane elastomers have been developed: polyester-based and polyether-based. At similar hardness, polyester-based urethane will exhibit better toughness, oil resistance, physical characteristics, and outstanding abrasion resistance. The polyether type has better hydrolytic stability

PLASTICS IDENTIFICATION CHART

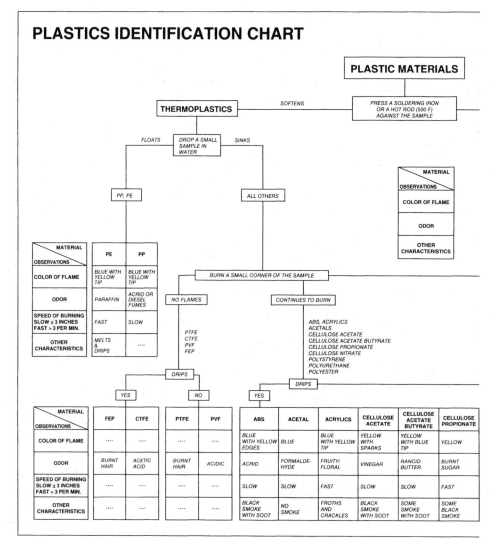

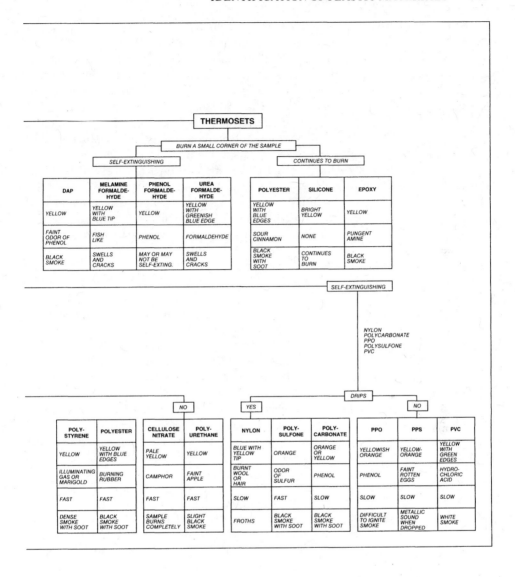

THERMOSETS

BURN A SMALL CORNER OF THE SAMPLE

SELF-EXTINGUISHING

	DAP	MELAMINE FORMALDE-HYDE	PHENOL FORMALDE-HYDE	UREA FORMALDE-HYDE
	YELLOW	YELLOW WITH BLUE TIP	YELLOW	YELLOW WITH GREENISH BLUE EDGE
	FAINT ODOR OF PHENOL	FISH LIKE	PHENOL	FORMALDEHYDE
	BLACK SMOKE	SWELLS AND CRACKS	MAY OR MAY NOT BE SELF-EXTING.	SWELLS AND CRACKS

CONTINUES TO BURN

POLYESTER	SILICONE	EPOXY
YELLOW WITH BLUE EDGES	BRIGHT YELLOW	YELLOW
SOUR CINNAMON	NONE	PUNGENT AMINE
BLACK SMOKE WITH SOOT	CONTINUES TO BURN	BLACK SMOKE

SELF-EXTINGUISHING

NYLON
POLYCARBONATE
PPO
POLYSULFONE
PVC

DRIPS

NO

	POLY-STYRENE	POLYESTER	CELLULOSE NITRATE	POLY-URETHANE
	YELLOW	YELLOW WITH BLUE EDGES	PALE YELLOW	YELLOW
	ILLUMINATING GAS OR MARIGOLD	BURNING RUBBER	CAMPHOR	FAINT APPLE
	FAST	FAST	FAST	FAST
	DENSE SMOKE WITH SOOT	BLACK SMOKE WITH SOOT	SAMPLE BURNS COMPLETELY	SLIGHT BLACK SMOKE

YES

NYLON	POLY-SULFONE	POLY-CARBONATE
BLUE WITH YELLOW TIP	ORANGE	ORANGE OR YELLOW
BURNT WOOL OR HAIR	ODOR OF SULFUR	PHENOL
SLOW	FAST	SLOW
FROTHS	BLACK SMOKE WITH SOOT	BLACK SMOKE WITH SOOT

NO

PPO	PPS	PVC
YELLOWISH ORANGE	YELLOW-ORANGE	YELLOW WITH GREEN EDGES
PHENOL	FAINT ROTTEN EGGS	HYDRO-CHLORIC ACID
SLOW	SLOW	SLOW
DIFFICULT TO IGNITE SMOKE	METALLIC SOUND WHEN DROPPED	WHITE SMOKE

and fungus resistance along with better low-temperature flexibility and higher resilience. The many formulations of urethanes result in elastomers ranging in hardness from 65 shore A to as high as 80 shore D. Elongation can be as high as 750% for the softer materials. These properties are accompanied by tensile strength ranging from 5000 to as high as 8000 PSI, by far the highest of any elastomeric material, thermoset or thermoplastic (9). TPUs service temperature ranges from −60°F to 270°F, and specific gravity ranges from 1.1 to 1.3. Key applications include wheels and casters, medical devices, automotive under-the-hood components, shoe bottoms, wire and cable jackets, O-rings, diaphragms, and seals.

Both types of TPUs dissolve in cyclohexanone, tetrahydrofuran, and pyridine. The polyether type burns with yellow flame with strong acrid odor and gray smoke. The material burns fast and completely and eventually chars. The polyester type burns slowly with yellow-orange flame. For positive identification, FTIR is most commonly used. FTIR as well as differential scanning calorimetry can differentiate between the two types of TPUs. C^{13} NMR has been proven to be useful in determining types of polyols.

C. Copolyesters (COPS)

The polyester elastomer, well known by the brand name Hytrel® manufactured by DuPont, are block copolymers composed of alternating hard and soft segments, polyalkylene terephthalate and polyalkylene ether, respectively (10). Hardness ranges from 35 to 72 shore D, which is closer to flexible thermoplastic than truly rubbery elastomers. COPS are considered high-performance elastomers due to their excellent toughness, impact resistance, wide usable temperature range of −80°F to 250°F, outstanding flex fatigue resistance, and excellent solvent resistance. COPS are often specified in applications requiring mechanical strength and durability along with flexibility. Examples include seals, belts, bushings, pump diaphragms, gears protective boots, hose and tubing, springs, and impact absorbing devices. COPS can be processed using conventional processing techniques, including injection molding, extrusion, blow molding, and rotational molding.

Polyester elastomer is soluble in cresol, concentrated sulfuric acid, and trifluoroacetic acid. Specific gravity ranges from 1.17 to 1.25. The elastomer burns very slowly with a yellow flame with blue tip. It gives off sooty smoke and a sweet characteristic odor. Infrared spectroscopy and DSC are preferred method for positive identification of the elastomer.

D. Olenfinics (Thermoplastic Elastomeric Olefins—TEOs)

TEOs, also referred to as thermoplastic polyolefin elastomers (TPOs), are a simple blend of two polymer systems, each with its own phase. The harder polymer system is a polyolefin—commonly polypropylene or polyvinyl chloride. The softer of this polymer is an elastomer, ethylene propylene rubber with PP or nitrile-butadiene with PVC, with little or no crosslinking. The fine dispersion of the two phases provides sufficient interphase surface contact for the attractive forces between the phases to be relatively strong (11). The molded or extruded materials have good tensile strength and elongation and excellent low-temperature flexibility. TEOs have the lowest specific gravity of all thermoplastic elastomers, with values ranging from 0.9 to 1.02. Hardness ranges from 50 shore A to 60 shore D. They have excellent weathering properties and good chemical resistance and are relatively inexpensive. The service temperature ranges from −75°F to 220°F. Typical applications include wire and cable insulations, automotive components, sporting goods, and flexible packaging. TEOs burn slowly with a blue flame with yellow tip and give off characteristic partition wax odor. The advanced methods for positive identification of the elastomer include FT-IR, DSC, and PY-GC. Solvent extraction technique is also used to extract certain additives used in elastomers prior to carrying out elastomer component identification.

E. Thermoplastic Vulcanizates (TPVs)

The TPVs are very similar to TEOs except that the elastomer phase is highly crosslinked and finely divided in a continuous matrix of polyolefin. The elastomeric phase is vulcanized during the manufacturing process, resulting in improved performance in areas such as compression set and thermal stability. TPVs exhibit the lowest compression set properties of all the TPE classes (12). They are available in a wide range of hardness and show excellent chemical resistance, colorability, oil resistance, and superior flex fatigue properties. TPVs can be processed using all conventional processing methods including injection molding, extrusion, blow molding, and calendering. They offer an attractive economical alternative to thermoset elastomers in a wide variety of markets including automotive, construction, appliance, medical, and sporting goods. Key applications are: industrial hose, caster wheels, gaskets, diaphragms, seals, and under-the-hood automotive components. TPV's specific gravity range is from 0.9 to 1.0. They burn with large yellow tip and blue base and black smoke. They burn fast and give off burning rubber odor. Advance identification techniques such as infrared spectroscopy, PY-GC, and DSC are commonly used for positive identification.

REFERENCES

1. E. I. DuPont Co., *Technical Bulletin: Identification of Thermoplastic Materials*, Wilmington, DE.
2. Coe, G. R., "Instrumental Methods of Polymer Analysis," *SPE ANTEC*, **23** (1977), pp. 496–499.
3. Waites, G. and Harrision, P., *Dictionary of Chemistry*, Cassell Wellington House, London, 1998, p. 139.
4. Cheremisinoff, N. P., *Polymer Characterization, Laboratory Techniques and Analysis*, Noyes Publications, Westwood, NJ, 1996, p. 60.
5. Waites, Reference 2.
6. Sax, N. I. and Lewis, R. J., *Condensed Chemical Dictionary*, Van Nostrand Reinhold, New York, 1987.
7. Portnoy, R., *Medical Plastics—Degradation Resistance and Failure Analysis*, Plastics Design Library, Norwich, NY, 1988, p. 22.
8. Marshall, J., "Thermoplastic Elastomers," *Modern Plastics Encyclopedia*, Nov. 1997.
9. Schwartz, S. and Goodman, S., *Plastics Materials and Processes*, Van Nostrand Reinhold, New York, 1982, p. 446.
10. Rader, C. P., Reference 1.
11. Rader, C. P., Reference 1.
12. Level, M. A., "Thermoplastic Elastomers," *Plastics Eng.*, January, 2003 p. 94.

GENERAL REFERENCES

"Identification of Plastics," *Modern Plastics Encyclopedia*, McGraw-Hill, New York, 1950, pp. 992–1001.

"How to Identify Plastics," *Western Plast. Mag.* (March 1966).

"Identification Chart," *Canad. Plast. Mag.* (1971).

Richardson, T. A., *Modern Industrial Plastics*, Howard W. Sams and Co., Inc., Indianapolis, IN, 1974, pp. 84–90.

Lever, A. E. and Rhys, J. A., *The Properties and Testing of Plastic Materials*, Temple Press, Feltham, England, 1968, pp. 269–275.

Haslam, J. and Willis, H. A., *Identification and Analysis of Plastics*, Iliffe Books, London, England, 1965.

Rodriguez, F., *Principles of Polymer Systems*, McGraw-Hill, New York, 1970, Chapter 15, pp. 464–475.

Kline, G. M., *Analytical Chemistry of Polymers*, Parts 1–3, Interscience, New York, 1962.

Beck, R. D., *Plastics Product Design*, Van Nostrand Reinhold, New York, 1970, pp. 432–440.

14

TESTING OF FOAM PLASTICS

14.1. INTRODUCTION

Cellular material (foam) is a generic term for materials containing many cells (either open, closed, or both) dispersed throughout the mass. The increasing use and popularity of foam plastics has created a need for developing test methods particularly suited to foamed plastics. For a long time, the standard test methods developed for solid plastics were employed to determine the properties of cellular plastics. The test methods had to be modified to a degree because of the lower overall strength of the cellular plastic materials. Such modifications and changes created numerous nonstandard test methods which in turn created more confusion among the designers and users of cellular plastics. Through a painstaking effort of ASTM committees, SPI, and material suppliers, many standard test methods have been developed. The majority of the test methods for cellular plastics are very similar to the ones already developed for noncellular plastics. However, some tests are developed to suit the particular needs of cellular plastics and are unique in that respect. The porosity test to measure the open-cell content of rigid foam plastics is one such test developed especially for foam plastics.

14.2. RIGID FOAM TEST METHODS

14.2.1. Density (ASTM D 1622, ISO 845)

The density of foam plastics is of considerable interest to parts designers since many important physical properties are related to foam density. The procedure to determine the density of cellular plastics is very simple. Basically, it requires conditioning a specimen of a shape whose volume can easily be calculated. The speci-

Handbook of Plastics Testing and Failure Analysis, Third Edition, by Vishu Shah
Copyright © 2007 by John Wiley & Sons, Inc.

men is weighed on a balance or a scale. Next, the specimen volume is calculated by measuring length, width, and height using a micrometer, a dial gauge, or a caliper. The density is calculated as follows:

$$\text{Density } (\text{lb/ft}^3) = \frac{\text{Weight of specimen (1b)}}{\text{Volume of specimen (ft}^3)}$$

The test is primarily intended for determination of apparent overall density and apparent core density of rigid cellular plastics. Apparent overall density is defined as the weight in air per unit volume of a sample, including all forming skins. Apparent core density is defined as the weight in air per unit volume of a sample, after all forming skin has been removed.

14.2.2. Cell Size

Cell size and cell orientation are very important since several physical properties of rigid cellular plastics depend upon them. For example, determining the water absorption and open-cell content requires knowledge of surface cell volume, which uses cell size values in the calculations.

This test method covers the determination of the apparent cell size of rigid cellular plastics by counting the number of cell-wall intersections in a specified distance. Two basic procedures are employed. Procedure A requires the preparation of a thin slice, not more than one half of the average cell diameter in thickness, that is mechanically stable. The cellular plastic specimen is cut to no more than one half the average cell diameter in thickness on a slicer and the shadowgraph is projected on the screen by the use of a cell-size scale slide assembly and a slide projector. The average chord length is obtained by counting the cells on cell-wall intersections and converting this value to average cell size by mathematical derivation.

Procedure B is intended for use with materials whose friable nature makes it difficult to obtain a thin slice for viewing. The cellular plastic specimen is sliced to provide a smooth surface. The cell walls are accented by the use of a marking pen. The average chord length is obtained by counting the cell-wall intersections and converting this value to average cell size by mathematical derivation.

14.2.3. Open-Cell Content (ASTM D 2856, ISO 4590-1981)

The porosity or percentage of open cells in a cellular plastic is determined by this method. This knowledge is important when specifying cellular plastics for flotation applications, where excessive porosity or high open-cell content can adversely affect flotation characteristics. A high percentage of closed cells, on the other hand, prevents escape of gases and improves insulation characteristics by promoting low thermal conductivity.

Since any conveniently sized specimen can only be obtained by some cutting operation, a fraction of closed cells will be opened during sample preparation and will be included as open cells. Three basic procedures are established to cope with this problem.

Procedure A. Corrects for cells opened during the sample preparation by measuring the cell diameter and allowing for the surface volume.

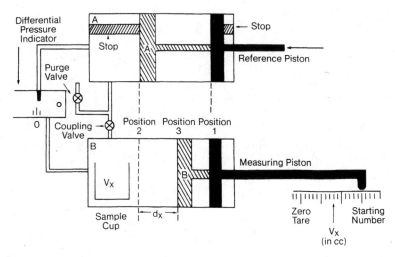

Figure 14-1. Schematic, air pychnometer. (Reprinted with permission of ASTM.)

Procedure B. Corrects for cells opened in sample preparation by cutting and exposing surface areas equal to the surface area of the original sample dimensions.

Procedure C. Does not correct for cells opened during sample preparation and gives good accuracy on highly open-celled materials. The accuracy decreases as the closed cell content increases and as the cell size increases.

The method is based on Boyle's law, which states that at a constant temperature, an increase in volume of a confined gas results in a proportional decrease in pressure. If a chamber size is increased equally with or without material present in the specimen chamber, the pressure drop will be less for the empty chamber. The extent of this difference and the actual volume of the material is a measure of the percentage of closed cells.

The apparatus used in the test is called an air pycnometer, and is schematically illustrated in Figure 14-1. It consists of two cylinders of equal volume with a specimen chamber provided in one. Pistons in both permit volume changes. When the volumes of both cylinders are altered, the volume change for the specimen-containing cylinder is smaller than for the empty chamber because of the presence of the sample. The extent of this difference is measured and a calculation is carried out to determine the open-cell content of the foam sample. A commercially available air pycnometer is shown in Figure 14-2.*

14.2.4. Compressive Properties (ASTM D 1621, ISO 844)

The test to determine compressive strength and compressive modulus of rigid cellular plastics is somewhat similar to the one developed for noncellular plastics and discussed in Chapter 2. The test is very useful in comparing the compressive strengths of various foam plastic formulations. It provides a standard method of obtaining data for research and development, quality control, and verifying specification. However, the test is not a direct indication of how the cellular plastic will

Figure 14-2. Air pychnometer.

behave under actual load over a period of time. Further tests for creep, fatigue, and impact resistance must be conducted when designing a cellular part for load-bearing applications.

The test is carried out using the test specimen of minimum height of 1 in. and maximum height no greater than the width or diameter of the specimen. A pre-conditioned test specimen is subjected to uniform loading using a compression testing machine. The crosshead movement is used as a measure of deflection. The test is continued until a yield point is reached or until the specimen has been compressed approximately 13% of its original thickness, whichever occurs first. From load-deflection curves such as those shown in Figures 14-3 and 14-4, compressive strength, compressive modulus and apparent modulus is calculated.

14.2.5. Tensile Properties (ASTM D 1623)

This method for determining tensile strength, modulus, and elongation of rigid cellular plastics is similar to the one used for noncellular plastics described in Chapter 2. The only variations are in the shape of the specimen and the method of preparation. Three basic types of test specimens are used. The type A specimen is preferred in cases where enough material exists to form the necessary specimen. Type B is used where only smaller specimens such as sandwich panels are available. Type C specimens are used to determine tensile adhesive properties of a cellular plastic to a substrate, as in a sandwich panel, or the bonding strength of a cellular plastic to a single substrate.

14.2.6. Shear Properties (ASTM C 273)

This test was developed specifically for studying the stress–strain behavior of sandwich constructions or cores when loaded in shear parallel to the plane of the facings. The test device illustrated in Figure 14-5 can be used for the

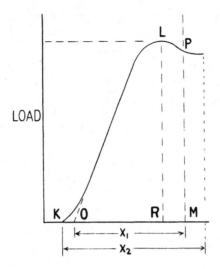

X$_1$ = PROCEDURE A: 10% CORE DEFORMATION
PROCEDURE B: 2% STRAIN
X$_2$ = DEFLECTION (APPROXIMATELY 13%)

Figure 14-3. Compressive strength (stress at yield point below 10 percent deformation). (Reprinted with permission of ASTM.)

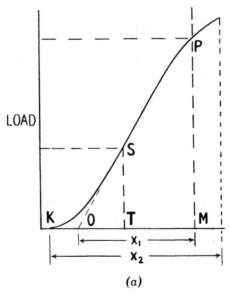

(a)

x$_1$ = Procedure A: 10% Core deformation
Procedure B: 2% Strain

x$_2$ = Deflection (approximately 13%)

Figure 14-4. Compressive strength (stress at 10 percent deformation). (Reprinted with permission of ASTM.)

shear testing of the complete sandwich panel or core materials alone. The load is applied to the ends of steel plates either in tension or compression in such a way that the load is distributed uniformly across the width of the specimen. The maximum load is applied for 3–6min. A stress–strain curve is plotted and from the initial slope, the shear modulus of the sandwich as a unit or core alone is calculated.

14.2.7. Flexural Properties (ASTM D 790, ISO 178)

Flexural strength at break, flexural yield strength, and modulus of elasticity of rigid cellular plastics can be determined by the procedures described in Chapter 2.

14.2.8. Dimensional Stability (ASTM D 2126, ISO 2796)

In recent years, cellular plastics have found numerous applications in the aerospace, electronics, and construction industries. It is important to know how cellular plastics

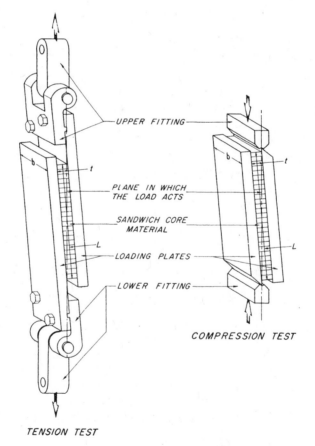

Figure 14-5. Test apparatus—shear property. (Reprinted with permission of ASTM.)

will behave under various conditions of temperature and humidity. Many insulation applications demand a high degree of stability for an extended period of time. The test developed for studying the response of rigid cellular plastics to thermal and aging stress provides a maximum use temperature as well as stability data at the specified temperature and humidity conditions. The test requires the use of a balance, an oven, a cold box, and gauges for accurate measurements of specimens. The test specimens are machined to specified dimensions and accurately measured. After appropriate conditioning, the specimens are subjected to one of the test conditions specified in Table 14-1. Final test measurements are determined after the specimens recover to room temperature. Visual examination and a dimensional check are performed. The maximum use temperature is determined by subjecting the specimens to successively higher temperatures and recording dimensional changes until an unacceptable dimensional change is measured.

14.2.9. Water Absorption (ASTM D 2842, ISO 2896)

Cellular plastics are used extensively in flotation applications because of their ability to maintain a low buoyancy factor. The buoyancy factor is directly affected by the amount of water a particular foam plastic will absorb. The test method developed to determine water absorption of rigid cellular plastics is fully described in the ASTM Standards Manual. Basically, the test consists of determining the volume of initial dry weight of the object and calculating the initial buoyancy force. The object is then immersed in water and, at the end of the immersion period, the final buoyancy force is measured with an underwater weighing assembly. The difference between the initial and final buoyancy force is the weight of water absorbed. This difference is expressed in terms of water absorbed per unit of specimen volume. The test results are seriously affected if proper steps are not followed closely and variables are not controlled carefully.

TABLE 14-1. Temperature and Relative Humidity Condition

Temperature (°C)	Relative Humidity
23 ± 2	50 ± 5
38 ± 2	97 ± 3
70 ± 2	97 ± 3
To be selected	95 ± 3
−40 ± 3	Ambient
−73 ± 3	Ambient
70 ± 2	Ambient
100 ± 2	Ambient
150 ± 2	Ambient
200 ± 2	Ambient
Temperatures and humidities selected for individual needs	

From ASTM D 2126. Reprinted with permission of ASTM.

14.2.10. Water Absorption of Core Materials for Structural Sandwich Constructions (ASTM C 272)

This method of determining the water absorption of core materials for structural sandwich constructions gives basic information about water absorption characteristics and helps to determine the effect of water absorption on mechanical, thermal, and electrical properties. A 3- × 3-in. specimen is cut from a sandwich panel and its edges smoothed. After properly conditioning and weighing the specimen, it is immersed completely in water at a specified temperature and time. After the immersion period, the specimen is removed, dried, and weighed. Water absorption is reported as water gained per cubic centimeter or specimen tested, or as percentage of weight gain. The test is also used as a control test for product uniformity.

14.2.11. Water Vapor Transmission (ASTM E 96)

The water vapor transmission test is used to determine values of water vapor transfer through permeable materials. Two basic methods, the desiccant method and the water method, have been developed. The desiccant method generally yields lower results than the water method.

The rate of water vapor transmission (WVT) is defined as the time rate of water vapor flow of a body between two specified parallel surfaces normal to the surfaces, under steady conditions through unit area, under conditions of test. Water vapor permeance is defined as the ratio of the WVT of a body between two specified parallel surfaces to the vapor pressure difference between the two surfaces. An accepted unit of permeance is perm. Water vapor permeability is defined as the product of permeance and thickness. The unit for permeability is perm-in.

In the desiccant method, the specimen is sealed to the open mouth of a test dish containing desiccant and the assembly is placed in a controlled atmosphere. The rate of water vapor transmission through the specimen into the desiccant is determined by periodic weighings of the dish assembly. Distilled water is used in the water method. The rate of water movement through the specimen is determined by periodically weighing the dish containing the distilled water and specimen. Figure 14-6 shows a commercial WVT analyzer.

14.2.12. Weathering Properties

The effect of the outdoor environment on foam plastics is not clearly known owing to a lack of data from either the material suppliers or the end users. In general, the resistance of foamed plastics to the outdoor environment is usually similar to that of the base polymer. The test methods described in Chapter 5 are applicable to foam plastics.

14.2.13. Thermal Conductivity (ASTM C 177)

Cellular plastics have the lowest thermal conductivity or K factor of any insulating material available today. This superior insulating ability depends upon many variable factors. Thermal conductivity of cellular plastics is greatly influenced by the blowing agent, cell size, closed-cell content, and density of the foam. Temperature and moisture are other factors that affect the thermal conductivity of cellular

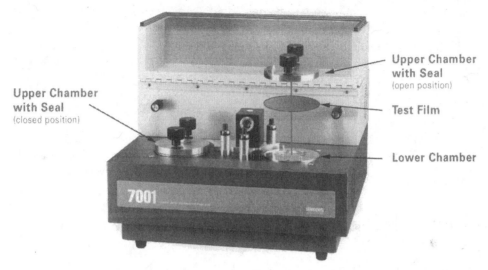

Figure 14-6. Commercial WVT analyzer. (Courtesy Illinois Instruments, Inc.)

plastics considerably. The method used to determine the thermal conductivity of cellular plastics is the same as that used for noncellular plastics. These techniques are described in detail in Chapter 3.

14.2.14. Flammability

Because of their superior insulating properties, foam plastics have found numerous applications in the construction and packaging industries. Along with increased use, a considerable amount of concern has been expressed in regard to the flammability of foam plastics. A detailed discussion of various tests used in the industry to determine the flammability of cellular and noncellular plastics and their limitations is included in Chapter 8.

14.2.15. Dielectric Constant and Dissipation Factor (ASTM D 1673)

Cellular plastics have unique advantages over conventional plastics in regard to electrical properties because part of the plastic material has been replaced by a gas that has a lower dielectric constant. Thus, cellular plastics have an improved dielectric constant, a distinct advantage in wire and cable insulation applications. Another advantage of cellular plastics is their low dissipation factor.

This method, although quite similar to the methods used for noncellular plastics, was developed especially for cellular plastics for several reasons. Cellular plastics have surfaces that preclude the use of conventional electrodes such as metal foils attached by petrolatum and similar adhesives. The greater thickness of the cellular plastic specimen does not lend itself to the attachment of conventional electrodes.

The dielectric constant and dissipation factor of flat sheets or slabs of both rigid and flexible cellular plastics can be determined by this method, at frequencies from 60 to 100 MHz. The basic apparatus consists of a bridge and a resonant circuit. Since cellular plastics do not have surfaces suitable for the attachment of conventional electrodes, prefabricated, rigid, metal plate electrodes must usually be employed for dielectric constant and dissipation factor tests. Such electrode systems may be either the direct-contact or the noncontacting type.

The specimens are conditioned using standard conditioning procedures. The specimens should be free of surface skin unless otherwise specified or agreed upon, because surface skin may affect the results considerably. A more detailed discussion of the procedures and factors affecting the test results is presented in Chapter 4.

Because of the basic nature of cellular plastics—inconsistency in thickness and foam densities—the reproducibility of the test results by these methods is considerably poorer than that expected from measurements on noncellular types of insultating materials.

14.3. FLEXIBLE CELLULAR MATERIALS TEST METHODS

14.3.1. Introduction

The test methods developed for testing flexible cellular plastics are quite different from those developed for rigid foams. For rigid cellular plastics, separate test methods were developed for specific properties. No such separate test methods relating to specific properties are developed for flexible cellular plastics. Instead, a series of test procedures that describe a variety of physical properties of a particular type of material are commonly used to test flexible cellular plastics.

In this section, an attempt is made to briefly describe all types of basic and suffix tests developed for flexible cellular plastics by the joint efforts of ASTM and the Society of the Plastics Industry. The individual tests for different forms and types of materials are also given along with a list of basic and suffix tests for each material.

Only a few of these tests apply to each different type of material. The test title and the applicable test are listed in Table 14-2.

14.3.2. Steam Autoclave Test

The steam autoclave test consists of exposing the foam specimens to a low-pressure steam autoclave at a prescribed temperature and time, and observing the changes in the physical properties of the specimen. The test specimens are exposed to steam following specified preconditioning. After the exposure period, specimens are properly dried in a dryair oven. The compression load deflection is obtained before and after exposure and percent change from original value is reported. The steam autoclave compression set value is also calculated.

14.3.3. Constant Deflection Compression Set Test

This test consists of deflecting the specimen under specified conditions of time and temperature and observing the effect on the thickness of the specimen. The test procedure requires the specimen to be compressed between two or more parallel plates to a specified deflection thickness. The entire assembly is placed in a mechanically convected air oven for a specified time and temperature. Following this exposure, the specimen is removed from the apparatus and the recovered thickness is measured. The constant deflection compression set, expressed as a percentage of the original thickness, is calculated.

TABLE 14-2. Test for Flexible Cellular Materials

	Slab Urethane Foams	Molded Urethane Foams	Slab, Bonded, and Molded Urethane Foams	Vinyl Polymers and Copolymers
ASTM designation	D 1564	D 2406	D 3574	D 1565
Steam autoclave test	X	X	X	
Constant deflection compression set test	X	X	X	X
Load–deflection tests Methods A and B	X	X	X	X
Air flow test	X		X	
Compression force– deflection test	X	X	X	
Dry heat test	X	X	X	
Fatigue test static and dynamic	X	X	X	X
Density test	X	X	X	X
Tear resistance test	X	X	X	
Tension test	X	X	X	
Resilience test	X		X	

14.3.4. Indentation Force Deflection Test

The load deflection test is performed two different ways:

Method A: Indentation to specified deflections

Method B: Indentation to specified loads

Method A. Indentation force deflection (IFD) consists of measuring the load necessary to produce deflection (generally 25–65 percent) in the foam product. The test is carried out by pushing a flat circular indentor foot down into the foam specimen and measuring the force on the foot at various compression amounts. The test is widely used in the cushioning and bedding industry. Higher IFD values indicate a stiffer foam.

Method B. Indentation residual deflection force (IRDF) consists of deflecting the specimen to a constant load rather than a constant thickness as in the IFD method. The specimen is compressed with a specified load for 1 min and the resulting thickness is noted. The IRDF value at the specified load is reported; IRDF values at different loads may also be obtained. Here, a higher IRDF value indicates a stiffer foam. This test is useful in determining how thick the padding should be under an average person sitting on a seat cushion.

14.3.5. Air Flow Test

The air flow test measures the ease with which air passes through a cellular structure. The resistance to air flow exhibited by the open cells in a flexible foam may be used as an indirect measurement of cell structure characteristics. The test consists of placing a flexible foam specimen in a cavity over a vacuum chamber and creating a specified constant air-pressure differential. The rate of flow of air required to maintain this pressure differential is the air flow value. The test is conducted two different ways: air flow parallel to foam rise and air flow perpendicular to foam rise. Air flow values are proportional to porosity in flexible foam.

14.3.6. Compression Force Deflection Test

The compression force deflection test consists of measuring the load necessary to produce a 50 percent compression over the entire top area of the foam specimen. This test differs from the IFD test described previously in that the flat compression foot used in the test is larger than the specimen. The compression foot is brought into contact with the specimen. The specimen is compressed 50 percent of its original thickness and after 1 min the final load is observed. The compression force deflection is recorded as the load required for a 50 percent compression of specimen area.

14.3.7. Dry Heat Test

The dry heat test consists of exposing foam specimens in an air-circulating oven and observing the effect on physical properties of the foam. The tensile properties are determined using the same procedure described in the tension test for flexible foams.

14.3.8. Fatigue Test

Fatigue tests are conducted four ways:

Method A: Static force loss test at constant deflection
Method B: Dynamic fatigue by the roller shear at constant force
Method C: Dynamic fatigue test by constant force pounding
Method D: Dynamic fatigue test for carpet cushion

Method A helps to determine:

1. A loss of force support
2. A loss of thickness
3. Structural breakdown by visual examination

The original thickness of the specimen is measured after preflexing. Next, the specimen is placed between two parallel plates with spacer bars to provide 75 percent deflection. The plates are clamped and held at 75 percent deflection for 22 hr at standard test conditions. Thirty minutes after the completion of the test, repeating the initial measurements using the six preflexes and the original thickness, the deflection for the final force reading is determined. The percent loss of thickness and of force deflection are calculated. Results of the visual examination indicating possible breakdown of cellular structure are also noted.

Method B tests the samples dynamically, at a constant force, deflecting the material both vertically and laterally. The fatigue test may be carried out by either procedure A, which uses 8000 cycles, or procedure B, which uses 20,000 cycles. The apparatus consists of a roller that moves back and forth over the specimen at a specified force. The final fatigue value is expressed as a total loss number. The total loss number is equal to the sum of percent losses at each different load. The visual examination results are also reported.

Method C—dynamic fatigue test by constant force pounding—is conducted to determine the loss of force support at 40 percent indentation force deflection (IFD), a loss in thickness and structural breakdown as assessed by visual inspection. The test evaluates the specimen by repeatedly deflecting the material by a flat horizontal indentation exerting a specified vertical force on the test specimen. This test may be conducted by using either 8000 or 80,000 cycles. The percent loss in thickness and percent loss of 40 percent IFD and results of visual examination are reported.

Method D, the dynamic fatigue test for carpet cushion, was developed to evaluate the specimen by repeatedly deflecting the carpet cushion by a rubber-covered roller exerting a specified force on the test specimens. The test determines the retention of force support at 65% compression force deflection (CFD), loss of thickness, and structural breakdown as determined by visual inspection.

14.3.9. Density Test

The density test determines the density of uncored foam by calculating the weight and volume of a regularly shaped specimen. The specimen is weighed on a balance and its dimensions are measured using calipers. The density of the specimen is calculated by dividing the weight by the volume of the specimen.

14.3.10. Tear-Resistance Test

This test was developed to determine the tear resistance of foam. The tear-resistance test, also known as the block method, uses the block specimen described in Figure 14-7. The block specimen is mounted by spreading the block so that each tab is held in the jaw of the tensile testing machine. The load is applied by separating the jaws at a specified speed. The force required to rupture the specimen is recorded. The tear resistance is calculated from the maximum load recorded and the average thickness of the specimen. The tear resistance is reported in newtons per meter.

14.3.11. Tension Test

This method is similar to the method described in Chapter 2 for the tensile testing of plastics. The test determines the effect of the application of tensile force to a foam specimen. Tensile stress, tensile strength, and ultimate elongation are also obtained. The test specimens are dumbbell-shaped and cut from a flat sheet with a die, such as is shown in Figure 14-8. The tensile strength is determined by placing the specimen in the jaws of the tensile testing machine and pulling it apart at a specified speed. The force at break divided by the original cross-sectional area of the specimen is the tensile strength.

14.3.12. Resilience Test

This test, also known as the ball rebound test, is very useful in evaluating the resilience of foams. The test consists of dropping a steel ball onto a foam specimen and noting the height of rebound. The apparatus, as shown in Figure 14-9, consists

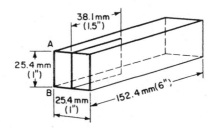

Figure 14-7. Block specimen—tear resistance test. (Reprinted with permission of ASTM.)

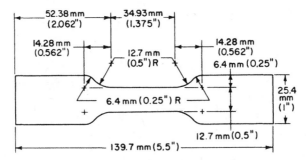

Figure 14-8. Die for cutting dumbbell-shaped specimen for tension test. (Reprinted with permission of ASTM.)

Figure 14-9. Resilience test apparatus.

TABLE 14-3. Foam Properties Chart

Type	Density (lb/ft³)	Tensile Strength (psi) (ASTM D 1623)	Compressive Strength at 10 Percent Deflection (psi) (ASTM D 1621)	Maximum Service Temperature (°F) Dry	Wet	Thermal Conductivity (Btu/ft²) (ASTM 177)	Coefficient of Linear Expansion (10^{-5}in./in./F) (ASTM D 696)	Dielectric Constant (ASTM D 1673)	Dissipation Factor at 28°C and 1 meg	WVT (perm-in.) (ASTM E 96)	Water Absorption, Percent by Volume (96hr) (ASTM D 2842)
ABS (acrylonitrile–butadiene–styrene) Injection molding type Pellets	40–56	2,000–4,000	2,300–3,700	176–180	—	0.58–2.1	3.7–9.5	—	—	—	0.4–0.6
Flame-retarded Pellets	45–55	1,500–3,000		180	—		5.1	—	—	—	—
Acrylic Boards	2.6–6.2	—	67–320ᵃ	220–230	—	0.22	2.9–3.2	1.90	0.0036	—	11.8–13.3
Cellulose acetate Boards and rods (rigid, closed cell foam)	6.0–8.0	170	125	350	—	0.31	2.5	1.12		—	13–17 at 100% R.H. 1.9–2.5 at 50% R.H.
Epoxy Rigid closed cell	5.0	51	90	350	—	0.26	—	$1.19 - 1.08$ at 10^6 and 10^{10}	—	—	—
Precast blocks, slabs, sheet	10.0	180	260	350	—	0.28	—	$1.36 - 1.24$ at 10^6 and 10^{10}	—	—	—
	20.0	650	1,080	350	—	0.32	—	$1.55 - 1.41$ at 10^6 and 10^{10}	—	—	—
Syntactic One-part free flowing powder	14.0–20.0	—	125	350–400	—	<0.5	—	1.38	0.006	—	—
Two-part moldable like damp sand	20.0	—	1,000	—	—	0.38	—	1.45	0.01	—	—
Rigid sheet or pack-in-place	23.0	—	2,100	500	—	0.36	—	—	—	—	1.8
Spray-applied Two-package systems (liquid)	1.8–2.0	26–31	13–17	160	—	0.11–0.12	—	—	—	—	—
Foam-in-place Two-package systems (liquid)	2–2.3	—	20–26	200	—	0.11–0.13	—	—	—	—	—
Ethylene copolymer Extruded sheet	35.0	600–800	—	130	—	—	—	—	—	—	<0.5

TABLE 14-3. *Continued*

Type	Density (lb/ft³)	Tensile Strength (psi) (ASTM D 1623)	Compressive Strength at 10 Percent Deflection (psi) (ASTM D 1621)	Maximum Service Temperature (°F) Dry	Wet	Thermal Conductivity (Btu/ft²) (ASTM 177)	Coefficient of Linear Expansion (10^{-5} in./in./F) (ASTM D 696)	Dielectric Constant (ASTM D 1673)	Dissipation Factor at 28°C and 1 meg	WVT (perm-in.) (ASTM E 96)	Water Absorption, Percent by Volume (96hr) (ASTM D 2842)
Ionomer Sheet, rod	2.0–20.0	57–105	24–15.2	150–155	—	0.27–0.34	—	1.5 at 10^6	0.003	0.34–2.00	0.40–1.0
Phenolic	⅓–1½	3–17	2–15	—	—	0.21–0.28	—	—	—	—	—
Foam-in-place Liquid resin	2–5	20–54	22.85	Continuous service at 300		0.20–0.22	0.5	—	—	2 lb/ft³ 5 lb/ft³ 2,074 g 1,844 g per day/m.²	13–51 at 100% R.H. 1–4 at 50% R.H.
	7–10	80–130	158–300	—		0.24–0.28	—	1.19–1.2	0.028–0.031	—	10–15 at 100% R.H.
	10–22	—	300–1,200	—		—	—	1.19–1.2	0.028–0.031	—	1–5 at 50% R.H.
Syntactic castable Two-component: liquid and paste	50–60	>1,000	8,000–13,000	275		1.0	10	2.1	0.03	—	>0.5
Phenylene oxide-based foamable resin Pellets	50	3,300	5,500	200		0.86ᶜ	3.8×10^{-5}	216	0.0017	—	—
Polybenzimidazole Blown Slabs	3–6	44–125	30–125	600		0.216	—	—	—	—	—
Syntactic Molding powder, slabs	15–38	200–2,000	700–4,310	600		0.42–0.78	—	1.82–2.40	—	—	—
Polycarbonate Pellets	50	5500	7,500	270		1.05ᶜ	2.5	220	0.001	—	—

Material											
Polyethylene											
Low-density foam											
Planks, rods, round. ovals. net	1.3–2.6	20–30	5	160–180		0.28–0.40	9.5–2.3	1.05 at 10^6	0.0002 at 10^9	0.40	<0.50
Noncrosslinked sheet	2.1–3.3	35–100	3	160–180	—	0.28–0.34	2.3	1.05 at 10^6	0.0002 at 10^9	0.20–<0.40	<0.50
Noncrosslinked tubing	2.1–3.3	—	—	160–180	—	0.26–0.28 at 40°F, mean	2.3	—	—	0.13	<0.50
Crosslinked sheet, rolls, cord	1.6–2.4	40–70	2.9–3.0	175–200	—	0.25–0.28	—	1.06 at 10^6	0.0002 at 10^9	<0.40	0.1 (24hr)
Intermediate density foam	3.6–4.4	70–110	4.3–14	180–200	—	0.30	—	1.07 at 10^6	0.0002 at 10^9	<0.40	<0.50
Crosslinked plank, sheet, rolls	5.5–7	110–210	2–18	180–200	—	0.32–0.34	—	1.15 at 10^6	0.0002 at 10^9	<0.40	<0.50
	9.0–10.5	210–300	15–30	180–200	—	0.34–0.40	—				<0.50

High-density foam
  Molded parts and shapes with
  solid integral skin
	25.0–50.0	1,200	1,300	230	230	0.92	4.18				0.22

Crosslinked foam Rolls, sheets	0.9–12.5	46–210	2.0–18.5	180		0.27–0.4	<13	1.1–1.55	0.002 – 0.0007		0.1–0.5
Polypropylene											
Low-density foam Roll, sheets	0.6	20–40	0.7	250		0.27		1.02 at 10^4	0.00006 at 10^4	—	0.02 (3hr)
High-density foam	35.0	1.600	2.100								
Molded parts and shapes with solid integral skin	—	4.2	—	—	—	—	—				
Crosslinked foam Sheets	3.0	118–147	175–1,200	275	275	0.27	—				0.5
Polystyrene											
Shapes, boards, and billets	1.0	21–28	13–18	165–175		0.26	3.0–4.0	1.06 – 1.02 at 10^3 to 10^6 cycles	0.0001– 0.0007 at 10^2 to 10^6 cycles	1.2–6.0	2–6
	2.0	42–68	35–45	165–175		0.24 at 70°F mean temp.	3.0–4.0			0.6–1.2	2–4
Molded from expandable sheets	5.0	148–172	85–130	165–175		0.246	3.0–4.0			0.4–0.6	2–4
	1.5–20	55–70	25–55 at 5%			0.21–0.29[b] 0.23–0.26[c]					
Extruded boards and billets	2.0–2.6	60–105 at 5%	25–60	165–175 (aged)		0.17–0.19[b]	3.0–4.0	<1.05 at 10^2 to 10^8 cycles	<0.0004 at 10^3 to 10^8 cycles	0.3–1.1	1.0
	2.0–5.0	180–200	100–180 at 5%			0.18–0.21[d]					
Extruded film and sheet	6.0	300–500	42.5	170–175 (aged)		0.24 at 70°F mean temperature		1.27	0.00011	1.50	
	8.0	400–700	52.5	175						1.25	
	10.0	600–1000	68.0	175				1.28	0.00015	1.00	

TABLE 14-3. *Continued*

Type	Density (lb/ft³)	Tensile Strength (psi) (ASTM D 1623)	Compressive Strength at 10 Percent Deflection (psi) (ASTM D 1621)	Maximum Service Temperature (°F) Dry	Wet	Thermal Conductivity (Btu/ft²) (ASTM 177)	Coefficient of Linear Expansion (10⁻⁵ in./in./F) (ASTM D 696)	Dielectric Constant (ASTM D 1673)	Dissipation Factor at 28°C and 1 meg	WVT (perm-in.) (ASTM E 96)	Water Absorption, Percent by Volume (96hr) (ASTM D 2842)
Polyurethane											
Rigid (closed cell)	1.3–3.0	15–95	15–60	180–250	—	$0.11–0.17^{r,f}$ 0.21	4–8	1.05	—	0.6–4.0	1.0–5.0
Molded parts; boards, blocks, slabs; pipe covering: one-shot, two- and three-package systems for foam-in-place; for spray, pour, or froth-pour techniques	4–8	90–290	70–275	200–250	—	$0.15–0.21^{c,f}$ $0.21–0.29^{e,h}$	4	1.10	0.0018	0.9–2.0	0.6–2.0
	9–12	230–450	290–550	250–275	—	$0.19–0.25^{c,f}$ $0.31–0.35^{g,h}$	4	1.2	0.0032	—	
	13–18	475–700	650–1,100	250–300	—	$0.26–0.34^{c,f}$ $0.36–0.40^{f,g}$	4	1.3	0.0055	—	
	19–25	775–1,300	1,200–2,000	250–300	—	$0.34–0.42^{d,f}$ $0.42–0.52^{c,f}$	4	1.4	—	—	0.2
	26–40	1,350–2,500	2,100–4,000	275–300	—	—	4	1.5	—	—	—
	41–70	3000–8000	5000–15000	300	—	0.57^k	4	—	—	0.08	—
Flexible—free rise (slabstock) Slabs, sheets, blocks, custom shapes	0.9–8.0	8–45	0.2–2.0 at 25%	150–175	—	0.2–0.25 at 2lb/cu. ft density	—	1.0–1.5	—	—	—
Flexible—molded (foam-in-place) Two- and three-package or one-shot systems for mixing on job	1.2–20.0	10–1,350	0.25–100 at 25%	150 25%	—	0.3 at 2lb/ cu. ft density	4	1.1	—	—	—
Isocyanurate foams											
Bun	1.5–3.0	15–50	20–60	300	—	0.105–0.17	4	—	—	3	1.5
Laminate	2.0–3.5	20–60	30–70	300	—	0.11–0.16	4	—	—	3	1.5
Pour	1.5–3.0	25–75	20–80	300	—	0.11–0.17	4	—	—	3	1.5
Spray	2.0–3.0	20–60	20–65	300	—	0.115–0.17	4	—	—	3	1.5
Semirigid (foamed-in-place)	1.2–26.0	20–1,350	20–2,100	175–350	—	0.11–0.30	—	—	—	—	—
Integral skin molded (flexible) Two- and three-package systems	25–65 (skin) 5–20 (core)	20–1,350	1–5 at 25%	150–175	—	—	—	2.2	—	—	—
Skinned molded (rigid) Two- and three-package systems	25–65 (skin) 3–30 (core)	100–2,700	40–3,000	150–250	—	0.12–0.80	4	2.5	—	—	—
Microcellular elastomer Two- and three-package systems	20–70	50–2,500 at 50% 5–100 at 10%	50–2,500 at 25%	200	150	—	—	—	—	—	—

Polyvinyl chloride									
Plastisol mechanically frothed									
Liquid or paste	13–60	—	—	150–175	Depends on density	—	—	—	—	—
Plastisol with blowing agent									
Liquid or paste	3–60	50–3,000	—	150–175	Depends on density	—	—	—	—	—
Flexible open cell									
Liquid or paste	10–up	10–200	—	125–225	Depends on density	—	—	—	—	—
Sheets and rolls; cored cushions; other molded shapes			0.5–40.0							
Flexible closed cell									
Sheets, tubes, and molded shapes	4.0–11	50–150 (D412)		130–150	0.24–0.28 at 70°F	—	—	—	0.20 max.	Nil
Rigid closed cell									
Boards and billets	2–4	1,000 and up	—	—	2.0 at 70°F	4.0–6.0	—	—	—	—
Silicone									
Liquid (10% closed cell)	9.6	(unrestricted)	—	400	—	—	1.42 at 10⁵	0.001 at 10⁵	—	—
Liquid (closed cell)	21.0–31.0	100–150	10 at 75%	500–650	0.36	—	1.3–1.4	<0.01	—	0.1
Sheet (open cell)	10.0	45	5 at 75%	500	0.6	—	1.2	0.007	—	—
Sheet (closed cell)	25–34	80–100	1.5–2.9	450	—	—	—	—	—	Maximum 5%[h]
Styrene–acrylonitrile									
Products or shapes molded from expandable beads; finished boards	0.5	—	1.5 at 5%	—	0.32	—	—	—	2.0–4.0	Nil
	0.8	20	6.0 at 5%	170–190	0.29	—	—	—	—	—
	1.0	30	6.0 at 5%	—	0.29	—	—	—	—	—
Urea formaldehyde									
Block, shred, foam-in-place	0.8–1.2	Poor	5	120	0.18–0.21	—	—	—	28–35	1.9

Note: 1.42 at 10^5; 0.001 at 10^5.

[a] At 5 percent.
[b] At 40°F.
[c] ASTM C 177.
[d] At 70°F.
[e] Blow with fluorocarbon.
[f] First number in each sequence is the value for unaged material; the number is the value after aging.
[g] Blown with CO_2.
[h] ASTM D 1056.

Reprinted with permission of McGraw-Hill Book Company, New York.

of a clear plastic tube with a series of circles inscribed on it to measure the rebound height directly in percentage. The ball is dropped so that it does not strike the tube on the drop or the rebound. The resilience value is the ratio of the rebound height to the original height expressed as a percent. The higher value of the resilience indicates more "lively" foam.

14.4. FOAM PROPERTIES

Table 14-3 (pp. 321–325) lists and compares important properties for both rigid and flexible foams.

GENERAL REFERENCES

Frisch, K. and Saunders, J., *Plastics Foams*, Marcel Dekker, New York, 1972.

Benning, C. J., *Plastics Foams*, Vols. I and II, Wiley-Interscience, New York, 1969.

Remington, W. J. and Priser, R., "A New Apparatus for Determining the Cell Structure of Cellular Materials," *Rubber World*, **183** (1958), pp. 261–264.

Harding, R. H., "Determination of Average Cell Volume in Foamed Plastics," *Mod. Plast.*, **37**(10) (1960), pp. 156–160.

Benning, C. J. and Nutter, J. I., "Filled Polyethylene Foams," *22nd SPE ANTEC*, Montreal, March 7–11, 1966 Reprints.

Benning, C. J., "Polyethylene Foam: I. Modified PE Foam Systems," *J. Cellular Plast.*, **3**(2) (February 1967), pp. 62–72.

Benning, C. J., "Polyethylene Foam, II. Mechanical Properties of Polyethylene Foams Prepared at High and Low Rates of Extrusion," *J. Cellular Plast.*, **3**(3) (March 1967), pp. 125–137.

15

FAILURE ANALYSIS

15.1. INTRODUCTION

The fundamental problem concerning the failures of parts made out of plastics materials is a lack of understanding the difference between the nature of relatively new polymeric materials and traditional materials such as metal, wood, and ceramics. Designers, processors, and end-users are all equally responsible in contributing to the problem. Merely copying a metal or wood product with some minor aesthetic changes can lead to premature and sometimes catastrophic failures. Designers are generally most familiar with metals and their behavior under load and varying conditions of temperature and environment. While designing metal parts, designers can rely on instantaneous stress–strain properties, and for most applications they can disregard the effect of temperature, environment, and long-term effect of load (creep). Plastics materials are viscoelastic in nature; unlike other materials, properties can vary considerably under the influence of temperature, load, environment, and presence of chemicals. For example, a well-designed part may perform its intended function for a very long time at room temperature and under normal load. The same part may fall apart quickly when exposed to extreme cold or hot environment, and the process can accelerate if the part is subjected to mechanical loading or exposed to chemical environment. Most often overlooked is the synergistic effect of all the factors such as temperature, creep, chemicals, ultraviolet (UV), and other environmental factors on plastic parts. Selecting the right material for a given application is one of the most important tasks for the product designers. Today, designers are faced with dealing with a large number of plastic materials and material grades. Choosing the right plastic for a new product from close to 50 major types of plastic materials and over 50,000 grades of commercial plastics from about 500 suppliers is not an easy task. Additionally, since the

Handbook of Plastics Testing and Failure Analysis, Third Edition, by Vishu Shah
Copyright © 2007 by John Wiley & Sons, Inc.

industry has not yet become standardized with regard to types of test specimen, specimen geometries, and test methods, making a true comparison is extremely difficult. The software such as Computer-Aided Material Preselection by Uniform Standards (CAMPUS) promotes the use of comparable data based on uniform global testing standards. A detailed discussion of CAMPUS is carried out in Chapter 20, entitled "Uniform Global Testing Standards." Designers must take into account all these factors prior to making final material selection in order to avoid premature failure arising from simply choosing the wrong material.

During processing, plastic materials are subjected to severe physical conditions involving elevated temperatures, high pressures, high shear rate, and chemical changes. Processor must follow the proper procedures and guidelines set forth by material suppliers and make sure that material is processed optimally in well-maintained equipment. Tool qualification, process validation, and documentation of standard operating procedures are important steps, and they should not be overlooked. Quite often, fundamentally sound processing techniques are compromised in order to produce aesthetically appealing parts and achieve faster cycle times and higher yields, resulting in unexpected failures. The end-users must also be educated by product manufacturers in the proper use of the plastics products and make sure that it is used for intended purpose. Plastics parts often fail prematurely because of the intentional or unintentional abuse by the consumers.

Finally, for years, companies have practiced over-the-wall approach (Figure 15-1) to new product development. The product development cycle begins with a marketing department generating a request to develop a new product with an outline of requirements related to the functionality and appearance. An industrial design group or an outside firm is commissioned to develop this product. At this point, the main emphasis is on basic functionality and aesthetic appeal of the product. No engineering calculations are carried out. Once this step is completed, the basic design is passed on to the product design group. The product design is scrutinized to make sure that the product will perform under various loads and environmental conditions. Prototyping and testing may also be carried out prior to passing the design over to the tooling group. A tooling group's primary responsibility is to develop a production tool to meet requirement. Once the tool is built, it is turned over to the molding department for high-volume production. In each successive step, there is very little communication between different groups and the responsibility is passed on from one group to the next. Apparent lack of communication with this approach does not allow various groups to take advantage of the valuable experience, knowledge, and expertise of the parties involved. Subsequent investigations of failed parts have pointed to the

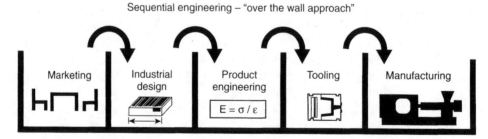

Figure 15-1. Over-the-wall approach to new product development. (Reprinted with permission of Hanser Publications.)

lack of communication between engineering, tooling and manufacturing groups. Many corporations have adopted the concurrent engineering approach to product development (Figure 15-2) to reduce a number of premature failures. This method promotes communication between all concerned parties right from the inception of the project and all the way up to the high volume production. Figure 15-3 shows the breakdown of major reasons behind plastics product failures.

Part failure is generally related to one of four key factors:

1. Material selection
2. Design

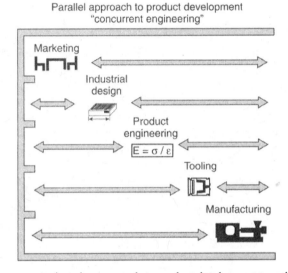

Figure 15-2. Concurrent engineering approach to product development to reduce number of premature failures. (Reprinted with permission of Hanser Publications.)

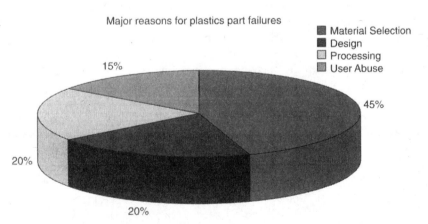

Figure 15-3. Breakdown of major reasons behind plastics product failures. (Reprinted with permission of RAPRA Technology Limited.)

3. Process
4. Service conditions

15.1.1. Material Selection

Failures arising from hasty material selection are not uncommon in plastics or any other industry. In an application that demands high-impact resistance, a high-impact material must be specified. If the material is to be used outdoors for a long period, an ultraviolet (UV)-resistant material must be specified. For proper material selection, careful planning, a thorough understanding of plastic materials, and reasonable prototype testing are required. Plastics are viscoelastic materials. Viscoelasticity is defined as the tendency of plastics to respond to stress as if they were a combination of elastic solids and viscous fluids. This property, possessed by all plastics to some degree, dictates that while plastics have solid-like characteristics such as elasticity, strength, and form stability, they also have liquid-like characteristics (such as flow) depending on time, temperature, rate, and amount of loading. This also means that unlike metals, ceramics, and other traditional materials, plastics do not exhibit a linear stress–strain relationship. Designers accustomed to working with metals and other materials often make the mistake of selecting and specifying incorrect plastic materials. It is this nonlinear relationship for plastics that makes an understanding of creep, stress relaxation, and fatigue properties extremely important.

Typically, for most designers the material selection process begins by reviewing the plastic material data sheets generally provided by the material suppliers. A misinterpretation of the data sheets is one of the most common reasons for selecting and specifying the wrong material, for a given application. First it is important to understand the purpose of a data sheet. Data sheets are useful only for comparing property values of different plastic materials such as the tensile strength of nylon versus polycarbonate or the impact strength of polystyrene versus ABS. Data sheets should be used for initial screenings of various materials. For example, if a designer is looking for a material that is strong and tough, he may start out by selecting materials whose reported values are higher than 7000-psi tensile strength and impact strength values of better than 1.0 ft-lb/in. and eliminating materials such as general-purpose polystyrene, polypropylene, and polyethylene. Data sheets are never meant to be used for engineering design and final or ultimate material selections. First, the reported data are generally derived from the short-term tests. Short-term tests, as the name suggests, are the tests conducted without consideration of time, and the values derived are instantaneous. Tensile test, izod impact test, and heat distortion temperature are the examples of such short-term tests. Data reported on data sheets are also derived from single-point measurements. These tests do not take into account the effect of time, temperature, environment, chemicals, and so on. A single number representing one point on a stress–strain curve cannot begin to convey plastics' behavior over a range of conditions. The standardized tests used to measure data sheet properties contain data measured in a laboratory under ideal conditions (as specified by ASTM or ISO standards) on standardized test specimens that bear little resemblance to the geometry of real-world parts. These tests likewise take place at temperatures, stress rates, and strain rates that rarely correspond to the real-world conditions (1).

The proper use of multi-point data for selecting the most appropriate plastic materials for the applications cannot be over emphasized. This point is well illustrated in a classic example of misinterpretation of published test data and the true meaning and usefulness of heat distortion temperature (HDT) values. As discussed in Chapter 3, the heat distortion

TABLE 15-1. HDT vs. Continuous-Use Temperature (UL Temperature Index)

Material	HDT	Continuous-Use Temperature
Ryton R-4 (polyphenylene sulfide)	>500°F	338°F
Ultem 4000 (polyetherimide)	412°F	122°F

temperature test is a short-term test conducted using standard test bars and laboratory conditions. The temperature values derived from this test for a particular plastic material is simply an indication of the temperature at which the test bar shall deform 0.010 in. under a specified load. The reported values are further distorted by factors such as residual stresses in the test bars, amount of load, and specimen thickness. This reported value is of limited practical importance and should not be used to select materials for applications requiring continuous exposure at elevated temperatures. Continuous-use temperature data such as UL temperature index is a better indication of how plastic materials will perform for an extended period at elevated temperatures. Table 15-1 shows temperature data derived from two different test methods. If a designer were to select the material solely based on published heat deflection temperature data without understanding the true meaning of the test, test limitations, and how the values are derived, the result could be disastrous. Note that the continuous-use temperature for a commercial grade of 40% glass-reinforced polyphenylene sulfide (PPS) is only 338°F, while the heat deflection temperature data derived from a short-term test is greater than 500°F, indicating that this material is not suitable in applications requiring continuous exposure to heat over 338°F.

Material Selection Using Multi-Point Data
As discussed, material selection difficulties stem from limited availability of multi-point data from the material suppliers. Data sheets with single-point measurement data are readily available. However, with a little effort, the designers can find multi-point data from the sources such as CAMPUS (2) and IDES (3) and from all leading material suppliers. Multi-point data are presented in the form of chart and graphs of shear modulus versus temperature, isochronous stress–strain curves, and creep data at a minimum of three different temperatures and four stress levels. While designing a product to withstand multiple impact loads, the designer must take into consideration the data generated from instrumented impact tests which can provide valuable information such as ductile-to-brittle transition and behavior of the specimen during the entire impact event. Modulus values are also often misinterpreted. The flexural modulus values which are derived from single-point measurement are frequently accepted as the indication of the stiffness of the material over a long period. Flexural modulus tests are conducted at a very low strain and generally represent only the linear portion of the stress–strain curve. The reported values do not correspond well with the actual use conditions, and they tend to overpredict the stiffness of the actual part. Plastic parts often fail due to the lack of consideration of creep values in material selection process. Plastics can creep or deform under a very small load at a very low strain, even at room temperature. Creep or apparent modulus data for the plastic materials over a long period at several temperatures should be evaluated.

Material Selection Process
The material selection should not be solely based on cost. A systematic approach to material selection process is necessary in order to select the best material for any application.

The proper material selection technique involves carefully defining the application require-ment in terms of mechanical, thermal, environmental, electrical, and chemical properties. In many instances, it makes sense to design a thinner wall part taking advantage of the stiffness-to-weight ratio offered by higher-priced, fast-cycling engineering materials (4). Many companies including material suppliers have developed software to assist in material selection simply by selecting application requirement in the order of importance. The material selection process starts with carefully defining the requirements and narrowing down the choices by the process of elimination. The designer must identify application requirements including mechanical, thermal, environmental, and chemical. All special needs such as outdoor UV exposure, light transmission, fatigue, creep, stress relaxation, and regulatory requirements must be considered. Processing techniques and assembly methods play a key role in selecting appropriate material and should be given consider-ation. Many plastics materials are susceptible to chemical attack, and therefore behavior of plastics material in chemical environment is one of the most important considerations in selecting material. No single property defines a material's ability to perform in a given chemical environment, and factors such as external or molded-in stresses, length of expo-sure, temperature, chemical concentration, and so on, should be carefully scrutinized. Some of the common pitfalls in material selection process are relying on published mate-rial property data, misinterpretation of data sheets, and blindly accepting material suppli-er's recommendations. Material property data sheets should only be used for screening various types and grades of materials and not for ultimate selection or engineering design. As discussed earlier, the reported data are generally derived from short-term tests and single-point measurements under laboratory conditions using standard test bars. The published values are generally higher and do not correlate well with actual use conditions. Such data do not take into account the effect of time, temperature, environment, and chemicals. Figure 15-4 shows a typical failure arising from improper material selection. In order to assist designers with the material selection process, material supplier have developed a comprehensive checklist such as one illustrated in Figure 15-5 (See p. 335. See also Appendix K for complete figure). Key considerations are as follows.

Mechanical Properties

- Tensile strength and modulus
- Flexural strength and modulus

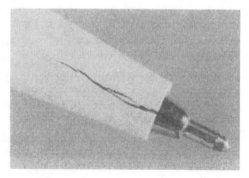

Figure 15-4. A typical failure arising from improper material selection. (Reprinted by permission of Hanser Publications.)

New Application Checklist

This checklist includes critical considerations for new part development.
Its use will help provide a more rapid and more accurate recommendation.

Name _____   Date _____

Customer _____   Part _____

_____   _____

_____   _____

Project timing _____

Driving force _____

Current product _____

Its performance _____

Comments _____

Part Function — *What is the part supposed to do?* _____

_____

_____

_____

Appearance

Clear

☐   water clear

☐   very clear

☐   generally clear,   maximum haze level: _____

☐   transparent color, maximum haze level: _____

Comments: _____

Opaque

☐   high gloss

☐   medium gloss

☐   low gloss

   ☐ from the plastic   ☐ from paint   ☐ from the mold

Comments: _____

Colors desired: _____

   ☐ from the plastic   ☐ from paint   ☐ from both

Criticality of color match: _____ %

☐ daylight   ☐ tungsten light   ☐ fluorescent light   ☐ all (no metamerism allowed)

Comments: _____

Critical appearance areas — *please attach sketch*

	None	Invisible	Minor	OK
gate blemishes	☐	☐	☐	☐
sink marks	☐	☐	☐	☐
weld lines	☐	☐	☐	☐

Comments: _____

_____

Critical structural areas — *please attach sketch*

Comments: _____

_____

_____

_____

Figure 15-5. New application checklist. See Appendix K (pp. 626–628) for complete figure. (Courtesy of Bayer Corporation.)

- Impact strength
- Compressive strength
- Fatigue endurance
- Creep
- Stress-relaxation

Both short- and long-term property data must be evaluated: Short-term data for quick comparison and screening of the candidates and long-term data for final material selection. Creep and stress relaxation data, which represent deformation under load over a long period, need to be scrutinized over the usable range of temperatures. Isochronous stress–strain curves are very useful for comparing different materials on an equal-time basis. Multi-point impact data obtained from instrumented impact test which provide more meaningful information such as energy at a given strain or total energy at break must be taken into account. Plastic parts often fail due to the lack of consideration of sudden loss of impact in a very cold environment. Multi-point low-temperature impact data, although generally not found on data sheets, is available from all major material suppliers.

Thermal Properties

As discussed earlier in the chapter, short-term values such as heat distortion temperature and Vicat softening point should only be used for initial screening. Meaningful values derived from continuous-use temperature and coefficient of thermal expansion test are more helpful for final material selection. Figure 15-6 shows the example of a failed part resulting from selecting incorrect material based on short-term thermal test data. Plastic materials tend to expand and contract anywhere from seven to ten times more than conventional materials like metals, wood, and ceramics. Designers must be well aware of this, and special consideration must be given if dissimilar materials are to be assembled. The thermal expansion differences can develop internal stresses from push–pull effect along with internal stresses and cause the parts to fail prematurely. The restraining of the tendency of a piping system to expand/contract can result in significant stress reactions in pipe and fittings, or between the piping and its supporting structure. The allowing of a moderate change in length of an installed piping system as a consequence of a temperature change is generally beneficial, regardless of the piping material, in that it tends to reduce

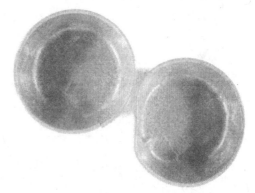

Figure 15-6. Example of a failed part resulting from selecting incorrect material based on short-term thermal test data.

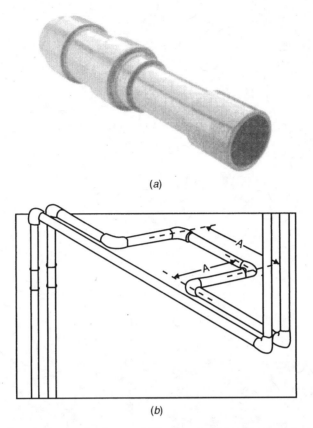

(a)

(b)

Figure 15-7. (*a*) A typical expansion loop and (*b*) expansion joint. These are installed to compensate for expansion and contraction.

and redistribute the stresses that are generated should the tendency for a dimensional change be fully restrained. Thus, allowing controlled expansion/contraction to take place in one part of a piping system is an accepted means to prevent added stresses to rise to levels in other parts of the system that could compromise the performance of, or cause damage to, the structural integrity of a piping component, or to the structure which supports the piping (5). Figure 15-7 illustrates a typical expansion loop and an expansion joint installed to compensate for expansion and contraction.

Exposure to Chemicals
One of the most important considerations in selecting the right material is its resistance to various chemicals. As discussed earlier, the resistance of plastics to various chemicals is dependent on time of contact with chemicals, temperature, molded-in or external stress, and concentration of the chemical. Part design and processing practices play a major role in a material's ability to withstand chemical attack. For example, the stress concentration factor increases significantly for the parts designed with radius-to-wall-thickness ratio of less than 0.4. As a rule, crystalline polymers are more resistant to chemicals when compared to amorphous polymers (Figure 15-8); therefore, if the application requires the parts to be constantly exposed to chemicals, crystalline materials should be given serious con-

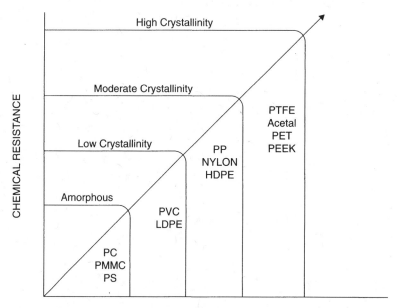

Figure 15-8. Effect of crystallinity on chemical resistance.

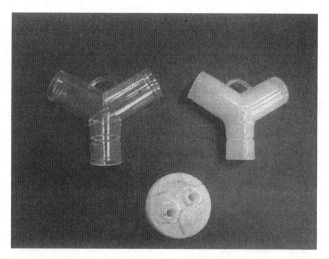

Figure 15-9. Premature part failure resulting from exposure to aggressive chemical.

sideration. Chemical exposure to plastic parts may result in physical degradation such as stress cracking, softening, swelling, discoloration, and chemical attack in terms of reaction of chemicals with polymers and loss of properties. Figure 15-9 shows premature failure of a part that was exposed to aggressive chemicals.

Environmental Considerations

Plastic materials are sensitive to environmental conditions. Environmental considerations include exposure to UV, IR, X-ray, high humidity, weather extremes, pollution from industrial chemicals, microorganisms, bacteria, fungus, and mold. The combined effect of various factors may be much more severe than any single factor and the degradation process in accelerated many times. It is very important to understand that the published test results do not include synergistic effects of various environmental factors, which almost always exist in real-life situations. Designers should consider exposing fabricated parts to environmental extremes much similar to the ones encountered during the actual use of the product.

Regulatory Approval Requirements
Material selection may be driven by the regulatory requirements put forth by agencies such as Underwriters Laboratories (UL), National Sanitation Foundation (NSF), and Food and Drug Administration (FDA) in terms of flammability, pressure ratings, and toxicological considerations.

Economics
As discussed earlier, material selection should not be driven by cost alone. The most logical approach calls for choosing three to four top candidates based on requirements and selecting one of them with economic considerations.

Other Considerations
Material selection process must also address processing considerations such as type of fabrication process, secondary operations, and component assembly.

15.1.2. Design

Proper material selection alone will not prevent a product from failing. While designing a plastic product, the designer must use the basic rules and guidelines provided by the material supplier for designing a particular part in that material. One must remember that with the exception of a few basic rules in designing plastic parts, the design criteria change from material to material as well as from application to application. Today, designers are challenged with multiple requirements while designing plastic parts. Major emphasis is on economics, functionality, manufacturability, and aesthetic appeal. Some compromise during the design process is inevitable; in some cases, trade-offs like these lead to premature failures. The most common mistakes made by designers when working in plastics are related to wall thickness, sharp corners, creep, draft, environmental compatibility, and placement of ribs. Failure arising from designing parts with sharp corners (insufficient radius) by far exceeds all other reasons for part failures. Maintaining uniform wall thickness is essential in keeping sink marks, voids, warpage, and, more importantly, areas of molded-in stresses to minimum. The viscoelastic nature of plastics materials as opposed to metals require the designer to pay special attention to creep and stress relaxation data. Plastics parts will deform under load over time, depending upon type of material, amount of load, length of time, and temperature. Design guides for proper plastic part design are readily available from material suppliers. Table 15-2 shows a typical part design checklist.

   Earlier in the chapter, we discussed the importance of concurrent engineering practices for a successful part design. The robust product development process that incorporates sound engineering is critical. This can be achieved by incorporating a systematic approach to developing a new product. This logical and scientific approach requires step-by-step

TABLE 15-2. Typical Part Design Checklist. (Courtesy of Bayer Corporation.)

Part Design Checklist

for injection molded engineering thermoplastics

Material Selection Requirements

☐ **LOADS** ___ Magnitude ___ Duration ___ Impact ___ Fatigue ___ Wear

☐ **ENVIRONMENT** ___ Temperature ___ Chemicals ___ Humidity ___ Cleaning agents
___ Lubricants ___ U.V. light

☐ **SPECIAL** ___ Transparency ___ Paintability ___ Warpage/Shrinkage
___ Plateability ___ Flammability ___ Cost ___ Agency test

Part Details Review

☐ **RADII** ___ Sharp corners ___ Ribs ___ Bosses ___ Lettering

☐ **WALL THICKNESS**
 • Material ___ Strength ___ Electrical ___ Flammability
 • Flow ___ Flow length ___ Too thin ___ Thin to thick
 ___ Picture framing ___ Orientation
 • Uniformity ___ Thick areas ___ Thin areas ___ Abrupt changes

☐ **RIBS** ___ Radii ___ Draft ___ Height ___ Spacing
 ___ Base thickness

☐ **BOSSES** ___ Radii ___ Draft ___ Inside diameter/outside diameter
 ___ Base thickness ___ Length/diameter
 ___ Proximity to load ___ Strength vs. load ___ Visual area

☐ **WELD LINES** ___ Draw polish ___ Texture depth

☐ **DRAFT** ___ Part geometry ___ Material ___ ½ degree (minimum)

☐ **TOLERANCE** ___ Tool design (across parting line, slides)

Assembly Considerations

☐ **PRESS FIT** ___ Tolerances ___ Long-term retention ___ Hoop stress
☐ **SNAP FIT** ___ Allowable strain ___ Assembly force
___ Tapered beam ___ Multiple assembly

☐ **SCREWS** ___ Thread-cutting vs. forming ___ Avoid countersinks
☐ **MOLDED**
THREAD ___ Avoid feather-edges, sharp corners and pipe threads

☐ **ULTRASONICS** ___ Energy director ___ Shear joint interference
___ Wall thickness ___ Hermetic seal
☐ **ADHESIVE &** ___ Shear vs. butt joint ___ Compatibility
SOLVENT BONDS ___ Trapped vapors

☐ **GENERAL** ___ Interfit tolerances ___ Stack tolerances ___ Thermal expansion
___ Care with rivets and molded-in inserts ___ Component compatibility

Mold Concerns

☐ **WARPAGE** ___ Cooling (corners) ___ Ejector placement
☐ **GATES** ___ Type ___ Size ___ Location
☐ **RUNNERS** ___ Size and shape ___ Sprue size ___ Balanced flow
___ Cold slug well ___ Sharp corners
☐ **GENERAL** ___ Draft ___ Part ejection ___ Avoid thin/long cores

progression in a definite order. Designers are cautioned not to skip any of the steps for economics reasons or time constraints.

Steps for Robust Part Design Process

Basic part design
Material selection
Structural analysis
Moldflow simulation and analysis
Rapid prototyping
Design review I
Single-cavity prototype
Design review II and tolerance analysis
Tooling protocol and mold cooling analysis
Mold construction phase with regular follow-ups
Sampling, pilot run, and establishing process parameter
Final part evaluation and acceptance

The prototyping aspect of the part design process is often overlooked due to cost and time constraints. Regardless of the medium chosen, the prototyping technique generates physical models that act as a primary means of communication between marketing, engineering, tooling, and manufacturing groups. The use of the prototype to describe the function, size, shape, feel, and look of a part inevitably leads to a major productive environment and a higher degree of interaction between the members of the product design team (6). Equally important is the use of structural analysis tools commonly known as finite element analysis (FEA) and process simulation techniques such as Moldflow® analysis. Computer simulations give designers an early indication of the weak areas and potential problems. Addressing these concerns prior to mold construction designers can avoid costly rework and untimely product failures. Oversimplification could be a real danger in such situations, because it may sway the results too far from the true picture. Often, the significant factors that affect failures are incorrectly considered or ignored. Proper differentiation between primary, secondary, and peak stresses must be made, since each has a separate failure mode that should be considered differently (7).

Many reasons for early product failures are attributed to poor part design. However, the following five reasons are the most prominent of all.

A. Lack of radius
B. Excessive wall thickness variations
C. Incorrect rib placement
D. Environmental compatibility
E. Lack of understanding the creep phenomenon

All plastics are notch-sensitive. Stress concentration resulting from sharp corners and lack of adequate radius tops the list of causes that contribute to the plastic part failures. Designers are constantly reminded to avoid sharp corners at all costs. Sharp corners introduce twofold problems in plastic parts. First, it increases stress concentration in terms of molded-in stresses which tend to reduce mechanical properties and even cause cata-strophic failures. Second, it impedes the flow of material and ejection of parts from the mold. This fact is illustrated in Figure 15-10, which shows the combined effect of stress concentration and molded-in stress due to the sharp corner at the base of two ribs. Residual tensile stresses are the highest at the base of the ribs due to the restricted shrinkage created by the metal core between the ribs which does not allow the part to shrink until ejected from the mold. Differential cooling of the thick sections at the intersection of the ribs with the nominal wall also contribute to the stresses. The addition of the high stress concentration factor at these sharp corners makes them vulnerable to even a small bending moment, resulting in failure at either point A or point B (7).

Stresses build rapidly in internal sharp corners of the part as shown in Figure 15-11, which illustrates the influence of a fillet radius on stress concentration. At a constant wall thickness, as radius increases, R/T also increases proportionally, thereby decreasing the

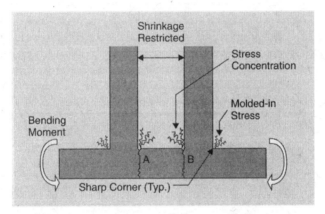

Figure 15-10. The combined effect of stress concentration and molded-in stress due to the sharp corner at the base of two ribs. (Courtesy of Bayer Corporation.)

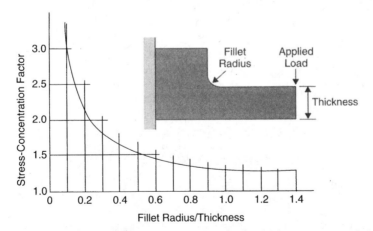

Figure 15-11. The influence of fillet radius on stress concentration. At a constant wall thickness, as radius increases, R/T also increases proportionally and thereby decreasing the stress concentration factor.

stress concentration factor. Even a slight increase in radius can reduce the stress concentration factor drastically. Conversely, not specifying enough fillet radius can be disastrous. As a rule, inside corner radii should be 30–50 percent of the nominal wall thickness, with a 0.020-in. radius as bare minimum. Outside corners should have radius equal to the inside corner plus the wall thickness. This practice allows wall thickness to be uniform at corners and reduces stress concentration. It is important to note that as the curve flattens out beyond an R/T ratio of 1.0, a further increase in fillet radius does not contribute significantly toward the improvement in stress concentration factor. In fact, too generous a radius can create a very thick wall section and possibility of increased molded-in stresses and voids. Effect of notch sensitivity on a very tough material like Polycarbonate is shown in Figure 15-12. There is a drastic reduction in the izod impact strength in a sample with a sharper notch (8). Figures 15-13 and 15-14 illustrate a typical part failure arising from lack of radius in the key areas of the part.

The fundamental rule for designing plastic parts is to maintain uniform wall thickness throughout the part. However, for most applications, complexity of the design requirements makes maintaining uniform wall thickness impractical. The designer must use well-established part design principles such as gradual transition between thick and thin walls to improve stress distribution, gating from thick to thin area, generous radius at intersections to promote flow, originating inner and outer radius from the same point to ensure uniform wall thickness through the corner, and so on. Excessive wall thickness

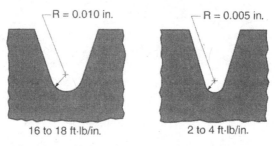

Figure 15-12. Effect of notch sensitivity on a very tough material like polycarbonate. There is a drastic reduction in the izod impact strength in a sample with a sharper notch. (Courtesy of Bayer Corporation.)

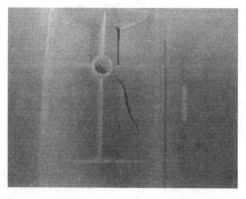

Figure 15-13. Typical part failure from lack of radius in key areas of the part.

Figure 15-14. Typical part failure from lack of radius in key areas of the part.

separation is perhaps the single largest cause of warpage, voids, and sink marks in thicker sections. Such variations lead to a high level of residual stresses. Residual stresses develop due to differential cooling and results in shrinkage differences between thick and thin sections. These internal stresses gradually lead to reduce mechanical performance, and stress parts are also more susceptible to chemical attacks. Warped parts present only aesthetic and minor functional issues in some cases. However, this problem is more severe in the case of assembled parts. Plastic parts are somewhat more ductile in nature and tend to give a little, making assembly of slightly warped parts possible. The constrained parts that are extremely stressed deform under load over time and eventually crack. The combination of internal stresses created by nonuniform wall thickness and external stresses from assembly accelerate the part failure process exponentially. Figure 15-15 shows failure resulting from excessive wall thickness variations.

As discussed earlier, the designer has the option to maintain wall thickness and still maintain the desired rigidity, strength, and structural integrity by incorporating ribs in the part design. The proper use of the ribs and correct rib placement makes the difference between a structurally strong part with uniform wall thickness and an extremely weak part prone to premature failures. Inadequate rib design generally results in high internal warpage, sink marks, voids, stresses, and molding and tooling issues. Rules for the proper rib design are as follows:

1. Make the rib thickness at its base equal to 50 percent of the adjacent wall thickness.
2. Height of the rib should be less than 300 percent of the wall thickness.
3. Radius of the base of the rib must be a minimum of 25 percent of the nominal wall thickness to avoid high stress concentration.
4. Distance between the ribs should be 200 percent of the nominal wall thickness.
5. All ribs should have a draft of 0.5 to 1.5 degrees.
6. Avoid free-standing ribs to minimize (a) air trapping in the blind hole and (b) the resulting burn marks and short shot.

Figure 15-15. Failure resulting from excessive wall thickness variations.

Guidelines for basic rib design are shown in Figure 15-16. Typical issues due to improper rib design are shown in Figure 15-17.

Statistics show that almost 50 percent of the failures of engineering plastics result from environmental degradation (9). Designers must take into account the effect of various environmental factors such as exposure to chemicals, ultraviolet rays, weather extreme, pollution, acid rain, moisture, and microorganisms on the end product. The major challenge is to predict long-term behavior from short-term laboratory or field exposures. The steps in the degradation process involves stress-enhanced absorption and concentration of the chemical molecules at acceptable microstructural sites. Localized plasticization then ensues, leading to crazing and subsequent crack development (9). For the most part, designers are well aware of the limitations of plastic material in terms of chemical compatibility and generally do a thorough job of investigating the effect of well-known solvents and other aggressive chemicals. The problem arises when the least suspected items such as O-rings, seals, gaskets, and trapped vapors from glues, adhesives, and solvents used in assembly—come in contact with the end product. A good example is the incompatibility of the well-known PVC plasticizer dioctyl phthalate (DOP) with polycarbonate (6). Figure 15-18 shows crazing induced in a polycarbonate barb by a PVC tube containing one such plasticizer. One of the major problems facing designers is the lack of chemical compatibility data along with the misinterpretation of published data. Published data are generally derived from immersion of a plastic specimen in a chemical environment for 24 hours suspended in a glass beaker. The ASTM D543 Immersion Test is discussed in Chapter 9 in further detail. Most polymers will undergo stress cracking when exposed to a certain chemical environment under high stress under a given period of time. Such cracking will occur even though some chemicals have no effect on unstressed part, and therefore the simple immersion of a test specimen is an inadequate measure of the chemical resistance of polymers. The combined effect of the molded-in or internal stresses, chemical concentration, temperature, exposure time, and external stresses can be devastating and usually bring about a catastrophic failure. Designers must also pay extra attention

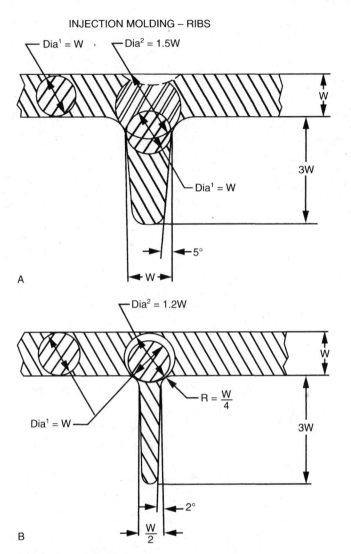

Figure 15-16. General Guidelines
- Height less that 300% of nominal wall thickness (3W)
- Thickness 40 to 60% of nominal wall thickness (W/2)
- Distance between ribs 200% nominal wall thickness (2W)
- Radius at the base of the rib 25 to 70% of nominal wall thickness (W/4)
- Draft .05 to 1.5 degrees
- Use lower values for better appearance and higher values for structural strength

to the parts molded with metal inserts since they tend to become stressed at the interface due to the coefficient of thermal expansion differences and poor molding practices. Cracking of the plastic around a metal insert due to chemical exposure is a common problem and is well illustrated in Figure 15-19. All metal inserts must be preheated to the same as the mold temperature prior to loading them into the mold to minimize molded-in stresses around the inserts. Detrimental effects of environmental factors on appearance and properties are discussed in Chapter 5.

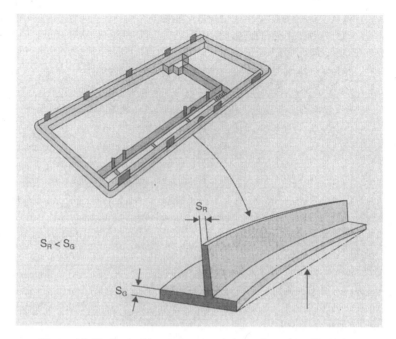

Figure 15-17. Typical issues (warpage) due to improper rib design.

Figure 15-18. Crazing in a polycarbonate barb resulting from contact with incompatible plasticizer used in PVC tubing. (Courtesy of Bayer Corporation.)

Creep is one of the most misunderstood and highly neglected phenomena by designers contributing to premature failure of plastic parts. This is discussed thoroughly in Section 15.1.1.

15.1.3. Process

After proper material selection and design, the responsibility shifts from the designer to the plastic processor. The most innovative design and a very careful material selection cannot

Figure 15-19. Cracks around a metal insert in a part. Heavy stress concentration in this area made the part susceptible to chemical attack.

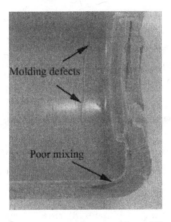

Figure 15-20. Part failure due to poor processing practices. (Courtesy of The Madison Group.)

make up for poor processing practices. Molded-in stresses, voids, weak weld lines, and moisture in the material are some of the most common causes for premature product failures. Ignoring sound processing techniques in order to produce aesthetically pleasing parts generally results in parts lacking physical quality and disappointing production levels (10).

The latest advancement in process control technology allows the processors to control the process with a high degree of reliability and also helps in record keeping should a product fail at a later date. Mold cavities should be fitted with date code inserts that allow changing of the date and other information easily with a screw driver and without taking the mold apart. Such records of processing parameters are invaluable to a person conducting failure analysis. Any assembly or secondary operation on processed part must be evaluated carefully to avoid failures. Many failures arise from stress cracking around metal inserts, drilled holes, and welded joints. Part failure resulting from poor processing practices is shown in Figure 15-20. Proper molding practice begins with a processor, making sure that the correct material is utilized. Historically, many plastic parts failures can be traced back to the processor simply using the wrong material. If mixing of color, additives, or regrind is required, the processor must make sure that the proper ratio is maintained. Appropriate use of automatic material loaders, color mixers, and blenders generally minimize the errors created by manual operation. If the material is hygroscopic in nature, a dehumidifying dryer and recommended drying procedures must be carried out. Poorly maintained and aging machinery cannot produce quality parts despite using the most advanced processing techniques. In the injection molding process, check ring assembly, screw, and barrels are considered to be the most critical components. Worn-out components can lead to material degradation, poor mixing of additives, nonhomogenized melt, underpacking, and other related issues. Processors often deviate from recommended processing parameters and optimum processing conditions to make up for the deficiencies in tooling and process equipment. Such practices are likely to produce physically inferior quality parts and result in untimely failures. For example, a molder may try to minimize the flash created by worn-out or poorly constructed tooling by underpacking and produce parts that are visually acceptable. However, underpacking can result in the formation of micro-voids that act as stress concentrators and crack initiation sites. Process-related failures generally arise from four major categories:

(a) Improper material drying
(b) Under- or overpacking
(c) Cold or overheated material
(d) Improper additives/regrind mixing and utilization

Drying

The majority of plastic materials absorb moisture to some degree. The degree of absorption depends strictly on the chemical structure of the material. Nonpolar materials such as polypropylene, polyethylene, polystyrene, PVC, and so on, have low affinity for moisture and only contain surface moisture. Simple hot air drying is sufficient in this case. Conversely, polar materials are hydrophilic in nature and therefore have a high affinity for moisture. Materials like nylons, polycarbonate, polyesters, and polyurethanes are considered hydroscopic and must be thoroughly dried using a dehumidifying dryer prior to processing. When exposed to high humidity, these materials get fully saturated with moisture and the moisture is trapped inside the resin pellet. This is illustrated is Figures 15-21a and 15-21b. Two things can happen if moisture is not removed completely prior to processing. Parts molded with inadequate drying can have voids and bubbles along with visible splay on the outside of the part. However, lack of visible splay does not necessarily validate proper drying. Moisture-laden parts molded from materials such as polyester may not show splay but can be full of voids within the walls of the parts. Voids and bubbles act as stress concentrators and reduce the ability to sustain load, thereby causing unexpected failures. The second and more severe problem arising from improper drying is one of hydrolysis. Plastic materials like nylons and polyesters that are produced using condensation polymerization technique are susceptible to hydrolysis. During the condensation polymerization process, water and macromolecules are formed. The water is then removed and the process repeats itself. The water molecules separated during the reaction must be constantly removed in order to allow the reaction to continue and form a very-long-chain molecule. If the material is processed wet, the condensation polymerization process can reverse itself (hydrolysis) and long-chain molecules will split into numerous smaller molecules (11). Henceforth, hydrolysis causes reduction is molecular weight, which in turn

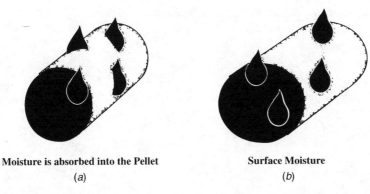

Moisture is absorbed into the Pellet

(a)

Surface Moisture

(b)

Figure 15-21. (*a*) Effect of moisture on hygroscopic resin pellet. (*b*) Effect of moisture on nonhygroscopic resin pellet.

results in drastic loss of physical properties and increase in melt viscosity. Note that this process is nonreversible, which means that if the molder molds hygroscopic material wet, the molded parts will be weak and brittle, and regrind generated from the parts and runners is degraded and cannot be reused. Loss of properties cannot be recovered by drying and reprocessing.

Packing

Molded plastic parts achieve their highest physical properties when the molecules are tightly packed. This is generally accomplished by injecting material into the mold, immediately followed by packing and holding under pressure until the gate seal has taken place. Maintaining sufficient pressure on the melt as it cools and solidifies is extremely critical due to the high volumetric shrinkage of plastic material. If adequate packing pressure is not maintained, molded plastic part will experience uneven shrinkage. Most significant effect of this shrinkage is residual stresses and warpage. By packing out the parts properly, the molder can also eliminate voids and bubbles. A failure analyst will routinely section-mold parts and view them under magnification to determine the presence of microbubbles and voids. Underpacking also creates weak weld lines, a major cause for failures around cored-out holes and bosses. Overpacking is equally detrimental to the quality of the molded parts. Overpacking results in highly stressed parts that are susceptible to warpage and mechanical failure and are likely to be attacked by chemical and solvents. Figure 15-22 shows a failed part as a result of voids created by underpacking.

Cold or Overheated Material. Optimum melt temperature is the key to molding physically strong and mechanically sound parts. Amorphous materials soften gradually and subsequently have a wide processing temperature range, while crystalline materials have a sharp melting point and a narrow processing range. Too often, molders rely on barrel temperature settings and do not make an effort to measure actual melt temperature. The actual melt temperature can be higher by as much as 50–70°F due to the frictional heat generated by the screw and temperature controller inconsistency. Additionally, material can pick up 20–30°F from the frictional heat generated from a high-speed injection through a very small nozzle into the mold cavity contributing to the degradation process. Material degradation resulting from overheating leads to molecular breakdown and loss

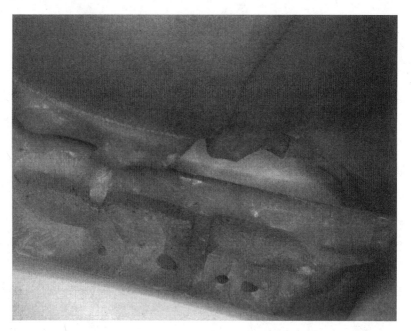

Figure 15-22. Part failure from voids as a result of underpacking. Voids generally lower the load-bearing capability of the part.

of physical properties. In the case of heat- and shear-sensitive materials like PVC, over-heating the melt brings about actual separation of hydrogen and chlorine molecules and formation hydrochloric acid. This triggers an exothermic reaction that accelerates the degradation process until a complete breakdown of the polymor all the way back to carbon occurs. Finally, overheating can also consume or deplete important additives like antioxi-dant and weaken the polymer. Conversely, too cold a melt does not allow the material to fuse properly and create a homogenized mix.

Improper Additive/Regrind Mixing and Utilization. Colors and additives are added to the base polymer for the aesthetic appeal, improve processibility, and enhance performance. Therefore, the proper mixing of the additives is very critical. Additives can generally be viewed in a broad sense as contaminants. Too much or too little can have a detrimental effect on the end product. For example, poor dispersion of antioxidant can result in embrittlement in polyolefin films. Regrind generated from the sprue, runners, and rejected parts are generally fed back into the machine at a predetermined virgin/regrind mix ratio. Controlled laboratory tests have shown declining physical properties with successive generation of regrind usage. Figure 15-23 shows the results of the experiment graphically. Along with declining properties, there is also marked depletion of the additives such as antioxidants and lubricants, which in turn create more processing and product perfor-mance issues. Cross-contamination resulting from grinding mistakes and inadequate cleaning of the grinder bins and hoppers in between jobs have contributed to many part failures. Proper level of regrind usage and through and uniform mixing is paramount to maintaining integrity of the product. Figure 15-24 shows a part failure resulting from excessive regrind usage.

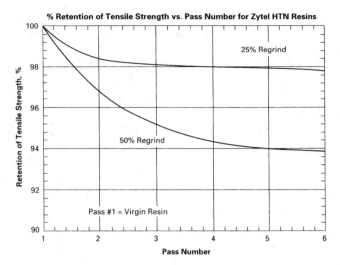

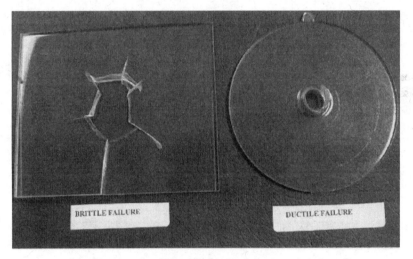

Figure 15-23. Declining physical properties in a part from successive generation of regrind usage. (Courtesy of DuPont, Nylon Zytel® HTN Molding Guide.)

Figure 15-24. Failure resulting from excessive regrind usage.

15.1.4. Service Conditions (Environmental Factors and User Abuse)

In spite of the built-in safety factor warning labels and user's instructions, failures arising from service conditions is quite common in the plastics industry. Five categories of unintentional service conditions are as follows.

1. Reasonable misuse.
2. Use of product beyond its intended lifetime.
3. Failure of product due to unstable service conditions.

4. Failure due to service condition beyond reasonable misuse.

5. Simultaneous application of two stresses operating synergistically (12).

During normal use, plastic parts encounter environmental conditions to varying degree. These conditions include weather extremes, radiation, pollution, and biological factors. Parts are also exposed to a variety of chemicals. Depending on the chemical concentration, amount of internal or external stress, and temperature level, the effect can range from slight to severe. The synergistic effect of chemicals, environment, and mechanical loads can bring about catastrophic failures in a relatively short time. Add to all this the human factor, a consumer who is unfamiliar with the limitations of the plastic products or is poorly instructed on how to properly install and/or use the product, premature product failure is inevitable. Some degree of reasonable misuse of the product is expected and unavoidable. Judging from the court ruling in product liability cases, it behooves the manufacturer to anticipate such reasonable misuse of the product by the consumer and design the product with additional safety margin. For example, a consumer may place plastic utensils or storage containers in the bottom rack of the dishwasher or in a microwave oven for extended period, use a plastic chair to reach the objects instead of a ladder, and expose products made out of inherently poor chemical resistant plastics to harmful solvents and chemicals. It is also reasonable to expect that a consumer will roll out a fully loaded trash container from the driveway and drop it from a 6-in.-high curb and put stress on the wheels. It is unreasonable to expect the consumer to lower the cart gently from the curb. This type of reasonable abuse must be factored-in by the manufacture. Many items fall under the category of "beyond reasonable misuse." Using a screwdriver handle as a chisel and placing a plastic pipe at the end of a wrench to gain extra leverage are the classic examples of user abuse and cannot be considered reasonable. Pipe-dope-containing linseed oil is known to attack products made from ABS. For years, plumbers and DIY consumers have used pipe dope as sealing agent for threaded components. Pool, spa, and sprinkler product manufacturers have made a conscious effort to warn consumers of the danger of using pipe-dope sealants in place of Teflon tape, yet the failures resulting from the misuse continue. Figure 15-25 shows a failed nut due to overtightening by the consumer using a pipe wrench in spite of clear warnings indicating hand tightening of the nut only. Many products fail because they

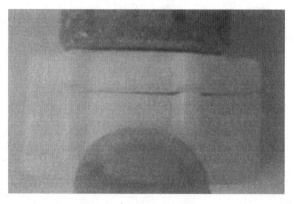

Figure 15-25. Failed nut due to overtightening. Nut from a water supply line. (Courtesy of The Madison Group.)

are used beyond intended lifetime. Disposable products such as cups, grocery bags, and stretch wraps are intended for one time use only.

Most of the stresses imposed on plastics products in service can be grouped under the headings of thermal, chemical, physical, biological, mechanical, and electrical (12). Effect of chemical, mechanical, thermal, environmental, and biological factors are discussed in the following sections.

15.2. TYPES OF FAILURES

15.2.1. Mechanical Failure

Mechanical failure arises from the applied external forces. When the forces exceed the yield strength of the material, they cause the products to deform, crack, or break into pieces. The force may have been applied in tension, compression, and impact for a short or a long period of time at varying temperatures and humidity conditions. There are many reasons why parts fail due to the mechanical force acting upon it. If we rule out thermal, chemical, and environmental reasons and strictly focus on mechanical failures, we find four major types of mechanical failures.

a. Brittle failures
b. Ductile failures
c. Fatigue failures
d. Creep and stress-relaxation-related failures

Brittle Failures
Brittle failures are characterized by a sudden and complete catastrophic failure in which rapid crack propagation is observed without appreciable plastic deformation. Brittle failures, once initiated, require no further energy for the crack to propagate. Material never reaches its yield stress when a part fails in a brittle manner. Notch-sensitive plastics such as polystyrene and acrylics are more prone to brittle failures. There is a direct relationship between the failure mode and the amount of stress in the part, rate of loading (strain rate), and temperature. At high rate of loading, which generally occurs under shock loading or high-speed impact conditions, a part is likely to break in a brittle manner. Low temperatures and high residual stresses tend to accelerate the onset of the brittle failure mode. In glassy amorphous polymers like polycarbonate and polyetherimide, brittle failures are almost always preceded by formation of one or more crazes under large triaxial fields. Crazes are similar to internal voids within a material and, once initiated, act as internal crack or defect. As loading continues, these cracks grow and often lead to brittle failure. Preexisting microcracks, which presumably develop during processing and subsequent cooling of the polymer, propagate and brittle failure occurs under the action of high localized stress at the crack tips. Plastic parts exposed to certain chemicals also transform crazing into cracks and brittle failures. The cracked surfaces of the part failed in a brittle manner are smooth and show very little, if any, torn or jagged surfaces much like the surfaces of a broken piece of a glass.

Ductile Failures
Ductile failures are characterized by gradual tearing of the surfaces when applied forces exceed the yield strength of the material. For the crack resulting from the ductile mode

of failure, additional energy must be provided to propagate the crack by some type of external loading. Ductile failure is slow and noncatastrophic in nature, and the failed specimen generally shows gross plastic deformation in terms of stress whitening, jagged and torn surfaces, necking (reduction in cross-sectional area), and some elongation.

Ductile to Brittle Transition

Generally, plastic materials are more brittle at lower temperatures and they become more ductile as the temperature increases. A strong and tough material may change its ductile behavior and break suddenly in a brittle manner at a low temperature and high strain rate. Of particular interest to the designers and the failure analysts is the point at which the transition from ductile to brittle mode occurs. For most polymers, there is an observable transition in impact energy and failure by brittle cleavage occurring below the transition temperature. Above the transition temperature, the material fails in a ductile manner with considerable deformation: Below the transition the failure is by rapid crack propagation with little deformation (15). Sharp notches and high strain rate lowers the ductile–brittle transition substantially as the temperature decreases. The presence of additives such as colorants also influences the ductile–brittle transition. High-speed impact tests (Section 2.7.3.E) have been developed to study and establish the point at which this ductile-to-brittle transition takes place.

Fatigue Failures

For all materials, the maximum stress that can be applied without failure in cyclic loading is less than that in static loading. Under the applied cyclic load, fatigue cracks initiate somewhere in the specimen and extend during cycling. Eventually, the crack expands to such an extent that the remaining material cannot support the stress and the part fails suddenly (14). Fatigue failures, quite similar to ductile failures, are noncatastrophic in nature and occur over time. The failure is from progressive localized permanent structural change that occurs in a material subjected to repeated or fluctuating stresses well below ultimate tensile strength. Fatigue fractures are caused by the simultaneous action of cyclic stresses, tensile stresses, and plastic strain, all three of which must be present. Cyclic stresses initiate a crack and tensile stresses propagate it (15). The number of cycles required to break the test specimen or a part depends upon the stress imposed on the part. If the stress is sufficiently low, the material will undergo many cycles prior to breakage and in some cases will never break. The fatigue endurance limit, which is defined as the stress below which the material will not fail, is a very important number from a design standpoint. A designer must design the product such that the maximum stresses the part will be subjected to are well below the fatigue endurance limit. This number is usually 20–35 percent of the static tensile strength. Like many other mechanical properties of plastics, fatigue life of the product is greatly influenced by the molecular weight, temperature, and notches in the product (16). Fatigue cracks may also start from microcracks and crazing, which tend to act as stress concentrators and crack initiators.

Creep/Stress Relaxation

Creep is a nonreversible deformation of material under load over time. Stress relaxation is a gradual decrease in stress with time under a constant deformation. Stated another way, creep is an increase in strain under constant stress, whereas stress relaxation is a decrease in stress under constant strain (17). Both of them play a key role in premature plastic product failures. As discussed earlier, designers often misunderstand or overlook the obvious difference between plastics and other traditional materials as it relates to the deformation under load over time and effect of temperature. Creep failure (creep rupture) occurs when polymer chains can no longer hold the applied load and stress reaches levels high enough for microcracks to form. In the case of stress relaxation, at a constant deformation the movement of the polymer chain reduces the force necessary for a given deformation. Stress relaxation rarely results in mechanical failure but can lead to a part losing functionality, as happens in the loosening of bottle caps and the failure of plastic springs (18). Figure 15-26 illustrates a creep-related failure occurring from ovetightening.

15.2.2. Thermal Failure

Four reasons responsible for majority of product failures resulting from thermal exposure of degradation are:

Temperature differences, extreme low to extreme high
Thermal expansion and contraction

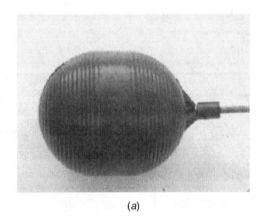

(a)

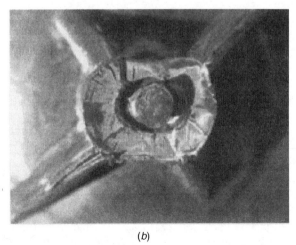

(b)

Figure 15-26. (*a*, *b*) Creep (deformation under load over time)-related failure due to overtightening. (Courtesy of The Madison Group.)

Thermal degradation

Misinterpretation of published heat resistance data

At abnormally high temperatures the product may warp, twist, melt, or even burn. Plastics tend to get brittle at low temperatures. Even the slightest amount of load may cause the product to crack or even shatter. Thermal degradation is basically a molecular deterioration of the polymer as a result of overheating. At high temperature the components of the long-chain backbone of the polymer can begin to separate, known as molecular scission, and react with one another to change the properties of the polymer (19). The thermal degradation can be the result of either (a) overheating during processing of the material or (b) exposing it to high temperatures in service. Thermal degradation generally involves changes to the molecular weight, and molecular weight distribution of the polymer and typical property changes include: reduced ductility, chalking, color changes, cracking, and overall reduction in physical properties. Depletion of the additives such as antioxidants, lubricants, and stabilizers due to overheating during processing is also documented as a major cause of product failures.

As stated earlier, plastics expand and contact anywhere from seven to ten times more than the conventional materials. Allowances must be made form such changes to avoid failures resulting from stresses induced by expansion or contraction. Some of the examples of the most common failures resulting from this effect are piping systems, parts molded with metal inserts, and constrained assembled components with no room for movement. The failures resulting from lack of understanding the real differences between published data for heat distortion temperatures and continuous-use temperatures have been discussed earlier in this chapter. Selecting and specifying materials for long-term high heat application solely based on HDT data is bound to result in premature failure.

15.2.3. Chemical Failure

Failure of plastics from exposure to chemicals occurs in two distinct ways: (1) a chemical attack from simply exposing plastics to incompatible chemicals and (2) a chemical attack from environmental stress cracking. A wide variety of chemicals—including solvents, greases, oils, cleaners, detergents, paints, coatings, and adhesives—can attack plastic parts. Even water at elevated temperatures can hydrolyze plastics and cause hydrolytic degradation in terms of loss of properties such as elongation and impact strength. Continual exposure to water will result in embrittlement. Hydrolytic degradation from inadequately drying the material prior to processing has been discussed in Section 15.1.3.

Chemical compatibility depends upon five factors: exposure time, temperature, chemical concentration, molded-in stresses, and externally applied stresses. The synergistic effect of these factors accelerates the process exponentially. Chemical exposure generally results in the softening of the polymer, crazing, and eventual cracking. As discussed in Chapter 9, Section 9.2, the published chemical compatibility data derived from emersion test should only be used to compare relative resistance of various plastics to a specific chemical reagent. Such data fail to take into account the effect of time, temperature, concentration, and residual stresses along with synergistic effect of these factors. Many plastic products fail due to material selection based on published chemical compatibility data. Product designers must test chemical compatibility separately for each application in all anticipated environmental conditions at various stress levels to avoid untimely failures. Figure 15-27 shows typical failure resulting from chemical attack.

Figure 15-27. Polystyrene part degraded from prolonged contact with gasoline. (Courtesy of The Madison Group.)

Environmental stress cracking (ESC) is a major cause of plastics products failures. Twenty-five percent of the part failures are attributed to environmental stress cracking. ESC is a phenomenon in which a plastic material is degraded by a chemical agent while under stress. It is a solvent-induced failure mode, where the synergistic effect of a chemical agent and mechanical stress result in cracking. It is important to understand that in the case of ESC the chemical agent does not cause direct chemical attack or molecular degradation. The mechanism is mainly physical, and the chemical penetrates into the molecular structure and interferes with the intermolecular forces binding the polymer chains leading to accelerated molecular disentanglement (20). For environmental stress cracking to occur, three things are necessary: polymer, chemical, and stress. The presence of chemical simply accelerates the process of stress cracking. The stress cracking without chemical degradation will eventually occur in the absence of an active fluid and will therefore eventually occur in air (21). Stress cracking begins with crazing in terms of microcracks and continues until a visible crack develops. The failure mode is characterized as brittle fracture, even in material that is otherwise ductile. The crack initiation sites are generally on the surface in the vicinity of highly stressed areas. The cracked surfaces appear somewhat shiny and smooth.

Multiple individual cracks leading to a large crack are also observed. In general, plastics with lower molecular weight, lower crystallinity, and amorphous structure are more susceptible to ESC. High-molecular-weight semicrystalline plastics offer the best environmental stress crack resistance. Figure 15-28 illustrates the progressive steps in failure caused by environmental stress cracking (22). Figure 15-29 show failed sprinkler housing as a result of environmental stress cracking. The stresses were generated by thread tightening and concentrated at the thread root to due relatively small radius. The requisite chemical agent responsible for the failure was identified as the ester-base oil with a commercial pipe dope sealant. Ester-based oils are known to product ESC failures within ABS resins. Pipe dope sealants have been widely recognized as deleterious to ABS pipes and fittings.

15.2.4. Environmental Failure

Plastics exposed to outdoor environments are susceptible to many types of detrimental factors. The effect can be anywhere from a mere loss of color, slight crazing, and cracking,

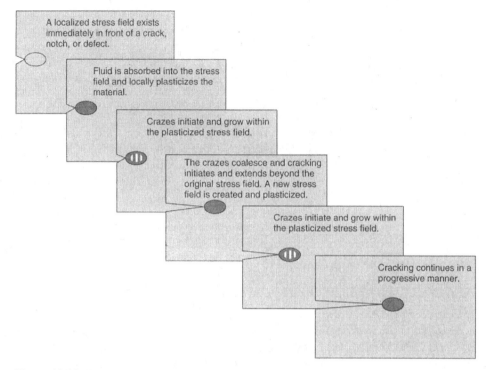

Figure 15-28. Progressive steps in failure due to environmental stress cracking. (Reprinted with permission of ASM International.)

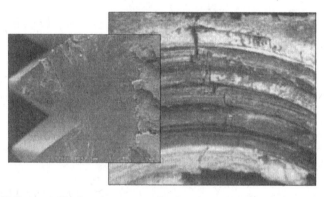

Figure 15-29. Failed sprinkler housing as a result of environmental stress cracking. (Reprinted by permission of ASM International.)

to a complete breakdown of the polymer structure. A typical product failure arising from weathering effects in shown in Figure 15-30.

Almost 50 percent of the failures of plastic parts result from environmental degradation. A large portion of the environmental degradation is attributed to (a) natural weathering and (b) exposure to solar radiation (which includes UV, IR, X-ray, etc.), pollution, industrial chemicals, and humidity. Other major contributing factors are microorganisms,

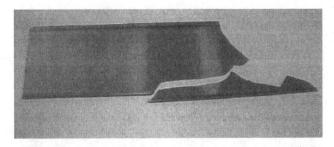

Figure 15-30. Failure occurring in a plastic part due to prolonged exposure and resulting degradation from UV rays. (Courtesy Consultek, LLC.) A full color version of this figure can be viewed on DVD included with this book.

such as bacteria, fungus, mold, water, thermal energy, and ionizing radiation resulting from gamma rays, X-rays, and electron beams. As discussed earlier, the synergistic effect of the mentioned factors is much more severe than the effect of any single factor and the degradation process accelerates many times.

The failure mode varies from simple discoloration to embrittlement in terms of microcracks, loss of impact, elongation, reduction in molecular weight, a complete breakdown of molecular structure, and severe cracking. The severity of the damage depends on the natural environment, geographic location, the type of plastic, and the concentration and duration of exposure. Predicting the long-term performance of plastic products when exposed to environmental conditions is a difficult task. Accelerated laboratory weathering tests tend to produce confusing and unrealistic results. Results from newly developed accelerated outdoor weathering tests are more realistic. Accelerated outdoor tests use solar tracking and/or concentrator systems to expose specimens to real outdoor weather while increasing the exposure to solar radiation. These methods can obtain useful results at least eight times faster than real-time weathering. Since natural weathering is invariably a combination of UV and oxidation from wetness, a nighttime spray cycle must be incorporated for more realistic results.

All products requiring long-term outdoor use must be protected by incorporating UV absorbers or UV stabilizers into the plastic base resins. The UV absorbers provide preferential absorption to most of the incident UV light and are able to dissipate the absorbed energy harmlessly. UV stabilizers, on the other hand, inhibit the bond rupture by chemical means or dissipate the energy to lower levels that do not attack the bonds. Protective coatings and paints are also very effective in extending the life of the products exposed to natural weathering. Figure 15-31 illustrates a decomposition of a swimming pool skimmer cover from continuous exposure to sunlight and moist environment.

15.3. ANALYZING FAILURES

The first step in analyzing any type of failure is to determine the cause of the failure. Before proceeding with any elaborate tests, some basic information regarding the product must be gathered. If the product is returned from the field, have the district manager or consumer give you basic information, such as the date of purchase, date of installation, date when the first failure encountered, geographic location, types of chemicals used with or around the product, and whether the product was used indoors or outdoors. All this

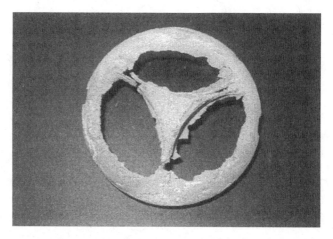

Figure 15-31. Decomposition of a swimming pool skimmer cover from continuous exposure to sunlight and moist environment.

information is very vital if one is to analyze the defective product proficiently. For example, if the report from the field along with the defective product indicates that a certain type of chemical was used with the product, one can easily check the chemical compatibility of the product or go one step further and simulate the actual-use condition using that same chemical. Record keeping also simplifies the task of failure analysis. A simple date code or cavity identification number will certainly enhance the traceability. Many types and styles of checklists to help analyze the failures have been developed. Table 15-3 illustrates a typical failure analysis checklist. Flow Chart 15-1 illustrates a procedure for methodically analyzing failures.

Ten basic methods are employed to analyze product failure.

1. Visual examination
2. Identification analysis
3. Stress analysis
4. Heat reversion technique
5. Microstructural analysis (microtoming)
6. Mechanical testing
7. Thermal analysis
8. Nondestructive testing (NDT) techniques
9. Fractography
10. Simulation testing

By zeroing in on the type of failure, one can easily select the appropriate method of failure analysis.

15.3.1. Visual Examination

Visual analysis is one of the most important aspects of failure analysis. A well-executed and carefully documented visual analysis forms the basis for other types of sophisticated

TABLE 15-3. Typical Failure Analysis Checklist

1. History and Description of the Failure
    1. Did it ever work?
    2. What changed?
    3. Field failure
    4. QC failure
    5. As-molded failure
    6. Percentage of failure
    7. Repeatable
    8. Location always the same

    What does it look like?
    1. Brittle
    2. Ductile
    3. Plastic deformation

    How bad is it?
    1. Minor
    2. Catastrophic

    What was the mode of failure?
    1. Flexure
    2. Tensile
    3. Compression
    4. Torsion
    5. Vibration
    6. Impact
    7. Creep

    How about area-crack origin?
    1. Crack origin
    2. Thin or thick wall
    3. Transition area
    4. Gate
    5. At a boss

2. Materials
    What about the resin?
    1. Grade and lot number
    2. Color
    3. Meet minimum specification
    4. Correct for the application
    5. Contains regrind
    6. How much?
    7. Is it contaminated?

3. Design
    1. Examinations for print to part variations
    2. Wall thicknesses and transitions
    3. Gate locations, knit lines
    4. Fillets, radii, and ribs
    5. Part deformation
    6. Snap-fit and press-fit
    7. Finite Element Analysis

4. Processing
    1. Tooling
    2. Proper gating and runner systems
    3. Cavity variations
    4. Review processing conditions and drying cycles
    5. Recycle
    6. Shot size, machine capacity
    7. Splay and gate blush
    8. Sinks and voids
    9. Burn marks, black specs
    10. Presence of mold releases

5. Secondary Operations of Assembly, Finishing, and Machining.

5.1 Assembly
    Bonding
    1. Performance concerns
    2. Strength
    3. Compatibility

    Welding
    1. Design
    2. Strength
    3. Sealing
    4. Burn marks on the part
    5. Equipments
    6. Process verification

TABLE 15-3. *Continued*

Fastening and inserting
1. Product selection
2. Process selection
3. Pilot hole size

4. Strength
5. Boss fracture

Staking
1. Process specifications
2. Strength

3. Equipments

Specialty
1. The effects of riveting

2. Special hardware

5.2 Finishing
Appliques
1. Adhesion
2. Application parameters

3. Performance

Color
1. Stability
2. Procedure

3. Compatibility

Metalization
1. Adhesion
2. Compatibility

3. Effectiveness
4. Environmental performance

Printing
1. Adhesion
2. Compatibility

3. Procedure

5.3 Machining
Induced thermal and mechanical stresses through operations such as
1. Cutting
2. Milling
3. Drilling

4. Annealing
5. Degreasing

6. Environmental
1. Shiny fracture surfaces
2. Chemical exposure
3. High or low heat exposure

4. Humidity
5. UV degradation
6. Friction and wear

7. Application Stresses
1. Impact
2. Weathering
3. Thermal

4. Handling
5. Storage
6. Electrical breakdown

Courtesy of GE Plastics.

tests, and it gives failure analyst a clue and a direction in which the rest of the analysis needs to be conducted. It is also the least expensive techniques, only requiring trained eyes, magnifying glass, and adequate lighting. First, the evidence should be handled very gently and carefully. Broken pieces and open surfaces must not be disturbed. Parts may have to be washed with special reagents at a later date to conduct chemical analysis to determine presence of unwanted stress inducing reagents. If parts are returned in several pieces, try to arrange them very loosely to form a complete part. This may give a clue

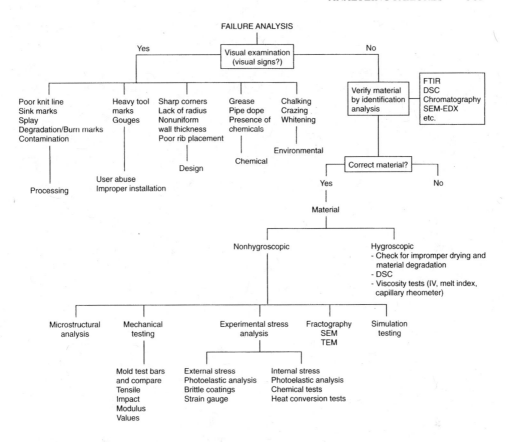

about the location of fracture initiation site. If sufficiently large sample of defective parts are available, try to look for a cavity numbers on the parts. Discovering that the majority of the defective parts come from one or two cavities can provide a clue. The defective parts should be compared with the good parts from the rest of the cavities. Areas to look for include gate size, gate location, sharp corners, missing radius, voids, and so on. A careful visual examination of the returned part can reveal many things. Excessive splay marks indicate that the material was not adequately dried before processing. The failure to remove moisture from hydroscopic materials can lower the overall physical properties of the molded article and in some cases even cause them to become brittle. The presence of foreign material and other contaminants is also detrimental and could have caused the part to fail. Burn marks on molded articles are easy to detect. They are usually brown streaks and black spots. These marks indicate the possibility of material degradation during processing causing the breakdown of molecular structure leading to overall reduction in the physical properties. Parts molded from PVC can be examined under black light (UV light) to investigate possible material degradation. A degraded or burned portion (chlorine molecules) generally fluoresces and is easy to detect. Sink marks and weak weld lines, readily visible on molded parts, represent poor processing practices and may contribute to part failure. A rule of thumb used to determine the quality of the weld line calls for an attempt to feel the depth of the open weld line with a fingernail. A well-formed weld line should not be felt with a fingernail.

Visual examination will also reveal the extent of consumer abuse. The presence of unusual chemicals, grease, pipe dope, and other substances may give some clues. Heavy marks and gouges could be the sign of excessively applied external force. Sharp corners, missing radii, nonuniform wall thickness, poor placement of ribs, and so on, may point to part failures related to design issues. Broken surfaces should be carefully examined under magnification. A smooth broken surface points to a brittle mode of failure, while a jagged, torn, or pulled surfaces with or without stress whitening indicates ductile failure. A smooth surface is also an indication of chemical attack and resulting brittle failure. Warped parts may indicate (a) high residual stresses in the part resulting from processing issues or (b) thermal problems arising from high heat exposure.

The defective part should also be cut in half using a sharp saw blade. The object here is to look for voids caused by trapped gas and excessive shrinkage, especially in thick sections during molding. A reduction in wall thickness and sharp notches created by such voids could be less than adequate for supporting compressive or tensile force or withstanding impact load and may cause part to fail. Lastly, if the product has failed because of exposure to UV rays and other environmental factors, a slight chalking, crazing, microscopic cracks, large readily visible cracks, or loss of color will be evident.

15.3.2. Identification Analysis

One of the main reasons for product failure is simply the use of the wrong material. When a defective product is returned from the field, material identification tests must be carried out to verify that the material used in the defective product is, in fact, the material specified on the product drawing. However, identifying the type of material is simply not enough. Since all plastic materials are supplied in a variety of grades with a broad range of properties, the grade of material must also be determined. A simple technique such as the melt index test can be carried out to confirm the grade of a particular type of material. The percentage of regrind material mixed with virgin material has a significant effect on the physical properties. Generally, the higher the level of regrind material mixed with virgin, the lower the physical properties. If, during processing, higher-than-recommended temperature and long residence time is used, chances are that the material is degraded. This degraded material, when reground and mixed with virgin material, can cause a significant reduction in overall properties. Unfortunately, the percentage of regrind used with virgin is almost impossible to determine by performing tests on the molded parts. However, a correlation between the melt index value and the part failure rate can be established by conducting a series of tests to determine the minimum or maximum acceptable melt index value (27). Alternatively, dilute solution viscosity tests can be used to determine the molecular weight of the polymers. By comparing the molecular weight of the known sample with the suspect, it is possible to determine if the material was degraded during processing. The degraded material will have lower molecular weight. Part failures due to impurities and contamination of virgin material are not that unusual. Material contamination often occurs during processing. A variety of purging materials are used to purge the previous material from the extruder barrel before using the new material. Not all of these purging materials are compatible. Such incompatibility can cause the loss of properties, brittleness, and delamination. Poor material handling practices contribute to the contamination resulting from regrind usage. In the vinyl compounding operation, failure to add key ingredients, such as an impact modifier, can result in premature part failure. Simple laboratory techniques cannot identify such impurities, contamination, or the absence of a key ingredient. More sophisticated techniques, such as chromatography, spectroscopy, and electron microscopy (SEM-EDX), must be employed. These methods not only can positively identify the basic material, but also can point out the type and level of impurities in most cases. Plastic materials are rarely manufactured without some type of additives. These additives play an important role in thermally or environmentally stabilizing the base polymer. Antioxidants, flame retardants, UV stabilizers, heat stabilizers, and lubricants are routinely added for enhancement. If for some reason these important additives are left out or depleted due to poor processing practices, plastic parts may fail prematurely. Techniques such as deformulation using complex HPLC or GPC followed by NMR, PY/MS, and FTIR of purified factions is employed for separation, identification, and quantification of ingredients. These techniques can also be used to reverse engineer a product or assembly. Advanced methods for detailed identification of additives and other ingredients are discussed in Chapter 13. Among many methods available for identification and analysis, for all practical purposes and cost containment, only four basic techniques are most frequently used for plastic materials:

1. FTIR
2. DSC
3. Ash content (burn-off test) or TGA
4. Viscosity tests

FTIR is one of the most powerful identification techniques available today for positive identification of polymeric materials. The test is conducted simply by passing an infrared beam through a small sample, whereby some of the infrared radiation is absorbed or transmitted. The spectrum resulting from this test acts like a fingerprint unique to a particular polymer. Since no two unique molecular structures produce the same infrared spectrum, the resulting spectra can be compared with known material spectra and material is positively identified. One of the biggest advantages of FTIR is speed, accuracy, and quantity of specimen required to obtain the results. Depending upon the sophistication and the depth of the spectra library used, one can identify materials, additives, and contaminants within minutes. Once the generic type of plastic material is identified (nylon, acetal, polycarbonate), further tests such as DSC can be carried out to identify specific type (nylon 6, nylon 66, nylon 612, and acetal homopolymer vs. copolymer) within the same generic family. The FTIR technique can also identify if something other than basic polymer such as a contaminant is present. If the failed product's base material is supposed to contain inorganic fillers and reinforcement such as talc, calcium carbonate, and glass fibers, the ash content test is very effective in determining and verifying a specified percentage. This is accomplished by simply burning off the polymer in a muffle furnace at 600°C and weighing the leftover ash. By viewing the ash content under a microscope, one can differentiate between glass and minerals. A more sophisticated and highly accurate test known as thermogravimetric analysis (TGA) can also be used for this purpose. This technique is useful when only a few milligrams of sample is available for testing. Once the material is properly identified, it can be labeled as one of the two types of materials, hygroscopic or nonhygroscopic. All hygroscopic materials must be dried thoroughly prior to processing to avoid resulting loss of properties and reduction in overall strength. Intrinsic viscosity, melt index, and capillary rheometer tests are conducted to determine the extent of reduction in molecular weight of the polymer due to hydrolytic and thermal degradation. The tests are discussed in-depth in Chapter 7.

15.3.3. Stress Analysis

Once the part failure resulting from poor molding practices or improper material usage through visual examination and material identification is ruled out, the next logical step is to carry out an experimental stress analysis.

Experimental stress analysis is one of the most versatile methods for analyzing parts for possible failure. This technique fills the gap between theoretical stress analysis conducted using finite element method and failure analysis. Performed on molded or prototype parts and models, experimental stress analysis not only complements the finite element method but also provides accurate insights into the total stress picture, including residual stresses in molded parts (23). The part can be externally stressed or can have residual or molded-in stresses. External stresses or molded-in stresses or a combination of both can cause a part to fail prematurely. Detection of residual stresses has a different meaning than evaluation of stresses due to applied forces. It is possible of course to see failure resulting from poor design, or underestimating of forces. These failures are usually detected in proof testing, or in early production. Residual stresses are altogether different: A molding process can generate residual stress just about anywhere, anytime. Residual stresses are responsible for reduction in strength, diminished temperature performance, and lower chemical resistance. Here, ongoing photoelastic inspection can prove extremely helpful, allowing early detection of defective molded parts or identification of failures in clear plastic products. It is important to understand that experimental stress analysis techniques do not measure stress, but instead measure strain. The measured strains are then converted into stresses by calculation, by taking into account the material properties of the product being tested (18). Five basic methods are used to conduct stress analysis:

> Photoelastic method
> Brittle-coating method
> Strain gauge method
> Chemical method
> Heat reversion technique

Photoelastic Method

The photoelastic method for experimental stress analysis is very popular among design engineers and failure analysts and has proved to be an extremely versatile, yet simple, technique.

If the parts to be analyzed are made out of one of the transparent materials, stress analysis is simple. All transparent plastics, being birefringent, lend themselves to photoelastic stress analysis. The transparent part is placed between two polarizing mediums and viewed from the opposite side of the light source. Photoelasticity is based on the theory that when polarized light passes through a material that is stressed, the light splits into two divergent polarized beams vibrating in different planes (x and y) along the direction of the principal stresses. This phenomenon, which results in two different indices of refraction, is known as *birefringence*. The two beams travel through the material at different speeds, depending on the stresses encountered and the thickness of the material. The relative distance between the two light waves (fast and slow) as they emerge from the material is called *retardation*. By rotating the second polarizing filter (analyzer), the user can control the amount (intensity) of light allowed to pass through. The components of the two light waves that do pass through at any given angle of analyzer rotation interfere with each other, resulting in a characteristic color spectrum. The intensity of colors displayed when a stressed transparent or translucent material is viewed under polarized light is modulated by the retardation. The observed colored fringes are simply level lines of constant stress along a fringe. The intensity of the colors diminishes as the retardation or fringe order increases. These color patterns, visible when using polarized light, can be used to observe and make a qualitative evaluation of stress in an object (25).

The fringe patterns are observed without applying external stress. This allows the observer to study the molded-in or residual stresses in the part. High fringe order indicates the area of high stress level, whereas low fringe order represents an unstressed area. Also, close spacing of fringes represents a high stress gradient. A uniform color indicates uniform stress in the part (26). Next, the part should be stressed by applying external force and simulating actual-use conditions. The areas of high stress concentration can be easily pinpointed by observing changes in fringe patterns brought forth by external stress. Figure 15-32 illustrates a typical stress pattern in a part. This type of evaluation is useful as a regular part of product inspection for quality control of transparent parts for any manufacturer to maintain product quality and consistency and to prevent failures. The method often reveals problems associated with many other process control parameters such as temperature, material, fill rate, design, and so on. Three basic measuring techniques are employed:

1. Simple observation of color pattern
2. Compensator method
3. Analyzer rotation method

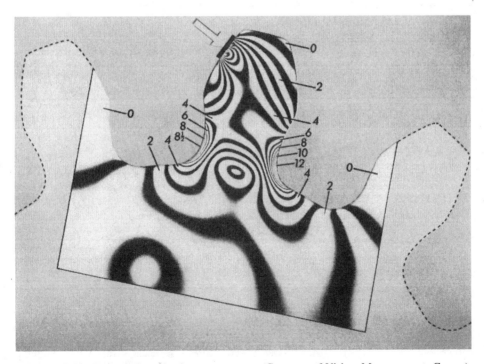

Figure 15-32. Typical photoelastic stress pattern. (Courtesy of Vishay Measurements Group.)

Simple observations of color pattern by viewing the part placed between two polarized sheets have limited capability. The results are highly subjective to interpretation, and they can only be used for qualitative measurements. The compensator method (ASTM D4093) uses a compensator (wedge), a calibrated hand-held device that optically adds retardation of equal but opposite sign to the sample. The operator adjusts the compensator position until a black fringe appears at the measurement point as shown in Figure 15-33. A scale on the compensator provides a quantitative reading of optical retardation. A more advanced analyzer rotation method is generally used to measure fractional levels of retardation. Injection-molded transparent parts used in medical devices are often annealed to relieve molded-in stresses. Photoelastic measurement techniques have been used to determine the effectiveness of annealing and also for the failure analysis of defective parts. Figure 15-34 shows residual stresses in medical packaging before and after annealing. Three-dimensional model analysis is used effectively to analyze the stress distribution on the surface and at any point inside the model. The method calls for subjecting the models prepared by machining or by stereolithography (SLA) to scaled-down forces while heated in an oven and then slowly cooled down to room temperature to freeze-in the photoelastic stress pattern. The model is cut into thin slices and photoelastically examined to determine the level of stresses.

Another technique known as the *photoelastic coating technique* can be used to photoelastically stress-analyze opaque plastic parts. The part to be analyzed is coated with a photoelastic coating, service loads are applied to the part, and the strains on the surface of the part are transmitted to the coating, which assumes the same strain condition as the part. The coating is then illuminated by polarized light from the reflection polariscope. With a digital compensator attached to the polariscope, quantitative stress analysis can be easily performed (27). In summary, photoelastic techniques can be used successfully for failure analysis of a defective product.

Observing the Areas of Stress Concentration. The location of the molded-in stress displayed by the patterns of birefringence rings is related to the type of material used and

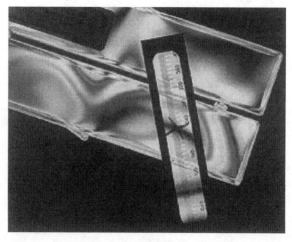

Figure 15-33. Compensator method for quantitative stress measurement. (Courtesy of Strainoptics, Inc.) A full color version of this figure can be viewed on DVD included with this book.

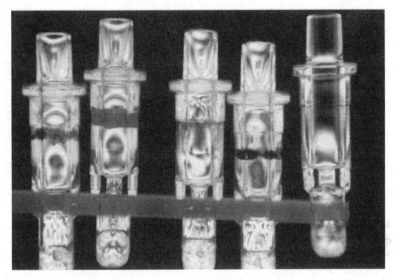

Figure 15-34. Residual stresses before and after annealing. A full color version of this figure can be viewed on DVD included with this book.

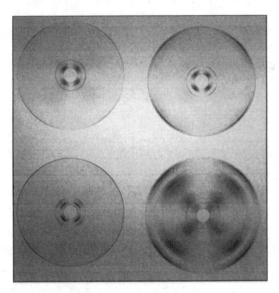

Figure 15-35. Birefringence comparison of four compact disks with varying degree of molded-in stresses. (Reprinted with permission of *Injection Molding.*)

molding conditions. Processing parameters such as pressure and temperature, will affect the location and/or the amount of molded-in stress. Figure 15-35 illustrates the birefringence comparison of four compact disks with varying degree of molded-in stress. The CD at the lower right as molded with general-purpose polystyrene; the other three are molded with high-flow optical grade of polycarbonate specifically formulated for compact disks. The compact disk in upper left corner shows that the molded-in stress is concentrated at the hub. The stress appears at the outer edges of the CD shown in upper right corner, where it is not critical since there are little or no data stored in these areas of CD. The compact disk in the lower left shows the lowest degree of molded-in stress of all three samples. This is represented by consistent clarity from center to all edges. The compact disk shown in the lower right represents the greatest amount of molded-in stress. The number of bands and their tight spacing indicate areas of high stress throughout the disk. The amount of molded-in stress is due to the high-molecular-weight general-purpose polystyrene used in manufacturing this CD and visibly explains why this material is not suitable for the use in manufacture of CDs.

Brittle-Coating Method

The brittle-coating method is yet another technique of conveniently measuring the localized stresses in a part. Brittle coatings are specially prepared lacquers that are usually applied by spraying on the actual part. The part is subjected to stress after air-drying the coating. The location of maximum strain and the direction of the principle strain are indicated by the small cracks that appear on the surface of the part as a result of external loading. Thus, the technique offers valuable information regarding the overall picture of the stress distribution over the surface of the part. Figure 15-36 shows a typical crack pattern in the coating after being strained. The data obtained from the brittle coating method can be used to determine the exact areas for strain gauge location and orientation, allowing precise measurement of the strain magnitude at points of maximum interest. They are also useful for the determination of stresses at stress concentration points that are too small or inconveniently located for installation of strain gauges (28). The brittle-coating technique, however, is not suitable for detailed quantitative analysis like photoelasticity as well as measurement of residual stresses. Sometimes it is necessary to apply an undercoating prior to the brittle coating to promote adhesion and to minimize compatibility problems. Perhaps the best way to use the information obtained from this technique is to pinpoint the location and direction of the stress in the part and determine the best place for mounting strain gages. Further discussion on this subject is found in the literature (29,30).

Strain Gauge Method

The electrical resistance strain gauge method is the most popular and widely accepted method for strain measurements. The strain gauge consists of a grid of strain-sensitive

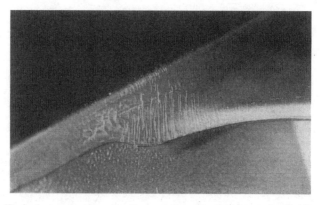

Figure 15-36. Typical stress pattern in the coating after being strained. (Courtesy of Bayer Corporation.)

metal foil bonded to a plastic backing material. When a conductor is subjected to a mechanical deformation, its electrical resistance changes proportionally. This principle is applied in the operation of a strain gauge. For strain measurements, the strain gauge is bonded to the surface of a part with a special adhesive and then connected electrically to a measuring instrument. When the test part is subjected to a load, the resulting strain produced on the surface of the part is transmitted to the foil grid. The strain in the grid causes a change in its length and cross section and produces a change in the resistivity of the grid material. This change in grid resistance, which is proportional to the strain, is then measured with a strain gauge recording instrument (31). In using strain gauges for failure analysis, care must be taken to test the adhesives for compatibility with particular plastics to avoid stress-cracking problems. Residual or molded-in stresses can be directly measured with strain gauges using the hole drilling method. This method involves measuring stress at a particular location, drilling a hole through the part to relieve the frozen-in stresses, and then remeasuring the stress. The difference between the two measurements is calculated as residual stress.

Chemical Method (Solvent Stress Analysis)

Most plastics, when exposed to certain chemicals while under stress, show stress cracking. This phenomenon is used in stress analysis of molded parts. Solvent stress analysis provides a quantitative means of determining stress levels in parts. The method is used to evaluate molded-in stress, which result from poor processing practices or other stresses created by metal inserts around hubs, screws, rivets, and other externally applied forces. Inherent stresses develop in molded parts as a result of injecting hot molten plastics material into a relatively cold mold. Such molecular orientation develops during the mold-filling phase as the melt is injected through the nozzle, runners, gates, and cavity. Internal stresses related to molecular orientation can lead to warpage or remain as internal stresses reducing the durability and environmental stress crack resistance of the molded part. Highly stressed parts are attacked by solvents above its critical stress level. The net effect is crazing and cracking, which can be visually seen on the part. Figure 15-37 illustrates the result of solvent attack in terms of crazing and cracking around the highly stressed mounting holes.

There are two factors that make polymers susceptible to chemical attack: basic chemical structure and polarity of the solvent and the polymer. A polymer with large free volume

Figure 15-37. The result of a solvent attack in terms of crazing and cracking around the highly stressed mounting holes.

such as polycarbonate allows solvents to penetrate into polymer matrix. Solvents penetrate and swell the polymer at preferential sites of stress concentration, making stress cracking or crazing possible. Polymer and solvent polarities also determine the degree of attraction or repulsion between them and help determine the sensitivity of a polymer to a particular solvent. If the solubility parameter of the solvent closely matches that of the polymer, the polymer will be attacked by the solvent. As the difference between parameters increases, resistance to failure under stress improves (32).

In one of the tests, the part is immersed in a mixture of glacial acetic acid and water for 2 min at 73°F and later inspected for cracks that occur where tensile stress at the surface is greater than the critical stress (33). The part may also be externally stressed to a pre-determined level and sprayed on with the chemical to determine critical stresses. Stress cracking curves for many types of plastics have been developed by material suppliers. If a defective product returned from the field appears to have stress-cracked, similar tests should be carried out to determine molded-in stresses as well as the effect of external loading by simulating end-use conditions. Failures of such types are seen often in parts where metal inserts are molded-in or inserted after molding. Three other tests—strain-resistance test, solvent stress-cracking resistance, and environmental stress-cracking resistance (ESCR)—are also employed to analyze failed parts. The acetone immersion test, to determine the quality of rigid PVC pipe and fittings as indicated by their reaction to immersion in anhydrous acetone, is very useful. An unfused PVC compound attacked by an anhydrous acetone causes the material to swell, flake, or completely disintegrate. A properly fused PVC compound is impervious to anhydrous acetone; only a minor swelling, if any, is observed. Defective PVC pipe or fittings returned from the field are subjected to this test for failure analysis.

A Sample Solvent Stress Analysis Report

EXECUTIVE SUMMARY. All injection-molded plastic parts have some degree of molded-in or frozen-in (residual stress). The degree of residual stress in the molded part depends upon part design, tooling design, and molding practices. Such residual stress at a very high level can increase sensitivity to environmental stress cracking, lower overall impact strength, and diminish the heat performance. This project was aimed at determining the amount of residual stress in various molded parts by conducting the Solvent Stress Analysis test. A secondary objective is to establish the minimum acceptable molded-in stress for these parts so that it can be used as quality assurance guidelines.

The test results showed a stress level of <1000 psi in all molded polycarbonate parts. Typical stress levels in properly designed, well-molded parts range from 900 to 1200 psi. The test results therefore indicate that the parts are designed and molded properly and should not show an adverse effect on long-term performance of the product.

PURPOSE. The purpose of conducting the Solvent Stress Analysis test was to determine the amount of molded-in stress present in various injection molded components. The assembly, after being subjected to the aging environment, had encountered functional issues. The general consensus was that functional issues may be due to the out-of-roundness condition created by exposing the assembly to high aging temperatures. If the molded parts were highly stressed, they would have a tendency to stress-relieve when exposed to higher aging temperature and subsequently show dimensional variations among various assembled components. The Solvent Stress Analysis test would determine the level of molded-in stress in the parts, and the results can be used to set a quality assurance minimum acceptable stress level guideline for the molder.

MATERIALS AND TEST PROCEDURE

Test:	Solvent Stress Analysis
Test Method:	GE Plastics Test Method (T-77)/Acetic Acid Immersion Test, ASTM 1939
Project #:	CP 1001
Customer:	Consultek, LLC
Date:	August 15, 2003
Material:	Polycarbonate and ABS
Test Condition:	23°C/50% RH
Sample:	Not required
Preparation:	Molded parts
Sample Types:	3 minutes per sample
Test Duration:	Methanol/Ethyl acetate

1. Various solutions are prepared to reflect critical stress values of 1200, 1100, 1000, 800, and 500 psi. An intermediate solution having a critical stress level of 1200 psi is used as a starting point. If no crazing is apparent, a more highly concentrated ethyl acetate solution is used and a test procedure is repeated using a fresh specimen.
2. A part is immersed and thoroughly wetted and kept wet for a 3-minute period in a given solution.
3. After 3 minutes, the part is withdrawn and placed in distilled water for rinsing.
4. The part is examined visually for signs of stress crazing.
5. If no crazing or cracking is apparent, a higher ethyl acetate concentration solution is used and the same procedure was repeated using a fresh part.
6. A minimum of three samples of each type is used in the test.

RESULTS

Critical Stess	Part A	Part B	Part C	Part D
Methanol/Ethyl Acetate				
1200 psi (71:29)	NSC	NSC	NSC	NSC
1100 psi (69:31)	NSC	NSC	NSC	NSC
1000 psi (67:33)	NSC	NSC	NSC	NSC
800 psi (63:37)	NSC	NSC	NSC	NSC
570 psi (50:50)	NSC	NSC	NSC	NSC

NSC, no stress cracking.

CONCLUSIONS AND RECOMMENDATIONS. Test results clearly show that the molded-in stress in vial parts are less than 1000 psi. Typical stress levels in properly designed, well-molded parts range from 900 to 1200 psi. The test results therefore indicate that the parts are designed and molded properly and should not show an adverse effect on long-term performance of the product.

In order to ensure that the parts are molded at a low and acceptable stress level, we recommend developing a quality assurance standard and a test procedure that can be used as ongoing as well as receiving inspection criteria. The maximum acceptable stress level in the part should be set at 1000 psi.

Heat Reversion Technique (ASTM F1057)

All plastic manufacturing processes introduce some degree of stress in the finished product. As stated previously, the stresses in molded parts are commonly referred to as molded-in (residual) stresses. By reversing the process, by reheating the molded or extruded product, the presence of stress can be determined. The test is conducted by simply placing the entire specimen or a portion of the specimen in a thermostatically controlled, circulating air oven and subjecting it to a predetermined temperature for a specified time. The specimens are visually examined for a variety of attributes. The degree and severity of warpage, blistering, wall separation, fish-scaling, and distortion in the gate area of molded parts indicate stress level. Stresses and molecular-orientation effects in the plastic material are relieved, and the plastic starts to revert to more stable form. The temperature at which this begins to occur is important. If changes start below the heat distortion temperature of the material, high levels of stress and flow orientation are indicated. The test has been significantly improved by new methods including the attachment of strain gauges to critical regions of the part to carefully monitor initial changes in the shape (34). ASTM F1057 describes the standard practice for estimating the quality of extruded PVC pipe by the heat reversion technique.

15.3.4. Microstructural Analysis (Microtoming)

Microtoming is a technique of slicing an ultra-thin section from a molded plastic part for microscopic examination. This technique has been used by biologists and metallurgists for years, but only in the last decade has this technique been used successfully as a valuable failure analysis tool.

Microtoming begins with the skillful slicing of an 8- to 10-μm-thick section from a part and mounting the slice on a transparent glass slide. The section is then examined under a light transmission microscope equipped with a polarizer for photoelastic analysis. A high-power (20× to 400×) microscope that will permit photographic recording of the structure in color is preferred. By examining the microstructure of a material, much useful information can be derived. For example, microstructural examination of a finished part that is too brittle may show that the melt temperature was either nonuniform or too low. The presence of unmelted particles is usually evident in such cases. Another reason for frequent failures of the injection-molded part is failure to apply sufficient time and pressure to freeze the gates. This causes the parts to be underpacked, which creates center-wall shrinkage voids. Figure 15-38 illustrates shrinkage voids. Voids tend to reduce the load-bearing capabilities and toughness of a part through the concentration of stress in a weak area. Contamination, indicated by abnormality in the microstructure, almost always creates some problems. Contamination caused by the mixing of different polymers can be detected through such analysis by carefully studying the differences in polymer structures. Quite often, poor pigment dispersion also causes parts to be brittle. This is readily observable through the microtoming technique. In order to achieve optimum properties, additives such as glass fibers and fillers must disperse properly. Microtoming a glass-fiber-reinforced plastic part reveals the degree of bonding of the glass fiber to the resin matrix as well as the dispersion

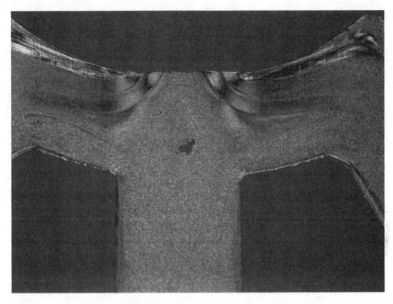

Figure 15-38. Shrinkage voids.

and orientation of glass fibers. Molded-in stresses as well as stresses resulting from external loading are readily observed under cross-polarized light because of changes in birefringence when the molecular structure is strained. The microtoming technique can also be applied to check the integrity of spin and ultrasonic or vibration welds (35).

15.3.5. Mechanical Testing

Defective product returned from the field is often subjected to a variety of mechanical tests to determine the integrity of the product. Two basic methods are employed. First one involves conducting mechanical tests such as tensile, impact, or compression on an actual part or small sample cut out from the part. The test results are then compared to the test results obtained from the retained samples. The second method requires grinding-up the defective parts and either compression or injection molding standard test bars and conducting mechanical tests. The test results are compared to the published data for the virgin material. The amount of material available for molding the test bars quite often precludes injection molding. Fatigue failure tests such as flexural fatigue or tensile fatigue can be employed to determine premature failure from cyclic loading.

15.3.6. Thermal Analysis

Thermal analysis techniques are used extensively in failure analysis. DSC is perhaps the most widely used technique, since it can determine key properties that affect the polymer's integrity. These are: melting point, heat of fusion, glass transition temperature, and degree of cure. As discussed earlier, since FTIR spectroscopy cannot differentiate between different materials within the same family of a polymer, DSC is used as a secondary means of identification. Different types of nylons (6 vs. 66), polyesters (PET vs. PBT), acetals

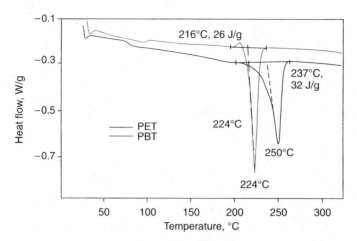

Figure 15-39. DSC thermogram showing the clear difference in melting point of PBT and PET polyesters. (Reprinted with permission of ASM International.)

(homopolymer vs. copolymer), and so on, are easily identified because all of these materials have distinct melting points. Figure 15-39 is a DSC thermogram showing the clear difference in melting point of PBT and PET polyesters (36). DSC also measures the heat of fusion, which is simply the energy required to melt the sample and is significant because it indicates the degree of crystallinity. For crystalline materials, level of crystallinity has a great effect on physical properties. Generally, higher levels of crystallinity result in improved tensile strength, increased stiffness, and superior chemical resistance. Failed parts are often analyzed for reduced level of crystallinity resulting from poor processing practices. Glass transition temperature (T_g) determined by DSC can provide information regarding material degradation and presence or absence of nucleating agent within the resin (36). Oxidative degradation can be monitored by the DSC method of oxidative induction time (OIT), which can determine level of antioxidant in the polymer. A reduced level of antioxidant can accelerate thermal degradation during processing, can make polymer susceptible to UV attack, and is one of the main reasons for the embrittlement in polyolefin parts. The glass transition temperature of nylon 6 can vary considerably, depending on the moisture content. DSC is used determine moisture content and its effect on polymer degradation (37). DSC can also evaluate thermoset plastics for degree of cure. If the molded component is undercured, the part exhibits a significant reduction in mechanical properties and may be likely to fail (38). TGA is often used to quantitative determination of the relative loading of additives such as fillers, reinforcement, plasticizers, chemical blowing agent, and so on. TMA is often overlooked as a failure analysis tool, but can provide insight into critical aspects of material properties as a part of such investigations. These aspects include coefficient of thermal expansion, glass transition temperature, and residual molded-in stresses (39). A detailed discussion on the use of TGA, TMA, and DSC techniques can be found in Chapter 7.

15.3.7. Nondestructive Testing (NDT) Techniques

NDT techniques are useful in determining the flaws, discontinuities, and joints. In simplest form, nondestructive testing involves measurement, weighing, visual examination, looking for surface imperfections, weak knit lines, and so on. Ultrasonic flaw detection testing, acoustic emission, and X-ray (radiographic NDT) are the most common techniques used for plastics component failure analysis.

15.3.8. Fractography (ASTM C1145)

Fractography is one of the most valuable tools available for failure analysis. It differs from the microstructural analysis technique using a microtome, an instrument that slices a very thin section from a part for microscopic examination. Fractogaphy is defined as the means and methods for characterizing a fractured specimen or component. The major use of fractography is to reveal the relationship between physical and mechanical processes involved in the fracture mechanism. The size of fracture characteristics will range from gross features, easily seen with unaided eyes, down to minute features just a few micrometers across. Light and electron microscopy are the two more common techniques used in fractography. An important advantage of electron microscopy over conventional light microscopy is that the depth of field in SEM is much higher; thus SEM can focus on all areas of a three-dimensional object identifying characteristic features such as striations or inclusions (40). By examining the exposed fractured surface and interpreting the crack pattern and fracture markings, one can determine the origin of the crack, direction of the crack propagation, failure mechanism, and stresses involved. On microscopic scale, all fractures fall into one of the two categories: ductile and brittle. Ductile fractures are characterized by material tearing and exhibit gross plastic deformation. Brittle fractures display little or no microscopically visible plastic deformation and require less energy to form. Fracture analysis usually involves the unexpected brittle failure of normally ductile material. In plastics there are four basic micro-fractographic features that clearly define the failure mechanism.

1. Branching
2. River marks
3. Wallner lines
4. Fatigue striations

Branching is observed on a failed part as the cracks that end before they reach the edge of the part away from the origin. Figure 15-40 illustrates branching.

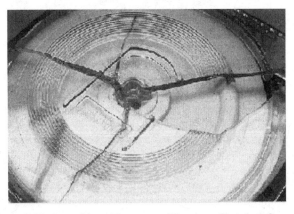

Figure 15-40. Branching. (Courtesy of Eastman Chemical Company.)

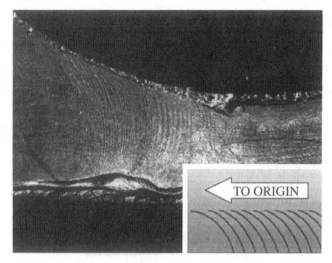

Figure 15-41. River markings pointing toward the fracture origin. (Courtesy of Eastman Chemical Company.)

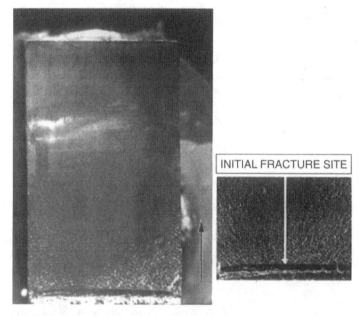

Figure 15-42. Wallner lines and initial fracture sites. (Courtesy of Eastman Chemical Company.)

River marks may be visible on fracture edge. The pattern shown in Figure 15-41 illustrates the river markings pointing toward the fracture origin. Wallner lines are wavelike bands that originate from the fracture origin and are very useful in pinpointing initial fracture sites. They are created by the interaction of a propagating crack front with a transverse elastic impulse, causing a momentary deviation of the crack front. Wallner lines are not typically observed in filled materials. Figure 15-42 indicates Wallner lines and an initial fracture site. Fatigue striations in plastics are similar to the ones observed in metal

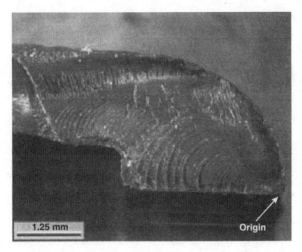

Figure 15-43. Fracture striations emanating from the fracture origin of the polycarbonate latch handle. (Courtesy of IMR Test Labs.)

fatigue fracture surfaces. Figure 15-43 illustrates fatigue striations emanating from the fracture origin of the polycarbonate latch handle.

15.3.9. Simulation Testing

In many instances, by simulating certain conditions such as exposing a part similar to the one that has failed to chemicals and/or other environmental factors, valuable knowledge concerning possible cause of failure can be gained.

15.4. SAMPLE OF A FAILURE ANALYSIS REPORT

A. Introduction

Failure of the ABS drain, waste, and vent (DWV) piping were encountered at an apartment complex. These failures were the result of circumferencial cracking of the pipe at the entrance of the fitting sockets and along the edge of the cement used to join the pipe to the fittings. An investigation was made into the problem to determine the cause of failure. This report covers the investigation.

B. Procedure

The overall investigation included visual inspection of pipe and fittings, testing of the pipe against the requirements of ASTM D 2661 (Standard Specification for ABS Schedule 40 DWV pipe), testing of the ABS plastics material against the requirements of ASTM D 1788 (Standard Specification for Rigid ABS plastics), and other specific evaluations. To assess the quality of ABS material, pieces of the pipe in question were compression molded into plaques. Since the amount of material available precluded injection molding, the specimens were prepared by compression molding. The material tests included deflection temperature under load (ASTM D 648), Izot impact (ASTM D 256), tensile stress and percent elongation (ASTM D 638), and specific gravity (ASTM D 792).

Additional properties determined included melt flow rate (ASTM D 1238), ash content (ASTM D 4218), and transition temperatures using a differential scanning calorimeter.

The monomer content of the polymer was determined using an FTIR spectrophotometer making 100 scans. The morphology of the polymer was examined through the use of transmission electron microscopy (TEM).

C. Results and Discussion

All the failures that have been reported or observed directly have been circumferencial in nature, at the edge of the cement layer at fitting socket entrances. The nature of these failures indicates that the failures are the result of solvent stress crack failure, that is, failure due to stress in the system which is accelerated by the presence of solvent from the cement.

The pipe did not meet all the requirements of ASTM specification D 2661. Minimum wall thickness and outside diameter values were below the specification minimums for some of the samples. All of the samples passed the minimum requirement for pipe stiffness but failed to meet the flattening requirement. All of the samples failed the impact test.

The test conducted on the materials indicated that they did not meet the minimum specification requirements. None of the materials met the minimum requirement for Izod impact. The deflection temperatures and the tensile stress at yield values were above the minimum requirement. All the samples had specific gravity significantly higher than the control. This is a result of the presence of pigmentation and other foreign particles, which would not be present in virgin pipe compound. One of the samples contained metallic particles. This was also reflected in higher ash content of the samples as compared to the control. The melt flow values were significantly higher than the control. The higher melt flow is indicative of lower molecular weight, probably from the excessive use of regrind material. The monomer content also varied from sample to sample, indicating that the virgin material was not used in manufacturing the pipe. One of the samples, upon close examination, showed three distinct layers. The outer and inner layers were made up of the same material while the center layer contained a different material. Both FTIR analysis and electron microscopic examination of the layers were conducted. The IR spectra for the two outer layers were significantly different from the center layer in terms of monomer content. The material in the center layer also had many foreign particles. The electron micrographs showed similar results.

D. Summary

The nature of the failures indicated that they were the result of solvent stress cracking. This is a well-recognized phenomenon, although the exact mechanism for such failure is not fully understood. This type of failure has not been encountered in pipe made from virgin, specification grade ABS. The failures are a direct result of using reprocessed materials, contrary to the requirements of the ASTM specification. The factors that contributed to the low environmental stress crack resistance were the distribution of the acrylonitrile, the rubber particle size, and the lower molecular weight.

E. Conclusions

1. The failures were caused by solvent stress cracking of the ABS pipe by the solvent in the cement.
2. The stress-crack resistance of the pipe was below that of good quality ABS pipe.
3. The reduced stress-crack resistance of the pipe was due to the use of materials with low stress-crack resistance.

4. The ABS pipe did not meet the requirements of ASTM D 2661, which was marked on the pipe.

5. The ABS material used to make the pipe did not meet the requirements of ASTM D 2661.

6. It is likely that the pipe will continue to fail.

15.5. CASE STUDIES

The following case studies can be found at the end of this chapter:

Case Study 1: Counterfeit Plastic Resin?
Case Study 2: Molded Part Defect Identification
Case Study 3: Stolen Formulations
Case Study 4: Failure of Plastic Plumbing Products
Case Study 5: Failure Analysis of a PVC Pipe
Case Study 6: Water Filter Housing Failure Analysis
Case Study 7: Miscellaneous Case Studies

REFERENCES

1. Ogando, J., "the Misunderstood Material," *Design News* (Feb. 2004).
2. Computer Aided Material Preselection by Uniform Standards, www.campusplastics.com
3. The Plastics Web, www.ides.com
4. Tobin, W., *Plastics Design Forum* (Jan./Feb. 1997), p. 48.
5. Plastics Pipe Institute, Technical Report, TR-21, "Thermal Expansion and Contraction in Plastic Piping System."
6. Malloy, R., *Plastic Part Design for Injection Molding*, Hanser Publications, Inc., Cincinnati, OH, 1994, pp. 125–130.
7. Mehta, K. S., "Designing Premature Failure Out of Injection Molded Parts," *SPE ANTEC*, **38** (1992).
8. Mehta, K. S., "Pandora's Medical Device—A Systematic Approach to Identify and Correct Part Design Problems," SPE, *Plastics South Conference Proceedings* (1986).
9. "Polymer Degradation" Article ID: 171, http://www.azom.com
10. "Plastics Product Failure," *Cycle Time Tips*, Vol. 11, Ashland, Inc., 2005.
11. Martes, S., "Why Dry? Why Brittle?," *Cycle Time Tips*, Vol. 33, Ashland, Inc.
12. Ezrin, M., *Plastics Failure Guide: Causes and Prevention*, Hanser Gardner Publications, Cincinnati, OH, 1996, p. 154.
13. Miller, p. 29.
14. Ibid, p. 32.
15. IMR test Lab article, www.imrtest.com
16. Basdekis, C. H., *ABS Plastics*, Reinhold Publishing Corporation, New York, 1964, p. 22.
17. "Technical Tidbits," *Bruch Wellman*, **2**(6), June 2000.
18. "Failure Mode of Plastics," www.zeusinc.com
19. "Introduction to Thermal Degradation," Zeus polymer minute, www.zeusinc.com

20. Jansen, J., "The Plastic Killer," *Adv. Mater. Processes* (June 2004), p. 50.

21. Wright, D., *Failure of Plastics and Rubber Products*, Rapra Technology Ltd., Shropshire, UK, 2001, p. 221.

22. Vogt, J. P., "Testing of Mechanical Integrity Assures Service Life of Plastics Parts," *Plastics Design & Processing* (March 1976), p. 13.

23. Mehta. K. S., "Experimental Stress Analysis for Molded Parts," *SPE ANTEC* (1986).

24. Corby, T. W. and Redner, A., "How to Use Experimental Stress Analysis," *Plastics Design Forum* (Jan./Feb. 1981).

25. *Fundamentals of Polariscope and Polarimeters*, Powerpoint Presentation, Strainoptics Technologies, Inc.

26. *Introduction to Stress Analysis by the Photoelastic Coating Technique*, Technical Bulletin IDCA-I, Photoelastic Division, Vishay Intertechnology, Inc.

27. *Reflection Polariscope*, Technical Bulletin S-103-A, Photoelastic Division, Vishay Intertechnology, Inc.

28. Hollman, J. P., *Experimental Methods for Engineers*, McGraw-Hill, New York, 1971, pp. 333–334.

29. Dally, J. W. and Riley, W. F., *Experimental Stress Analysis*, McGraw-Hill, New York, 1965.

30. Durelli, A. J., Phillips, E. A., and Tsao, C. H., *Introduction to Theoretical and Experimental Analysis of Stress and Strain*, McGraw-Hill, New York, 1958.

31. Measurement Group, Technical Bulletin, Vishay Intertechnology, Inc.

32. "Polycarbonate: Testing for Chemical Resistance Under Stress," *Injection Molding*, April, 1996.

33. Vogt, Reference 27, p. 12.

34. Levy, S., "Product Testing, Insurance Against Failure," *Plastics Design Forum* (July–Aug. 1984), p. 83.

35. Sessions, M. L., *Microtoming, Engineering Design with Dupont Plastics*, E. I. Dupont Co., Wilmington, DE, 1977, p. 12.

36. Jansen, J., "Plastic Component Failure Analysis," *Adv. Mater. Processes* (May 2001), p. 58.

37. "Thermal Analysis," Application Examples, Mattler-Toledo, Technical Literature.

38. Jansen, Reference 41, p. 59.

39. Jansen, Reference 41, p. 57.

40. Zamanzadeh, M., Gibbon, D., and Larkin, E., *A Reexamination of Failure Analysis*, MATCO Associates, Inc., Pittsburgh, PA, 2004.

GENERAL REFERENCES

Dally, J. W. and Riley, W. F., *Experimental Stress Analysis*, McGraw-Hill, New York, 1965.

Hetenyi, M., *Handhook of Experimental Stress Analysis*, John Wiley & Sons, New York, 1950.

Kuske, A. and Robertson, G., *Photoelastic Stress Analysis*, John Wiley & Sons, New York, 1974, pp. 263–274.

Bell, R. G. and Cook, D. C., "Microtoming, an Emerging Tool for Analyzing Polymer Structures," *Plast. Eng.* (Aug. 1979), p. 18.

Frocht, M. M., *Photoelasticity*, John Wiley & Sons, New York, 1941.

Leven, M. M., *Photoelasticity*, Pergamon Press, New York, 1963.

GE Plastics, Technical Literature, www.geplastics.com

CASE STUDY 1: COUNTERFEIT PLASTIC RESIN?*

The Case

Company ABC approached Polymer Solutions Incorporated for us to determine if any differences existed between two different polyacetal samples, labeled as Sample A and Sample B. Sample A was a product from a major supplier of polyacetal. Sample B was purported to be an innovative, direct replacement for the Sample A plastic product. The supplier of Sample B claimed to have developed a novel synthesis process and, as a result, was attracting substantial capital from investors.

Due to several sources of credible information, it became highly doubtful that a novel plastic was being produced. Instead, it was suspected that the Sample A product was being purchased and repackaged as being representative of pilot scale quantities of Sample B. If Sample A and Sample B were produced in different facilities using different reaction parameters then some differences between the two samples would result. Polymer Solutions Inc. was asked to compare the chemical compositions and key chemical and physical properties of the two polyacetal resins to determine if Sample B was "counterfeit." It was concluded that there were no differences between Sample A and Sample B so Sample B was essentially a counterfeit plastic resin.

The Approach

In order to arrive at this conclusion, Polymer Solutions Incorporated used several analytical techniques to compare and contrast the two samples:

- Melt rheology
- Nuclear magnetic resonance (NMR) spectroscopy
- Fourier transform infrared (FTIR) spectrometry
- Extraction and additive analysis
- Capillary gas chromatography (GC)

The melt rheological analysis probed the two polyacetal resin samples for differences in their melt viscosities as a function of shear rate. The instrument used was a TA Instruments AR-1000 controlled stress rheometer with a parallel plate geometry (Figure 15-1-1).

*Case study prepared by J. Brooks, J. Todd, and J. Rancourt, Polymer Solutions Incorporated, Blacksburg, VA 24060.

Figure 15-1-1. TA Instruments AR-1000 controlled stress rheometer.

Duplicate viscosity versus shear rate flow curves are compared for each resin in the plot shown in Figure 15-1-2.

The flow curves show a plateau in the low shear rate range below about 1 sec^{-1} and shear thinning behavior from about 5 sec^{-1} and above. Qualitatively, the flow curves of the two samples were very similar in terms of the viscosity in the low-shear plateau region, the location of the transition to shear-thinning behavior, and the slope of the flow curve in the shear-thinning region.

Quantitative comparisons were made by fitting the Williamson model to the flow curve data. The zero-shear viscosity and rate index are two parameters obtained from the Williamson model fits. The zero-shear viscosity is the viscosity value extrapolated to infinitely low shear rate, while the rate index is the slope of the viscosity versus shear rate relationship in the shear-thinning region at high shear rates. These two parameters are very sensitive to differences in the molecular weight distributions of the resins. Table 15-1-1 lists the zero-shear viscosities and rate indexes of the two resin samples.

TABLE 15-1-1. Zero-Shear Viscosities (Pa·sec)

	Zero-Shear Viscosity (Pa·sec)		Rate Index	
	Sample A	Sample B	Sample A	Sample B
Run 1	5000	4669	0.8261	0.7712
Run 2	4863	4411	0.7912	0.7763
Average	4932	4540	0.8087	0.7738
Std. Dev.	97	182	0.0247	0.0036
T-test	0.1156		0.1864	

Very minor differences were seen in the zero-shear viscosities and rate indexes between the two resin samples. Statistically, these differences were not significant.

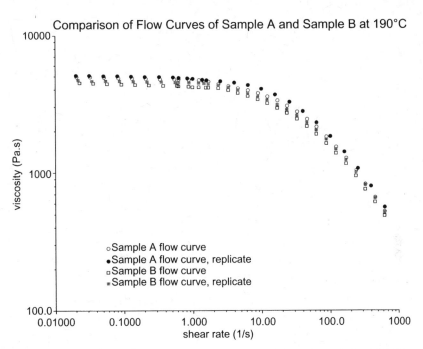

Figure 15-1-2. Comparison of flow curves of Sample A and Sample B at 190°C.

End-Group Analysis and Molecular Weight by NMR Spectroscopy

The two polyacetal resin samples were dissolved, as-received, in deuterated dimethylsulf-oxide at 150°C. The proton NMR spectra of the two samples are compared in Figures 15-1-3 and 15-1-4.

The NMR peak indicated by the arrow was determined to be due to the methyl proton resonances of the acetate end caps of the acetal polymer chains. The number average molecular weight of each sample was calculated from the area under this peak, assuming each polymer chain has two acetate end groups. The calculated number average molecular weight (M_n) and degree of polymerization (DP_n) values were as follows:

Resin	DP_n	M_n (g/mol)
Sample B	3330	100,000
Sample A	3200	96,000

These number average molecular weight values seem quite reasonable for a high-viscosity grade polyacetal engineering resin. The M_n and DP_n values were not significantly different between the two resin samples. The difference between the molecular weight values of the two resins is only about 4%. The error associated with the NMR determination of the end groups is expected to be at least as large as 4%, if not larger.

The many minor peaks in the NMR spectra, which represent additives and impurities in the resins, were very similar for the two resins in terms of occurrence, chemical shift (x-axis position), and intensity. This is strong evidence that the same, or nearly identical, additive package was also used in each resin.

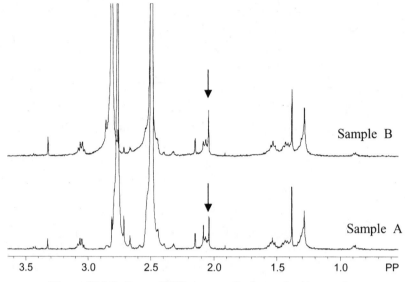

Figure 15-1-3. Proton NMR spectra of polyacetal resin samples.

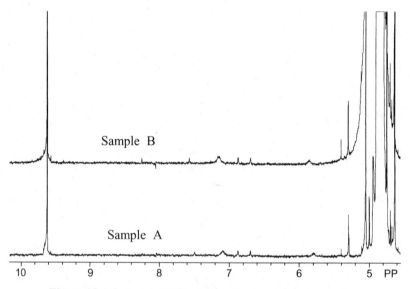

Figure 15-1-4. Proton NMR spectra of polyacetal resin samples.

The spectral peak at approximately 2.75 to 2.85 ppm is due to residual protons in the NMR solvent, dimethyl sulfoxide (DMSO), and is not due to any difference between the two resin samples. Because of the interaction between the NMR solvent and water, the chemical shift of the DMSO peak is very sensitive to water content and temperature. Therefore the difference in the location of this peak between the two spectra was most likely the result of a small difference in the water content or temperature of the two NMR samples during acquisition of the NMR spectra. Water is likely responsible for at least a part of the broad peak near 4.8 to 5.0 ppm.

FTIR Analysis of Pressed Polymer Films (Figure 15-1-5)

Samples of each polyacetal resin as received were pressed into films between heated chrome-plated steel plates at 190°C. The approximately 0.25-mm-thick films were analyzed by transmission-mode FTIR spectroscopy. The FTIR spectra of the two resin films are shown in Figure 15-1-6.

The major infrared absorbance bands are off-scale because relatively thick films were used. Of greatest interest are the minor absorbance bands that contain information regarding the type and concentration of additives in the resins. When the two spectra are overlaid, the minor absorbance bands match up almost exactly, as shown in Figure 15-1-7.

This is very strong evidence that the additive package is identical for the Sample A and Sample B resins.

Extraction and Analysis of Additives

The Sample A and Sample B polyacetal resin samples were dissolved in boiling ethylene glycol (EG) at 200°C and then precipitated in tetrahydrofuran (THF). A fine, gel-like

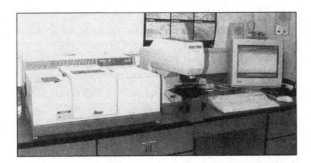

Figure 15-1-5. FTIR analyzer.

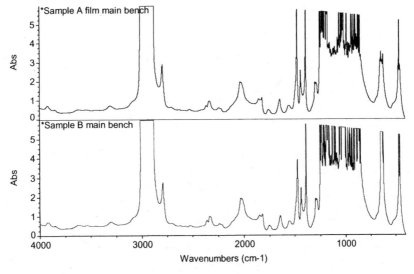

Figure 15-1-6. FTIR spectra of the two resin films.

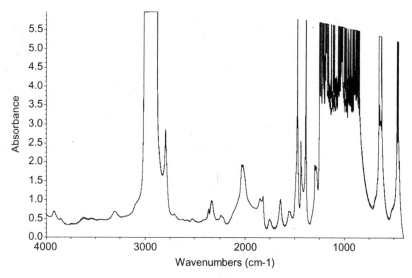

Figure 15-1-7. FTIR spectra with infrared absorbance bands.

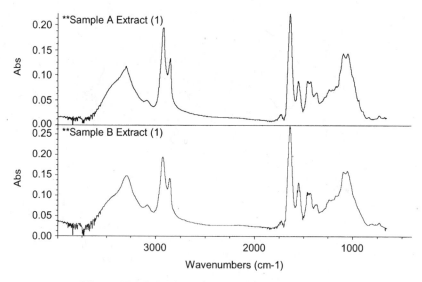

Figure 15-1-8. ATR-FTIR spectra of filtered extracts.

precipitate of acetal resin was formed. This procedure ensured a high recovery of additives from the resin. The EG/THF extracts were filtered through 0.2-μm-pore-size membrane filters. The filtered extracts were evaporated on glass slides in an oven at 130°C until films were formed on the slides. The film was analyzed by attenuated total reflectance Fourier transform infrared spectroscopy (ATR-FTIR). The ATR-FTIR spectra of the extracts are shown in Figure 15-1-8.

The overlay of the two spectra is shown in Figure 15-1-9.

As seen in the ATR-FTIR spectra of the extracts, the absorbance bands are nearly identical in terms of wavenumber (x-axis) and intensity (y-axis) for the two extracts. This

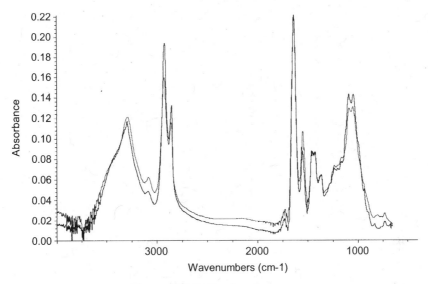

Figure 15-1-9. Overlay of spectra.

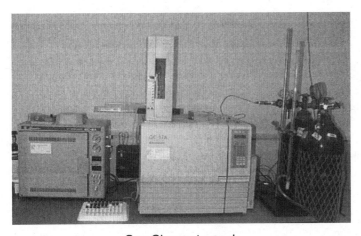

Gas Chromatograph

Figure 15-1-10. Gas chromatograph.

spectral data set is additional very strong evidence that the additive package is identical between the two polyacetal resins.

The absorbance bands present in the FTIR spectra of the extracts suggest that the specialty additive package was the same for Sample A and Sample B.

The extracts of the two resin samples were analyzed by gas chromatography on a nonpolar capillary GC column (Figure 15-1-10). The major stabilizer additive was not sufficiently volatile to be detected by gas chromatography. The GC analysis is most sensitive to minor components of the extract. The GC chromatograms of the extracts are overlaid in Figure 15-1-11.

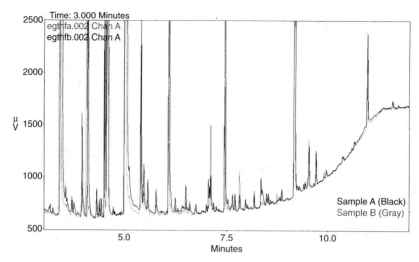

Figure 15-1-11. GC chromatograms of extracts.

As seen in the GC chromatograms, the peaks present were very similar for both extracts. The GC chromatogram represents a chemical "fingerprint" of each resin sample. Each chemical compound will elute at a specific retention time on a given GC column under a specific set of operating conditions. Without knowing the identities of the chemical compounds responsible for each peak in the chromatogram, it is possible to compare two samples for similarities or differences. GC analysis is employed in forensic science in such a manner to trace the origin of materials found at a crime scene, for example. And, fraud is definitely a crime!

There were some very slight differences in peak heights between the two extracts, but the vast majority of peaks that were present in the Sample B extract were also present in the Sample A extract. These peaks may represent residual initiators and other impurities from the polymerization process, additives and/or impurities in the additives, polymer degradation products, impurities in the extraction solvents, and contaminants transferred from laboratory glassware. The extractions of the two resins were performed in parallel. The extraction solvents and glassware preparation were kept the same for both resin samples. The fact that the same peaks were present for both resin samples indicates that the two resins are chemically very similar, even down to minor additives and impurities. This is strong evidence that the two resins originated from the same source, or were produced using an identical process and raw materials and additives that came from the same origin.

Conclusion

From the data sets it was concluded that Sample A and Sample B are indistinguishable. The so-called "novel synthesis process" appears to actually be a repackaging of "conventional" polyacetal that is produced in large quantities by a well-established chemical manufacturer.

CASE STUDY 2: MOLDED PART DEFECT IDENTIFICATION*

Polymer Solutions Incorporated received a molded silicone medical part that contained a small discoloration on the tip. It was of interest to identify the composition of this defect region in order to discover its origin. Testing showed that the defect was composed of an alloy of cobalt, chromium, and molybdenum, embedded in a silicone matrix. This alloy was present in another process in the same manufacturing facility and contaminated the silicone part.

The Approach

In order to arrive at this conclusion, Polymer Solutions Incorporated used two main analytical techniques to examine the defect region:

- Optical microscopy
- Scanning electron microscopy (SEM) (Figure 15-2-0) coupled with energy dispersive spectroscopy (EDS)

Initially, the defect was viewed and documented using an optical microscope. A white dashed oval in Figure 15-2-1 encompasses the location of the contaminants.

In order to characterize the defect region, scanning electron microscopy (SEM) coupled with energy dispersive spectroscopy (EDS) was utilized. This analytical technique is ideal for determining the presence of inorganic species, especially in very small spot sizes. SEM magnifies the region of interest while EDS acquires spectra of the chemical elements present in the magnified section. One can also map the spatial location of the chemical elements that are detected.

The sample was analyzed using the SEM, which is shown at the right. A reference spectrum was acquired of a silicone region where there were no defects (Figure 15-2-2). Silicon, oxygen, and aluminum were detected. The presence of silicon and oxygen were generated from the silicone medical part, and aluminum may have derived from interference from the SEM vacuum chamber components or the sample mount.

*Case Study prepared by B. Starr and J. Rancourt, Polymer Solutions Incorporated, Blacksburg, Virginia 24060.

Figure 15-2-0. Scanning electron microscope (SEM).

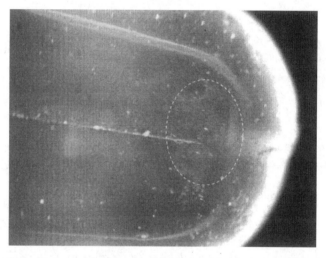

Figure 15-2-1. Optical micrograph of the tip of silicone medical part containing the defect region (32× magnification).

The defect region was then located under the SEM (Figure 15-2-3). Many small particles were embedded in the surface of the silicone. EDS analysis of a wide area of the defect region (Figure 15-2-4) indicated the presence of silicon, oxygen, aluminum, cobalt, chromium, and molybdenum.

Next, one of the particles in the defect region was isolated for spot mode analysis. Spot mode EDS analysis allowed a spectrum to be obtained from a very specific location on a defect particle. High concentrations of cobalt, chromium, and molybdenum were detected (Figure 15-2-5) compared to the wide scan (Figure 15-2-4).

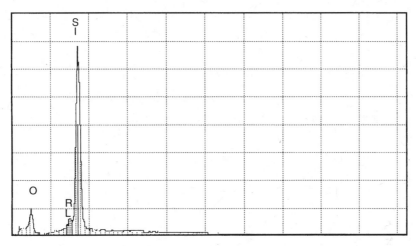

Figure 15-2-2. EDS spectrum of reference region of silicone.

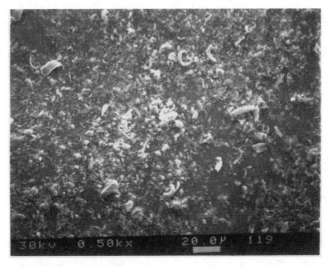

Figure 15-2-3. SEM micrograph of defect region at 500× magnification.

In order to illustrate the spatial distribution of the elements in the defect region, elemental maps were acquired in the defect region (Figure 15-2-6). The elemental maps (Figure 15-2-7) illustrate the spatial position of the chemical elements that were detected and relate directly to the SEM micrograph (Figure 15-2-6) that was obtained prior to mapping. The maps show that the particle in Figure 15-2-6 is composed of cobalt, chromium, and molybdenum and is embedded in a silicone matrix. The distribution of these three metals in the particle was consistent with an alloy of cobalt, chromium, and molybdenum. Table 15-2-1 lists the atomic concentrations of each chemical element that was detected in the reference region and a defect particle.

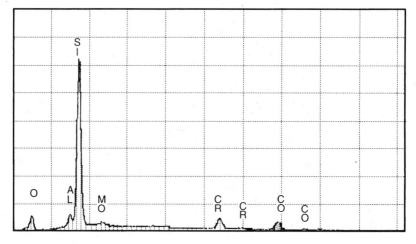

Figure 15-2-4. EDS spectrum of defect region (wide scan).

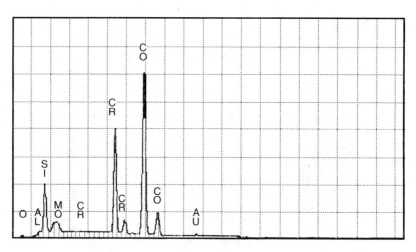

Figure 15-2-5. EDS spot mode analysis of a defect particle.

TABLE 15-2-1. Atomic Concentrations (%) of Elements Detected

Element Detected	Reference Spectrum	Defect Particle Spot Mode
Silicon	30.66	25.68
Oxygen	66.36	0.76
Aluminum	2.99	2.29
Cobalt	ND[a]	45.22
Chromium	ND[a]	20.44
Molybdenum	ND[a]	5.60

[a]ND, not detected.

Figure 15-2-6. SEM micrograph of defect particle at 2000× magnification.

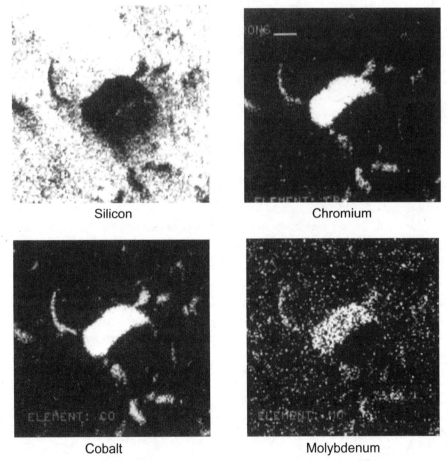

Silicon

Chromium

Cobalt

Molybdenum

Figure 15-2-7. Elemental maps depicting elements detected in regions corresponding to Figure 15-2-6.

CASE STUDY 3: STOLEN FORMULATIONS*

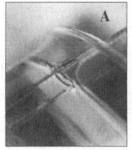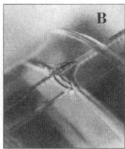

A former employee of Company A was discharged, started his own company (Company B), and began producing a specialty coating. Company A, however, suspected that the employee had actually stolen the proprietary formula for this coating. Polymer Solutions Incorporated (PSI) was asked to analyze this coating from Company A and the corresponding coating from Company B to objectively document their similarities and differences.

After extensive analysis, PSI determined that the likelihood of the formula being identical was very high. The corresponding coatings from the two companies showed virtually no difference. The chances of this occurring by coincidence were astronomical.

The Approach

In order to arrive at the conclusion of chemical similarity, Polymer Solutions Incorporated used two major analytical techniques to compare and contrast the coating samples:

- Fourier transform infrared (FTIR) spectrometry (Figure 15-3-0)
- Capillary gas chromatography (GC)

Approximately 10 mL of each coating sample was placed in a glass test tube and centrifuged until the solids began to settle and the first two or more millimeters from the top of the liquid in the tube was clear of solid particles. This took approximately 15–20 minutes. This centrifuging was done in order to separate the liquid fraction of the coating formulations from solid particles. Talc and silica particles present in the formulations could interfere with the Fourier transform infrared spectroscopic analysis.

Two or three drops of the clear liquid from the top of each test tube were placed on a glass microscope slide for the FTIR spectroscopic analysis. Approximately 40–50 μL of the clear liquid from the top of each test tube was placed in a 2-mL glass autosampler vial and diluted with approximately 1.8 mL of chloroform for the GC analysis.

Fourier Transform Infrared Spectroscopy (FTIR)
The FTIR spectra of the coating formulations were acquired. An infrared spectrum provides a rather unique fingerprint of an organic chemical or mixture of chemicals as a result

*Case study prepared by J. Todd and J. Rancourt, Polymer Solutions Incorporated, Blacksburg, Virginia 24060.

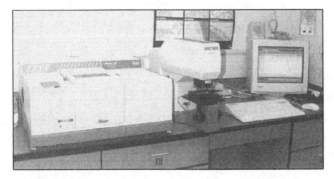

Figure 15-3-0. Fourier transform infrared spectrometer.

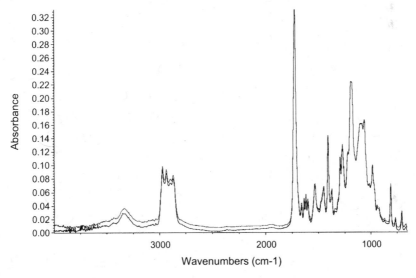

Figure 15-3-1. Comparison of duplicate infrared spectra acquired from the Sample A1.

of absorbance values (peaks) that have specific intensities (the y-axis) at specific wavelengths (the x-axis). Each peak in the spectrum represents the absorption of energy by a specific chemical bond or functional group. Every chemical compound or mixture of compounds has its own unique combination of chemical bonds and functional groups and therefore will provide a rather unique infrared spectrum.

Duplicate infrared spectra were acquired from each coating formulation sample. A comparison of the duplicate infrared spectra is shown in Figure 15-3-1. As seen in Figure 15-3-1, the duplicate spectra for the same sample are nearly identical, as expected.

A comparison of the spectra from samples A1 and B1 is shown in Figure 15-3-2. As seen in this overlay plot, the two spectra are nearly identical in terms of the locations (x-axis) and intensities (y-axis) of spectral peaks. This is strong evidence that these two coating formulations contain the same components because the positions of the peaks along the x-axis are identical. The nearly identical peak intensity for corresponding peaks indicates that the concentration of each component is very similar in the two coating formulations. If the concentrations of the various components were significantly different

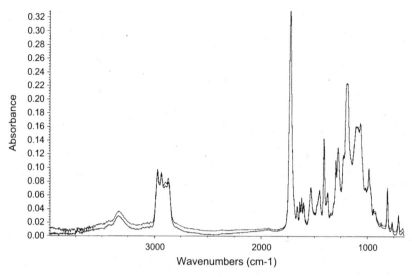

Figure 15-3-2. Comparison of the infrared spectra acquired from Sample A1. and Sample B1.

between the two formulations, then there would be noticeable differences in peak intensities (*y*-axis values) between the two spectra in Figure 15-3-2. Slight variations in peak intensities could be the result of (1) slight variations in chemical composition between the two coating samples or (2) the inherent variability of the infrared spectroscopic technique.

Infrared spectra can be compared to libraries of other infrared spectra for similarities. Various comparison algorithms have been developed for the purpose of quantitatively rating the similarity between two infrared spectra. Comparison algorithms generate a numerical score based on the similarities of the intensities of the various peaks in the two infrared spectra. Comparison algorithms are typically used to search libraries of infrared spectra in order to find one or more spectra that most closely match the sample spectrum. A match score is generated that quantifies the degree of similarity between each library spectrum and the sample spectrum. The higher the score, the better the match. A score of 100 indicates a perfect match.

The infrared spectra acquired from the four coating samples, which included two duplicate spectra from each coating sample for a total of eight spectra, were placed in a spectral search library. A comparison algorithm was used to rate the degree of similarity between each spectrum and the eight spectra in the search library. The match scores are shown in Table 15-3-1. The lower the score, the less similar the two spectra are. Generally, a score of 90 or higher is considered to be a good match.

All of the match scores ranged between 96.97 and 100.00. The scores of 100.00, indicating perfect matches, occurred only when spectra were compared to themselves. The highest match scores between different spectra occurred most of the time between the two duplicate spectra acquired from the same sample. This was the case for the A2, B1, and B2 samples. For the A1 sample, the B1 spectrum actually provided a higher match score than the duplicate A1 spectrum. In other words, the difference between duplicate spectra of the A1 sample was greater than the difference between the A1 sample spectrum and the B1 spectrum. The match scores between the A1 and B1 spectra ranged between 98.85

TABLE 15-3-1. Spectral Match Scores for Duplicate Infrared Spectra Acquired from the Four Coating Samples

	A1(1)	A1(2)	A2(1)	A2(2)	B1(1)	B1(2)	B2(1)	B2(2)
A1(1)	100.00	98.01	97.73	96.97	99.02	98.85	97.79	97.69
A1(2)		100.00	97.65	97.90	99.23	99.30	97.85	97.95
A2(1)			100.00	99.61	98.37	98.33	99.73	99.74
A2(2)				100.00	98.26	98.36	99.58	99.69
B1(1)					100.00	99.68	98.53	98.57
B1(2)						100.00	98.52	98.56
B2(1)							100.00	99.78
B2(2)								100.00

TABLE 15-3-2. Parameters for the Gas Chromatographic Analysis

Parameter	Value
Instrument	Shimadzu GC-17A with flame ionizaiton detector (FID)
Column	Restek Rtx-5 capillary GC column
	15-m length $\times$ 0.25-mm diameter $\times$ 0.25-μm film thickness
Temperature program	(1) Hold at 75°C for 1 minute
	(2) Heat to 300°C at 15°C per minute
	(3) Hold at 300°C for 9 minutes
Injection technique	Split injection, 100:1 split ratio
Injected volume	1 μL

and 99.30. These are considered to be very high match scores. This indicates a very high degree of similarity between the A1 and B1 coating formulations.

Gas Chromatography (GC)

The four coating samples were analyzed by gas chromatography (GC) on a capillary column, as shown to the right. Gas chromatography separates volatile chemical compounds based on their differing rates of travel through a column through which a gas stream is flowing. A capillary GC column is a very long and narrow column that provides a high degree of resolution of the different chemical compounds in a mixture. The term *resolution* refers to the ability to separate and distinguish one compound from another. Two compounds that are retained in the GC column for the same amount of time are said to not be resolved, while compounds that exit the column at different times are said to be resolved. A concentration-sensitive detector positioned at the outlet of the GC column measures the amount of each compound in a mixture as it exits the column. Specific compounds will yield peaks on the x-axis (retention time) having intensities (y-axis) or peak areas that are proportional to their concentrations in the sample. Nonvolatile components will not yield peaks and different chemical compounds can yield peaks having identical or nearly identical retention times.

The parameters for the gas chromatographic analysis are presented in Table: 15-3-2.

Samples of the individual components used in the Company A coating formulations were provided to Polymer Solutions Incorporated. The samples listed Table 15-3-3 were

TABLE 15-3-3. Samples of Individual Components Used in Company A Coating Formulations

Abbreviation	Chemical Compound
AM1	Acrylate monomer 1
AM2	Acrylate monomer 2
AM3	Acrylate monomer 3
AM4	Acrylate monomer 4
AO	Acrylate oligomer
PI1	Photoinitiator 1
PI2	Photoinitiator 2
VM	Viscosity modifier

diluted with chloroform and analyzed by gas chromatography in order to establish retention times for each of the volatile components in the coating formulations. It should be noted that the AO acrylate oligomer and the VM viscosity modifier were not volatile enough to be analyzed by gas chromatography; however, these two samples contained minor volatile compounds that were detectable by GC. Not listed in Table 15-3-3 were a wax-coated silica powder and a talc powder that were used in the A2 coating formulation. The wax-coated silica and talc additives were not analyzed by GC because they were not soluble in any solvent and thus could not be injected into the GC instrument.

The gas chromatograms of a pair of coating samples are shown in Figures 15-3-3 and 15-3-4. The first five minutes of the chromatograms are not shown, because the only major peaks occurring in this region were associated with the chloroform used to dilute the coating samples. Likewise, the last 5 minutes of the chromatograms were not shown because no significant peaks eluted during this time. The peaks associated with the major volatile components are labeled in each chromatogram.

The chromatograms for the pair of coating samples are very similar in terms of the retention times and heights of the peaks that are present. This indicates that the chemical compositions are very similar for the coating samples. Approximately 31 different peaks were detected in the GC chromatograms of the coating samples. These peaks include the major volatile components listed in Table 15-3-3, as well as minor components present in each of the major coating formulation components. These minor compounds represent impurities present in individual components of the coating formulations.

The concentrations of each GC peak, calculated as a percentage of the total area of all peaks detected in the GC chromatogram, are compared in Table 15-3-4 for the A1 and B1 coating formulations. In general, each peak in the GC chromatogram represents a unique chemical compound. Sometimes two or more compounds will have the same or very similar retention times, so a GC peak can represent more than one compound. Thus, the 31 peaks listed in Table 15-3-4 represent at least 31 different volatile chemical compounds present in the coating samples. Of these 31 peaks, 6 represent major components in the coating formulation. These are listed in boldface type in Table 15-3-4.

The remaining peaks, listed in parentheses, represent impurities and minor additives in the major components of the coating formulations.

The high level of similarity in area percent values between the A and B coating formulations are remarkable, even for the minor peaks in the chromatograms. This informa-

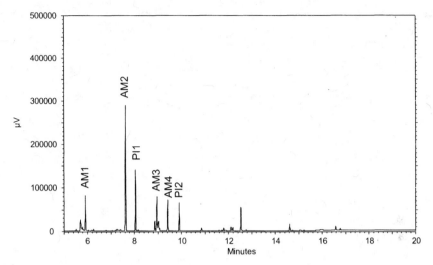

Figure 15-3-3. Gas chromatogram of the A1 coating sample.

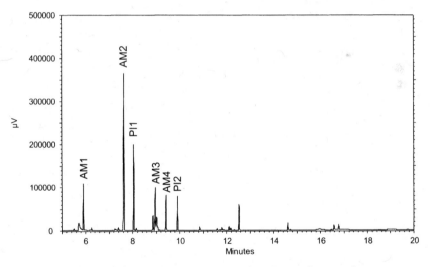

Figure 15-3-4. Gas chromatogram of the B1 coating sample.

tion indicates that the chemical compositions of the volatile fractions of the A1 and B1 coatings are almost identical.

Comparisons of the area percents of the major peaks in the GC chromatograms are shown in Table 15-3-5. The differences in area percents ranged between 0.1% and 1.5% for the A1 and B1 samples, again showing a very high degree of similarity between the A and B coating formulations. It should be noted that the GC analysis does not measure the concentration of nonvolatile components in the coating samples.

TABLE 15-3-4. Comparison of the Concentrations of Volatile Compounds in the A1 and B1 Coating Formulations as a Percentage of the Total Peak Area as Determined by Gas Chromatography

Peak No.	Peak ID or Origin*	Retention Time (minutes)	Area Percent A1	Area Percent B1
1	(AM1)	5.52	0.37	0.34
2	(AO)	5.70	6.05	5.64
3		5.79	0.41	0.00
4	**AM1**	5.91	7.04	7.15
5		6.20	0.15	0.00
6	(AM1)	6.26	0.31	0.63
7		6.78	0.00	0.00
8	(AM3)	7.25	0.47	0.31
9		7.29	0.36	0.30
10		7.40	0.45	0.63
11	**AM2**	7.63	29.00	30.50
12	**PI1**	8.05	13.15	14.25
13		8.16	0.25	0.37
14	**AM3**	8.96	15.04	15.72
15	(VM)	9.13	0.00	0.13
16	**AM4**	9.42	6.40	5.70
17	**PI2**	9.91	5.96	5.60
18	(AM2)	10.85	0.70	0.69
19		11.59	0.15	0.32
20	(AM3)	11.70	0.16	0.17
21	(AM3)	11.80	0.67	0.56
22	(AM3)	11.85	0.29	0.25
23		12.11	1.32	0.97
24	(AM4)	12.19	0.77	0.43
25	(AO)	12.53	5.68	4.69
26		12.76	0.25	0.00
27	(AM2)	14.61	1.59	1.31
28	(AM3)	16.00	1.19	1.01
29	(AM4)	16.57	1.21	1.12
30	(AO)	16.77	0.61	1.21
31		19.74	0.00	0.00
	Totals:		**100.00**	**100.00**

Entries in boldface type indicate the major peak associated with each component in the coating formulation. Entries in parentheses indicate minor peaks that represent minor additives or impurities in that component.

TABLE 15-3-5. Comparison of the Major GC Peaks of the A1 and B1 Coating Samples

Peak ID or Origin	Retention Time (minutes)	Area Percent A1	Area Percent B1	Area Percent Difference
(AO)	5.70	6.0	5.6	0.4
AM1	5.91	7.0	7.2	−0.1
AM2	7.63	29.0	30.5	−1.5
PI1	8.05	13.2	14.2	−1.1
AM3	8.96	15.0	15.7	−0.7
AM4	9.42	6.4	5.7	0.7
PI2	9.91	6.0	5.6	0.4
(AO)	12.53	5.7	4.7	1.0

CASE STUDY 4: FAILURE OF PLASTIC PLUMBING PRODUCTS*

Abstract

Failures of plastic components are being seen more often in industrial, household, and commercial settings. Many of these failures involve the transport of water and cause significant damage when they occur. These failures can be caused by improper material specification, bad design, overloading, or incorrect molding conditions. Issues such as chemical resistance, environmental deterioration, geometric sensitivity, temperature dependence, and aging are at times overlooked.

Background of Plastics

The unique properties of plastic make it one of the most sought-after materials in the world today. Their low weight, ability to be easily shaped or molded, low cost, rigidity or flexibility, and the ability to insulate are only a few of the characteristics that make plastics a popular option when designing a new product or replacing an existing nonplastic one. Products such as vinyl siding, water bottles and filters, computer housings, polyvinyl chloride (PVC) piping, disposable packaging, automotive body panels and engine components are all products that have successfully been made from plastic.

Plastic has been in use for over a millennia. South American Indians used natural rubber in the manufacture of waterproof containers, shoes and torches. Thanks to the development of new plastics and processing equipment, application of plastic in every industry has exploded over the last 50 years. Nearly everyone can think of at least one product that was once made of metal or wood that is now plastic.

Failure of Plastics

Plastics are a unique material that can be ductile under one condition, and with a relatively small change in conditions be brittle. A common example of this is seen with Silly Putty®. If one pulls this material slowly it can be stretched almost indefinitely (ductile). However, if one pulls quickly on this material it breaks (brittle). Decreasing the temperature of the Silly Putty®, decreases the stretching rate at which it becomes brittle. In essence, plastics have important properties that are temperature- and time (rate)-dependent—a characteristic that is not seen with other materials. This is an important consideration when designing a part out of plastic. If the engineer does not take this into account, a plastic part may be able to absorb loads under one condition, but completely fail under another condition—two conditions that might possibly be relatively close to one another.

In a similar way that nearly everyone knows of a plastic part that was once metal or wood, nearly everyone can recall a plastic part that has failed. Failure of plastic can be caused by a number of reasons. It is important to determine why a failure occurred. Was it due to the plastic not performing up to specification? Was the failure the result of an incorrect design? Was the plastic introduced to an unexpected environmental change?

*Case study prepared by *Paul Gramann, Antoine Rios and Bruce Davis, The Madison Group: PPRC, Madison, WI 53719.*

Possibly, cracking of the plastic was designed to do such that under certain conditions. One such example is a Formula 1® racecar crashing into the side of a wall. When an impact like this occurs the plastic body panels and other components are designed to absorb the energy and release it as they crack; keeping the energy away from the driver. This could be viewed as a plastic failure; however, the driver that walks away from this high-speed crash would probably have a different opinion.

Common causes of plastic failure, which is far from an exhaustive list, are:

- Environmental stress cracking
- Chemical attack
- Ionizing radiation attack
- Oxidation
- Improper processing conditions
- Incorrect material properties

Of recent interest are disinfectants that are used to kill bacteria in water. Some investigators have reported chemicals such as chlorine and chloramines to be detrimental to some plastics used in plumbing applications.

Analyzing Plastic Failures

There are several steps that can be taken while analyzing a plastic failure. Certainly, the first step is to determine the actual cause of failure. This is usually not a trivial step and can take great deal of time and resources. This step begins with collecting as much historical information as possible. This includes information such as the age of the product, when and under what conditions did the failure occur, previous conditions that the part may have been exposed to, what stress or loads was the part under, did any of these conditions recently change. This information could be critical to help determine the cause of the failure. For example, if one finds that a plastic component was recently cleaned with a solvent that the plastic is not resistant too, this would be investigated further to determine its role in the failure.

The next step to determine the cause of failure is a visual examination of the failed part. Cosmetically displeasing features on the part may point to improper processing conditions. Burn marks on the part may indicate degradation during processing. Degradation of the polymer's molecular structure may adversely affect its mechanical properties. Likewise, sink marks on the part may also indicate improper conditions were used during the making of the part.

Visual examination will also disclose the extent of consumer abuse. Gouges and deep marks in the plastic show signs of possible excessive use of the part. A foreign substance, such as a liquid, on the part should be investigated for possibly chemical attack on the plastic.

Identification of the plastic recipe is an important step in most failure analyses. This includes identifying and quantifying the plastic resin, fillers, additives and reinforcements. This is done due to the fact that a primary reason for a plastic part to fail is merely because the wrong material was used for the application. In a similar fashion, the failure may have been caused by the lack of a material, such as, an impact modifier to increase ductility. There are several techniques that can be employed to accurately determine the type of plastic and all of its constituents (1). Moreover, these techniques can also give an idea of the processing history that the plastic experienced.

Failure of a PVC Pipe

Plastic pipe, tubing, and other profiles are a popular alternative to copper, steel, aluminum and other materials. These products are used in a wide variety of industries including, building & construction, automotive, consumer goods, lawn & garden, windows & doors, furniture, plumbing, and electrical. One of the most widely used materials for these products is polyvinyl chloride or commonly known as PVC. This material is popular in these industries because of the wide range of properties that can be obtained, depending on the additives that are mixed with it. PVC can be made to have high strength, rigidity and hardness, good electrical properties, and high chemical resistance, as well as to be self-extinguishing—all this at a relatively inexpensive price. However, depending if a plasticizer additive is used with the PVC, along with what kind and how much, the characteristics of the final part can be dramatically altered to have high-impact strength with relatively low hardness and rigidity.

An unplasticized PVC pipe is quite rigid with high strength and good chemical resistance. These properties make it attractive for use in above or below ground plumbing applications. However, a very important change in properties occurs as the temperature gets colder: The impact strength of PVC drastically changes for the worse. This means that at low temperatures the ability of PVC to dissipate the energy from a sudden blow is limited and may result in part failure. The best way to describe this phenomenon, apart from demonstrating the impact of several PVC pipes at different temperatures, is to graph the impact strength of PVC as a function of temperature, shown in Figure 15-4-1. The most interesting part of this graph, the area that can explain many plastic failures, is boxed out in gray. Here, you will see a dramatic decrease in impact strength as the temperature gets colder: The part is becoming increasingly more brittle. The impact strength is four times less at −10°C than at 20°C—a temperature range that is easily experienced in many

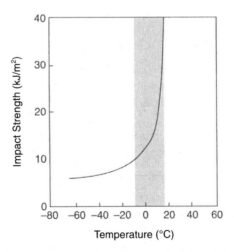

Figure 15-4-1. Impact strength of PVC as a function of temperature.

regions of the world. This phenomenon is one that is not seen with everyday metals and is commonly overlooked when designing with plastics.

One can improve this situation by using additives, in this case a plasticizer, that ultimately moves the graph to the left and gives the part a high impact strength at a much lower temperature. However, the gain in one property usually means the loss of other properties, in this case, the loss of stiffness. Figure 15-4-2 shows the modulus (stiffness) of PVC as a function of temperature (solid line). The dashed line indicates the temperature at which the stiffness will decrease dramatically, approximately 50°C for this PVC. For many uses, 50°C is a temperature that the product would never experience; however, if an additive is used to increase the impact strength (as described above), then this graph will also move to the left, lowering the temperature at which the stiffness is lost.

A compromise must be made for how much, if any, additive is to be used for the application and environment that the product will be used. In the case of PVC pipe, high mechanical strength, rigidity, hardness and high chemical resistance is required at the lowest cost. Plasticizing additives typically add to the cost of a product and are not used in pipe production. Other additives, such as calcium carbonate, can reduce costs. Unfortunately, these cost-reducing additives can make the product even more brittle causing the impact strength graph, shown in Figure 15-4-1, to move to the right, making the product more brittle and more susceptible to failure.

An example of a failed PVC pipe is shown in the Figure 15-4-3. A visual inspection of the part indicates that this was a brittle failure as opposed to a ductile failure. Many brittle failures occur very quickly, whereas ductile failures will typically occur over a longer period of time—for example, Silly Putty®. It was revealed that the pipe was in a cold condition of approximately −5.0°C. The pipe was in an environment where the temperature was low enough that it became very brittle. A force, which could be caused by an external blow or from internal pipe pressure, became to great and the part failed catastrophically. The cause of failure may not be because the engineer specified the wrong

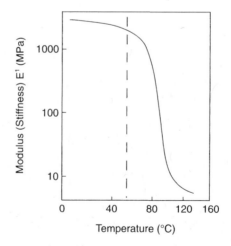

Figure 15-4-2. Stiffness of PVC as a function of temperature.

Figure 15-4-3. Failed PVC pipe.

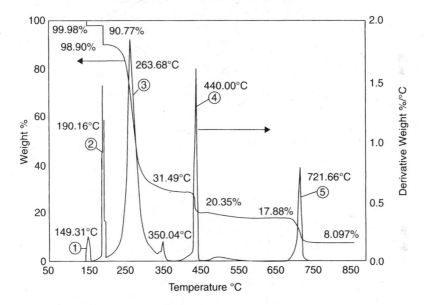

TGA Analysis of a PVC sample. (1) Volatiles: humidity, monomers, solvents
(2) DOP plasticizer (3) HCL formation (4) carbon-carbon scission
(5) CO_2 formation

Figure 15-4-4. TGA of a PVC compound (2).

pipe for the job or environment, but because of the formulation of the material or the processing conditions at the production plant were wrong.

To determine if the pipe had the correct formulation a wide variety of material tests can be performed. One such test is the thermogravimetric analyzer (TGA). This device is often used to identify the components of a plastic part. It works by gradually heating a small sample of the plastic to a very high temperature. At different temperatures the compounds of the plastic will decompose. The TGA accurately records the change of weight with respect to the temperature. Figure 15-4-4 shows an example of a TGA test on PVC sample (2). Here, the decomposition of the different compounds can be seen along with the percent weight lose. Using data from an extensive library, the decomposition peaks are matched with known materials to decompose at the same exact temperature.

Figure 15-4-5. Steel braided hose with cracked plastic nut on the end.

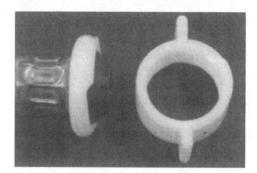

Figure 15-4-6. Cracked plastic nut on the end.

Figure 15-4-7. Close-up of failed surface of plastic nut.

Failure of a Plumbing Nut

Plastic nuts are commonly used to fasten water transport hoses to toilets, sinks, and other plumbing products. Figures 15-4-5 and 15-4-6 show a typical failure of a plastic water nut at the end of steel braided hose. The nut shows a circumferential crack that extends completely around the nut. The circumferential crack is located at the last or bottom-most thread inside of the nut. This is the area where the stresses are higher during tightening after the nut has topped out with the mating piece.

Figure 15-4-7 shows a close-up of the fracture surface. The initiation of this crack appears* to be at the end of the thread and at the root of the thread. The smooth and sharp

---

*This can be confirmed by microscopy analysis using a scanning electron microscope

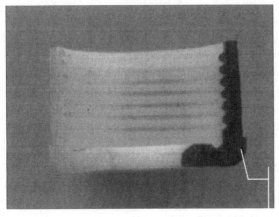

Decrease in sidewall thickness at high stress region

Figure 15-4-8. Cross section of a water nut.

nature of the fracture indicates that this is a brittle failure. In this case, brittle failure with this plastic material is a consequence of slow crack growth leading to a sudden catastrophic failure. The most likely cause of the initial crack was the tightening force during assembly or post-adjustment of the nut. The stress during tightening causes a stress higher than the yield stress of the material. How the part handles this stress depends on a number of things—material properties being one of them. The higher stress leads to microcracks throughout the cross section. These cracks grew slowly over time as the tightening stress was released leading to catastrophic failure of the nut.

It is very likely that the material used for this nut has lower yield strength than what is required to prevent it from failing under normal installation and operating conditions. Further, the material exhibited brittleness when ductility would have been preferred. Figure 15-4-8 shows a cross section of an unfailed connector. The cross-section shown is from a connector that is very similar, if not identical, to the one that has failed. The wall thickness where the crack occurred is relatively thin. There is a reduction in thickness just below this high stress area. A better design may one that has an increased wall thickness past the high stress region, instead of decreasing it at the high stress region as shown in the figure. A material that is inherently more ductile or better handle the stress involved with this application should also be investigated.

REFERENCES

1. Sepe, M., "Materials Troubleshooting," a chapter in *Injection Molding Handbook*, edited by Osswald, T. A., Turng, T. and Gramann, P. J., Hanser, Munich (2002).
2. Osswald, T. and Menges. G., *Material Science of Polymers for Engineers*, Hanser, Munich, 1995.

CASE STUDY 5: FAILURE ANALYSIS OF A PVC PIPE*

Plastic pipe, tubing, and other profiles are an extremely popular alternative to copper, steel, aluminum and other materials. In fact, by 2003 it is predicted that 33% of all US pipe production will be made with plastic. Plastic pipes, tubing, and profiles are used in a wide variety of industries including building & construction, automotive, consumer goods, lawn & garden, windows & doors, furniture, plumbing, and electrical. One of the most widely used materials for these products is polyvinylchloride, commonly known as PVC. This material is popular in these industries because of the wide range of properties that can be obtained, depending on the additives that are mixed with it. PVC can be made to have high strength, rigidity and hardness, good electrical properties, and high chemical resistance, as well as to be self-extinguishing—all this at a relatively inexpensive price. However, depending if a plasticizer additive is used with the PVC, along with what kind and how much, the characteristics of the final part can be dramatically altered to have high impact strength with relatively low hardness and rigidity.

An unplasticized PVC pipe, shown in Figure 15-5-1, is quite rigid with high strength and good chemical resistance. These properties make it attractive for use in above- or below-ground plumbing applications. However, a very important change in property occurs as the temperature gets colder: The impact strength of PVC drastically changes for the worse. This means that at low temperatures the ability of PVC to dissipate the energy from a sudden blow is limited and may result in part failure. The best way to describe this phenomenon, apart from demonstrating the impact of several PVC pipes at different temperatures, is to graph the impact strength of PVC as a function of temperature, shown in Figure 15-5-2. The most interesting part of this graph, the area that can explain many plastic failures, is boxed out in gray. Here, you will see a dramatic decrease in impact strength as the temperature gets colder: The part is becoming increasingly more brittle. The impact strength is four times less at −10°C than at 20°C—a temperature range that is easily experienced in many regions of the United States. This phenomenon is one that is not seen with everyday metals and is commonly overlooked when designing with plastics.

One can improve this situation by using additives, in this case a plasticizer, that ultimately moves the graph to the left and gives the part a high impact strength at a much lower temperature. However, the gain in one property usually means the loss of other properties, in this case, the loss of stiffness. Figure 15-5-3 shows the modulus (stiffness) of PVC as a function of temperature (solid line). The dashed line indicates the temperature at which the modulus will decrease dramatically, approximately 50°C for this PVC. For many uses, 50°C is a temperature that the product would never experience; however, if an additive is used to increase the impact strength (as described above), then this graph will also move to the left, lowering the temperature at which the stiffness is lost.

Thus, a compromise must be made for how much, if any, additive is to be used for the application and environment that the product will be used. In the case of PVC pipe, high mechanical strength, rigidity, hardness, and high chemical resistance is required at the lowest cost. Plasticizing additives typically add to the cost of a product and are not used in pipe production. Other additives, such as calcium carbonate, can reduce costs.

---

*Reprinted with permission of The Madison Group. *Note*: The analysis and its write-up are property of The Madison Group and cannot be copied and/or distributed in anyway without prior permission from The Madison Group.

Figure 15-5-1. Unplasticized PVC pipe.

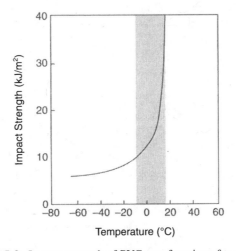

Figure 15-5-2. Impact strength of PVC as a function of temperature.

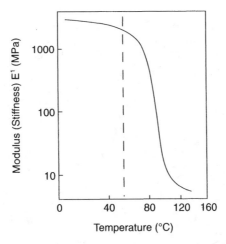

Figure 15-5-3. Modulus (stiffness) of PVC as a function of temperature.

Figure 15-5-4. Failed PVC pipe.

Unfortunately, these cost-reducing additives typically make the product even more brittle, causing the impact strength graph shown in Figure 15-5-2 to move to the right, making the product more brittle and more susceptible to failure.

An example of a failed PVC pipe is shown in Figure 15-5-4. To determine the cause of failure, a variety of techniques can be used:

- Visual inspection of the failed part
- Structural finite element analysis
- Dynamic finite element analysis
- Material evaluation
- Process evaluation

A visual inspection of the part indicates that this was a brittle failure as opposed to a ductile failure. Many brittle failures occur very quickly, whereas, ductile failures will typically occur over a longer period of time. It was revealed that the pipe was in a cold condition of approximately −5.0°C. The pipe was in an environment where the temperature was low enough that it became very brittle. A force, which could be caused by an external blow or from internal pipe pressure, became too great and the part failed catastrophically. The cause of failure may not be because the engineer specified the wrong pipe for the job or environment, but because of the formulation of the material or the processing conditions at the production plant were wrong.

To determine if the pipe had the correct formulation, a wide variety of material tests can be performed. One such test is the thermogravimetric analyzer (TGA). This device is often used to identify the components of a plastic part. It works by gradually heating a small sample of the plastic to a very high temperature. At different temperatures the compounds of the plastic will decompose. The TGA accurately records the change of weight with respect to the temperature. Figure 15-5-5 shows an example of a TGA test on PVC sample. Here, the decomposition of the different compounds can be seen along with the percent weight lose. Using data from an extensive library, the decomposition peaks are matched with known materials to decompose at the same exact temperature.

To establish the mode and forces of failure, along with providing confidence that a failure took place in the manner that was determined, a finite element analysis (FEA) can be made. This type of analysis allows the engineer to place the part in a realistic environment under normal to extreme conditions and observe what happens to the part—if failure occurs. Figure 15-5-6 and Figure 15-5-7 shows the predict failure of a pipe caused by an extreme internal pressure using FEA.

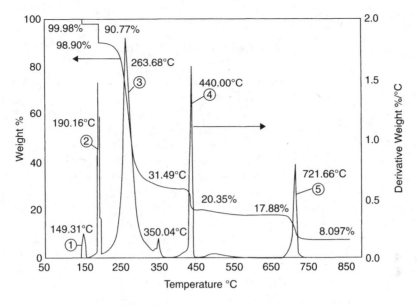

TGA Analysis of a PVC sample. (1) Volatiles: humidity, monomers, solvents
(2) DOP plasticizer (3) HCL formation (4) carbon-carbon scission
(5) CO_2 formation

Figure 15-5-5. Thermogravimetric analyzer (TGA) test on PVC sample.

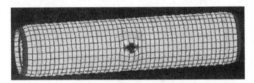

Figure 15-5-6. Failure of a pipe caused by an extreme internal pressure using finite element analysis (FEA)—view from pipe side. A full color version of this figure can be viewed on DVD included with this book.

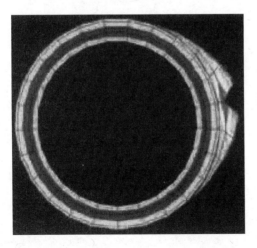

Figure 15-5-7. Failure of a pipe caused by an extreme internal pressure using finite element analysis (FEA)—view looking into pipe. A full color version of this figure can be viewed on DVD included with this book.

CASE STUDY 6: WATER FILTER HOUSING FAILURE ANALYSIS*

A water filter housing was received for failure analysis. The filter's housing shows a failure located at the housing's bottom cap (Figures 15-6-1 and 15-6-2). The failure appears as a circumferential crack that completely separated the bottom cap from the housing. This failure caused extensive water damage in the property where it was installed.

Analyzing the stresses and forces on the filter's housing during operation, two main sources of stresses are identified. There is a stress originating at the threads of the housing, along with a stress caused by the internal water pressure. The stress at the threads is produced while tightening the filter housing to seal it to the filter's base. A structural analysis (see Figure 15-6-3) was performed to determine which areas are exposed to high stress. The maximum stress (reddish color) occurs at the inner corner of the bottom cap. This is the exact place where the failure occurred on the analyzed housing. The higher stress is at the smallest radius (stress concentration point) and where the housing's wall is thinnest.

Further inspection of the housing reveals defects that originated during molding. As seen in Figure 15-6-3 and Figure 15-6-4, there are molding ripples and poor material mixing. The molding ripples are generated during mold filling, and they can be the source of stress concentrations that can lead to crack initiation. Poor mixing during processing leads to material inhomogeneities that can be the source of stress concentrations, local material weaknesses, and local material degradation. The molding defects identified here are contributing factors for failure to occur. Other important factor is material selection. There are thousands of material grades to choose for a given application. It is important

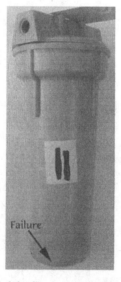

Figure 15-6-1. Cracked water filter housing.

Figure 15-6-2. Close-up of cracked water filter housing.

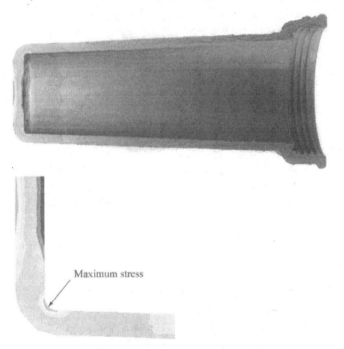

Figure 15-6-3. Structural analysis determining high stress exposure.

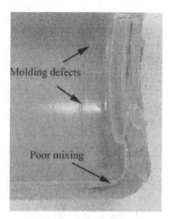

Figure 15-6-4. Molding defects: molding ripples and poor material mixing.

to have the right knowledge about the application to choose the appropriate material. A balance between material cost and performance must be obtained to avoid failure. A material such as the one used to mold this filter can include fillers (such as talc). The use of fillers reduces material costs, but at the same time reduces material strength. Material processing is also an important issue. Molders prefer to use low-mold-temperature and low-viscosity materials to obtain the fastest processing time possible. A material with low viscosity allows faster mold filling, but it is usually of lower molecular weight that reduces the mechanical performance of the part. A colder mold can be the cause of numerous filling problems. It is usually the combination of several factors that leads to failure of a plastic part. It is important to consider all these when properly designing a plastic product.

CASE STUDIES 7: MISCELLANEOUS CASE STUDIES*

The following are a series of failure analyses performed by Microbac's Hauser Laboratories Division, which should demonstrate the practical application of the above-described failure analysis techniques.

Case Study A: Product Misuse

Hauser received three samples of 12-in. nominal ABS Truss pipe from a State's Utility District that used the pipe in a sanitary sewer application. An ABS truss pipe consists of a coextruded ABS shell that is filled with Portland cement. Upon receipt, the samples were inspected and found to contain an axially oriented area of damaged ABS, such that under ring compression the pipe had buckled. This buckling reportedly occurred at the 6 o'clock position in the sewer system, which is consistent with a gravity flow system. The ABS appeared to be damaged due to chemical attack and also displayed evidence of softening, possibly due to exposure to solvents. The softened ABS had subsequently altered shape and dimensions. A main component of the failure analysis was heated headspace gas chromatography (mass spectroscopy) HHGC-MS. The HHGC-MS test results indicated the presence of perchloroethylene in quantities greater than 1000 ppm in the wall of the ABS at the 6 and 12 o'clock positions. Trichloroethene was also found. It is known that perchloroethylene is used for dry cleaning, and the damage to the sewer system was isolated to an area directly downstream from a commercial dry-cleaning operation (Figure 15-7-1). Subsequent testing by the municipality confirmed that the dry cleaner was discharging solvents directly to the sanitary sewer system and was found liable for repairs.

Case Study B: Poor Fabrication

The client submitted numerous samples of solvent cemented 4-in. nominal schedule 40 bell/spigot joints In PVC pipe. The joints were from a system that was used to transport a sodium hypochlorite solution. The system owner and user reported that the joints had leaked in service at an ambient system pressure of approximately 80 psi. The joints were sectioned and inspected. The inspection results were then compared to the requirements of an ASTM specification, which describes how to correctly assemble PVC solvent cement joints. Numerous deficiencies were found, notably lack of complete insertion, inadequate fusion (possibly due to lack of primer use), incomplete coverage, and possible contamination of the glue (sand/dirt/debris). The system installer was required to return to the jobsite and make repairs without additional cost to the owner/user.

Case Study C: Inadequate Design

Hauser received two samples of 2-in. nominal solvent cement PVC tee fittings that had failed in service. The tees were portions of an amusement park ride where water jets were constantly cycled on and off. The tees branched vertically from the 2-in. supply line for approximately 6 in., where the branch then changed direction 90 degrees (from the axis of the supply line). The system operated using recycled (potable) water. Hauser performed failure analysis, and in the process he determined that the fitting samples displayed adequate workmanship of manufacture and assembly, possessed adequate material strength, and were manufactured from an appropriate material. The fractures indicated that a stress

*For more information, contact Dr. Kim Baughman, Director of Development, Microbac Laboratories, Inc., kbaughman@microbac.com.

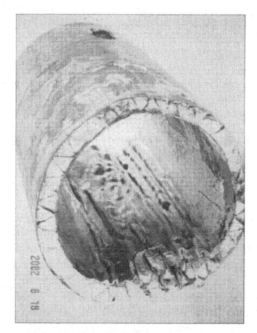

Figure 15-7-1. Section of ABS truss pipe affected by long-term exposure to dry-cleaning solvents.

Figure 15-7-2. Cyclic fatigue fracture of PVC tee used in water spraying system.

field was present which caused fracture that could only have come from operation of the system. This stress field ultimately caused a fatigue-type failure in the fittings due to torsional loading of the fittings through the 6-in. lever arm, caused by water hammer as the water was cycled on and off repeatedly (Figure 15-7-2). The failures could have been eliminated by using a significantly more robust fitting and/or thrust blocking of the tee and lever arm of the pipe branch to reduce or eliminate the stress.

Case Study D: Material Substitution

The manufacturer of a golf club bag properly designed all components on their newest product. One of the design changes was the incorporation of a plastic three-point harness. The Asian manufacturer molded and submitted prototypes for approval. These prototypes were tested and performed exceptionally well. The first production lot was received and given away to professional golfers on tour. In a matter of weeks the company received a number of complaints that the harnesses were coming apart.

The initial fix was to mold-in additional washers around the rivets to prevent rivet pullout, but the field failures continued. DSC and FTIR analysis of the harnesses revealed that the prototypes were made from acetal, and the production harnesses were made from nylon 6. The reason for this error was that the molder selected what they believed was the best material for the application during the prototyping stage and the customer assumed that they were molding nylon. During a subsequent conversation between the two parties, the design company referred to the "nylon" harnesses. Please note that the material type was *never* specified in writing. The part is now molded in acetal, and no additional field failures have been observed.

Case Study E: Environmental Stress Cracking

A client manufacturing escalators with a special design was seeing field failures of a yellow polycarbonate part that was screwed into the steel of the escalator step tread to make the leading edge of the step more visible to the passengers. The edge piece was becoming brittle and cracking after three months of service. This was only happening at one installation. Analysis of the failed parts by ICP revealed a high level of chlorine that was not present in as-manufactured samples. It was determined that a manufacturer of escalator cleaning equipment had recently introduced a new cleaner that contained a high level of chlorinated hydrocarbons. The combination of chlorinated hydrocarbons, stress at the screw holes, and a fluid reservoir formed by the recess for the screw head spelled disaster for the polycarbonate. Subsequent examination revealed that the plastic part was cracking everywhere—it simply failed at the screws first. Changing the cleaner resolved the issue.

Case Study F: Poor Design

The customer submitted a fondue pot, manufactured in China and sold in Europe, for failure analysis after the plastic handles pulled off of the pot during use. The customer had filled the pot with vegetable oil and heated the oil to boiling, in accordance with the photograph on the front of the box that the set was packaged in. While moving the pot from the stove to the table both handles came off the pot. Hot oil spilled onto, and severely burned, one person. The handles were connected to the pot by molded-in screw-on metal inserts. These inserts had large barbs that should have held the handles in place. Visual inspection of the handles showed clear signs of melting. DSC analysis of the part revealed a glass transition at −100°C. FTIR analysis determined that the part was styrene butadiene copolymer. The Vicat softening temperature of the handle was measured as 103°C. The physical properties of this material were totally inappropriate for the application. While we have cited this case as poor design, it may well be a material substitution failure, because it seems highly unlikely that a design engineer would make such an error.

Case Study G: Poor Manufacturing Practices

The client submitted a sample of polyethylene (PE) natural gas pipe that had failed in service after approximately 13 years in service by slow crack growth through the wall of the pipe. The sample was documented, and the fracture was opened for inspection. Upon inspection, a material was found molded into the pipe wall. Optical microscopy indicated that the molded-in contamination compromised the stress-carrying capacity of the pipe wall, and it also allowed a fracture to initiate and propagate. When tested by FTIR, the material was found to be cellulose. Silicone grease was also found in close proximity. A greasy shop rag had found its way into the extruder during pipe manufacture, ultimately causing failure of the pipeline.

16

QUALITY CONTROL

16.1. INTRODUCTION

Quality, as defined by the *American Heritage Dictionary*, is a characteristic or attribute of something, property, or a feature. According to Juran (1), "Quality is a universal concept, applicable to all goods and services, more appropriately described as fitness for use." Fitness for use is the extent to which the product successfully serves the purpose of the user during usage. Quality control is the regulatory process through which we measure actual quality performance, compare it with standards, and act on the difference (2).

Quality control is the means by which every step in the production process receives the attention required to assure that all parts and end products meet the desired specifications. Some parts are so complicated that if a production machine introduces a minor variance, the end unit cannot be assembled. This means that the manufacturer must make a determination of every critical step of the manufacturing process. Then, he must decide at which point inspections will be made and what controls will be established in order to produce parts and end products with all of the desired specifications.

In these days of rules and regulations, the only way any manufacturer can survive and be profitable is to have a firm grasp and a clear understanding of the science of quality control. For a plastics manufacturer or processor, the challenge is unique. The majority of materials are newly developed and are not precise in their composition. Manufacturing processes and procedures are different from conventional techniques; products made from such new materials have no previous history. Finally, the rapid growth of the plastics industry has crated severe problems in training new people.

Handbook of Plastics Testing and Failure Analysis, Third Edition, by Vishu Shah
Copyright © 2007 by John Wiley & Sons, Inc.

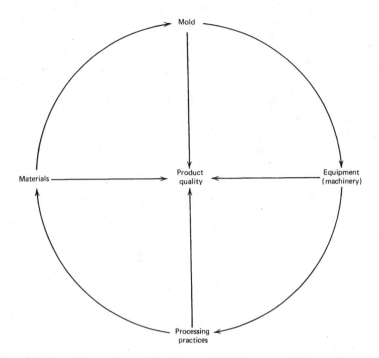

Figure 16-1. Interdependence of major variables in controlling product quality.

A well-established quality control system serves many useful purposes. First, it keeps the present customer happy, which in turn attracts new business. Second, it allows one to meet all regulatory and contractual obligations. More important, the system acts as a signaling device for any unforeseen problems and thereby reduces costly rejects. In the case of plastics, controlling the quality of the product is not a simple matter of inspecting and testing the product as it comes off the machine or assembly line. Many variables and unknowns, such as post-mold shrinkage, play an important role in controlling the ultimate quality of the product. Figure 16-1 graphically illustrates the interdependence of the major variables and their effect on product quality. In this chapter, we discuss statistical quality control, an important aspect of quality control, and the quality control system in general.

16.2. STATISTICAL QUALITY CONTROL

Statistical quality control (SQC) is relatively new. The original concepts were developed by Walter A. Shewart of Bell Telephone Laboratories in the early 1930s. Before the development of SQC methods, the only way to assure the quality of the product was to inspect 100 percent of the goods produced. Thus, the quality was "inspected into" the product. No consideration was given to the concept of controlling the process to produce a good part. The statistical quality control method emphasizes controlling the process by examining the product rather than controlling the product alone. Simply stated, the more adequately the manufacturing process is controlled, the less the probability of scrap and rejects (3).

There are two main facets of statistical quality control. One of them is the use of process control charts for in-process manufacturing operations. These charts, also referred to as "variables control charts" or "attributes control charts," are aimed at evaluating present as well as future performance. The other facet of statistical quality control is acceptance inspection or acceptance sampling. This technique forms the basis for scientifically evaluating past performance and accepting or rejecting the product.

16.2.1. Process Control Charts

Ideally, every manufacturer would like to see the manufacturing process controlled to a degree where he could produce identical parts day in and day out. In the real world, this is not possible. The difficulties in controlling the process arise from inherent variations caused by equipment and tool wear, operator skill, and manufacturing environment. Statistical quality control techniques, through the use of control charts, allow us to achieve the following:

1. To statistically determine whether the process is in or out of control
2. If the process begins to drift out of control, it signals a need for investigation and correction
3. To pinpoint the cause of the process variation
4. To reduce inspection cost
5. To improve the quality of the product and reduce the reject and scrap rate
6. To establish a stable, smooth-running, and predictable process
7. To study the capability of a particular process
8. To check suitability of specifications

In recent years, the growing trend in the plastics industry has been toward automation and high-speed production. Many new high-speed, fast-cycling injection-molding machines and high-throughput extruders have been developed. A 64-cavity mold in a machine operating at a 3-sec cycle and producing over 70,000 parts/hr is not uncommon. In contrast, a single-cavity mold in a 3000-ton machine producing a part weighing over 100 lb is also not uncommon. Some large extruders which produce 8- and 10-in. diameter pipe can consume as much as 2000 lb/hr. It is obvious that in such a high-production environment, one cannot allow the process to drift out of control or produce high scrap. Statistical quality control plays a very important role in controlling such processes in a highly competitive market where a high scrap and reject rate is totally unacceptable.

A. Variables Control Charts ($\bar{x}$ and R Charts)

Before we discuss how control charts are used in the plastics manufacturing operation, it is necessary to understand a few basic terms.

$\bar{x}$ *(Average, Arithmetic Mean).* The mean value of a sample of observations of the variable x.

$$\bar{x} = \frac{\Sigma x}{n}$$

For example, the individual weights of bottle caps are 7, 3, 9, 5, and 6 g. To compute the average ($\bar{x}$), sum the weights (x) and divide the sum by the total number (n) of bottle caps.

$$\bar{x} = \frac{7+3+9+5+6}{5} = 6$$

R (Range). The range is the difference between the highest and the lowest individual values in the group. The range calculated from the previous example would be six. $\bar{R}$ (pronounced R-bar) is the average of a series of R values.

Figure 16-2a illustrates a typical variables control chart ($\bar{x}$ and R chart). The center line of the $\bar{x}$ chart represents the average of a series of $\bar{x}$ values.

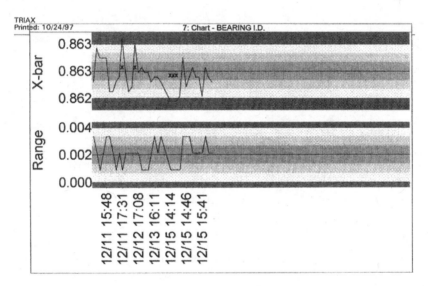

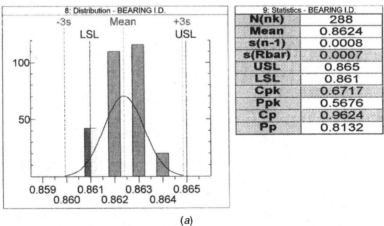

(a)

Figure 16-2. (a) Typical variables controls chart.

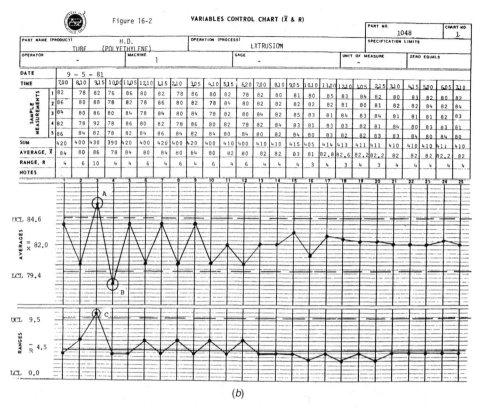

Figure 16-2. (*b*) Variable controls chart.

This value is symbolized by $\bar{\bar{x}}$ (pronounced *x* double-bar). The center line of the *R* chart represents the average of the ranges of the samples. The top and the bottom lines of both charts represent the upper control limit (UCL) and the lower control limit (LCL), respectively, Generally, the control limits are three standard deviations (3σ) above and below the center line. This means that the probability of measurements falling between ± 3 standard deviations is 99.73 percent of the observed values. Stated another way, between ± 3 standard deviations from the mean, one expects to find 99.73 percent of all the observed values. The values represented in Figure 16-2*b* are derived from the following example.

Example. Let us assume that we have just installed an extruder to make a 2-in.-inside-diameter tubing. The nominal wall thickness of 82 mil is to be maintained. As in the case of any manufacturing process, variations are expected. These variations are expected, owing to a number of factors such as variations in material, melt temperature, normal wear in the die, extruder speed, and tube puller speed. The quality control inspector is required to cut five pieces of tubing approximately every hour, measuring wall thickness and computing the average weight. Figure 16-2*b* also lists the wall thickness of the samples taken on Monday morning, soon after starting the extruder. The wall thicknesses are recorded in mils for the next

24 hr. The following calculations are carried out so that the variables control chart for $\bar{x}$ and R can be plotted.

$$\bar{\bar{x}} = \frac{\text{Sum of the means of the samples}}{\text{Sum of the number of sample means}} \qquad \bar{R} = \frac{\text{Sum of sample ranges}}{\text{Sum of the number of sample ranges}}$$

$$= \frac{\Sigma\bar{x}}{n} = \frac{2050}{25} = 82 \qquad\qquad = \frac{\Sigma R}{n} = \frac{113}{25} = 4.52$$

The upper control limit (UCL) and lower control limit (LCL) of the $\bar{x}$ chart is computed by:

$$\text{UCL} = \bar{\bar{x}} + A_2\bar{R} \qquad\qquad \text{LCL} = \bar{\bar{x}} - A_2\bar{R}$$
$$= 82 + 0.577(4.52) = 84.6 \qquad = 82 - 0.577(4.52) = 79.4$$

A_2 is a factor used in calculating the control limits for the $\bar{x}$ chart. Table 16-1 lists the factors for calculating the control limits for the average and range charts. The n in the table refers to the number of items in the sample. In this example, the number of items in the sample is five.

The upper and lower control limit for the range chart is calculated in a very similar manner.

$$\text{UCL} = D_4\bar{R} \qquad\qquad \text{LCL} = D_3\bar{R}$$
$$= 2.115(4.52) = 9.55 \qquad\qquad = 0(4.52) = 0$$

TABLE 16-1. Factors for Control Charts

Number of Items in Sample, n	Chart for Averages Factors for Control Limits A_2	Chart for Ranges Factors for Central Line d_2	Factors for Control Limits D_3	D_4
2	1.880	1.128	0	3.267
3	1.023	1.693	0	2.575
4	0.729	2.059	0	2.282
5	0.577	2.326	0	2.115
6	0.483	2.534	0	2.004
7	0.419	2.704	0.076	1.924
8	0.373	2.847	0.136	1.864
9	0.337	2.970	0.184	1.816
10	0.308	3.078	0.223	1.777
11	0.285	3.173	0.256	1.744
12	0.266	3.258	0.284	1.716
13	0.249	3.336	0.308	1.692
14	0.235	3.407	0.329	1.671
15	0.223	3.472	0.348	1.652

From B. Mason, "Statistical Techniques in Business and Economics." Reprinted with permission of Richard D. Irvin, Inc.

Here again, the D_4 and D_3 values derived from Table 16-1 are the factors used to compute the control limits for the range chart. From this information we now can chart $\bar{x}$ and R values as shown in Figure 16-2b.

Interpretation. By surveying the average chart, we can safely conclude that if a sample of five tubes is randomly picked and measured, the arithmetic mean wall thickness will fall between 79.4 and 84.6 mil about 99.73 percent of the time. A similar conclusion regarding the range chart can also be made. The average chart also reveals that at two points marked A and B, the process went outside the limits of the expected variability. This variation was attributed to the instability in the process due to the start-up. As is evident from the chart, the process stabilized itself with time and variations minimized. By looking at the range chart, we can see that the process was out of control at point C (during start-up), but stabilized later on.

Figure 16-3 shows the control chart of the same process a few days later. Up to point D in the chart, the process was well within the control limits. Beyond point D, the process seems to have gone completely out of control. A little investigation revealed that a high-density polyethylene material (with a slightly different melt index value) from a new supplier was introduced at point D. The difference in the extrusion characteristics of this new material accounted for such gross variation. Figure 16-4 illustrates a typical $\bar{x}$ chart showing a trend. The assignable cause

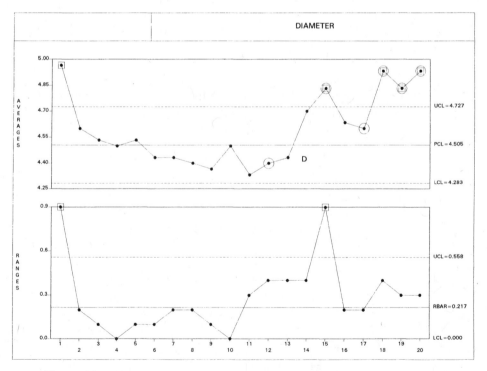

Figure 16-3. Chart showing process out of control due to material variation.

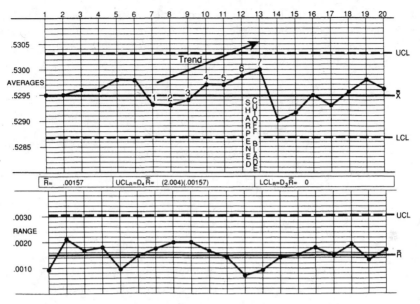

Figure 16-4. $\bar{x}$ chart showing a trend.

producing such a trend was traced down to the tooling wear. A steady but gradual increase in the wall thickness was recorded over a long period.

From the foregoing discussion, it is clear that a thorough understanding of the process equipment, equipment operators, process conditions, and tooling condition is required so that control charts can be interpreted intelligently and a meaningful conclusion can be drawn.

B. Control Charts for Attributes

Attribute is defined as a characteristic or property that is appraised in terms of whether it does or does not exist with respect to a given requirement (4). In preparing control charts for attributes, the only item of concern is the presence or absence of a given characteristic or defect. The most common inspection tools for attributes are Go and No-Go gauges. Unlike variable charts, attributes charts do not require actual measurements. Instead, a simple measurement in terms of percent defective or number of defects/unit is made. Some common examples of attributes are the presence or absence of flash, sink marks or shorts in a molded article, molded-in threads acceptable or unacceptable as measured by a Go–No-Go thread gauge, and the color of the molded article visually matches the control or it does not match. These types of charts are inevitable, especially when the item is produced in extremely large quantities, making individual measurements impractical, time-consuming, and very expensive. Control charts for attributes are also explored when the measurement technique is destructive. One of the biggest advantages of such charts is their simplicity and ease of understanding by all levels of personnel.

P Charts. One of the most commonly used control charts for attributes is the "percent defective" or *P* chart. *P* charts are prepared by obtaining a series of appropriately sized samples. A sample size of 50 or 100 is the most convenient. Next, the number of defective samples is counted and recorded. From this information, the percent defective in a sample (P), average (mean) percent defective ($\bar{P}$), upper control limit, and lower control limit can be calculated as follows:

$$P = \frac{\text{Number of defective units in sample}}{\text{Sample size}}$$

$$\bar{P} = \frac{\text{Sum of percent defective}}{\text{Total number of samples}}$$

$$\text{UCL} = \bar{P} + 3\sqrt{\frac{\bar{P}(1 - \bar{P})}{n}}$$

$$\text{LCL} = \bar{P} - 3\sqrt{\frac{\bar{P}(1 - \bar{P})}{n}}$$

If a negative value (less than zero) of the lower control limit is obtained, it should be considered as zero on the chart since the percent defective cannot be less than zero.

C Charts. The C chart is also known as the "C-bar" chart and is a special type of attributes control chart. Unlike the P chart, which portrays *percent defective*, the C chart uses the number of defects per unit. For example, a black speck and a deep sink mark on a molded part are two defects per part for C chart calculation but one part defective for P chart purposes.

The upper and lower control limits of a C chart are calculated as follows:

$$\text{Control limit} = \bar{C} \pm 3\sqrt{C}$$

$\bar{C}$ is the center line of the control chart.

16.2.2. Acceptance Sampling

Acceptance sampling is a widely used and accepted statistical quality control technique. The acceptance sampling technique calls for selecting a sample randomly from a lot and deciding whether to accept or reject the lot, based on the number of defective items found in the sample.

Contrary to the belief that 100 percent inspection means 100 percent quality, 100 percent inspection is not only impractical but also unreliable, ineffective, and impossible in the case of destructive testing. Acceptance sampling drastically reduces the cost of inspection and in most cases provides better quality assurance than 100 percent inspection due to lower inspector fatigue and boredom. At this point, it is important to understand that the acceptance sampling technique does not "control quality" but it helps in determining the course of action. Based on the acceptance sampling plan, one may decide to accept or reject a particular lot.

A. Sampling Theory

The need for an acceptance sampling plan arises from another need to compromise between the consumer's demand for a perfect quality product and the manufacturer's inability to provide a perfect quality product owing to process limitations and variations. Since 0 percent defective product quality is not possible, a compromise in terms of some absolute value of quality greater than 0 percent defective must be made. Theoretically, an ideal sampling plan is one that rejects all lots worse than the standard and accepts all lots equal to or better than the standard. Figure 16-5 illustrates one such ideal sampling plan diagram. According to this plan, all groups of products greater than 5 percent defective would be rejected and less than 5 percent defective would be accepted. Such an ideal sampling plan, which can distinguish between acceptable and rejectable lots 100 percent of the time, is not possible. Even 100 percent inspection is not capable of achieving perfect discrimination between good and bad lots. In other words, there is always a chance that good lots may be rejected and bad lots accepted with any type of sampling plan. The smaller the samples size, the greater the risk of making an erroneous judgment. The probability of acceptance is high with a good quality product and it becomes less and less so as the product gets worse. This fact forms the basis for a curve known as the operating characteristic curve. The curve is also commonly referred to as the OC curve. An OC curve is a plot of the quality of an incoming lot in percent defective against the probability that a lot will be accepted when sampled according to the plan. The lots that contain 0 percent defective will be accepted 100 percent of the time; conversely, the lots that are 100 percent defective will never be accepted. Depending upon the sampling plan and the quality of the incoming lots, the probability of acceptance (or rejection) increases or decreases. A typical

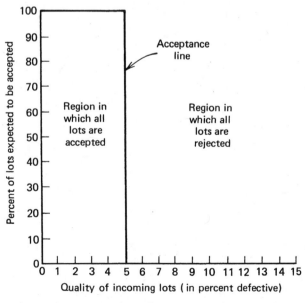

Figure 16-5. Ideal sampling plan.

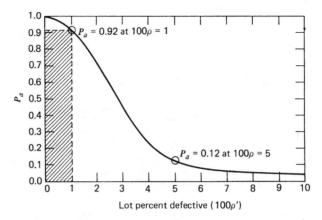

Figure 16-6. A typical operating characteristic curve. (Reprinted with permission from Richard D. Irwin, Inc.)

OC curve of a sample size (n) of 100 and acceptance number (c) of 2 is illustrated in Figure 16-6. In this case, $n = 100$ and $c = 2$ means that for a sampling plan that calls for a sample size of 100, the maximum allowable number of defects in the sample is two. The OC curve in Figure 16-6 also indicates that if a lot is 1 percent defective, the lot will be accepted under this sampling plan ($n = 100$, $c = 2$) 92 percent of the time. Also, if the lot is 5 percent defective, it will be accepted under this sampling plan 12 percent of the time. Obviously, a distinctly different OC curve for each sampling plan with its won characteristic risk pattern can be plotted. There are two points on the OC curve that describe, relatively, the acceptable and nonacceptable region. These two are known as producer's risk (α) and consumer's risk (β). Producer's risk is defined as the risk or probability of rejecting a good lot. Every producer would like to keep this risk as small as possible. The consumer's risk is defined as the risk or probability of accepting a poor lot. This is the probability of accepting a poor lot which is unsatisfactory and sending it out to the consumer. The consumer may reject the lot quite often if the consumer's risk of a sampling plan is kept high. A balance between the producer's risk and the consumer's risk must be attained for economics reasons.

B. Types of Sampling Plans
The three basic sampling plans that are most frequently used are:

1. Single sampling plan
2. Double sampling plan
3. Multiple sampling plan

Single Sampling Plan. This type of sampling plan is used when the results of a single sample from an inspection lot are conclusive in determining its acceptability. The lot is accepted if the number of defectives found in the sample is equal to or less than the acceptance number AC or C. Similarly, the lot is rejected if the number of defectives found in the sample is equal to or greater than the rejection number RE or r.

Double Sampling Plan. A double sampling plan is used if the inspection of the first sample leads to the decision to accept, reject, or take a second sample. After inspection of the second sample, if necessary, a decision to accept or reject is made. Figure 16-7 shows a typical double sampling plan. Figure 16-8 illustrates a switching plan for sampling inspection.

Multiple Sampling Plan. A multiple sampling plan is an extension of a double sampling plan. As long as the number of defectives falls between acceptance and rejections numbers, the inspection is continued.

C. Classification of Sampling Plans

Sampling plans fall into three major classes:

AQL sampling plan
LTPD sampling plan
AOQL sampling plan

A detailed discussion of all three classes of sampling plans is beyond the scope of this book. Our discussion will be confined to the AQL sampling plan.

Lot Size	Sample	Sample size	Cumulative sample size	1.0%	
				AC	RE
2–8	First	2	2	0	1
9–15	First	3	3	0	1
16–25	First	5	5	0	1
26–50	First	8	8	0	1
51–90	First	13	13	0	1
91–150	First	20	20	0	1
151–280	First Second	20 20	20 40	0 1	2 2
281–500	First Second	32 32	32 64	0 1	2 2
501–1200	First Second	50 50	50 100	0 3	3 4
1201–3200	First Second	80 80	80 160	1 4	4 5
3201–10,000	First Second	125 125	125 250	2 6	5 7
10,001–35,000	First Second	200 200	200 400	3 8	7 9
35,001–150,000	First Second	315 315	315 630	5 12	9 13

Figure 16-7. Double sampling plan.

Switching Rules for ANSI Z1.4 System

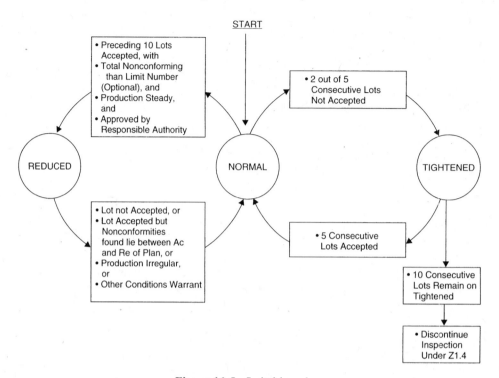

Figure 16-8. Switching plan.

D. AQL Sampling Plan

The acceptable quality level (AQL) is defined as the maximum percent defective that for acceptance sampling can be considered acceptable as a process average. A 4 percent AQL sampling plan will regularly accept 4 percent defective product. AQL sampling plans are designed to protect the supplier from having good lots rejected. The consumer's risk of accepting bad lots can only be determined by studying the OC curve for the AQL sampling plans. The specified producer's risk in an AQL plan is generally 5 percent, meaning a plan that will accept lots of AQL quality with a stated probability of 95 percent. For example, a 2 percent AQL plan will reject 2 percent defective material 5 percent of the time or accept it 95 percent of the time.

AQL sampling plans are the most widely accepted and used plans throughout the industry and government. The best-known published source of such AQL sampling plans has been the government publication titled *Sampling Procedures and Tables for Inspection by Attributes MIL-STD-105 E.* Since 1995 ANSI/ASOC Z1-4 has replaced MIL-STD-105E with minor modifications.

E. Use of Sampling Tables

The following procedure is recommended for using the sampling tables:

1. Determine the lot size.
2. Select the inspection level as specified. If the inspection level is unspecified, normal inspection level II should be used. Some companies use past quality history as a guideline for selecting the inspection level. For example, if the supplier has mostly submitted "good" lots in the past, level I (reduced inspection) is used. Level III (tightened inspection) is used if lots have been mostly "bad." Special inspection levels S-1 through S-4 may also be used for expensive as well as destructive-type inspections or for parts made with repetitive-type processes.
3. Determine the sample size code letter from the lot size information using Table 16-2.
4. Select the sampling plan as specified. If unspecified, the single sampling plan should be used.
5. From the information regarding the inspection level, sample size code letter, sampling plan, and specified AQL value, determine the acceptance and rejection number. If the AQL value is unspecified, start with a 2.5 percent defective AQL value.

Example. Given the following information, determine the sample size and acceptance and rejection numbers using (1) single and (2) double sampling plans.

Lot size: 1000
Inspection level: Normal
AQL: 1.0

TABLE 16-2. Sample Size and Code Letters

Lot or Batch Size	Special Inspection Levels				General Inspection Levels		
	S-1	S-2	S-3	S-4	I	II	III
2–8	A	A	A	A	A	A	B
9–15	A	A	A	A	A	B	C
16–25	A	A	B	B	B	C	D
26–50	A	B	B	C	C	D	E
51–90	B	B	C	C	C	E	F
91–150	B	B	C	D	D	F	G
151–280	B	C	D	E	E	G	H
281–500	B	C	D	E	F	H	J
501–1,200	C	C	E	F	G	J	K
1,201–3,200	C	D	E	G	H	K	L
3,201–10,000	C	D	F	G	J	L	M
10,001–35,000	C	D	F	H	K	M	N
35,001–150,000	D	E	G	J	L	N	P
150,001–500,000	D	E	G	J	M	P	Q
500,001 and over	D	E	H	K	N	Q	R

Solution. From Table 16-2, note that the 1000-piece lot size corresponds to code letter "J" for normal inspection level II.

1. Using Table 16-3 (single sampling plan for normal inspection), we find that for code letter "J" and AQL of 1.0, the sample size is 80 with acceptance number AC = 2 and rejection number RE = 3. Thus, the course of action should be accept—if the number of defectives is 2 or less—or reject—if the number of defectives is 3 or more.

2. Using Table 16-4 (for double sampling plan for normal inspection), we find that for the code letter "J" and AQL of 1.0, the first sample size is 50 pieces and acceptance number AC = 0 and rejection number RE = 3.

The second sample size is 50 pieces and the acceptance number (for both samples) is AC = 3. The rejection number (for both samples) is RE = 4. The following procedure should be followed.

Select 50 pieces from the lot and accept—if the number of defectives is 0; reject—if the number of defectives is 3 or more. If the number of defectives falls between 0 and 3, let's say 2, select another 50 pieces and accept—if the number of defectives in both samples is 3 or less; reject—if the number of defectives in both samples is 4 or more.

F. LTPD Sampling Plan

The lot tolerance percent defective (LTPD) sampling plan is defined as an allowable percentage defective, a figure which may be considered a borderline distinction between a satisfactory lot and an unsatisfactory one. A 2 percent LTPD sampling plan will regularly reject 2 percent defective product. The specified consumer's risk in an LTPD sampling plan is generally 10 percent, meaning a plan that will reject lots of LTPD quality with a stated probability of 90 percent. For example, a 6.0 LTPD plan will accept 6 percent defective material 10 percent of the time and reject it 90 percent of the time.

G. AOQL Sampling Plan

Average outgoing quality (AOQ) is defined as the average quality of outgoing products including all accepted lots plus all rejected lots after the rejected lots have been effectively 100 percent inspected and all defectives have been replaced by nondefectives. Average outgoing quality limit (AOQL) is defined as the maximum of AOQ for all possible incoming qualities for a given sampling inspection plan. In other words, AOQL is the worst average quality that can exist in the long run in the outgoing product. For example, a 2.5 AOQL plan assures us that in the long run the accepted material will not be more than 2.5 percent defective. AOQL sampling plans are designed to protect the consumer with specified risk. They offer a low probability of acceptance if the product quality exceeds the required AOQL.

16.2.3. Process Capability

A. Introduction

Process capability is a measure of variation of a process and its ability to consistently produce components within specification. Process capability can be

TABLE 16-3. Single Sampling Plan for Normal Inspection (Master Table)

Sample Size Code Letter	Sample Size	Acceptable Quality Levels (Normal Inspection) 0.010	0.015	0.025	0.040	0.065	0.10	0.15	0.25	0.40	0.65	1.0	1.5	2.5	4.0	6.5	10	15	25	40	65	100	150	250	400	650	1000
		Ac Re	Ac Re	Ac Re	Ac Re	Ac Re	Ac Re	Ac Re	Ac Re	Ac Re	Ac Re	Ac Re	Ac Re	Ac Re	Ac Re	Ac Re	Ac Re	Ac Re	Ac Re	Ac Re	Ac Re	Ac Re	Ac Re	Ac Re	Ac Re	Ac Re	Ac Re
A	2	↓	↓	↓	↓	↓	↓	↓	↓	↓	↓	↓	↓	↓	↓	0 1	↓	↓	1 2	2 3	3 4	5 6	7 8	10 11	14 15	21 22	30 31
B	3	↓	↓	↓	↓	↓	↓	↓	↓	↓	↓	↓	↓	↓	0 1	↑	↓	1 2	2 3	3 4	5 6	7 8	10 11	14 15	21 22	30 31	44 45
C	5	↓	↓	↓	↓	↓	↓	↓	↓	↓	↓	↓	↓	0 1	↑	↓	1 2	2 3	3 4	5 6	7 8	10 11	14 15	21 22	30 31	44 45	↑
D	8	↓	↓	↓	↓	↓	↓	↓	↓	↓	↓	↓	0 1	↑	↓	1 2	2 3	3 4	5 6	7 8	10 11	14 15	21 22	30 31	44 45	↑	↑
E	13	↓	↓	↓	↓	↓	↓	↓	↓	↓	↓	0 1	↑	↓	1 2	2 3	3 4	5 6	7 8	10 11	14 15	21 22	30 31	44 45	↑	↑	↑
F	20	↓	↓	↓	↓	↓	↓	↓	↓	↓	0 1	↑	↓	1 2	2 3	3 4	5 6	7 8	10 11	14 15	21 22	↑	↑	↑	↑	↑	↑
G	32	↓	↓	↓	↓	↓	↓	↓	↓	0 1	↑	↓	1 2	2 3	3 4	5 6	7 8	10 11	14 15	21 22	↑	↑	↑	↑	↑	↑	↑
H	50	↓	↓	↓	↓	↓	↓	↓	0 1	↑	↓	1 2	2 3	3 4	5 6	7 8	10 11	14 15	21 22	↑	↑	↑	↑	↑	↑	↑	↑
J	80	↓	↓	↓	↓	↓	↓	0 1	↑	↓	1 2	2 3	3 4	5 6	7 8	10 11	14 15	21 22	↑	↑	↑	↑	↑	↑	↑	↑	↑
K	125	↓	↓	↓	↓	↓	0 1	↑	↓	1 2	2 3	3 4	5 6	7 8	10 11	14 15	21 22	↑	↑	↑	↑	↑	↑	↑	↑	↑	↑
L	200	↓	↓	↓	↓	0 1	↑	↓	1 2	2 3	3 4	5 6	7 8	10 11	14 15	21 22	↑	↑	↑	↑	↑	↑	↑	↑	↑	↑	↑
M	315	↓	↓	↓	0 1	↑	↓	1 2	2 3	3 4	5 6	7 8	10 11	14 15	21 22	↑	↑	↑	↑	↑	↑	↑	↑	↑	↑	↑	↑
N	500	↓	↓	0 1	↑	↓	1 2	2 3	3 4	5 6	7 8	10 11	14 15	21 22	↑	↑	↑	↑	↑	↑	↑	↑	↑	↑	↑	↑	↑
P	800	↓	0 1	↑	↓	1 2	2 3	3 4	5 6	7 v8	10 11	14 15	21 22	↑	↑	↑	↑	↑	↑	↑	↑	↑	↑	↑	↑	↑	↑
Q	1250	0 1	↑	↓	1 2	2 3	3 4	5 6	7 v8	10 11	14 15	21 22	↑	↑	↑	↑	↑	↑	↑	↑	↑	↑	↑	↑	↑	↑	↑
R	2000	↑	↑	1 2	2 3	3 4	5 6	7 8	10 11	14 15	21 22	↑	↑	↑	↑	↑	↑	↑	↑	↑	↑	↑	↑	↑	↑	↑	↑

↓ = Use first sampling plan below arrow. If sample size equals, or exceeds, lot or batch size, do 100 percent inspection.

↑ = Use first sampling plan above arrow.

Ac = Acceptance number.

Re = Rejection number.

TABLE 16-4. Double Sampling Plan for Normal Inspections

Acceptable Quality Levels (Normal Inspection)

Sample size code-letter	Sample	Sample size	Cumulative sample size	0.010		0.015		0.025		0.040		0.065		0.10		0.15		0.25		0.40		0.65		1.0		1.5		2.5		4.0		6.5		10		15		25		40		65		100		150		250		400		650		1000	
				Ac	Re	Ac	Re	Ac	Re	Ac	Re	Ac	Re	Ac	Re	Ac	Re	Ac	Re	Ac	Re	Ac	Re	Ac	Re	Ac	Re	Ac	Re	Ac	Re	Ac	Re	Ac	Re	Ac	Re	Ac	Re	Ac	Re	Ac	Re	Ac	Re	Ac	Re	Ac	Re	Ac	Re	Ac	Re
A																														↓		*		↑																						
B	First	2	2																											*		↓		0	2	0	3	1	4	2	5	3	7	5	9	7	11	11	16	17	22	25	31
	Second	2	4																															1	2	3	4	4	5	6	7	8	9	12	13	18	19	26	27	37	38	56	57
C	First	3	3																									*		↓		0	2	0	3	1	4	2	5	3	7	5	9	7	11	11	16	17	22	25	31	↑	
	Second	3	6																													1	2	3	4	4	5	6	7	8	9	12	13	18	19	26	27	37	38	56	57		
D	First	5	5																							*		↓		0	2	0	3	1	4	2	5	3	7	5	9	7	11	11	16	17	22	25	31	↑			
	Second	5	10																									1	2	3	4	4	5	6	7	8	9	12	13	18	19	26	27	37	38	56	57				
E	First	8	8																					*		↓		0	2	0	3	1	4	2	5	3	7	5	9	7	11	11	16	17	22	25	31	↑					
	Second	8	16																							1	2	3	4	4	5	6	7	8	9	12	13	18	19	26	27	37	38	56	57						
F	First	13	13																			*		↓		0	2	0	3	1	4	2	5	3	7	5	9	7	11	11	16	17	22	25	31	↑							
	Second	13	26																					1	2	3	4	4	5	6	7	8	9	12	13	18	19	26	27	37	38	56	57								
G	First	20	20																	*		↓		0	2	0	3	1	4	2	5	3	7	5	9	7	11	11	16	17	22	25	31	↑									
	Second	20	40																			1	2	3	4	4	5	6	7	8	9	12	13	18	19	26	27	37	38	56	57										
H	First	32	32															*		↓		0	2	0	3	1	4	2	5	3	7	5	9	7	11	11	16	17	22	25	31	↑											
	Second	32	64																	1	2	3	4	4	5	6	7	8	9	12	13	18	19	26	27	37	38	56	57												
J	First	50	50													*		↓		0	2	0	3	1	4	2	5	3	7	5	9	7	11	11	16	17	22	25	31	↑													
	Second	50	100															1	2	3	4	4	5	6	7	8	9	12	13	18	19	26	27	37	38	56	57														
K	First	80	80											*		↓		0	2	0	3	1	4	2	5	3	7	5	9	7	11	11	16	17	22	25	31	↑															
	Second	80	160													1	2	3	4	4	5	6	7	8	9	12	13	18	19	26	27	37	38	56	57																
L	First	125	125									*		↓		0	2	0	3	1	4	2	5	3	7	5	9	7	11	11	16	17	22	25	31	↑																	
	Second	125	250											1	2	3	4	4	5	6	7	8	9	12	13	18	19	26	27	37	38	56	57																		
M	First	200	200							*		↓		0	2	0	3	1	4	2	5	3	7	5	9	7	11	11	16	17	22	25	31	↑																			
	Second	200	400									1	2	3	4	4	5	6	7	8	9	12	13	18	19	26	27	37	38	56	57																				
N	First	315	315					*		↓		0	2	0	3	1	4	2	5	3	7	5	9	7	11	11	16	17	22	25	31	↑																					
	Second	315	630							1	2	3	4	4	5	6	7	8	9	12	13	18	19	26	27	37	38	56	57																						
P	First	500	500			*		↓		0	2	0	3	1	4	2	5	3	7	5	9	7	11	11	16	17	22	25	31	↑																							
	Second	500	1000					1	2	3	4	4	5	6	7	8	9	12	13	18	19	26	27	37	38	56	57																								
Q	First	800	800	*		↓		0	2	0	3	1	4	2	5	3	7	5	9	7	11	11	16	17	22	25	31	↑																									
	Second	800	1600			1	2	3	4	4	5	6	7	8	9	12	13	18	19	26	27	37	38	56	57																										
R	First	1250	1250	*	↓		0	2	0	3	1	4	2	5	3	7	5	9	7	11	11	16	17	22	25	31	↑																										
	Second	1250	2500			1	2	3	4	4	5	6	7	8	9	12	13	18	19	26	27	37	38	56	57																										

↓ = Use first sampling plan below arrow. If sample size equals or exceeds lot or batch size, do 100 percent inspection.

↑ = Use first sampling plan above arrow.

Ac = Acceptance number.

determined only when a process is in the state of statistical control. This is the condition describing a process from which all assignable causes of variation have been eliminated and only common causes remain. It is evidence on a control chart by the absence of points beyond the control limits and by the absence of nonrandom patterns and trends within the control limits.

When a process has been brought in control using control charts, we only know that the process is producing uniformly the same quality, but there is no guarantee that the process is producing good quality. Therefore, the next step is to check the process against the specifications.

A process that is not in control has no definable capability. Therefore, we should first bring the process in control before we try to measure its capability.

B. Process Capability Studies

These studies provide a preliminary assessment of the potential of the process to produce products that meet specifications. Since these studies are of short duration, they cannot provide information about long-term process performance. Process capability studies are conducted using variable data on the applicable sample size taken from a true production run. The data is gathered in subgroups of consecutive units (typically five unites per subgroup) from throughout the production run and are analyzed using an X-bar, R chart, or other appropriate control chart. When the chart shows neither any points out of control nor any evidence of trends, the process capability may then be determined by using one of the several statistical capability analysis techniques.

When performing a process capability study using attribute data, the process must demonstrate that it can produce 300 consecutive parts with no defects for all attributes being studied.

C. Process Capability Index, C_{pk}

C_{pk} is the process capability index, a statistical ratio that compares the process capability to the product-tolerance band. The capability index indicates whether the process will produce units within the tolerance limits.

C_{pk} = the lesser of

$$\text{(a)} \quad \frac{\text{USL} - \bar{X}}{3\,\text{Sigma}} \qquad \text{(b)} \quad \frac{\bar{X} - \text{LSL}}{3\,\text{Sigma}}$$

A C_{pk} of 1 means statistically that 2700 defective parts will be present for every million parts produced. Even though a C_{pk} of 1.5 is considered to be excellent quality, the manufacturers are constantly challenged to meet a C_{pk} of 2, which translates to only 3.4 defects per million. Generally speaking, a process is deemed acceptable if the C_p and C_{pk} indices of the characteristic are 1.33, but this is the minimum acceptable level. The goal, but by no means the end point, is $C_{pk} = 2.0$.

When the above criteria are not met, 100 percent inspection must be implemented. The manufacturing engineer should immediately determine the reasons for not meeting the criteria and, wherever practical, revise the process accordingly. The 100 percent inspection or approved alternate must be continued unitl capability is demonstrated or unitl an engineering authorization or change is approved by the product engineering department.

TABLE 16-5. Reaction Table

	Capability Indices Value (C$_{pk}$)		
Process—in control	Less than 1.00 C_{pk} 100% Inspect	1.00–1.33 C_{pk} Accept parts	Greater than 1.33 C_{pk} Accept parts
Process—out of control	Continue 100% inspection	Sort components 100% since last "in control" sample	Accept parts

Identify and Correct Special Cause	
Process out of control and one or more points in the sample outside the tolerance	Identify and correct special cause; sort components 100% since last "in control" sample

Since 100 percent inspection is generally both inefficient from a cost standpoint and ineffective in screening out nonconforming products, it should normally be considered only as an emergency measure and should not be established as a permanent feature of the process. The above criteria are minimum acceptable levels and should be considered as starting points for continuous improvement.

After control limits have been developed and process performance has been demonstrated, decisions on process actions and part acceptance must be made according to Table 16-5.

16.2.4. Computerized Data Acquisition and Analysis

Data acquisition and analysis is no longer a tedious process with the use of personal computer and Windows-based SPC/SQC software. The only action required of an operator is to input measurements either manually using a keyboard or automatically using a gauge connected to the computer. The data can be analyzed with a few simple key strokes. The program displays the result in terms of X-bar and R charts, sigma, process capability index, histogram, and a variety of other statistical data. Visual defects are analyzed using attributes data option and results are displayed in terms of pareto or control charts. Figure 16-9a illustrates a typical screen from commercially available SPC/SQC software.

16.3. INTRODUCTION TO STATISTICAL PROCESS CONTROL

16.3.1. Purpose of SPC

The purpose of statistical process control (SPC) is to determine whether a process is "in control" and to give an alert when the process is out of control. An in-control process produces consistent output. If the output of a process is in control, we can predict future output. If the output of a process is not in control, we cannot predict future output. If a process is in control, we do nothing if the process is capable of meeting specification limits (tolerances). If a process is in control and not capable

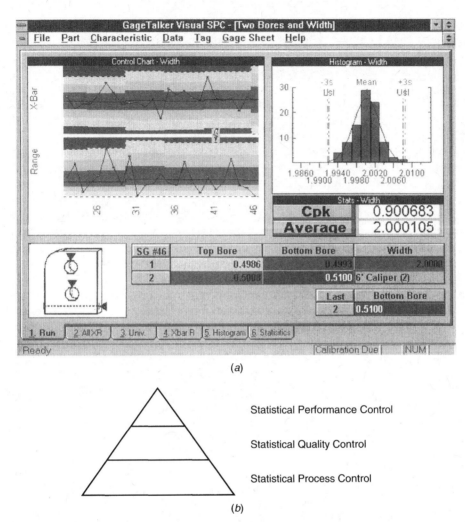

Figure 16-9. (*a*) Computerized SPC/SQC software. (Courtesy of Gagetalker CimWorks.) (*b*) Three different aspects or levels of statistical process control.

of meeting specification limits, we must (i) improve the process, (ii) change to a more capable process, or (iii) relax the specification limits. If a process is out of control, we should investigate the out-of-control condition. If it truly is an out-of-control condition instead of a naturally occurring event, we should stop the process and fix the out-of-control condition and take steps to ensure that it does not occur again. In essence, SPC tells us when to take action on a process and when to do nothing.

16.3.2. In-Control Versus Out-of-Control Conditions

A process is in control if the output of the process approximates a normal distribution. If the output of a process does not approximate a normal distribution, it is called an out-of-control condition or an out-of-control process. There are three general categories of out-of-conditions.

Out-of-Control Case #1: Outliers

The first out-of-control condition is when the process generates part parameters outside of some predetermined distance from the center of the output distribution. These limits are sometimes referred to as the upper process limits (UPL) and the lower process limits (LPL). Some authors refer to these limits as the upper and lower natural process limits. The selection of these limits is arbitrary. The current custom is to select plus and minus three sigma limits. If the distribution of the process output is a normal distribution, then 99.74% of the process will typically be between the upper and lower three sigma process limits. In some cases, such as for extremely stable processes, it is more practical to use 3.5 or 4.0 sigma limits instead of 3.0 sigma limits. This minimizes the number of false out-of-control signals.

No measuring or control system is perfect. Trade-offs are made between the cost and time spent evaluating the output of the process versus the benefits gained. An in-control process will generate data points beyond the upper and lower process limits 0.26% of the time. When this occurs, the user typically evaluates the situation to determine whether these outlier data points beyond the three sigma limits were caused by natural variation from common causes or by special causes. If the outlier data points were caused by common cause variation, they can be ignored. If they were caused by special cause variation, the special cause or causes of variation should be removed immediately. If the process output is from a production line, production is usually stopped and halted until special causes of variation are removed. Management is usually viewed as having primary responsibility for reducing common cause variation. Operators and leads are usually viewed as having primary responsibility for eliminating and preventing special cause variation.

Out-of-Control Case #2: Non-normal Distribution

The second out-of-control condition is when the process generates a distribution of part parameters that is inconsistent with the shape of a normal distribution. The shape of a normal distribution is typically referred to as a bell-shaped curve. A flat-shaped distribution between the three sigma limits is called a uniform distribution. The flat shape of a uniform distribution is different from the bell shape of a normal distribution. A uniform distribution of process output would be, by definition, an out-of-control condition. There are many types of conditions where the shape of the process output distribution does not approximate that of a normal distribution.

Out-of-Control Case #3: Time Dependencies

The third out-out-control condition is when the process output exhibits some time-dependent characteristic. This could be a certain number of constantly increasing (or decreasing) values that indicate that a trend is taking place. If a trend is occurring, the process can be re-centered.

16.3.3. SPC, Process Capability and Quality

In a manufacturing context, quality means (i) producing parts at design target, (ii) with minimum variable, and (iii) within specification limits. SPC is used to determine how well the process meets the first of these three requirements. Process capability studies are used to determine how well the process meets the third requirement. SPC measures the average output of the process (X-bar). X-bar is compared to design target to determine how closely the process output is centered relative to the design target. SPC also measures the variability of the output of the process. Typical measures of variability are range (R) and standard deviation (s). R or s is evaluated for stability in the same fashion that X-bar is evaluated for stability. Variability (R or s) should be in-control before X-bar is evaluated.

Process capability studies are used to determine what fraction of the tolerance band the process output consumes. Cp measures the fraction with consideration of where the process is centered. Cpk measures the fraction to the closest specification limit if the process is not centered. Cp and Cpk assume that the process is in control. Ppk is similar to Cpk but does not assume that the process is in-control.

16.3.4. SPC Techniques

SPC techniques can be either numerical or graphical techniques or a combination of the two. Computer software is commonplace and is used almost universally to do the necessary computations.

16.3.5. SPC Levels

"SPC" has evolved into a generic term that more precisely includes, at least for manufactured parts, three different aspects or levels of statistical process control as shown in Figure 16.9b. The top level is statistical performance control (SPFC). SPC in this context is used to determine the consistency with which the part (or assembly) performs and the average performance level. The performance metric is some measure of how well the part (or assembly) performs its intended function. Performance parameters can include flow rates, power amplification ratio, or any other parameter that the part is intended to perform. The intermediate level is statistical quality control (SQC). SQC in this context determines the consistency with which the different part characteristics, such as dimensions, are produced and the average dimensional value. The lowest, and most fundamental level, is statistical process control (SPC). SPC in this context determines the consistency with which the different process settings, such as pressures, temperatures, times, and speeds, are maintained and the average value of each of these settings.

The next comments in this section assume that process capability studies are done along with the SPC studies. Initially, parts and assemblies were evaluated using SPFC to see if they met performance requirements. After all, performance is the quality that the customer sees and feels. SPFC was typically done in addition to SQC. The next step in the evolution of SPC was based on the premise that if the design engineer had a good design, then the part would perform as intended if the part was manufactured according to the design specification. Consequently, the focus of SPC shifted, via SQC, to analyzing part dimensions. Stated another way, the premise was that if the part was built according to the specification, then there would be no need to do performance testing or SPFC. This approach saved time, cost, and effort.

The next step in the evolution of SPC was based on the premise that if the process settings could be maintained within prescribed limits, then the parts would be manufactured according to the engineering design specification. Consequently, the focus of SPC was to measure and analyze the process settings such as pressures, temperatures, times, and speeds.

Feedback control systems are frequently incorporated in process equipment to hold the process setting at a predetermined value and to minimize variability. With SPC, the premise is that if the process settings will be consistent and will be maintained within predetermined process limits, then all parts would be produced in conformance with the design specification. This would eliminate the need to do SQC on the parts and would also eliminate the need to do SPFC.

16.4. QUALITY CONTROL SYSTEM

16.4.1 Raw Material Quality Control

Any well-established quality control system begins with control of purchased material. Such a system assures one that the purchased material, in fact, meets the specified requirements. In most cases, processors rely on material suppliers to provide the same quality material time after time. If the end product or the particular process employed to make this end product is sensitive to changes in material quality and uniformity, such reliance on material suppliers may prove costly in the long run. The steps involved in setting up a good raw material quality control system are:

1. *Supplier Selection.* The first step in setting up such a system is to select a reputable supplier of material. Items to check for are past history, industry reputation, and future commitment. The supplier's ability to verify the quality of the material he is supplying should be investigated. One must also look at the quality and the type of supplier's manufacturing and test facility, the frequency of testing, quality control procedures and quantity, and more important, the quality of personnel. These considerations are of the utmost importance when a purchase of material from a custom compounding house or from a totally unknown supplier is considered. If the product liability risk is high, it may behoove you to consider requiring material suppliers to certify that the material meets the minimum requirements specified in the material specifications.

2. *Receiving Inspection.* Many types of tests have been devised for testing raw materials. Depending upon the severity of the need for inspection, the types of tests selected may vary from being basic and simple to sophisticated and complex. Some of the most common basic tests are the melt index test, specific gravity, bulk density, spiral flow test, and viscosity tests. Gel permeation chromatography, infrared analysis, thermal analysis, and rheometry are some of the more elaborate raw material quality control tests. These tests are discussed in detail in Chapter 7. Some processors also choose to mold test bars from a small sample of raw material and conduct physical tests such as tensile, impact, and flexural tests and then evaluate the results to see if they meet the preestablished specifications.

16.4.2. Process Quality Control

In-process quality control serves the basic purpose of providing assurance that the product continues to meet the specified requirements. By employing the process control chart techniques of statistical quality control, we are able to continuously monitor the process and determine whether the process is in or out of control. Patrol or floor inspection gives the inspector an opportunity to verify the visual and dimensional conformity of the processed parts.

16.4.3. Product Quality Control

There are two major areas of interest in product quality control. One is the receiving inspection, where a product manufactured by an outside vendor is inspected when it is received. The other is the outgoing lot inspection in which the product manufactured in-house is inspected prior to shipping. Here again, the principles of statistical quality control are applied. A sampling plan is selected based on the requirements and the AQL, LPTD, or AOQL value is specified. The product quality control involves visual inspection, measurement inspection, and in some cases, actual product testing. More and more emphasis is being put on end-product testing. Preferably, the test will simulate actual use, since a part that is aesthetically appealing and well within the specified tolerance only gives a partial indication of overall part quality.

16.4.4 Visual Standards

One of the reasons for the tremendous success of plastic products in the consumer market is that products made from plastic materials are aesthetically more appealing in terms of color and feel than products made from other materials. The majority of quality control systems fail to recognize the importance of visual standards or guidelines. Quite often, too much emphasis is placed on measurement and testing of the product and not enough on the visual standards.

Visual defects, such as orange peel, sinks, and cold flow, are quite common among fabricated plastic parts. These defects are not usually encountered in parts made from other materials. Furthermore, the terminology that prevails in the plastics industry to describe visual defects is totally different than the terminology used for conventional materials. Identifying visual defects is not only necessary for controlling the aesthetic quality of fabricated parts but is also necessary in assessing the overall quality and strength of the part. For example, a visual defect such as splay marks on the part indicates the presence of moisture in the material, brown streaks indicate the beginning of material degradation, both of which can lower the overall properties of the material.

A visual standards manual must include the basic definitions and explanations of recurring visual defects along with proper illustrations. Since it is difficult to qualify the visual defects in terms of actual measurements, such as gauging a diameter or wall thickness, some guidelines and accept–reject criteria must be established. Figures 16-10 and 16-11 illustrate typical pages from a visual standards manual. A typical visual defects summary chart is shown in Figure 16-12*a*. The

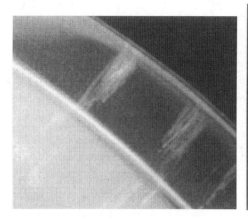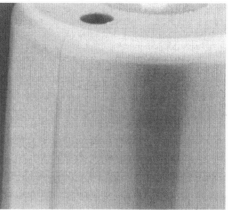

Figure 16-10. Visual defect.

ORANGE PEEL

Preferred

1. No uneven surface texture on molded part resembling orange peel.

Acceptable

1. A slight orange peel on inside of an external part.
2. A slight orange peel on outside of an internal part.
3. Orange peel not to exceed 1/4″ in diameter on the part.

Reject

1. Orange peel on sealing surfaces.
2. Excessive orange peel, greater than 1/4″ in diameter on molded part.

Figure 16-11. Criteria for accepting or rejecting the part.

Molders Division of The Society of Plastics Industry (SPI) has developed cosmetic specifications of injection-molded parts. The purpose of this standard is to provide quantitative definitions and recommended methods of inspection and measurement of the cosmetic quality attributes in the absence of customer provided specifications. The specification addresses the cosmetic quality of molded plastic parts and related post-molding activities. Figure 16-12*b* shows one of the cosmetic specifications.

16.4.5. Mold (Tool) Control

The quality of the molded part is only as good as the mold that produces that part. New equipment, skilled operators, or good molding practices cannot make up for a defective or worn out mold. In spite of this proven fact, the majority of

The Society of the Plastics Industry
Cosmetic Specifications
of Injection Molded Parts

GRADE 2 - (Low grade polish, textured, clear translucent)

SURFACE	A	B	C
WELDLINES & BLUSH	Limits for BLUSH and WELDLINES are established in agreement with customer and held based upon limit samples		
SINK	None allowed.	.005" max.	.015" max.
SPECKS & BUBBLES ACCEPT IF -	Less than or equal to .010". No closer than 1 inch.	Less than or equal to .010". No closer than 1 inch	Less than or equal to .015". No closer than 1 inch.
ALLOWABLE RANGE	.010" - .030". Allow 1 per 16 inch^2 (4" x 4").	.010" - .030". Allow 2 per 16 inch^2 (4" x 4"). No closer than 2 inches.	.015" - .040". Allow 3 per 16 inch^2 (4" x 4"). No closer than 1 inch.
REJECT IF -	Greater than .030".	Greater than .030".	Greater than .040".
SCRATCHES ACCEPT IF -	.150" or less in length. Allow 1 per 16 inch^2 (4" x 4").	.200" or less in length. Allow 1 per 16 inch^2 (4" x 4").	.300" or less in length. Allow 3 per 16 inch^2 (4" x 4"). No closer than 1 inch.
REJECT IF -	Greater than .150"	Greater than .200"	Greater than .300"
SPLAY	None Allowed	None Allowed	Refer to limit samples
BURNS	"	"	or note exceptions in
GAS MARKS	"	"	Quality Plan.
MARBLING	"	"	"
ORANGE PEEL	"	"	"
NON-UNIFORM TEXTURE	"	"	"
PITTING	"	"	"
CRACKING	"	"	"
CRAZING	"	"	"
DELAMINATION	"	"	"
COLD SLUGS	"	"	"

(a)

Figure 16-12. (*a*) Visual defect summary chart. (From "Cosmetic Specifications of Injection Molded Parts," reprinted with permission of Society Plastics Industry, Inc.) (*b*) Cosmetic specifications. (From "Cosmetic Specifications of Injection Molded Parts," reprinted with permission of Society of Plastics Industry, Inc.)

manufacturers often fail to recognize the importance of effective mold-control systems.

A good mold-control system starts with proper inventory control and adequate mold storage facilities. Documentation is the key word. From the inception of the mold, every little detail regarding the particulars of the mold must be logged. A mold information form such as the one shown in Figure 16-13 can be used. Once

The Society of the Plastics Industry
Cosmetic Specifications

Cutomer:_____

Part Name: _____   Part Number:_____

Authorized by:_____

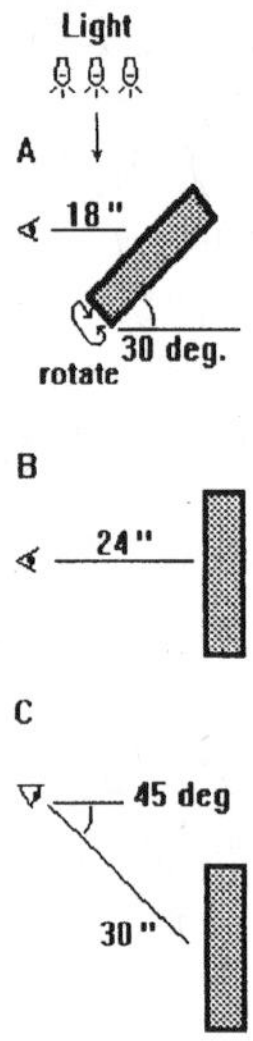

Light
₽ ₽ ₽

A ↓

◁ _18"_

30 deg.
rotate

GRADE 2 - (Low grade polish, textured, clear translucent)

SURFACE	A	B	C
WELDLINES & BLUSH	Limits for BLUSH and WELDLINES are established in agreement with customer and held based upon limit samples		

B

◁ _24"_

C

▽ _45 deg_

_30"_

		A	B	C
SINK		None allowed.	.005" max.	.015" max.
SPECKS & BUBBLES	ACCEPT IF -	Less than or equal to .010". No closer than 1 inch.	Less than or equal to .010". No closer than 1 inch	Less than or equal to .015". No closer than 1 inch.
	ALLOWABLE RANGE	.010" - .030". Allow 1 per 16 inch^2 (4" x 4").	.010" - .030". Allow 2 per 16 inch^2 (4" x 4"). No closer than 2 inches.	.015" - .040". Allow 3 per 16 inch^2 (4" x 4"). No closer than 1 inch.
	REJECT IF -	Greater than .030".	Greater than .030".	Greater than .040".
SCRATCHES	ACCEPT IF -	.150" or less in length. Allow 1 per 16 inch^2 (4" x 4").	.200" or less in length. Allow 1 per 16 inch^2 (4" x 4").	.300" or less in length. Allow 3 per 16 inch^2 (4" x 4"). No closer than 1 inch.
	REJECT IF -	Greater than .150"	Greater than .200"	Greater than .300"
SPLAY		None Allowed	None Allowed	Refer to limit samples
BURNS		"	"	or note exceptions in
GAS MARKS		"	"	Quality Plan.
MARBLING		"	"	"
ORANGE PEEL		"	"	"
NON-UNIFORM TEXTURE		"	"	"
PITTING		"	"	"
CRACKING		"	"	"
CRAZING		"	"	"
DELAMINATION		"	"	"
COLD SLUGS		"	"	"

(b)

Figure 16-12. *Continued*

the mold is released for production, the first article should be performed, all dimensions carefully logged, and samples retained for future reference. A mold history record card such as the one shown in Figure 16-14 should be generated and kept up to date. Any mold modification, however minor, must be recorded along with routine maintenance and repair data on the history card. The mold service schedule should be based upon the number of parts produced from the mold.

Mold Information

Mold Number:_____

Company:_____

Mold Maker:_____

Date Mfg.:_____

No.	Part Number	Model No. Desc.	Material	Supplier/Grade

☐ 2 Plate   ☐ 3 Plate   ☐ Stacked   ☐ Hot Runner
☐ Hot Manifold   ☐ Insulated Runner   ☐ Std. Unit Die   ☐ Other

Mold Base Type:_____ H

Mold Base Size:_____ L X _____ W X _____ H

Insert Size:_____ L X _____ W X _____ H

Mold Material:_____

Insert Material:_____

Number of Cavities:_____

Mold Hardness:_____

Machine Size		Nozzle Radius	Locating Ring Diameter	Sprue Bushing Size
Tons	OZ's			

Figure 16-13. Mold information form. (Courtesy of Performance Engineered Products, Inc.).

16.4.6. Workmanship Standards

Workmanship standards are essential to the smooth and successful operation of any quality control system. They are considered the best means of achieving quality workmanship.

Workmanship standards are nothing more than a simplified guide, explaining through the use of drawings, sketches, and photographs, the proper method of carrying out the specified task. The task may consist of simply deburring or hot stamp-

Type of Gate: _____

Type of KO: _____

Unwinding: _____

Core Pull: _____

Limit Switches: _____

Mold Transducer: _____

Mold Finish: _____

Plating Information: _____

Hot Sprue Bushing: _____

Hot Runner: _____

Insulated Runner: _____

_____

Top Half	GPM	Bottom Half	GPM
Cavity: _____		Cavity: _____	
Core: _____		Core: _____	
Other: _____		Other: _____	

Mold Accesory Information: _____

_____

Scheduled Maintenance Record

Maintenance	Date

Gaging

Part Number	Gage Type	Serial Number

Figure 16-13. *Continued*

ing the parts or assembling them using solvent cementing techniques. The majority of workmanship standards provide preferred, acceptable, and reject criteria. Figure 16-15 illustrates a typical workmanship standard.

Some of the most obvious advantages of such workmanship standards are a lower reject rate, elimination of unnecessary rework, early detection of defects, and reduced risk of rejecting good parts. Proper implementation of workmanship

Mold History Record Card

Mold No: _____   Print No: _____
Part No: _____   Mold Maker: _____
Company: _____   Date Mfg: _____

Mold Movement Record

Date	From	To	Authorization	Remarks

Mold Modification Record

Date	Modification No.	Reason

Figure 16-14. Mold history record card. (Courtesy of Performance Engineered Products, Inc.).

Critical Variables Data Record

Dimension	Tolerance		Measured	Remarks	Date
	+	–			

Maintenance/Repair History

Date	P. O. No.	No. of Parts Run	Description of Repair/Maintenance	Cost	Mold Shop

Figure 16-14. *Continued*

workmanship
standards

SOLDERLESS CONNECTIONS
Crimped Pins - Multi-Wire

MAGNIFICATION 5X

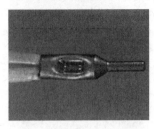

PREFERRED

1. Insulation has been stripped evenly and terminates flush with the rear of the contact barrel.

2. Conductors are bottomed in the support well.

3. Crimp indent is well formed and centered.

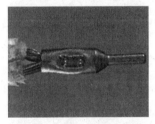

ACCEPTABLE

1. Insulation terminates within 1/16 inch of the rear of the contact barrel.

2. Insulation trim is slightly uneven.

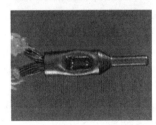

ACCEPTABLE MINIMUM

1. Exposed bare wire is maximum but does not exceed 1X insulated wire diameter plus 1/16 inch or a maximum gap of 1/8 inch, whichever is less, from the rear of the contact barrel. (Above tolerance includes allowance for latent shrinkage of the wire insulation).

REJECT

1. Exposed bare wire exceeds maximum tolerance specified above.

2. Conductors are not bottomed in the barrel.

3. Crimp indent is too low.

Figure 16-15. A typical workmanship standard.

standards can improve quality and reliability by eliminating the time lost in rework because of varied interpretation and personal opinions.

16.4.7. Documentation

Documentation is the heart of the quality control system. The data compiled through the documentation of test results, dimensional measurements, process capability studies, and sampling inspection can be used for statistical analysis. Many private and governmental agencies require that proper documents and records be retained for certain minimum time periods. For example, the nuclear industry requires that records be maintained for 40 years (5). Retention of such records can also provide necessary proof in the courtroom in case of product liability suits.

A "good" documentation system is one that is easy to implement, easy to understand, and easy to maintain. The records documenting inspection must indicate the characteristics observed, number of observations made, number and types of discrepancies, final dispositions, inspector identification, and most important, date of documentation.

Lastly, without proper records, it would be practically impossible to trace the reason for product failure. By carefully studying the dimensional measurement records of a part, one can also identify the equipment as well as the tool wear.

16.4.8. Quality Assurance Manual

A quality control system without a quality assurance manual describing in detail a quality assurance program cannot function adequately. The sole purpose of a quality assurance manual is to provide clear and precise written instructions and procedures so that there can be no misunderstanding and confusion between different organizations within and outside the company. Whenever possible, oral instruction should be avoided, because it can only result in misinterpretation and gross distortion of the message. The following is a broad outline of a typical quality assurance manual.

1. Mission statement
2. An organization chart describing the responsibilities of each individual in the organization
3. Function and responsibility of the quality control organization
4. Material review board (MRB) function and corrective action procedures
5. Receiving inspection procedures
6. In-process inspection procedures (first article inspection procedures)
7. Shipping inspection procedures
8. Disposition guidelines
9. Quality audit program. (Under this program, the Quality Manager or Supervisor periodically makes unannounced checks on inspected units, accuracy of gauges and test equipment, etc., to verify the accuracy as well as adequacy of quality assurance systems)
10. Procedures for handling customer returns
11. Gauge and test equipment calibration and maintenance procedure
12. Mold control (tool and die) program
13. Miscellaneous test procedures
14. Method of recording inspection data and exhibit of sample forms
15. Retention of records and documents
16. Defective material rework and reinspection procedures

17. Visual standards
18. Workmanship standards

16.5. GENERAL

16.5.1. Quality Control and Machine Operators

Although not a part of the quality control team, machine operators usually play a very important and direct role in controlling the quality of the parts. A smart quality control manager places part of the burden of inspecting the quality of the parts on machine operators. This also helps to lower the inspection cost by reducing the work load of inspectors and improving the overall quality. Such a program of machine operator participation in quality control of the product cannot be successful without the proper training of the participants. Machine operators must be trained to look for visual defects such as flow marks, shorts, and flash, to trim the gate without leaving protrusions from the part or gouging into the part, to use a go-no-go gauge, or to perform a simple test between the cycles. An illustration or sample of preferred maximum acceptable and unacceptable parts should be provided to the machine operator for the sake of better understanding and comparison purposes.

16.6. SUPPLIER CERTIFICATION

In the past 10 years, an increasing number of businesses and major corporations are requiring suppliers to be certified. The trend toward supplier certification has evolved from the renewed interest in quality assurance and greater competition from abroad.

Suppliers are essential to the success of any business, none can succeed without quality materials and services. Businesses must develop working relationship with suppliers. A byproduct of this new approach is a certification program that builds from a partnership concept as opposed to mandating certification and associated "partnership" objective. Certification, properly done, requires a major team effort and commitment from both parties. Any well-developed supplier certification program starts with a clear, well-defined purpose and objective. Figure 16-16 illustrates a typical supplier selection process. The next step is to establish criteria for supplier selection along with a comprehensive supplier performance evaluation process. Figure 16-17 illustrates a typical evaluation form used for such evaluations. A certification team, consisting of members of various departments such as quality assurance, purchasing, engineering, research and development, and production would visit the supplier to evaluate supplier capabilities and establish guidelines for working relationship. Many companies also require the supplier to undergo a self-audit prior to the certification team audit. After a series of discussions, programs for self-improvement, and corrective measures the certification team would proceed with certification. A formal presentation of "certified supplier" plaque is made by a high-ranking officer of the company. Guidelines are generally established for yearly recertification.

```
┌─────────────────────────────────────────────┐
│          Supplier Selection Process          │
└─────────────────────────────────────────────┘
```

The first indicator of a supplier's potential to be certified
is management commitment and understanding.

Levels	Management Traits	Quality Organization	Company Status
1	Lack of understanding	No one present	Unaware of what causes poor quality
2	Sees value of change, but no commitment of money	Fire fighting mode; symptoms treated, not causes	Constant quality problems are present
3	Willing to change, support and learn	Becomes involved in prevention	Commitment to continuous improvement
4	Management is participating	Quality effectively controls process	Shift to defect prevention
5	Management is part of improvement team	Zero-defects is the only acceptable method	Supplier certification is a way of life

Figure 16-16. Supplier selection process for certification.

16.7.1. New Technologies

Part characteristics can be separated into three categories. This first is dimensional and weight characteristics. The second is a material characteristic such as tensile strength. The third is a performance characteristic such as flow rate, angular velocity, or some other metric that relates to how well the part or assembly performs for the customer.

Prior technology SPC and process capability studies are based on the assumption that part characteristics are not related. This is not the case for many, but not all, manufacturing processes. Part characteristics are related (correlated) for many processes. To the extent that this is true, doing SPC and process capability analysis on all part characteristics is redundant, wasteful, and unnecessarily time-consuming.

The remainder of this discussion will focus on dimensional part characteristics, but correlations also exist to material and performance characteristics for many parts and processes.

Plastic injection molding is an example where part dimensions are related. If a plastic injection molded part has five critical dimensions and is manufactured in a 16-cavity mold, the molding press will produce 80 dimensions for each machine cycle (shot). It is possible to determine the single predictor dimension that is the statistically best predictor of the remaining 79 dimensions.

The correlations between dimensions are shown in the form of regression lines on a correlation chart. If there are 80 dimensions, then there will be 79 charts showing the relationship between each of the 79 remaining dimensions and the

SUPPLIER PERFORMANCE EVALUATION

SUPPLIER: A0001 ANY SUPPLIER USA   DATE: 01/01/99
TOTAL SHIPMENTS: 70

FACTORS	RATING		WEIGHT		WEIGHTED RATING
Quality					
(Total Shipments Which Conform to Spec)	69	x	20%	=	13.80
Quality					
(Total Shipments Received Without Cautions)	69	x	15%	=	10.35
Quality					
(Total Shipments Received Without Rejection Other than Spec)	70	x	20%	=	14.00
Certificate of Analysis (Total Received on or Before Day of Delivery)	68	x	20%	=	13.60
On-Time Delivery (Total on Time)	68	x	15%	=	10.20
Freight Damage (Total Shipments Received Free of Error Due to Freight Damage)	70	x	10%	=	7.00
			Overall Rating:		68.95

SCALE CALCULATION: (Overall Rating x 100)/Total Shipments=**98.50**

99-100	Excellent	93-95	Fair	89-Below	Unsatisfactory
96-98	Good	90-92	Marginal		

Figure 16-17. Supplier performance evaluation form.

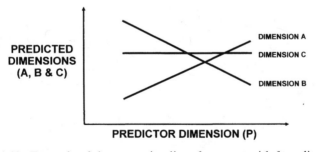

Figure 16-18. Example of the regression lines for a part with four dimensions.

predictor dimension. Figure 16-18 shows an example of the regression lines for a part with four dimensions.

Determining the predictor dimension that is the statistically best predictor of all other dimensions can be (and usually is) very computationally intensive. It is not unusual to have to calculate 20,000–30,000 correlation coefficients to determine

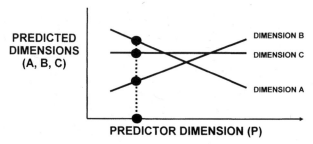

Figure 16-19. The relationships between part dimensions are consistent and predictable, irrespective of changes in press settings.

the best predictor dimension. Fortunately, computer software called Correlation Master™ is available from Algoryx, Inc™ (6) to quickly and easily perform the computations and generate the correlation charts.

The relationships are determined by either using existing manufacturing data for parts that have been produced or by inducing variation into the manufacturing process for new parts. No change is required to the manufacturing process.

When the predictor dimension is known, all other dimensions are also known. It is then not necessary to measure or do SPC or process capability analysis on all of the predicted dimensions as shown in Figure 16-19.

Many large and small companies are now saving money and improving profitability as a result of using this new technology to accomplish Lean SPC™.

In a similar fashion, many companies perform process capability (Cpk) analyses. Returning to our example of a plastic injection molded part with 5 critical dimensions that is manufactured in a 16-cavity mold, one would do 80 process capability analyses with prior technology. With the Lean Process Capability™ technology afforded by Correlation Master, one would have to do only process capability analysis on the predictor dimension.

In this example, use of Correlation Master would save 99% of the measurement costs, 99% of the SPC analysis costs and 99% of the process capability analysis costs. Further, the costs of destructive measurement can be eliminated in many cases.

This new technology has been successfully applied to the following processes:

- Plastic injection molding
- Rubber injection molding
- Rubber transfer molding
- Plastic extrusion
- Cold heading
- Hot heading
- Sheet metal punching
- Sheet metal forming

- Plating during semiconductor wafer fabrication
- Etching during semiconductor wafer fabrication

Additional studies are underway in the areas of metal injection molding (MIM) and metal extrusion.

Many additional benefits occur when specification target values and tolerances are superimposed on the correlation charts as shown in Figures 16-20 and 16-21. This is particularly true during the development phase of new parts.

Tooling changes and design tolerance relaxations are accomplished in a single-step instead of many steps. This is key because the tooling changes and design tolerance relaxations are accomplished independently of operator press settings (Figure 16-22).

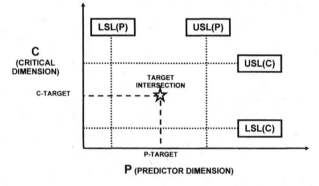

Figure 16-20. Unconstrained critical dimensions never have to be measured when the predictor dimension is conforming.

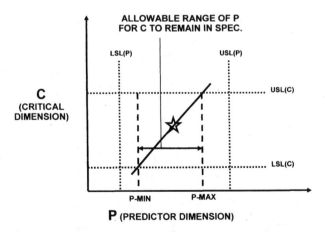

Figure 16-21. Constraining critical dimensions never have to be measured when predictor is between P-max and P-min.

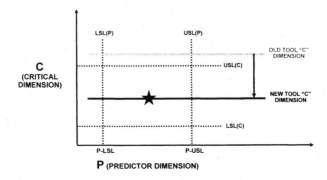

Figure 16-22. Design tolerance relaxation can determined in a single step.

In the past, determining tooling changes and design tolerance relaxations has been problematic. Any changes made could be undone by the operator or process engineer changing process settings. The tooling changes and design tolerance relaxations generated by Correlation Master are determined independently of operator press settings. This is a huge boon to toolmakers and design engineers.

The Correlation Master software also enables the process engineer to "dial in" process settings to the optimum value that will maximize quality and minimize scrap and reworks.

Finally, the Correlation Master software also identifies dimensions where the design tolerances can be tightened without reducing the producibility of the part.

These capabilities of the Correlation Master Software increase process knowledge, greatly accelerate time-to-market, and reduce conflict between the design, tooling, process and quality engineers.

REFERENCES

1. Juran, J. M., *Quality Control Handbook*, McGraw-Hill, New York, 1974, p. 2–2.
2. *Ibid.*, p. 2–11.
3. Deming, L. M., *Quality Control for Plastics Engineers*, Reinhold, New York, 1957, p. 1.
4. "Quality Assurance Terms and Definitions," *Military Standard Mil-Std-109B*, U. S. Govt. Printing Office, Washington, D. C., 1969.
5. "Nondestructive Inspection and Quality Control," *Metals Handbook*, **11**, American Society for Metals, Philadelphia, PA, p. 408.
6. Algoryx, Inc.™, 750 S. Bundy Drive, #304, Los Angeles, CA 90049 310-820-0987 www. algoryx.com.

GENERAL REFERENCES

Grant, E. L., *Statistical Quality Control*, McGraw-Hill, New York, 1974.

Juran, J. M., *Quality Control Handbook*, McGraw-Hill, New York, 1974.

Mason, R. D., *Statistical Techniques in Business and Economics*, Irvin, Homewood, IL, 1974.

Dodge, H. F. and Romig, H. G., *Sampling Inspection Tables—Single and Double Sampling*, John Wiley & Sons, New York, 1944.

Kenney, C. W. and Andrew, D. E., *Inspection and Gaging*, Industrial Press, New York, 1977.

Statistical Quality Control Handbook, Western Electric Company, New York, 1956.

"Sampling Procedures and Tables for Inspection by Attributes" ANSI/ASQC 21.4—1993, American Society for Quality Control (ASQC), Milwaukee, WI.

17

PRODUCT LIABILITIES AND TESTING

17.1. INTRODUCTION

Between 1970 and 1980, one phrase caught the attention of the manufactures and suppliers more than any other phrase: product liability. The Consumer Product Safety Commission reports that the number of product liability suits increased 983% from 1974 to 1988. In 1991 it was reported by insurance industry that $1.6 billion was paid out in product liability losses. They also reported that an additional $1.1 billion was spent on legal costs (1). The cost of liability insurance has increased substantially in the last five years. Virtually every industry has been plagued with liability suits and the plastics industry is certainly no exception.

The entire concept of product liability suits emerged from the total lack of concern regarding product safety that characterized product designers and manufacturers. The Occupational Safety and Health Administration (OSHA) reports the loss of over 10,000 lives each year and over 2 million disabling injuries annually (2). Recently, a material supplier and a plastic fittings manufacturer and distributor were sued by an angry consumer because plastic fittings failed prematurely, flooding an entire building and ruining expensive furnishings and carpeting. In another case, a molding machinery manufacturer was sued because a machine operator lost his right hand while trying to free a part from the mold and accidentally tripped the switch that closed the mold. Yet another classic illustration of a plastic product liability case involved a small manufacturer of PVC handles that were incapable of handing high heat. The handle was softened by high temperature, exposing live electrical contacts and electrocuting a person.

From the foregoing discussion it is clear that product liability and product safety are interrelated. The majority of manufacturers, especially the smaller ones, allow

Handbook of Plastics Testing and Failure Analysis, Third Edition, by Vishu Shah
Copyright © 2007 by John Wiley & Sons, Inc.

themselves to become the target of such product liability suits by thinking that they are impervious to such suits. The fact of the matter is that product liability involves everyone: material manufacturers, product designers, fabricators, sellers, and installers. The manufacturer may be held liable if:

1. The product is defective in design and is not suitable for its intended use.
2. The product is manufactured defective and proper testing and inspection was not carried out.
3. The product lacks adequate labeling and warnings.
4. The product is unsafely packaged.
5. The proper records of product sale, distribution, and manufacturer are not kept up to date.
6. The proper records of failure and customer complaints are not maintained (3).

What can a machine manufacturer do to avoid expensive lawsuits? What are the steps a manufacturer of a product must take before placing the product into the hands of somewhat novice customers? How can a molder who is merely providing a service to the industry protect himself from unknowingly getting involved in such product liability problems? How many ways can a design engineer design a product with all the safeguards built-in without affecting the product's originality, cost, and his or her creativity? How can a material supplier prevent getting sued because a product made from his material failed because of product design problems and not material quality?

Obviously, there is no single answer to all of these questions. The following is a general guideline everyone should follow to steer clear of unwarranted product liability suits.

17.2. PRODUCT/EQUIPMENT DESIGN CONSIDERATIONS

The product or equipment design engineer is often considered a prime mover of product safety. The key factors to be considered in designing a safe product or equipment are consumer ignorance, manufacturing mishaps, and deliberate misuse. A team of designers must review the design individually and collectively from different viewpoints and all possible angles. If a company is not large enough to staff a team of designers, so outside consulting firm should be allowed to review the design. Product insurance representatives should also be consulted during preliminary design since many insurance companies have engineers on staff as safety consultants. The design engineer should also be familiar with all standards and regulations concerning his or her product. Some of the other minor design considerations are selecting components with a high degree of reliability, designing systems to permit ready access for operating, repairing, and replacing components, designing equipment that takes into consideration the capabilities and limitations of operators, designing components that are incapable of being revised or improperly installed, and anticipating all possible environmental and chemical hazards (4).

17.3. PACKAGING CONSIDERATIONS

Packaging in a broad sense is defined as the outer shell of the product. Since plastics are used extensively as an encloser material for products such as appliances, electrical equipment, liquid chemicals, food, and beverages, the product safety and liability considerations are of extreme interest. The product enclosures should be designed to be tamperproof to prevent the insides from being exposed to persons unfamiliar with the potential hazards. The packaging material should be tested for toxicity, chemical compatibility, and environmental resistance. Identification labels indicating product name, model number, serial and lot number, date code, and manufacturing code have been found useful in making products safer.

17.4. INSTRUCTIONS, WARNING LABELS, AND TRAINING

One of the major lines of defense against the product liability suits is providing adequate instructions, warning labels, and training to the consumer as well as to the installer and service persons. All products cannot be made 100 percent safe. There will always be some degree of risk involved in handling certain products or machinery. Therefore, designers of the product or equipment must take into account the safety aspect and come up with a systematic procedure to deal with well-designed but inherently hazardous products. First, a clear, concise, but easy to read instruction manual is in order. The writer of such an instruction manual must take into account possible misinterpretation by the reader and must be aware of the consequences in case this happens.

The machinery or product manufacturer should not only comply with government or industry regulations regarding warning labels, but also place warning labels on his own wherever it is deemed necessary to prevent accidents. The warnings should not only "warn" but also indicate the consequences of disregarding the warning (5). Machinery manufacturers as well as fabricators can prevent the majority of accidents by implementing a proper training program for machine operators, installers, maintenance personnel, and foremen. Developing a safety training program on specific machines that the operators will be working with, having safety refresher courses, and distributing safety bulletins are a few of the most useful suggestions that have proven very successful (6).

17.5. TESTING AND RECORDKEEPING

One of the most powerful weapons any manufacturer trying to steer clear of product liability suits can have is a comprehensive testing and recordkeeping program. In many cases, quality control records may be the only defense the manufacturer may have in a courtroom.

The quality control testing should start with an inspection of the raw material and components as they are received. Whenever possible, the supplier should be required to meet military or other industry specifications. If the raw material or components are to be used in a potentially hazardous product, the supplier should be requested to provide certification along with each shipment. In-process testing is equally important. Here again, the quality control and destructive or nondestructive testing requirements should be set in accordance with industry standards. In the case of machine manufacturers, a thorough preshipment inspection should help eliminate any unexpected surprises. In some cases, it is advisable to retain an independent testing laboratory. The data generated by the independent test laboratories is often found to be more useful and convincing than self-generated data in courtroom defense. Four good reasons outlined below make a strong case for independent testing for product liability (7).

1. *Objectivity.* A manufacturer may be too close to his own products to maintain an impartial, unbiased viewpoint regarding their safety features. An outside safety engineer can look at the products impassionately, pointing out unsafe features that may escape an internal review.
2. *Exposure.* Independent engineering and testing laboratories make it their business to keep abreast of current specifications and safety laws as well as proposed safety legislation.
3. *Independence.* Since accountability is becoming a more important aspect of product liability, a documented report containing solid evidence of testing and fail-safe analysis by an independent firm carries much weight in establishing the intent of the manufacturer to design and make safe products before they reach the market.
4. *Anticipation.* Product testing before injury and/or product liability suits is good preventive medicine.

The other important task for a manufacturer is recordkeeping. A well-organized recordkeeping policy accomplishes many objectives. First, it establishes that the company is taking reasonable precautions to produce safe products. Second, in the event of an accident or injury, it allows the manufacturer to backtrack and pinpoint the cause of failure. Third, with the help of records, the manufacturer can prove that his product did in fact meet the minimum requirements. It is advisable to keep records of design, manufacturing, inspection, quality control, and testing procedures and results on file for at least five years. The retention period should be based upon individual product need. The records should also include material specification, suppliers, serial and lot numbers, and customer names (8). Other useful documents include the company's safety policy manual, design changes, failure reports, marketing and shipping records, and advertisement records.

17.6. SAFETY STANDARDS ORGANIZATIONS

Appendix E lists the organizations responsible for setting standards for safety.

REFERENCES

1. Goodden, R. L., *Preventing and Handling Product Liability,* Marcel Dekker, Inc., New York, 1996.

2. Kolb, J. and Ross, S. S., *Product Safety and Liability,* McGraw-Hill, New York, 1980, p. 4.

3. *Ibid.,* p. 12.

4. *Ibid.,* p. 330.

5. *Ibid.,* p. 208.

6. Allchin, T., "Product Liability in Plastics Industry," *SPE Pacific Tech. Conf.* (PACTEC). **57**, 1975, p. 57.

7. Zavita, Reference 1, p. 34.

8. Fountas, N., "Product Liability—Prepare Now for Judgment Day." *Plast. World* (Feb. 1978), p. 68.

18

NONDESTRUCTIVE TESTING AND MEASUREMENTS

18.1. INTRODUCTION

The term nondestructive testing is applied to tests or measurements carried out without harming or altering the properties of the part. Quite often, in order to make measurements or to study certain characteristics of a part, it becomes necessary to destroy the integrity of the part. For example, it is practically impossible to measure wall thickness of certain areas of 6-in. PVC pipe fitting without actually cutting the part. Similarly, shrinkage voids in a 3-in-diameter extruded Teflon* rod can only be found by cutting the rod in several places. Such destructive techniques are not only very expensive, but are also time consuming. Nondestructive testing methods allow one to determine flaws, imperfections, and nonuniformities without destroying the part. Nondestructive tests range from simple visual examination, weighing, and hardness measurements to more complex electrical and ultrasonic tests. The discussion on nondestructive testing this chapter is confined to ultrasonic, gamma backscatter, beta transmission, optical laser, X-ray fluorescence, and Hall effect measurement and testing techniques. Visual examination, hardness measurements, and electrical measurement techniques are discussed in other chapters.

18.2. ULTRASONIC

Ultrasonic testing is one of the most widely used methods for nondestructive inspection. In plastics, the primary application is the detection of discontinuities and measurement of thickness. Ultrasonic techniques can also be used for determining the moisture content

*Teflon is a registered trademark of the DuPont Company.

Handbook of Plastics Testing and Failure Analysis, Third Edition, by Vishu Shah
Copyright © 2007 by John Wiley & Sons, Inc.

of plastics, studying the joint integrity of a solvent-welded plastic pipe and fittings, and testing welded seams in plastic plates (1).

The term ultrasonic, in a broad sense, is applied to describe sound with a frequency above 20,000 cycles/sec. Commercial ultrasonic testing equipment generally employs the testing frequency in the range from 0.75 to 20 MHz. To provide a basis for understanding the ultrasonic system and how it operates, it is necessary to introduce the following terms:

Frequency Generator. A device that imposes a short burst of high-frequency alternating voltage on a transducer.

Transducer. A transducer or a probe is a device that emits a beam of ultrasonic waves when bursts of alternating voltage are applied to it. An ultrasonic transducer is comprised of piezoelectric material. Piezoelectric material is material that vibrates mechanically under a varying electric potential and develops electrical potentials under mechanical strain, thus transforming electrical energy into mechanical energy and vice versa (2). As the name implies, an electrical charge is developed by a piezoelectric crystal when pressure is applied to it and reverse is also true. The most commonly encountered piezoelectric materials are quartz, lithium sulfate, and artificial ceramic materials such as barium titanate.

Many different types of ultrasonic transducers are available, differing in diameter of the probe, frequency, and frequency bandwidth. Each transducer has a characteristic resonant frequency at which ultrasonic waves are most effectively generated and received. Narrow bandwidth transducers are capable of penetrating deep, as well as detecting small flaws. However, these transducers do a poor job of separating echos. Broad bandwidth transducers exhibit excellent echo separation but poor flaw detection and penetration (3). Transducers of a frequency range of 2–5 MHz are most common. For plastic materials, transducers in the range of 1–2 MHz seem to yield the best results.

Couplants. Air, being one of the worst transmitters of sound waves at high frequencies owing to a lack of impedance matching between air and most solids, must be replaced by a suitable coupling agent between the transducer and the material being tested. Many different types of liquids have been used as coupling agents. Glycerine seems to have the highest acoustic impedance. However, oil is the most commonly used couplant. Grease, petroleum jelly, and pastes can also be used as couplants, although a wetting agent must be added to increase wettability as well as viscosity. Some couplants have a tendency to react with the test specimen material and therefore chemical compatibility of the couplant should be studied prior to application. Couplants that are difficult to remove from test specimens should also be avoided. Any type of contamination between the test specimen and the transducer can seriously affect the thickness measurements, especially in the case of thin films. Therefore, it is absolutely necessary to remove these contaminants before applying the couplant (4). The basic sequence of operation in any ultrasonic measurement system is:

1. Generation of ultrasonic frequency by means of a transducer
2. Use of a coupling agent (couplant), such as oil or water, to help transmit the ultrasonic waves into the material
3. Detection of the ultrasonic energy after it has been modified by the material

4. Displaying of the energy by means of a recorder, cathode-ray tube, or other devices

The three basic ultrasonic measurement techniques most widely used today are:

1. Pulse echo
2. Transmission
3. Resonance

18.2.1. Pulse-Echo Technique

The pulse-echo technique is the most popular of the three basic ultrasonic, nondestructive testing techniques. The pulse-echo technique is very useful in detecting flaws and for thickness measurement.

Figure 18-1 shows the principle of the pulse-echo technique. The initial pulse of ultrasonic energy from a transducer is introduced into the test specimen through the couplant. This sound wave travels through the thickness of the specimen until a reflecting surface is encountered, at which time the sound wave reflects back to the transducer. This is called the back-wall echo. If the wave encounters a flaw in its path, the flaw acts as a reflecting surface and the wave is reflected back to the transducer. The echo in this case is referred to as a flaw echo. In both cases, the reflected wave travels back to the transducer, causing the transducer element to vibrate and induce an electrical energy that is normally amplified and displayed onto a CRT or other such device. The echo wave coming from the back wall of the specimen is marked by its transit time from the transducer to the back wall and return. Similarly, the transit time for the flaw echo can also be determined by this technique. Since transit time corresponds to the thickness of the specimen, it is quite possible to calculate the thickness of the specimen using simple computer logic. Figure 18-2 illustrates a typical commercially available pulse-echo instrument. One other technique, known as the immersion test technique, has generated tremendous interest among the manufactures who are in favor of automated inspection techniques. In the immersion technique, as shown in Figure 18-3, the specimen is completely immersed in the liquid

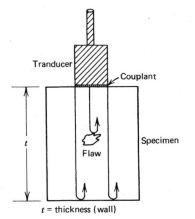

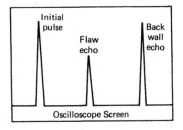

Figure 18-1. Pulse-echo technique.

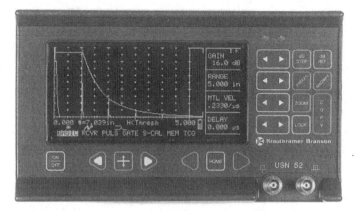

Figure 18-2. Instrument for detecting flaws using pulse-echo technique. (Courtesy of Kraut Kramer-Branson Inc.)

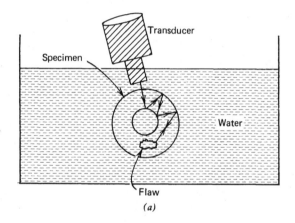

(a)

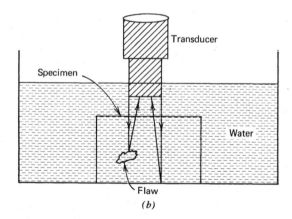

(b)

Figure 18-3. Immersion technique.

(usually water), which in turn acts as a couplant. The transducer is mounted in a fixture that is moved across the specimen, which passes continuously under the fixture. The sound beam can be directed either perpendicular to or at an angle to the surface of the test piece. The water between the transducer and the specimen acts as a couplant. In this manner, a uniform and nonabrasive coupling is achieved (5). The immersion technique is very useful in the automatic inspection of pipe, sheet, rods, and plates for detecting flaws as the pieces are being extruded under water.

18.2.2. Transmission Technique

In this technique, the intensity of ultrasound is measured after it has passed through the specimen (6). The transmission technique requires two transducers, one to transmit the sound waves and one to receive them. Figure 18-4 illustrates the basic concept of the transmission technique and flaw detection using the technique. The transmission testing can be done either by direct beams or reflected beams. In either case, the flaws are detected by comparing the intensity of ultrasound transmitted through the test specimen with the intensity transmitted through a reference standard made of the same material. The best results are achieved by using the immersion technique since this technique provides uniform and efficient coupling between transducers and test specimen. The main application of the transmission technique is in detecting flaws in laminated plastic sheets.

18.2.3. Resonance Technique

This method is primarily useful for measuring the thickness of the specimen. This is accomplished by determining the resonant frequencies of a test specimen. The detailed discussion of this technique is found in the literature (7).

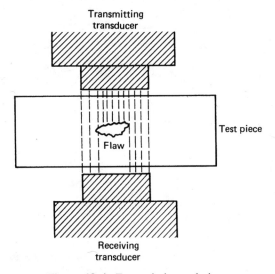

Figure 18-4. Transmission technique.

18.3. APPLICATION OF ULTRASONIC NDT IN PLASTICS

Ultrasonic nondestructive testing (NDT) has gained popularity in the past decade along with the growth of the plastic industry and along with an increasing emphasis placed on automation and material saving. Two major areas in which ultrasonic testing concepts are applied extensively are flaw detection and thickness measurement. The pulse-echo technique is used to detect a flaw such as voids and bubbles in an extruded rod of rather expensive materials such as Teflon and nylon. The flaw detection unit and other auxilliary equipment can be programmed so that the specific portion of the rod with a flew is automatically cut off and discarded without disturbing the continuous extrusion process. The transmission technique is commonly used to detect flaws in laminates. Thickness measurement by ultrasonic equipment is simple, reliable, and fast. This NDT technique simplifies the wall thickness measurement of parts with hard-to-reach areas and complex part geometry. Automated wall thickness measurement and control of large diameter extruded pipe is accomplished by using the immersion technique. An ultrasonic sensing unit is placed in a cooling tank to continuously monitor wall thickness. In the event of an out-of-control condition, a closed-loop feedback control system is activated and corrections are made to bring the wall thickness closer to the set point. Many such systems are commercially available. Figure 18-5 illustrates one such system based on using an ultrasonic

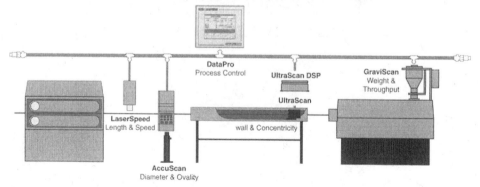

Figure 18-5. Wall thickness measurement system. (Courtesy of Lasermike.)

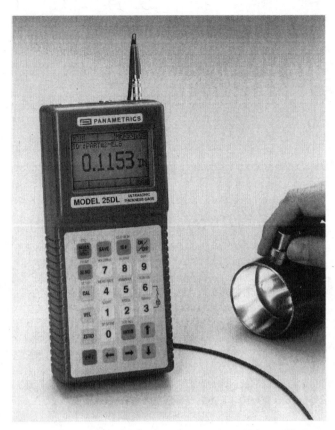

Figure 18-6. Thickness tester, using ultrasonic measurement technique. (Courtesy of Panametrics Corporation.)

gage and a laser-based scanner. Also illustrated in Figure 18-6 is a wall thickness measurement instrument. The ultrasonic NDT technique is used extensively by gas companies to examine the integrity of plastic pipe socket joints after they have been solvent cemented together (8). Ultrasonic measurements can also be used for determining the moisture content of plastic. In materials like nylons, the attenuation and the acoustic velocity change with the change in moisture content (9). The use of ultrasonics in testing reinforced plastics (10) and missiles and rockets (11) has been discussed.

18.4. GAMMA BACKSCATTER

Compton photon backscatter, commonly known as the gamma backscatter (GBS) gauging technique, allows one-sided measurement of film, sheet, pipe, composites, coatings, and laminations. The GBS gauge operates on the principle of Compton photon backscatting. When gamma rays, or photons, are directed at material to be measured, many of the photons are scattered back, losing some of their energy in the process. The backscattered photons strike scintillator detectors, which produces flashes of visible light. The flashes are passed through a photomultiplier tube (PTM) and converted to electons and amplified. Output from the PMT is a train of pulses whose height is proportional to the energy of

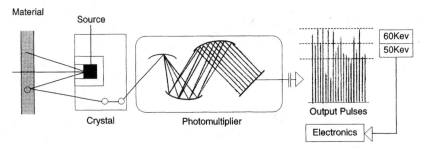

Figure 18-7. Principle of operation—gamma backscatter gauging technique. (Courtesy of NDC Systems.)

Figure 18-8. Measuring system using a gamma backscatter gauge. (Courtesy of NDC Systems.)

the detected gamma photons and whose pulse rate is proportional to the mass of the material being measured. Figure 18-7 graphically illustrates the principle of operation. Figure 18-8 illustrates a commercially available gamma backscatter system. The data collected from the gauge can either be displayed on a simple CRT or transported to a sophisticated computerized measurement system for statistical analysis and closed-loop control of various parameters.

18.5. BETA TRANSMISSION

This measurement technique is based upon the absorption of beta particles as they interact with the web. On either side of the web, a source head and detector head is mounted. As the mass of the web increases, more beta particles are absorbed by the web, resulting in

Figure 18-9. Beta transmission sensor. (Courtesy of NDC Systems.)

a reduced detector output. This output is subsequently linearized and converted to a very precise measurement of web basis weight or thickness.

Beta transmission sensors are used extensively for on-line thickness or weight measurement of continuous web processes. The sensors can also be mounted on a scanning platform for high-speed bidirectional scanning. A commercially available beta transmission sensor is shown in Figure 18-9.

18.6. SCANNING LASER

This noncontact gauging consists of a helium/neon laser that is projected in a straight line with virtually no diffusion. This laser beam is directed to rotating mirrors which in turn "scan" the laser beam through a collimator lens in a parallel path across the measurement area. The laser beam is projected toward the receiving lens. The receiving lens focuses the light onto a photo detector. A part placed in the beam casts a shadow, which becomes the starting point for the measurement. Through the edge-sensing process, the laser light signal entering the receiver is used to calculate the distance between the shadow edges. Dimensional data is instantly displayed and can be transmitted to a computer for further processing.

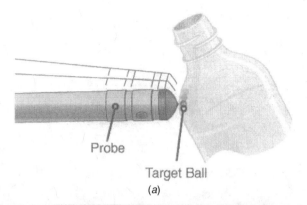

Probe

Target Ball

(a)

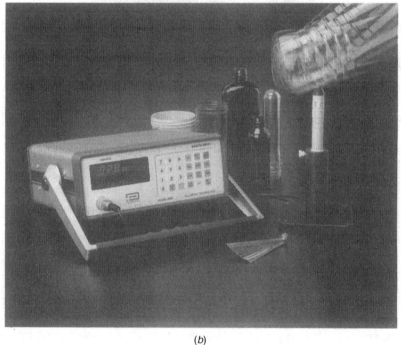

(b)

Figure 18-10. Wall thickness measurement using Hall effect gauging. (Courtesy of Panametrics Corporation.)

This state of the are technology is extremely flexible and accurate, allowing a variety of precise measurement including single diameter, multiple diameter, ovality, and dual axis measurement. The technique is widely used for in-process dimensional monitoring and control as well as sample inspection of products such as plastics, wire, monofilament, hose, tubing, and fiber-optic fibers.

18.7. X-RAY FLUORESCENCE

This technique is generally used for measurement of magnetic coatings on recording tape, floppy disks, and other metallic coatings on plastics. These probes are capable of measuring and controlling coating thickness to 0.01 µm.

18.8. HALL EFFECT

The thickness measurement technique known as the Hall effect uses a magnetic field applied at right angles to a conductor carrying a current. The combination induces a voltage in the other direction. The change in magnetic field and resulting induced voltage is proportional to the wall thickness.

The wall thickness measurement is done by simply placing a Hall probe on one side of the product to be measured and placing a ferromagnetic target, usually a small steel ball on the other side of the product. The thickness measurement is read directly off the digital panel meter. This method provides fast, accurate, and reliable thickness measurement without the use of messy couplant, and variations in material sound velocity resulting from changes in temperature usually encountered during ultrasonic thickness measurement. Figures 18-10a and 18-10b illustrate a commercially available thickness measurement unit based on the Hall effect principle.

REFERENCES

1. Krautkramer, J. and Krautkramer, H., *Ultrasonic Testing of Materials,* Springer-Verlag, New York, 1969.
2. Ostrofsky, B., "Ultrasonic Inspection of Welds," *Welding J.* (March 1965), p. 97-5.
3. "Nondestructive Inspection and Quality Control," *Metals Handbook.* Vol. II, American Society for Metals, Metals Park, OH, 1976, p. 179.
4. Krautkramer, Reference 1, p. 152.
5. *Ibid.,* p. 238.
6. *Ibid.,* p. 141.
7. *Ibid.,* p. 131.
8. *Nondestructive Examination of Plastic Pipe Socket Joints,* NDT Application Report. No J-714, Branson Instrument Company, Stamford, CN.
9. Krautkramer, Reference 1, p. 436.
10. Hastings, C. H., Lopilato, S. A., and Lynnworth, L. C., *Ultrasonic Inspection of Reinforced Plastics and Resin–Ceramic Composites. Nondestructive Test,* **19** (1961), pp. 340–346.
11. "Symposium on Recent Developments in Nondestructive Testing of Missiles and Rockets," *ASTM Spec. Tech. Pub. No. 350* (1963).

SUGGESTED READING

Lamble, J. H., *Principle and Practice of Nondestructive Testing,* Wiley-Interscience, New York, 1962.

LeGrand, R., "Nondestructive Testing Methods," *Machinist* (Sept. 1946), p. 893.

"Symposium on Nondestructive Testing." *ASTM Spec. Tech. Publ. No. 149* (1953).

Hitt, W. C. and Ramsey, J. B., "Ultrasonic Inspection of Plastics," *Rubber and Plastics Age,* **44**(4) (Apr. 1963), p. 411.

Seaman, R. E., "Ultrasonic Inspection by Pulsed Transmissions," *Br. Plast.,* **29**(7) (July 1956), p. 262.

Baumeister, G. B., "Production Testing of Bonding Materials with Ultrasonics," *ASTM Bull. No. 204* (Feb. 1955), p. 50.

Miller, N. B. and Boruff, V. H., "Adhesive Bonds Tested Ultrasonically," *Adhesives Age 6* (June 1963), p. 32.

Zurbrick, J. R., "Nondestructive Testing of Glass-Fiber Reinforced Plastics: Key to Composition Characterization and Design Properties Prediction," *S.P.E. J.,* **24**(9) (Sept. 1968), p. 56.

Hatfield, P., "Ultrasonic Measurements in High Polymers," *Research*, **9** (Oct. 1956), p. 388.

Coggeshall, A. D., "Nondestructive Quality Control Tests on Finished Reinforced Plastic Parts," *Plast. Tech.* (Dec. 1969), p. 43.

Bray, D. E. and McBride, D., *Nondestructive Testing Techniques,* Wiley-Interscience, New York, 1992.

GENERAL REFERENCES

Sales and Technical literature on "Gamma Backscatter, Beta Transmission, X-Ray Fluorescence," NDC Systems. Irvindale, CA.

Sales and Technical literature on "Hall Effect Technique," Panametrics. Waltham, MA.

Sales and Technical literature on "Laser Scanning," Lasermike Inc., Dayton, OH.

19

PROFESSIONAL AND TESTING ORGANIZATIONS

19.1. AMERICAN NATIONAL STANDARDS INSTITUTE (ANSI)

In 1918, when ANSI was founded, standardization activities were just beginning in the United States. Many groups were developing standards and their interests and activities overlapped. The standards they produced often duplicated or conflicted with each other. The result was the waste of manpower, money, and considerable confusion. Five professional/technical societies and three government departments decided a coordinator was needed and created ANSI to handle the job.

ANSI is a federation of standards competents from commerce and industry, professional, trade, consumer, and labor organizations, and government. ANSI, in cooperation with these federation participants,

1. Identifies the needs for standards and sets priorities for their completion.
2. Assigns development work to competent and willing organizations.
3. Sees to it that public interests, including those of the consumer, are protected and represented.
4. Supplies standards writing organizations with effective procedures and management services to ensure efficient use of their manpower and financial resources and timely development of standards.
5. Follows up to assure that needed standards are developed on time.

Another role is to approve standards as American National Standards when they meet consensus requirements. It approves a standard only when it has verified evidence presented by a standards developer that those affected by the standard have reached

Handbook of Plastics Testing and Failure Analysis, Third Edition, by Vishu Shah
Copyright © 2007 by John Wiley & Sons, Inc.

substantial agreement on its provisions. ANSI's other major roles are to represent American interests in nongovernmental international standards work, to make national and international standards available, and to inform the public of the existence of these standards.

19.2. ASTM INTERNATIONAL

ASTM International is one of the largest voluntary standards development organizations in the world—a trusted source for technical standards for materials, products, systems, and services. Known for their high technical quality and market relevancy, ASTM International standards have an important role in the information infrastructure that guides design, manufacturing, and trade in the global economy.

ASTM International, originally known as the American Society for Testing and Materials (ASTM), was founded in 1898. ASTM operates through more than 130 main technical committees with 1550 subcommittees. These committees function in prescribed fields under regulations that ensure balanced representation among producers, users, and general interest participants. Membership in ASTM is open to all concerned with the fields in which ASTM is active. An ASTM standard represents a common viewpoint of those parties concerned with its provisions, namely, producers, users, and general interest groups. It is intended to aid industry, government agencies, and the general public. The use of an ASTM standard is purely voluntary. It is recognized that, for certain work or in certain regions, ASTM specifications may be either more or less restrictive than needed. The existence of an ASTM standard does not preclude anyone from manufacturing, marketing, or purchasing products, or using products, processes, or procedures not conforming to the standard. Because ASTM standards are subject to periodic reviews and revision, it is recommended that all serious users obtain the latest revision. A new edition of the Book of Standards is issued annually. On the average, about 30 percent of each part is new or revised. Standards developed at ASTM are the work of over 30,000 ASTM members. These technical experts represent producers, users, consumers, government, and academia from over 100 countries.

ASTM Committee D20 on Plastics was formed in 1937. D20 meets three times a year in March, July and November with approximately 200 members attending over three days of technical meetings. The committee, with current membership of approximately 780, currently has jurisdiction of over 470 standards, published in the *Annual Book of ASTM Standards*, Section 8 on Plastics. These standards have and continue to play a preeminent role in all aspects important to the effective utilization of plastics, including specimen preparation, material specifications and testing methodologies. D20 also houses the US Technical Advisory Group is ISO Technical Committee 61 on Plastics, developing US positions and advocating them in the ISO arena. D20 sponsors two technical and professional training courses: one on "Testing Techniques of Plastics" and one on "Instrumental Analysis of Polymers."

19.3. FOOD AND DRUG ADMINISTRATION (FDA)

The Food and Drug Administration, first established in 1931, is an American government agency of the U.S. Department of Health and Human Services. The FDA's activities are directed toward protecting the health of the nation against impure and unsafe foods, drugs, cosmetics, and other potential hazards.

The plastics industry is mainly concerned with the Bureau of Foods, which conducts research and develops standards on the composition, quality, nutrition, and safety of foods, food additives, colors, and cosmetics, and conducts research designed to improve the detection, prevention, and control of contamination. The FDA is concerned about indirect additives. Indirect additives are those substances capable of migrating into food from contacting plastic materials. Extensive tests are carried out by the FDA before issuing a safety clearance to any plastic material that is to be used in food contact applications. Plastics used in medical devices are tested with extreme caution by the FDA's Bureau of Medical Devices, which develops FDA policy regarding the safety and effectiveness of medical devices.

Field operations for the enforcement of the laws under the jurisdiction of the FDA are carried out by 11 regional field offices, 22 district offices, and 124 resident inspection posts.

19.4. NATIONAL INSTITUTE OF STANDARDS AND TECHNOLOGY (NIST)

The National Bureau of Standards was established by an act of Congress in March 1901. The NBS was renamed the National Institute of Standard and Technology in 1988. The bureau's overall goal is to strengthen and advance the nation's science and technology and to facilitate their effective application for public benefit.

The bureau conducts research and provides a basis for the nation's physical measurement system, scientific and technological services for industry and government, a technical basis for increasing productivity and innovation, promoting international competitiveness in American industry, maintaining equity in trade and technical services, and promoting public safety. The bureau's technical work is performed by the National Measurement Laboratory, the National Engineering Laboratory, and the Institute for Computer Sciences and Technology.

19.5. NATIONAL ELECTRICAL MANUFACTURERS ASSOCIATION (NEMA)

The National Electrical Manufacturers Association was founded in 1926. This 600-member association consists of manufacturers of equipment and apparatus for the generation, transmission, distribution, and utilization of electric power. The membership is limited to corporations, firms, and individuals actively engaged in the manufacture of products included within the product scope of NEMA product subdivisions.

NEMA develops product standards covering such matters as nomenclature, ratings, performance, testing, and dimensions. NEMA is also actively involved in developing National Electrical Safety Codes and advocating their acceptance by state and local authorities. Along with a monthly news bulletin, NEMA also publishes manuals, guidebooks, and other material on wiring, installation of equipment, lighting, and standards. The majority of NEMA standardization activity is in cooperation with other national organizations. The manufacturers of wires and cables, insulating materials, conduits, ducts, and fittings are required to adhere to NEMA standards by state and local authorities.

19.6. NATIONAL FIRE PROTECTION ASSOCIATION (NFPA)

The National Fire Protection Association was founded in 1896 with the objective of developing, publishing, and disseminating standards intended to minimize the possibility and effect of fire and explosion. The NFPA's membership consists of individuals from business and industry, fire service, health care, insurance, and educational and government institutions. The NFPA conducts fire safety education programs for the general public and provides information on fire protection and prevention. Also provided by the association is field service by specialists on flammable liquids, electricity, gases, and marine problems.

Each year, statistics on causes and occupancies of fires and deaths resulting from fire are compiled and published. The NFPA sponsors seminars on the Life Safety Codes, National Electrical Code, industrial fire protection, hazardous materials, transportation emergencies, and other related topics. The NFPA also conducts research programs on delivery systems for public fire protection, arson, residential fire sprinkler systems, and other subjects. NFPA publications include *National Fire Codes Annual, Fire Protection Handbook, Fire Journal,* and *Fire Technology.*

19.7. NATIONAL SANITATION FOUNDATION (NSF)

The National Sanitation Foundation, more commonly known as NSF, is an independent, nonprofit environmental organization of scientists, engineers, technicians, educators, and analysts. NSF frequently serves as a trusted neutral agency for government, industry, and consumers, helping them to resolve differences and unite in achieving solutions to problems of the environment.

At NSF, a great deal of work is done on the development and implementation of NSF standards and criteria for health-related equipment. Standard No. 1, concerning soda fountain and luncheonette equipment, was adopted in 1952. Since then, many new standards have been developed and successfully implemented. The majority of NSF standards relate to water treatment and purification equipment, products for swimming pool applications, plastic pipe for potable water as well as drain, waste, and vent (DWV) uses, plumbing components for mobile homes and recreational vehicles, laboratory furniture, hospital cabinets, polyethylene refuse bags and containers, aerobic waste treatment plants, and other products related to environmental quality.

Manufactures of equipment, materials, and products that conform to NSF standards are included in official listings and these producers are authorized to place the NSF seal on their products. Representatives from NSF regularly visit the plants of manufacturers to make certain that products bearing the NSF seal do indeed fulfill applicable NSF standards.

19.8. SOCIETY OF PLASTICS ENGINEERS (SPE)

The Society of Plastics Engineers was founded in 1942 with the objective of promoting scientific and engineering knowledge relating to plastics. SPE is a professional society of plastics scientists, engineers, educators, students, and others interested in the design, development, production, and utilization of plastics materials, products, and equipment.

SPE currently has over 37,000 members scattered among its 91 sections. The individual sections as well as the SPE main body arranges and conducts monthly meetings, conferences, educational seminars, and plant tours throughout the year. SPE also publishes *Plastics Engineering, Polymer Engineering and Science, Plastics Composites*, and the *Journal of Vinyl Technology*. The society presents a number of awards each year, encompassing all levels of the organization, section, division, committee, and international. SPE divisions of interest are color and appearance, injection molding, extrusion, electrical and electronics, thermoforming, engineering properties and structure, vinyl plastics, blow molding, medical plastics, plastics in building, decorating, mold making, and mold design.

19.9. SOCIETY OF PLASTICS INDUSTRY (SPI)

The Society of Plastics Industry is a major society, whose membership consists of manufacturers and processors of plastics materials and equipment. The society has four major operating units consisting of the Eastern Section, the Midwest Section. the New England Section, and the Western Section. SPI's Public Affairs Committee concentrates on coordinating and managing the response of the plastics industry to issues like toxicology, combustibility, solid waste, and energy. The Plastic Pipe Institute is one of the most active divisions, promoting the proper use of plastic pipes by establishing standards, test procedures, and specifications. The Epoxy Resin Formulators Division has published over 30 test procedures and technical specifications. Risk management, safety standards, productivity, and quality are a few of the major programs undertaken by the machinery division. SPI's other divisions include Expanded Polystyrene Division, Fluoropolymers Division, Furniture Division, International Division, Plastic Bottle Institute, Machinery Division, Molders Division, Mold Makers Division, Plastic Beverage Container Division, Plastic Packaging Strategy Group, Polymeric Materials Producers Division, Polyurethane Division, Reinforced Plastic/Composites Institute, Structural Foam Division, Vinyl Siding Institute, and Vinyl Formulators Division.

The National Plastics Exposition and Conference, held every three years by the Society of Plastic Industry, is one of the largest plastic shows in the world. SPI works very closely with other organizations such as ASTM and ANSI to develop new test methods, standards, and specifications.

19.10. UNDERWRITERS LABORATORIES (UL)

Underwriters Laboratories, founded in 1894, is chartered as a not-for-profit organization to establish, maintain, and operate laboratories for the investigation of materials, devices, products, equipment, constructions, methods, and systems with respect to hazards affecting life and property.

There are five testing facilities in the United States and over 200 inspection centers. More than 700 engineers and 500 inspectors conduct tests and follow-up investigations to insure that potential hazards are evaluated and proper safeguards provided. UL has six basic services it offers to manufacturers, inspection authorities, or government officials. These are product listing service, classification service, component recognition service, certificate service, inspection service, and fact finding and research.

UL's Engineering Services department is in charge of evaluating individual plastics and other products using plastics as components. Engineering Services evaluates consumer products such as TV sets, power tools, appliances, and industrial and commercial electrical equipment and components. In order for a plastic material to be recognized by UL, it must pass a variety of UL tests including the UL 94 flammability test and the UL 746 series short- and long-term property evaluation tests. When a plastic material is granted "Recognized Component" status, a yellow card is issued. The card contains precise identification of the material including supplier, product designation, color, and its UL 94 flammability classification at one or more thicknesses. Also included are many of the property values, such as relative temperature index, hot wire ignition, high-current arc ignition, and arc resistance. These data also appear in the Plastics Recognized Component Directory.

UL publishes the names of the companies who have demonstrated the ability to provide a product conforming to the established requirements, upon successful completion of the investigation and after agreement of the terms and conditions of the listing and follow-up service. Listing signifies that production samples of the product have been found to comply with the requirements, and that the manufacturer is authorized to use the UL Listing Mark on the listed products which comply with the requirements.

UL's consumer advisory council was formed to advise UL in establishing levels of safety for consumer products, to provide UL with additional user field experience and failure information in the field of product safety, and to aid in educating the general public in the limitations and safe use of specific consumer products.

19.11. INTERNATIONAL ORGANIZATION FOR STANDERDIZATION (ISO)

ISO is a network of the national standards institutes of 148 countries, on the basis of one member per country, with a Central Secretariat in Geneva, Switzerland, that coordinates the system.

ISO is a nongovernmental organization: Its members are not, as is the case in the United Nations system, delegations of national governments. Nevertheless, ISO occupies a special position between the public and private sectors. This is because, on the one hand, many of its member institutes are part of the governmental structure of their countries, or are mandated by their government. On the other hand, other members have their roots uniquely in the private sector, having been set up by national partnerships of industry associations.

Therefore, ISO is able to act as a bridging organization in which a consensus can be reached on solutions that meet both the requirements of business and the broader needs of society, such as the needs of stakeholder groups like consumers and users (www.ISO.Org.).

20

UNIFORM GLOBAL TESTING STANDARDS

20.1. INTRODUCTION

The initial interest in the Uniform Global Testing Standards has been driven in large part by American manufacturers as well as the U.S. government. The globalization of the economy has forced the government and business community alike to focus on international standards. Over the past few years, there has been a growing demand to adopt ISO/IEC standards. There are many factors driving the trend toward uniform global ISO test methods (1):

1. The European Union (EU) and most other European nations have replaced their national standards with ISO test methods.
2. The Japanese Plastics Industry Federation is converting to ISO test methods.
3. USCAR (United States Council for Automotive Research) has mandated conversion from ASTM to ISO test methods.
4. U.S. based multinational companies such as Xerox have instituted worldwide material specifications based on ISO tests. Similar interest has been expressed by other multinational companies in electronics, medical, appliance, and consumer industries.
5. The U.S. government is also moving toward international standards.
6. ASTM is continuing to harmonize its plastics material standards and test methods to international standards.
7. Plastics materials manufacturers are slowly converting to ISO test methods.

Handbook of Plastics Testing and Failure Analysis, Third Edition, by Vishu Shah
Copyright © 2007 by John Wiley & Sons, Inc.

One cannot simply overlook the numerous benefits derived from standardization. Standardization of plastics testing on a global basis will help resin manufacturers and customers gain a more equitable access to expanding markets, consistency of product quality, reduced long-term costs, and improved communication. The use of one set of uniform, global standards for testing materials facilitates comparison and selection by ensuring that reported property data is much more consistent, regardless of the supplier or where the resin is produced (2).

20.2. ISO/IEC STANDARDS

The International Standards Organization (ISO), founded in 1947, is a worldwide federation of 120 national standards bodies at present representing each participating country. Each participating country is allowed one official member in ISO. The International Electrotechnical Commission (IEC), founded in 1906, has 50 national electrotechnical committees. The participants meet under the auspices of ISO and IEC to develop standards which can be accepted as international standards. The scope of ISO covers standardization in all fields except electrical and electronic engineering, including some areas of telecommunications, which are the responsibility of IEC. The objective of ISO and IEC is to promote the development of standardization and related activities in the world to facilitate the international exchange of goods and services and to develop cooperation in the spheres of intellectual, scientific, technological, and economic activity (3).

The ISO/IEC standards are uniform global standards for selecting, molding, and testing test specimens under provisions established by the technical committees. The ISO and IEC standards for testing plastic materials offer specific testing procedures based upon a comprehensive approach for generation of comparable data which reduces variability in preparing and testing specimens used to determine plastics material properties. The standards incorporate details which have been negotiated among and agreed upon by ISO member national standards bodies to ensure acceptability worldwide. Figure 20-1 illustrates important milestones in the trend toward global standardization (4).

20.3. ISO AND ASTM

While two thirds of the world is racing toward adopting International Standards with global manufacturing and marketing in mind, the United States is moving at a slow speed to achieve this goal and join with other industrial nations of the world. The demand put forward by the multinational U.S. automotive industry and other multinational corporations or material suppliers has finally started the ball rolling. The wisdom of U.S. companies adopting the ISO standards in place of current ASTM standards has been questioned by many traditionalists. The answer is quite simple. Both ISO and IEC test methods have gained international acceptance because of their organizational structure. Since the membership of the ISO is comprised of the national standards bodies of some 120 countries, and standards are developed through a balloting process, one vote per national body, it ensures consensus and agreement worldwide. Conversely, since the ASTM membership is primarily domestic, it is generally perceived to be an American standards organization.

Trend Towards Global Standardization Milestones

1947	1951	Mid-60s	1970	1979	Mid-80s	1987	1989	1991	1992	1993	1994	1995	1997
• International Organization for Standardization (ISO) founded in Geneva, Switzerland.	• ISO forms Technical Committee 61 to develop global materials testing standards for plastics. • U.S. establishes Technical Advisory Group (TAG) to ISO-TC61 within ANSI to represent U.S. positions on ISO ballots.	• ASTM begins call for the use of SI units in standards specifications.	• ASTM forms Sub-Committee D20.61 which becomes the new home for the U.S. TAG to ISO-TC61 so that U.S. positions on ISO ballots could be established using a process based upon consensus.	• British Standards Institute (BSI) produces quality systems standard BS5750 for the defense equipment sector of the U.K.	• CEN begins standardization based upon ISO test methods.	• Ford Motor Company notifies resin suppliers that new specifications would use only ISO test standards. • General Motors requests use of ISO standards in ASTM documents.	• ASTM Committee D20.94.02 establishes task group on automotive needs, becoming the basis for the formation of the USCAR (United States Council for Automotive Research) plastics committee.	• ASTM publishes D1999 document, referencing ISO specimens and equivalent ASTM and ISO test methods.	• ASTM harmonizes test methods for acetal with ISO protocols. • ASTM harmonizes test methods for acrylics with ISO protocols. • Resin suppliers in Europe (Hoechst, BASF, Hüls and Bayer) form database called "CAMPUS" (Computer Aided Material Preselection by Uniform Standards) using standards established by ISO and IEC (International Electrotechnical Commission), and offer license to all resin suppliers. • Europe forms Common Market through Community of European Nations (CEN) driving need for standardization to facilitate trade and commerce. • SPI's Polymeric Materials Producers Division (PMPD) calls for members to adopt ISO test methods as a means to ensure access to global markets and prevent the use of regional or national standards as barriers. • Big Three automakers formally establish USCAR (United States Council for Automotive Research) to address the need for precompetitive cooperation, with initial focus on standardization.	• USCAR agrees to promote use of SI system and ISO test bars under SAE J1639 test standard. • USCAR leads move to standardize test methods for nylon based upon ISO test methods published through SAE. • ISO publishes test standards for single-point data (ISO 10350).	• ISO publishes test standards for multipoint data: mechanical properties. • ISO publishes test standards for multipoint data: thermal and processing properties. • U.S. suppliers of engineering resins initiate and/or continue conversion to ISO test methods. • SPI publishes the U.S. plastics industry's first white paper on ISO testing methods, calling for uniform global materials test methods based upon ISO/IEC standards. • GM notifies resin suppliers that nylon specifications must be based on ISO test standards by January 1996. • "The Need for Uniform Global Materials Testing Standards" is subject of industry forum sponsored by SPI at National Plastics Exposition, Chicago. • CAMPUS version 3.0 rolled out in U.S. at NPE. • Japan Plastics Industry Federation (JPIF) begins considering adoption of ISO test methods. • U.S. ratifies NAFTA, GATT.	• CAMPUS licensees in Europe grow to 26, add 6 in U.S.	• ASTM to use only SI units in standards.

Figure 20-1. Trend toward global standardization. (Courtesy of Ticona Corporation.)

The ISO has been very successful in bringing together the interest of producers, users, governments, and the scientific community in the development of international standards. ASTM committee D20 by contrast is represented by material manufactures, end users and general interest groups. Table 20-1 lists participating members of ISO TC61 on plastics.

ASTM is "internationalizing" its protocols with ISO methodology as support for the goal to have one set of uniform global standards. Standards specifications for acetal and acrylic resins have been rewritten to require ISO test methods. ASTM has committed to revising the remaining standards to make them equivalent to ISO methods as they come up for review. ASTM is also playing a significant role in the development of the actual ISO/IEC standards through its Technical Advisory Group (TAG) to the ISO Technical Committee on Plastics. The ASTM group members develop and present the U.S. industry's position on all new and revised ISO test protocols for plastics, and are effective in influencing the development of standards within ISO (5). Table 20-2 illustrates the comparison between ISO and ASTM test methods.

TABLE 20-1. Participating Members of ISO TC61 on Plastics

Country	National Standards Organization	
Belgium	Institut Belge de Normalisation	IBN
Canada	Standards Council of Canada	SCC
Peoples Republic of China	China State Bureau of Technical Supervision	CSBTS
Colombia	Instituto Colombiano de Norms Technicas	ICONTEC
Czech Republic	Czech Office for Standards, Metrology and Testing	COSMT
Finland	Finnish Standards Association	SFS
France	Association Francaise de Normalisation	AFNOR
Germany	Deutsches Institut für Normung	DIN
Hungary	Magyar Szabványügyi Hivatal	MSZH
India	Bureau of Indian Standards	BIS
Islamic Republic of Iran	Institute of Standards and Industrial Research of Iran	ISIRI
Italy	Ente Nazionale Italiano di Unificazione	UNI
Japan	Japanese Industrial Standards Committee	JISC
Republic of Korea	Korean Industrial Advancement Administration	KIAA
Netherlands	Nederlands Normalisatie-Instituut	NNI
Philippines	Bureau of Product Standards	BPS
Poland	Polish Committee for Standardization	PKN
Romania	Institutul Roman de Standardizare	IRS
Russian Federation	Committee of the Russian Federation for Standardization, Metrology and Certification	GOST R
Slovakia	Slovak Office of Standards, Metrology and Testing	UNMS
Spain	Associacion Espanola de Normalizacion y Certificacion	AENOR
Sweden	SIS—Standardiseringen i Sverige	SIS
Switzerland	Schweizerische Normen-Vereinigung	SNV
United Kingdom	British Standards Institution	BSI
United States of America	American National Standards Institute	ANSI

Reprinted with permission of Society of Plastics Industry Inc.

TABLE 20-2. Cross-Reference List of ASTM/ISO Plastics Standards

Similar ASTM Standard	Standard Topic	Similar ISO Standard
	Specimen Preparation	
D0955	Measuring Shrinkage from Mold Dimensions of Molded Thermoplastics	294-4
D3419	In-Line Screw-Injection Molding of Test Specimens from Thermosetting Compounds	10724
D3641	Injection Molding Test Specimens of Thermoplastic Molding and Extrusion Materials	294-1,2,3
D4703	Compression Molding Thermoplastic Materials into Test Specimens, Plaques, or Sheets	293
D524	Compression Molding Test Specimens of Thermosetting Molding Compounds	95
D6289	Measuring Shrinkage from Mold Dimensions of Molded Thermosetting Plastics	2577
	Mechanical Properties	
D0256	Determining the Pendulum Impact Resistance of Notched Specimens of Plastics	180
D0638	Tensile Properties of Plastics	527-1,2
D0695	Compressive Properties of Rigid Plastics	604
D0785	Rockwell Hardness of Plastics and Electrical Insulating Materials	2039-2
D0790	Flexural Properties of Unreinforced and Reinforced Plastics and Electrical Insulating Materials	178
D0882	Tensile Properties of Thin Plastic Sheeting	527-3
D1043	Stiffness Properties of Plastics as a Function of Temperature by Means of a Torsion Test	458-1
D1044	Resistance of Transparent Plastics to Surface Abrasion	9352
D1078	Tensile Properties of Plastics by Use of Microtensile Specimens	6239
D1822	Tensile-Impact Energy to Break Plastics and Electrical Insulating Materials	8256
D1894	Static and Kinetic Coefficients of Friction of Plastic Film and Sheeting	6601
D1922	Propagation Tear Resistance of Plastic Film and Thin Sheeting by Pendulum Method	6383-2
D1938	Tear Propagation Resistance of Plastic Film and Thin Sheeting by a Single Tear Method	6383-1
D2990	Tensile, Compressive, and Flexural Creep and Creep-Rupture of Plastics	899-1,2
D3763	High-Speed Puncture Properties of Plastics Using Load and Displacement Sensors	6603-2
D4065	Determining and Reporting Dynamic Mechanical Properties of Plastics	6721-1
D4092	Dynamic Mechanical Measurements on Plastics	6721

TABLE 20-2. *Continued*

Similar ASTM Standard	Standard Topic	Similar ISO Standard
D4440	Rheological Measurement of Polymer Melts Using Dynamic Mechanical Procedures	6721-10
D5023	Measuring the Dynamic Mechanical Properties of Plastics Using Three Point Bending	6721-3
D5026	Measuring the Dynamic Mechanical Properties of Plastics in Tension	6721-5
D5045	Plane-Strain Fracture Toughness and Strain Energy Release Rate of Plastic Materials	572
D5083	Tensile Properties of Reinforced Thermosetting Plastics Using Straight-Sided Specimens	3268
D5279	Measuring the Dynamic Mechanical Properties of Plastics in Torsion	6721
	Thermoplastic Materials	
D1239	Resistance of Plastic Films to Extraction by Chemicals	6427
D1693	Environmental Stress-Cracking of Ethylene Plastics	4599
D1928	Preparation of Compression-Molded Polyethylene Test Sheets and Test Specimens	293 and 1872-2
D2581	Polybutylene (PB) Plastics Molding and Extrusion Materials	8986-1 and 8986-2
D2765	Determination of Gel Content and Swell Ratio of Crosslinked Ethylene Plastics	10147
D2951	Resistance of Types III and IV Polyethylene Plastics to Thermal Stress-Cracking	TR1837
D4020	Ultra-High-Molecular-Weight Polyethylene Molding and Extrusion Materials	11542-1&2
D4976	Polyethylene Plastics Molding and Extrusion Materials	1872-1&2
D0788	System For Poly(Methyl Methacrylate) (PMMA) Molding and Extrusion Compounds	FDIS 8257-1 ISO 8257-2
D4802	Poly(Methyl Methacrylate) Acrylic Plastic Sheet	ISO 7823-1 and -2
D3965	Rigid Acrylonitrile-Butadiene-Styrene (ABS) Compounds for Pipe and Fittings	7245
D4203	Styrene Acrylonitrile (SAN) Injection and Extrusion Materials	4894-1 4894-2
D4594	Polystyrene Molding and Extrusion Materials (PS)	1622-1&2 2897-1&2
D4634	Styrene-Maleic Anhydride Materials (S/MA)	10366-1 10366-2
D4673	Acrylonitrile-Butadiene-Styrene (ABS) Molding and Extrusion Materials	2580-1 2580-2
D3012	Thermal Oxidative Stability of Propylene Plastics Using a Blaxial Rotator	4577
D4101	Propylene Plastic Injection and Extrusion Materials	1873-1&2

TABLE 20-2. *Continued*

Similar ASTM Standard	Standard Topic	Similar ISO Standard
D5857	Propylene Plastic Injection and Extrusion Materials for International Use	1873-1&2
D1203	Volatile Loss From Plastics Using Activated Carbon Methods	176
D1243	Dilute Solution Viscosity of Vinyl Chloride Polymers	DIS 1628-2
D1705	Particle Size Analysis of Powdered Polymers and Copolymers of Vinyl Chloride	1624
D1755	Poly(Vinyl Chloride) Resins	1264 (pH only)
D1784	Rigid Poly(Vinyl Chloride) (PVC) Compounds and Chlorinated Poly (Vinyl Chloride) (CPVC) Compounds	1163-1 1163-2
D1823	Apparent Viscosity of Plastisols and Organosols at High Shear Rates by Extrusion Viscometer	3219
D1824	Apparent Viscosity of Plastisols and Organosols at Low Shear Rates by Brookfield Viscometer	2555
D2115	Oven Heat Stability of Poly (Vinyl Chloride) Compositions	305
D2287	Nonrigid Vinyl Choloride Polymer and Copolymer Molding and Extrusion Compounds	2896-1 2898-2
D3030	Volatile Matter (Including Water) of Vinyl Chloride Resins	1269
D3367	Plasticizer Sorption of Poly (Vinyl Chloride) Resins Under Applied Centrifugal Force	4608
D422	Thermal Stability of Poly (Vinyl Chloride) (PVC) Resin	305
D0789	Determination of Relative Viscosity, Melting Point, and Moisture Content of Polyamide (PA)	960 (water only)
D4066	Nylon Injection and Extrusion Materials	1874-1 1874-2
D1430	Polychlorotrifluoroethylene (PCTFE) Plastics	12086-1 12086-2
D1710	Polytetrafluoroethylene (PTFE) Basic Shapes, Rod, and Heavy-Walled Tubing	13000-1 13000-2
D2116	Fep-fluorocarbon Molding and Extrusion Materials	12086-1 12086-2
D3159	Modified ETFE-Fluoropolymer Molding and Extrusion Materials	12086-1 12086-2
D3275	E-CTFE-Fluoroplastic Molding, Extrusion, and Coating Materials	12086-1 12086-2
D3307	PFA-Fluorocarbon Molding and Extrusion Materials	12086-1 12086-2
D3308	PTFE Resin Skived Tape	13000-1 13000-2
D4441	Aqueous Dispepsions of Polytetrafluoroethylene	12086-1 12086-2

TABLE 20-2. *Continued*

Similar ASTM Standard	Standard Topic	Similar ISO Standard
D4591	Determining Temperatures and Heats of Transitions of Fluoropolymers by Differential Scanning Calorimetry	12086-1 12086-2
D4894	Polytetrafluoroethylene (PTFE) Granular Molding and Ram Extrusion Materials	12086-1 12086-2
D4895	Polytetrafluoroethylene (PTFE) Resins Produced from Dispersion	12086-1 12086-2
D5575	System and Basis for Specification of Copolymers of Vinylidene Fluoride (VDF) with Other Fluorinated Monomers	12086-1 12086-2
D5675	Fluoropolymer Micro Powders	12086-1 12086-2
D6040	SG to Test Methods for Unsintered PTFE Extruded Film or Sheet	12086-1 12086-2
D3935	Polycarbonate (PC) Unfilled and Reinforced Material	7391-1 7391-2
D4349	Poylphenylene Ether (PPE) Materials	DIS 15103-1 DIS 15103-2
D4181	Aoetal (POM) Molding and Extrusion Materials	DIS 9988-1 DIS 9988-2
D4507	Thermoplastic Polyester (TPES) Materials	7792-1 7792-2
D4550	Thermoplastic Elastomer-Ether-Ester (TEEE)	14910-1 14910-2
D5927	Thermoplastic Polyester (TPES) Injection and Extrusion Materials Based on ISO Test Methods	7729-1 7729-2
	Reinforced Thermosetting Plastics	
D2584	Ignition Loss of Cured Reinforced Resins	1172
D2734	Void Content of Reinforced Plastics	7822
D3846	In-Plane Shear Strength of Reinforced Plastics	4585(?)
D3914	In-Plane Shear Strength of Pultruded Glass-Reinforced Plastic Rod	4585(?)
	Film and Sheeting	
D1709	Impact Resistance of Plastic Film by the Free-Falling Dart Method	7765-1,2
D2578	Wetting Tension of Polyethylene and Polypropylene Films	8296
D2732	Unrestrained Linear Thermal Shrinkage of Plastic Film and Sheeting	11501
D3354	Blocking Load of Plastic Film by the Parallel Plate Method	11502
D4321	Package Yield of Plastic Film	4591

TABLE 20-2. *Continued*

Similar ASTM Standard	Standard Topic	Similar ISO Standard
	Cellular Plastics and Elastomers	
D1056	Flexible Cellular Materials-Sponge or Expanded Rubber	6916-1
D1621	Compressive Properties of Rigid Cellular Plastics	844
D1622	Apparent Density of Rigid Cellular Plastics	845
D1623	Tensile and Tensile Adhesion Properties of Rigid Cellular Plastics	1926
D2126	Response of Rigid Cellular Plastics to Thermal and Humid Aging	2796
D2842	Water Absorption of Rigid Cellular Plastics	2896
D2856	Open-Cell Content of Rigid Cellular Plastics by the Air Pycnometer	4590
D3453	Flexible Cellular Materials-Urethane for Furniture and Automotive Cushioning, Bedding, and Similar Applications	5999
D3574-A	Flexible Cellular Materials—Slab, Bonded, and Molded Urethane Foams	845
3574-B1	Flexible Cellular Materials—Slab, Bonded, and Molded Urethane Foams	2439
3574-B2	Flexible Cellular Materials—Slab, Bonded, and Molded Urethane Foams	2439
3574-C	Flexible Cellular Materials—Slab, Bonded, and Molded Urethane Foams	3386-1
3574-D	Flexible Cellular Materials—Slab, Bonded, and Molded Urethane Foams	1856
3574-E	Flexible Cellular Materials—Slab, Bonded, and Molded Urethane Foams	1798
3574-F	Flexible Cellular Materials—Slab, Bonded, and Molded Urethane Foams	8067
3574-G	Flexible Cellular Materials—Slab, Bonded, and Molded Urethane Foams	7231
3574-H	Flexible Cellular Materials—Slab, Bonded, and Molded Urethane Foams	8307
3574-13	Flexible Cellular Materials—Slab, Bonded, and Molded Urethane Foams	3385
3574-14	Flexible Cellular Materials—Slab, Bonded, and Molded Urethane Foams	None
3574-J	Flexible Cellular Materials—Slab, Bonded, and Molded Urethane Foams	2440
3574-K	Flexible Cellular Materials—Slab, Bonded, and Molded Urethane Foams	2440
D3575	Flexible Cellular Materials Made From Olefin Polymers	7214
D3576	Rubber Cellular Cushion Used For Carpet or Rug Underlay	5999

TABLE 20-2. *Continued*

Similar ASTM Standard	Standard Topic	Similar ISO Standard
D3748	Evaluating High Density Rigid Cellular Thermoplastics	9054
D4274-A	Testing Polyurethane Polyol Raw Materials: Determination of Hydroxyl Numbers of Polyols	CD 14900
D4274-B	Testing Polyurethane Polyol Raw Materials: Determination of Hydroxyl Numbers of Polyols	CD 14900
D4274-C	Testing Polyurethane Polyol Raw Materials: Determination of Hydroxyl Numbers of Polyols	CD 14900
D4669	Polyurethane Raw Materials: Determination of Specific Gravity of Isocyanates	10349-11
D4660	Polyurethane Raw Materials: Determination of Isomer Content of Isocyanates	TR9372
D4661	Polyurethane Raw Materials: Determination of Total Chlorine in Isocyanates	4615
D4661	Polyurethane Raw Materials: Determination of Total Chlorine in Isocyanates	7725
D4669	Polyurethane Raw Materials: Determination of Specific Gravity of Polyols	10349-11
D4672	Polyurethane Raw Materials: Determination of Water Content of Polyols	CD14897
D4876	Polyurethane Raw Materials: Determination of Acidity of Crude or Modified Isocyanates	CD 14898
D4877	Polyurethane Raw Materials: Determination of APHA Color in Isocyanates	6271
D4878	Polyurethane Raw Materials: Determination of Viscosity of Polyols	3219
D4889	Polyurethane Raw Materials: Determination of Viscosity of Crude or Modified Isocyanates	3104
D4890-B	Polyurethane Raw Materials: Determination of Gardner and APHA Color of Polyols	6271
D5155	Polyurethane Raw Materials: Determination of the Isocyanate Content of Aromatic Isccyanates	CD 14896
D6099	PU RM: Acidity in Moderate to High Acidity Aromatic ISOS	CD 14898

Thermal Properties

D0635	Rate of Burning and/or Extent and Time of Burning of Self-Supporting Plastics in a Horizontal Position	1210
D0648	Deflection Temperature of Plastics Under Flexural Load	75-1,-2
D0746	Brittleness Temperature of Plastics and Elastomers by Impact	974
D1238	Flow Rates of Thermoplastics by Extrusion Plastometer	1133
D1525	Vicat Softening Temperature of Plastics	306
D1790	Brittleness Temperature of Plastic Sheeting by Impact	8570
D1929	Ignition Properties of Plastics	871

TABLE 20-2. *Continued*

Similar ASTM Standard	Standard Topic	Similar ISO Standard
D2863	Measuring the Minimum Oxygen Concentration to Support Candle-like Combustion of Plastics (Oxygen Index)	4589-2
D3417	Heats of Fusion and Crystallization of Polymers by Thermal Analysis	11357-3
D3801	Measuring the Comparative Extinguishing Characteristics of Solid Plastics in a Vertical Position	1210B
D3835	Determination of Properties of Polymeric Materials by Means of a Capillary Rheometer	11443
D4804	Determining the Flammability Characteristics of Non-Rigid Solid Plastics	9773
D4986	Horizontal Burning Characteristics of Cellular Polymeric Materials	9772
D5025	A Laboratory Burner Used for Small-Scale Burning Tests on Plastic Materials	IEC 695-11-3 and 10093
D5048	Measuring the Comparative Burning Characteristics and Resistance to Burn-Through of Solid Plastics Using 125-mm Flame	10351
D5207	Calibration of 20 and 125 mm Test Flames for Small-Scale Burning Tests on Plastic Materials	IEC 695-11-3,-4

Optical Properties

D0542	Index of Refraction of Transparent Organic Plastics	489
D1003	Haze and Luminous Transmittance of Transparent Plastics	14782/13468

Analytical Methods

D0494	Acetone Extraction of Phenolic Molded or Laminated Products	308/59
D1505	Density of Plastics by the Density-Gradient Technique	1183
D1601	Dilute Solution Viscosity of Ethylene Polymers	1191
D1603	Carbon Black in Olefin Plastics	6964
D1921	Particle Size (Sieve Analysis) of Plastic Materials	4610
D2124	Analysis of Components in Poly(Vinyl Chloride) Compounds Using an Infrared Spectrophotometric Technique	1265
D2857	Dilute Solution Viscosity of Polymers	307, 1628-1
D3749	Residual Vinyl Chloride Monomer in Poly (Vinyl Chloride) Homopolymer Resins by Gas Chromatographic Headspace Technique	4601
D4322	Residual Acrylonitrile Monomer in Styrene-Acrylonitrile Copolymers and Nitrile Rubber by Headspace Gas Chromatography	4581
D5830	Ash Content in Thermoplastics	3451-1, 6427

TABLE 20-2. *Continued*

Similar ASTM Standard	Standard Topic	Similar ISO Standard
	Degradable Plastics	
D5209	Determining the Aerobic Biodegradation of Plastic Materials in the Presence of Municipal Sewer Sludge	14852
D5210	Determining the Aerobic Biodegradation of Plastic Materials in the Presence of Municipal Sewer Sludge	CD 14853
D5271	Determining the Aerobic Biodegradation of Plastic Materials in an Activated-Sludge-Wastewater-Treatment System	14851
D5338	Determining Aerobic Biodegradation of Plastic Materials Under Controlled Composting Conditions	14855
D5511	Determining Anaerobic Biodegradation of Plastic Materials Under High-Solids Anaerobic-Digestion Conditions	CD 15985

20.4. TEST DATA ACQUISITION AND REPORTING

To minimize the differences between the test results, generally caused by variations in test specimens and test conditions, ISO has developed a series of specifications:

ISO 10350: 1993	The acquisition and presentation of comparable single-point data.
ISO 11403-1: 1994	The acquisition and presentation of comparable multi-point data. Part 1—Mechanical properties.
ISO 11403-2: 1994	The acquisition and presentation of comparable multi-point data. Part 2—Thermal and processing properties.

20.4.1. Mold Design and Construction

ISO 3167 specifically deals with the production of multipurpose test specimens. It requires the use of a balanced, two-cavity mold with a defined gate design to prepare test specimens used to measure properties of plastics materials. Two identical cavities with large gates and a balanced runner system promotes filling and packing consistently and uniform orientation, minimizes shear effects, and produces uniform specimen. Figure 20-2 illustrates differences between old style, multicavity, unbalanced family mold and ISO test specimens (6).

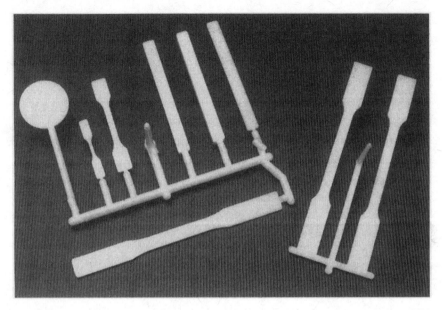

Figure 20-2. Differences between ASTM and ISO test specimens. (Courtesy of Ticona Corporation.)

20.4.2. Test Specimen Dimensions

The use of the multipurpose test specimen for a variety of primary tests such as tensile, flexural, charpy impact, and HDT brings about the uniformity in test results. This approach reduces the variables typically associated with specimen preparation, thereby ensuring more reliable, reproducible, and comparable data.

20.4.3. Process Parameters

ISO 294-1, -2, -3, and -4 underscore the importance of uniform processing parameters in order to produce consistent test specimens time after time. Since fill rate affects the mechanical and physical properties significantly, the molding procedure specifies average injection velocity rather than injection pressure and time. Melt and mold temperatures, as well as injection velocity for the particular product family, is found in the appropriate ISO material standards. This approach minimizes differences due to viscosity and shear sensitivity of a particular material (7).

20.4.4. Injection Molding Machine

Machine size and process control can affect resin degradation and shot-to-shot consistency. ISO 294 protocols specify limitations on barrel capacity in proportion to shot size. ISO also recommends using a molding machine capable of controlling injection velocity independent of injection pressure (8).

Just as the importance of uniform data acquisition cannot be overlooked, the uniform data reporting must also be given equal importance. ISO requires that all test data be reported in SI units. Table 20-3 lists test conditions and the format for presentation of single-point data according to ISO 10350.

TABLE 20-3. Test Conditions and Format for Presentation of Single-Point Data According to ISO 10350: 1993

Property	Standard	Specimen Type Dimension in mm	Unit	Test Conditions and Supplementary Instructions
Rheological Properties				
Melt mass-flow rate	ISO 1133	Molding compound	g/10 min	At test temperature and test load specified in Part 2 of material designation standards
Melt volume-flow rate	ISO 1133	Molding compound	cm³/10 min	
Molding shrinkage	ISO 2577		%	Thermosetting materials only
Mechanical Properties (at 23°C/50% RH, unless noted)				
Tensile modulus	ISO 527-1 and 527-2		MPa	Testing speed 1 mm/min; between 0.05% and 0.25% strain
Stress at yield		ISO 3167	MPa	Testing speed 50 mm/min; if strain at yield or break > 10%
Strain at yield			%	
Nominal strain at break			%	At 50 mm/min; only if no yield at ≤ 50% strain
Stress at 50% strain			MPa	Testing speed 5 mm/min; only if strain at break < 10%
Stress at break			MPa	
Strain at break			%	
Tensile creep modulus	ISO 899-1	ISO 3167	MPa	At 1 hr }
Tensile creep modulus			MPa	At 1000 hr } strain ≤ 0.5%
Flexural modulus	ISO 178	80 × 10 × 4	MPa	Test speed 2 mm/min
Flexural strength			MPa	
Charpy impact strength	ISO 179/1eU		kJ/m²	At +23°C; edgewise impact
Charpy impact strength			kJ/m²	At −30°C; edgewise impact
Charpy notched impact strength	ISO 179/1eA	80 × 10 × 4	kJ/m²	At +23°C; edgewise impact
Charpy notched impact strength			kJ/m²	At −30°C; edgewise impact
Tensile impact strength	ISO 8256		kJ/m²	At +23°C; Double V 45° notches; $r = 1$ mm; record if fracture cannot be observed with notched Charpy test.

Thermal Properties

Property	Standard	Specimen	Units	Conditions
Melting temperature	ISO 3146	Molding compound	°C	Method C (DSC or DTA) at 10°C/min
Glass transition temperature	IEC 1006	Molding compound	°C	Method A (DSC or DTA) at 10°C/min
Temperature of deflection under load	ISO 75-1 and 75-2	110 × 10 × 4 or 80 × 10 × 4	°C °C	0.45 MPa } for less rigid materials 1.8 MPa } for both soft and rigid materials 8.0 MPa } for rigid materials only
Vicat softening temperature	ISO 306	10 × 10 × 4	°C	Heating rate 50 K/hr. Load 50 N
Coefficient of linear thermal expansion—parallel —normal	TMA	Prepared from ISO 3167	1/K 1/K	Mean secant value over the temperature range 23°C to 55°
Flammability	ISO 1210	125 × 13 × 1.6 Additional thickness 150 × 150 × 3 Additional thickness	Steps Steps Steps Steps	Indicate class from the following sequence: HB, V-2, V-1, V-0 Indicate 5VA, and 5VB
Limiting oxygen index	ISO 4589	80 × 10 × 4	%	Method A; top surface ignition

Electrical Properties

Property	Standard	Specimen	Units	Conditions
Relative permittivity	IEC 250	2-mm plate	—	Frequency 100 HZ and 1 MHz
Dissipation factor			—	
Volume resistivity	IEC 93		Ω × cm	Measurement with contact electrodes Voltage 100 V
Surface resisivity			Ω	
Dielectric strength	IEC 243-1		kV/mm	Use electrode configuration 25-mm/75-mm coaxial cylinders. Immersion in IEC 296 transformer oil. Short-time test
Comparative tracking index	IEC 112	≥15 × ≥15 × 4	—	Use solution A

Other Properties

Property	Standard	Specimen	Units	Conditions
Water uptake	ISO 62	2-mm plate	% %	Saturation value in water at 23°C Saturation value at 23°C/50%RH
Density	ISO 1183	≥10 × ≥10 × 4	kg/m³	

Reprinted with permission of Society of Plastics Industry Inc.

20.5. COMPUTER-AIDED MATERIAL PRESELECTION BY UNIFORM STANDARDS (CAMPUS)

First introduced in Europe in 1988, CAMPUS is fast becoming a worldwide standard in database software for plastics materials. CAMPUS promotes use of comparable data based on uniform global testing standards and differentiates its effort from the proliferating group of commercial material databases (8). The software, reportedly developed with the plight of the product engineer in mind, is now distributed throughout the world free of charge in five different languages by more than 40 leading material suppliers.

All suppliers to the CAMPUS database must adhere to ISO standard methods. Unlike the other commercial databases, which may obtain data from many different sources using different testing methods, CAMPUS assures the end user of uniform data acquisition and reporting. The PC-based Windows version program is extremely user friendly. To select any material, the user simply determines the material properties needed for the application and enters the required material property value or range of values necessary for application. The program searches the database and displays the acceptable material candidate (7). Figures 20-3 and 20-4 illustrate a typical page from CAMPUS.

The latest version of CAMPUS includes more sophisticated design information, notably "multi-point" data in the form of chart and graphs of shear modulus versus temperature, isochronous stress–strain curves, and secant modulus versus strain. These contrast with elementary "single-point" data on mechanical properties such as tensile strength, flex modules, or Izod impact strength. Multi-point design data is rarely present in standard data sheets (9). The user can also select and view properties for comparison, sort according to specific requirement, switch between SI and US units, choose and plot any pairs of properties for listed products, and overlay up to 10 curves of different products. The software allows the user to select a processing method, such as injection or extrusion, for a

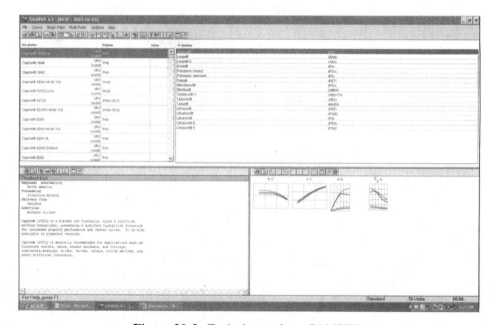

Figure 20-3. Typical page from CAMPUS.

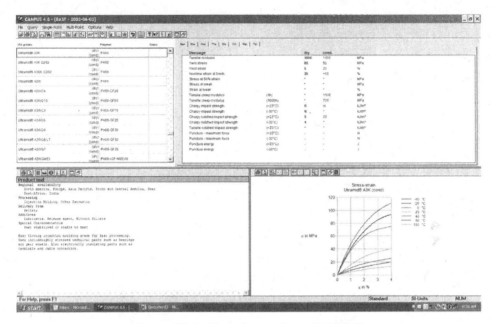

Figure 20-4. Typical page from CAMPUS.

given material and obtain a text display of processing information along with application examples and agency recognition.

The most useful feature of the CAMPUS software comes in the form of a separate merge program that allows the end user to superimpose stress–strain or viscosity curves and conduct side-by-side comparisons of the properties of materials manufactured by CAMPUS member suppliers (10).

20.6. PLASTICS MATERIALS DATABASES

In the last 10 years, the number of resin grades on the market has more than tripled. The shear number of available plastics, estimated at about 15,000, makes it difficult to determine the most appropriate material for a given application. The difficulty in selecting a material, moreover, is compounded by the fact that the test data may not sufficiently represent the behavior of plastics material because the test data provided in technical data sheets from the material supplier do not always reflect real-world conditions (11).

Plastics materials databases offer solutions to the problem of sorting through the maze of material data sheets with the help of a personal computer. However, except for CAMPUS, all the currently available materials databases do not provide material selection by uniform standards. All databases are programmed to allow the user to search for a material by properties, generic name, trade mark, applications, and manufacturer. The majority include a comparative format for a side-by-side comparison of material properties from the same or different suppliers. One of the versions also provides scanned manufacturer data sheets and additional material and marketing information.

Table 20-4 is a compilation of currently available plastics material databases.

TABLE 20-4. Plastic Materials Database Directory

Company	Product	Materials	Data Types	Web siite
CAMPUS	Computer Aided Material Preselection by Uniform Standards	Over 90% of Engineering Plastics	Properties, Comparable Data from Different Suppliers	www.campus.com
GE Plastics	Engineering Design Database	GE Materials	Properties, Ratings, Design Data, Chemical Resistance	www.geplastics.com
IDES	Material Property Database	Over 60,000 Plastic Material Datasheets	Properties	www.ides.com
MatWeb	Material Property Database	Metals, Plastics, Ceramics and Composites	Properties	www.matweb.com
Plastics Technology	Material Database	Plastic Materials	Properties Based on Search Criteria	www.ptonline.com
Bayer	Material and Design Database	Plastics and Elastomers	Properties, Design Data	www.bayermaterialscience.com
Dupont	Material and Design Database	Dupont Materials	Properties, Design Data	www.plastics.dupont.com

Name	Type	Description	Data	URL
Ticona	Material and Design Database	Ticona Materials	Properties, Design Data	www.ticona.com
M-Base Engineering	Material Database	Plastic Materials	Properties	www.m-base.de
Solvay	Material and Design Database	Solvay Materials	Properties, Design Data	www.solvayplastics.com
Teijin	Material and Design Database	Materials and Films	Properties, Design Data	www.teijinkasei.com
Plasticsusa.com	Material Database	Materials and General information	Properties, Trade Name Directory	www.plasticsusa.com
UL IQ for Plastics	Material Database	60,000 Grades of UL-Recognized Plastics	Properties, Rankings	www.ul.com/plastics
Plastics Design Library RAPRA	Material Database Bibliographic and Directory Database	Plastics, Rubber and Other Materials Rubber, Plastics, Adhesives, and Polymeric Composites	Properties	www.knovel.com www.rapra.net

REFERENCES

1. "Uniform Global Material Testing Standards—A Primer for Managers," SPI Polymeric Materials Producers Division, SPI, Washington, D.C., 1996, p. 1.
2. *Ibid.*, p. 2.
3. "Uniform Global Material Testing Standards—A Technical Primer," SPI, Washington, D.C., 1996, p. 3.
4. Wigotsky, V, "The Road to Standardization," *Plastics Eng.*, April 1995, p. 22.
5. Reference 1, p. 3.
6. Gabriele, M., "Global Standards Could Resolve Inconsistencies," *Plastics Technol.*, June 1993, p. 48.
7. Reference 1, p. 4.
8. Renstrom, R., "ISO draft new guidelines for plastics design," *Plastics News*, September 8, 1997, p. 5.
9. Leonard, L., "Comparable Data for Plastic Materials—Help is on the way," *Plastics Design Forum*, Jan.–Feb. 1993, p. 37.
10. Gabriele, Reference 6, p. 53.
11. Renstrom, Reference 8, p. 5.

APPENDIX A

INDEX OF TEST EQUIPMENT MANUFACTURERS

Abrasion

Taber Industries
Qualitest USA
Benz Material Testing Instruments
Zwick USA

Accelerated Weathering

Atlas Material Testing Technology,
   LLC
Q-Panel Company

Arc Resistance

Amprobe Instruments
CEAST USA Inc.

Bubble Viscometer

BYK-Gardner USA

Burst Strength Tester

Applied Test Systems, Inc.
Testing Machines, Inc.

Colorimeter/Spectrophotometer

BYK-Gardner USA
Datacolor
GretagMacbeth Corp.
Testing Machines, Inc.
X-Rite Inc.
Konica Minolta
HunterLab

Compressive Strength and Modulus

CEAST USA Inc.
Chatillon
Instron Corp.
MTS Systems Corp.
Testing Machines, Inc.
Tinius Olsen Testing Machine Co.
Zwick USA

Handbook of Plastics Testing and Failure Analysis, Third Edition, by Vishu Shah
Copyright © 2007 by John Wiley & Sons, Inc.

Creep Properties

Applied Test Systems, Inc.
Ceast USA Inc.
Instron Corporation
MTS Systems Corp.
Tinius Olsen Testing Machine
    Company

Cup Viscosity Test

BYK-Gardner USA
Fisher Scientific Co.

Density

CEAST USA Inc.
Qualitest USA
Ohaus Scale Corporation
Techne Inc.
Zwick USA

Dielectric Strength, Dielectric Constant

Biddle Sawyer Corp.
MTI Corp.
CEAST USA Inc.

Drop Impact Test

Applied Test Systems
BYK-Gardner USA
Tinius Olsen Testing Machine Co.
CEAST USA Inc.
Benz Material Testing Instruments

DSC

DuPont Co.
Haake
Mettler-Toledo Inc.
Perkin-Elmer Corp.
TA Instruments
Shimadzu Scientific instruments

Durometer/Rockwell Hardness

Instron Corp.
Mitutoyo
Rex Gauge Co.
Testing Machines, Inc.
Qualitest USA
Kernco Instruments Co.
Benz Material Testing
    Instruments
Zwick USA

Elongation

CEAST USA Inc.
Chatillon
Instron Corp.
MTS Systems Corp.
Testing Machines, Inc.
Tinius Olsen Testing Machines,
    Co.
Zwick USA

EMI/RFI Shielding

Enthone Inc.

Fatigue Failure

Fatigue Dynamics, Inc.
Instron Corp.
Tinius Olsen Testing Machine
    Co.
CEAST USA Inc.
Qualitest USA
Benz Material Testing
    Instruments

Flammability

CEAST USA Inc.
SGS U.S. Testing Co.
Testing Machines, Inc.
Fire Testing Technology, Ltd.

Flexural Modulus/Strength

Chatillon
Instron Corp.
MTS Systems Corp.
Testing Machines, Inc.
Tinius Olsen Testing Machine Co.
Zwick USA

Gel Point/Gel Time

CEAST USA Inc.
Shyodu Instrument Co.
Sunshine Instrument Co.
Techne, Inc.
Testing Machines, Inc.
BYK-Gardner USA

Glossmeter

BYK-Gardner USA
Datacolor
Konica Minolta

GPC

Waters Associates, Inc.
Viscotek

Hardness

Instron Corp.
Mitutoyo
Shore Instruments & Mfg. Co.
Qualitest USA
Benz Material Testing Instruments
Zwick USA

Haze

BYK-Gardner USA

HDT

CEAST USA Inc.
Testing Machines, Inc.
Tinius Olsen Testing Machines, Inc.

Impact Strength

BYK-Gardner USA
CEAST USA Inc.
Instron Corp.
Testing Machines, Inc.
Tinius Olsen Testing Machines, Inc.
Qualitest USA
Zwick USA

Inherent/Intrinsic Viscosity

Fisher Scientific Co.

IR Analysis

Infrared Engineering Inc.
Nicolet Instruments
The Foxboro Co.

Melt Index

CEAST USA Inc.
Goettfert
Instron Corp.
Kayeness, Inc.
Testing Machines, Inc.
Tinius Olsen Testing Machine, Inc.
Benz Material Testing Instruments
Zwick USA

Oxygen Index

CEAST USA Inc.
Testing Machines, Inc.
Fire Testing Technology Ltd.

Moisture Analyzer

Arizona Instrument Corp.
Newport Scientific Inc.
Omnimark Instrument Corp.
Zarad Inc.

NDT

GE Inspection Technologies, LP.
Panametrics, Corporation
NDC.
Lasermike.

Polarizer/Photoelasticity

Measurements Group Inc.
Strainoptic Technologies, Inc.

Refractometer

Reichert, Inc.
Kernco Instruments Co.

Rheometer

Brookfield Engineering
CEAST USA Inc.
Goettfert
Haake
Dynisco
TA Instruments
Tinius Olsen Testing Machine Co.
Benz Material Testing Instruments

Smoke Generations

Newport Scientific.
Fire Testing Technology Limited

Specific Gravity

Ohaus Scale Corp.
Techne, Inc.
Testing Machines, Inc.
Qualitest USA
Benz Material Testing Instruments

Stress Relaxation

CEAST USA Inc.
Instron Corp.

Stress/Strain

CEAST USA Inc.
Instron Corp.
MTS System
Tinius Olsen Testing Machine Co.
Zwick USA

Tensile Impact Strength

Instron Corp.
Testing Machines, Inc.
Tinius Olsen Testing Machine Co.

Tensile Strength

CEAST USA Inc.
Instron Corp.
MTS Systems Corp.
Testing Machines, Inc.
Tinius Olsen Testing Machine Co.
Benz Material Testing Instruments
Zwick USA

Test Specimen

Lansen Mold Co., Inc.
Benz Material Testing Instruments
Master Precision Products, Inc.

Thermal Conductivity

Anter Corporation

Torque Tester

Mountz, Inc.

TGA

DuPont Co.
Haake
Mettler-Toledo Inc.
Perkin-Elmer Corp.
TA Instruments
Shimadzu Scientific instruments

TMA

DuPont Co.

Haake

Mettler-Toledo Inc.

Perkin-Elmer Corp.

TA Instruments

Shimadzu Scientific instruments

Vicat Softening Point

Atlas Material Testing Technology

Testing Machines, Inc.

Tinius Olsen Testing Machine Co.

CEAST USA Inc.

Viscosity

Brabender C.W. Instruments, Inc.

Brookfield Engineering Laboratories, Inc.

BYK-Gardner USA

Fisher Scientific

Haake, Inc.

Testing Machines, Inc.

Benz Material Testing Instruments

Weathering

Atlas Material Testing Technology

Q-Panel Co.

ALPHABETICAL INDEX OF COMPANIES ADDRESSES, PHONE NUMBERS, AND WEBSITES

Anter Corporation

1700 Universal Road

Pittsburg, PA 15235

(415) 795-6410

www.anter.com

Atlas Material Testing Technology

4114 N. Ravenswood

Chicago, IL 60613

(773) 327-4520

www.altas-mts.com

Applied Test Systems, Inc.

348 Newcastle Road,

Butler, PA 16001

(724) 283-1212

www.atspa.com

Arizona Instrument

1912 West 4th Street

Tempe, AZ 85040-1941

(800) 290-1414

www.azic.com

Linberg/Blue M

308 Ridgefield court

Asheville, NC 28806

(828) 658-2711

www.lindberg-bluem.com

Brabender, C.W. Instruments, Inc.

50 E. Wesley Street

S. Hackensack, NJ 07606

(201) 343-8425

www.cwbrabender.com

Benz Materials Testing Instruments

73 Maplehurst Avenue

Providence, RI 02908

(401) 331-5650

www.benztesters.com

Brookfield Engineering Laboratories, Inc.

11 Commence Boulevard

Middleboro, MA 02346

(508) 946-6200 or (800) 628-8139

www.brookfieldengineering.com

BYK-Gardner USA
Rivers Park II
9104 Guilford Road
Columbia, MD 21046-2729
(800) 343-7721 or (301) 483-6500
www.byk-gardner.com

Biddle Sawyer Corp.
21 Penn Plaza
360 West 31st Street
New York, NY 10001
(212) 736-1580
www.biddlesawyer.com

Ceast USA Inc.
4816 Sirus Lane
Charlotte NC 28208
(704) 423-0042
www.ceast.com

Fisher Scientific
2000 Park Lane
600 Business Center Drive
Pittsburgh, PA 15205
(412) 490-7286 or (800) 926-0505
www.l.fisherci.com

Fatigue Dynamics, Inc.
965 Decker Road
Walled Lake, MI 48390
(248) 669-6100 or (800) 394-7568
www.fdinc.com

Thermo Haake USA
5225 Verona Road
Madison, WI 53711
(608) 327-6777
www.thermohaake.com

Instron Corp.
100 Royal Street
Canton, MA 02021
(617) 828-2500 or (800) 564-8374
www.instron.com

Dynisco Polymer Test
Westgate II, 730 Hemlock Road
Morgantown, PA 19543
(508) 541-9400
www.dynisco.com

MTS Systems Corp.
1400 Technology Drive
Eden Prairie, MI 55344
(800) 944-1687
www.mts.com

Ohaus Corp.
19A Chapin Road
P.O. Box 2033
Pine Brook, NJ 07058
(973) 377-9000
www.ohaus.com

Omnimark Instrument Corporation
1711 West University Drive
Suite 159
Tempe, AZ 85281
(602) 784-2200
www.omniwww.com

PerkinElmer Instruments
710 Bridgeport Avenue
Shelton, CT 06484
(800) 762-4000 or (203) 925-4600
www.perkinelmer.com

Measurements Group Inc
P.O. Box 27777
Raleigh, NC 27611
(919) 365-3800
www.measurementsgroup.com

Q-Panel Lab Products
800 Canterbury Road
Cleveland, OH 44145
(440) 835-8700
www.q-panel.com

Rex Gauge Co.
1250 Busch Parkway
Buffalo Grove, IL 60089
(800) 927-3982 or (847) 465-9009
www.durometer.com

Shyodu Instrument Co.
6351 Old Tipton Road
Millington, TN 38053
(901) 872-6894
www.shyodu.com

Strainoptics Technologies, Inc.
108 West Montgomery Avenue
North Wales, PA 19454
(215) 661-0100
www.strainoptics.com

Sunshine Scientific Instruments, Inc.
2200 Michener Street, Suite 23
Philadelphia, PA 19115
(215) 673-5600 or (800) 343-1199
www.measurebetter.com

Techne, Inc.
3 Terri Lane, Suite 10
Burlington, NJ 08016
(800) 225-9243
www.techneusa.com

Testing Machines, Inc.
2 Fleetwood Court
Ronkonkoma, NY 11779
(631) 439-5400 or (800) 678-3221
www.testingmachines.com

Tinius Olsen Testing Machine Co., Inc.
1065 Easton Road
P.O. Box 1009
Horsham, PA 19044
(215) 675-7100
www.tiniusolsen.com

Mountz, Inc.
1080 North, 11th Street
San Jose, CA 95112
(408) 292-2214
www.etorque.com

SGS-U.S. Testing Corporation
291 Fairfield Avenue
Fairfield, NJ 07004
(973) 575-5252 or (800) 777-8378
www.ustestingsgsna.com

Newport Scientific Inc.
8246-E Sandy Court
Jessup, MD 20794-9632
(301) 498-6700
www.newport.scientific.com

Thermo Electron Corporation
Process Instruments Division
501 90th Avenue N.w.
Minneapolis, MN 55433
(763) 783-2500
www.thermo.com

Goettfert Inc.
488 Lakeshore Parkway
Rock Hill, SC 29730
(803) 324-3883
www.goettfert.com

Taber Industries
P.O. Box 164
455 Bryant Street
North Tonawanda, NY 14120
(716) 694-4000
www.taberindustries.com

GE Inspection Technologies, LP
50 Industrial Park Road
Lewiston, PA 17044
(866) 243-2638 or (717) 242-0327
www.geinspectiontechnologies@ae.ge.com

Datacolor
5 Princess Road
Lawrenceville, NJ 08648
(609) 924-2189
www.datacolor.com

X-Rite
3100 44th Street S.W.
Grandville, MI 49418
(616) 534-7667 or (800) 248-9748
www.x-rite.com

Amprobe Advanced Tests Products
3270 Executive Way
Miramar, FL 33025
(800) 327-5060
www.amprobe.com

Panametrics Inc.
48 Woerd Avenue
Waltham, MA 02453
(800) 225-8330 or (781) 899-3900
www.panametrics-ndt.com

Enthone Inc.
P.O. Box 1900
350 Frontage Road
West Haven, CT 06516
(203) 934-8611
www.enthone.com

Mattler-Toledo Inc.
1900 Polaris Parkway
Columbus, OH 43240
(614) 438-4511
www.mt.com

TA Instruments Inc.
109 Lukens Drive
New Castle, DE 19720
(302) 427-4000
www.tainst.com

Fire Testing Technology, LTD.
Charlwoods Road, East Grinstead
West Sussex, RH19 2HL, UK
44 (0) 1342-323600
www.fire-testing.com

Qualitest USA
Cross Roads One Center
8201 Peters Road, Suite 1000
Plantation, FL 33324
(877) 884-8378
www.qualitest-inc.com

Zared
380 Country Oak Lane
Inverness, Illinois, 60067
(800) 633-7507
www.zared.com

NDC Infrared Engineering
5314 North Irwindale Avenue
Irwindale, California 91706
(626) 960-3300
www.ndcinfrared.com

Konica Minolta
Instrument Systems Division
725 Darlington Avenue
Mahwah, NJ 07430
(888) 473-2636
www.konicaminolta.us

Lansen Mold Co., Inc.
Main Street
Berkshire, MA 01224
(413) 443-5328
www.lansenmold.com

Master Precision Products, Inc.
1212 Fairplains Street
Greenville, MI 48838
(800) 354-3170
www.masterprecision.com

Zwick USA
1620 Cobb International Blvd., Suite 1
Kennesaw, GA 30152
Phone: 770-420-6555
www.zwick.com

Viscotek
15600 West Hardy Road
Houston, TX 77060
(800)-375-5966
www.viscotek.com

APPENDIX B

ABBREVIATIONS: POLYMERIC MATERIALS

ABS	Acrylonitrile–butadiene–styrene
AN	Acrylonitrile
CA	Cellulose acetate
CAB	Cellulose acetate butyrate
CAP	Cellulose acetate propionate
CN	Cellulose nitrate
CP	Cellulose propionate
CPE	Chlorinated polyethylene
CPVC	Chlorinated polyvinyl chloride
CTFE	Chlorotrifluoroethylene
DAP	Diallyl phthalate
EC	Ethyl cellulose
ECTFE	Poly(ethylene–chlorotrifluoroethylene)
EP	Epoxy
EPDM	Ethylene–propylene–diene monomer
EPR	Ethylene propylene rubber
EPS	Expanded polystyrene
ETFE	Ethylene/tetrafluoroethylene copolymer
EVA	Ethylene–vinyl acetate
FEP	Perfluoro(ethylene–propylene) copolymer
FRP	Fiberglass-reinforced polyester
HDPE	High-density polyethylene
HIPS	High-impact polystyrene
HMWPE	High-molecular-weight polyethylene
LDPE	Low-density polyethylene

Handbook of Plastics Testing and Failure Analysis, Third Edition, by Vishu Shah
Copyright © 2007 by John Wiley & Sons, Inc.

MF	Melamine–formaldehyde
PA	Polyamide
PAPI	Polymethylene polyphenyl isocyanate
PB	Polybutylene
PBT	Polybutylene terephthalate (thermoplastic polyester)
PC	Polycarbonate
PE	Polyethylene
PES	Polyether sulfone
PET	Polyethylene terephthalate
PF	Phenol–formaldehyde
PFA	Polyfluoro alkoxy
PI	Polyimide
PMMA	Polymethyl methacrylate
PP	Polypropylene
PPO	Polyphenylene oxide
PS	Polystyrene
PSO	Polysulfone
PTFE	Polytetrafluoroethylene
PTMT	Polytetramethylene terephthalate (thermoplastic polyester)
PU	Polyurethane
PVA	Polyvinyl alcohol
PVAC	Polyvinyl acetate
PVC	Polyvinyl chloride
PVDC	Polyvinylidene chloride
PVDF	Polyvinylidene fluoride
PVF	Polyvinyl fluoride
TFE	Polytetrafluoroethylene
SAN	Styrene–acrylonitrile
SI	Silicone
TPE	Thermoplastic elastomers
TPX	Polymethylpentene
UF	Urea formaldehyde
UHMWPE	Ultrahigh-molecular-weight polyethylene
UPVC	Unplasticized polyvinyl chloride

APPENDIX C

GLOSSARY

A

Abrasion resistance: The ability of a material to withstand mechanical action such as rubbing, scraping, or erosion that tends to progressively remove material from its surface.

Accelerated aging: A test procedure in which conditions are intensified in order to reduce the time required to obtain a deteriorating effect similar to one resulting from normal service conditions.

Accelerated weathering: A test procedure in which the normal weathering conditions are accelerated by means of a device (machine).

Aging: The process of exposing plastics to natural or artificial environmental conditions for a prolonged period of time.

Amorphous polymers: Polymeric materials that have no definite order or crystallinity. The polymer molecules are arranged in completely random fashion. Examples of amorphous plastics are polystyrene, PVC, and PMMA.

Apparent density (Bulk density): The weight of the unit volume of material including voids (air) inherent in the material as tested.

Arc resistance: The ability of a plastic material to resist the action of a high-voltage electrical arc, usually stated in terms of time required to render the material electrically conductive.

B

Birefringence (double refraction): Birefringence is the difference between index of refraction of light in two directions of vibration.

Handbook of Plastics Testing and Failure Analysis, Third Edition, by Vishu Shah
Copyright © 2007 by John Wiley & Sons, Inc.

Brittle failure: The failure resulting from the inability of a material to absorb energy, resulting in instant fracture upon mechanical loading.

Brittleness temperature: The temperature at which plastics and elastomers exhibit brittle failure under impact conditions.

Brookfield viscometer: The Brookfield viscometer is the most widely used instrument for measuring the viscosity of plastisols and other liquids of a thixotropic nature. The instrument measures shearing stress on a spindle rotating at a definite, constant speed, while immersed in the sample. The degree of spindle lag is indicated on a rotating dial. This reading, multiplied by a conversion factor based on spindle size and rotational speed, gives a value for viscosity in centipoises. By taking measurements at different rotational speeds, an indication of the degree of thixotropy of the sample is obtained.

Bubble viscometer: In a bubble viscometer, a transparent liquid streams downward in the ring-shaped zone between the glass wall of a sealed tube and a rising air bubble. The rate at which the air bubbles rise, under controlled conditions and within certain limits, is a direct measure of kinematic viscosity of streaming liquids.

Bulk density: *See* Apparent density.

Bulk factor: The ratio of the volume of any given quantity of the loose plastic material to the volume of the same quantity of the material after molding or forming. Bulk factor is a measure of volume change that may be expected in fabrication.

Burst strength: The internal pressure required to break a pressure vessel such as a pipe or fitting. The pressure (and therefore the burst strength) varies with the rate of pressure build-up and the time during which the pressure is held.

C

Capillary rheometer: An instrument for measuring the flow properties of polymer melts. It is comprised of a capillary tube of specified diameter and length, means for applying desired pressures to force the molten polymer through the capillary, means for maintaining the desired temperature of the apparatus, and means for measuring differential pressures and flow rates. The data obtained from capillary rheometers is usually presented as graphs of shear stress against shear rate at constant temperature.

Cellular plastics: *See* Foamed plastics.

Chalking: A whitish, powdery residue on the surface of a material caused by material degradation (usually form weather).

Charpy impact test: A destructive test of impact resistance, consisting of placing the specimen in a horizontal position between two supports, then striking the specimen with a pendulum striker swung from a fixed height. The magnitude of the blow is increased until the specimen breaks. The result is expressed in in.-lb or ft-lb of energy.

Chroma (saturation): The attribute of color perception that expresses the degree of departure from gray of the same lightness.

CIE (Commission Internationale De L'Eclairage): The international a commission on illuminants responsible for establishing standard illuminants.

Coefficient of thermal expansion: The fractional change in length or volume of a material for unit change in temperature.

Colorimeter: An instrument for matching colors with results approximately the same as those of visual inspection, but more consistently.

Compressive strength: The maximum load sustained by a test specimen in a compressive test divided by the original cross section area of the specimen.

Conditioning: Subjecting a material (or test specimens) to standard environmental and/or stress history prior to testing.

Continuous use temperature: The maximum temperature at which material may be subjected to continuous use without fear of premature thermal degradation.

Crazing: An undesirable defect in plastic articles, characterized by distinct surface cracks or minute frostlike internal cracks, resulting from stresses within the article. Such stresses may result from molding shrinkage, machining, flexing, impact shocks, temperature changes, or action of solvents.

Creep: Because of its viscoelastic nature, a plastic subjected to a load for a period of time tends to deform more than it would from the same load released immediately after application, and the degree of this deformation is dependent on the load duration. Creep is the permanent deformation resulting from prolonged application of stress below the elastic limit. Creep at room temperature is sometimes called cold flow.

Creep modulus (apparent modulus): The ratio of initial applied stress to creep strain.

Creep rupture strength: Stress required to cause fracture in a creep test within a specified time.

Crosslinking: Applied to polymer molecules, the setting up of chemical links between the molecular chains. When extensive, as in most thermosetting resins, crosslinking makes one infusible super-molecule of all the chains. Crosslinking can be achieved by irradiation with high-energy electron beams or by chemical crosslinking agents such as organic peroxides.

Crystallinity: A state of molecular structure in some resins attributed to the existence of solid crystals with a definite geometric form. Such structures are characterized by uniformity and compactness.

Cup flow test: Test for measuring the flow properties of thermosetting materials. A standard mold is charged with preweighed material, and the mold is closed using sufficient pressure to form a required cup. Minimum pressures required to mold a standard cup and the time required to close the mold fully are determined.

Cup viscosity test: A test for making flow comparisons under strictly comparable conditions. The cup viscosity test employs a cup-shaped gravity device that permits the timed flow of a known volume of liquid passing through an orifice located at the bottom of the cup.

D

Density: Weight per unit volume of a material expressed in grams per cubic centimeter, pounds per cubic foot and so on.

Dielectric constant (permittivity): The ratio of the capacitance of a given configuration of electrodes with a material as the dielectric to the capacitance of the same electrode configuration with a vacuum (or air for most practical purposes) as the dielectric.

Dielectric strength: The electric voltage gradient at which an insulating material is broken down or "arced through" in volts per millimeter of thickness.

Differential scanning calorimetry (DSC): DSC is a thermal analysis technique that measures the quantity of energy absorbed or evolved (given off) by a specimen in calories as its temperature is changed.

Dimensional stability: Ability of a plastic part to retain the precise shape in which it was molded, fabricated, or cast.

Dissipation factor: The ratio of the conductance of a capacitor in which the material is dielectric to its susceptance, or the ratio of its parallel reactance to its parallel resistance. Most plastics have a low dissipation factor, a desirable property because it minimizes the waste of electrical energy as heat.

Drop impact test: Impact resistance test in which the predetermined weight is allowed to fall freely onto the specimen from varying heights. The energy absorbed by the specimen is measured and expressed in in.-lb or ft-lb.

Ductility: Extent to which a material can sustain plastic deformation without fracturing.

Durometer hardness: Measure of the indentation hardness of plastics. It is the extent to which a spring-loaded steel indenter protrudes beyond the pressure foot into the material.

E

Elongation: The increase in length of a test specimen produced by a tensile load. Higher elongation indicates higher ductility.

Embrittlement: Reduction in ductility due to physical or chemical changes.

Environmental stress cracking: The susceptibility of a thermoplastic article to crack or craze formation under the influence of certain chemicals and stress.

Extensometer: Instrument for measuring changes in linear dimensions; also called a strain gauge.

Extrusion plastometer (rheometer): A type of viscometer used for determining the melt index of a polymer. It consists of a vertical cylinder with two longitudinal bored holes (one for measuring temperature and one for containing the specimen, the latter having an orifice of stipulated diameter at the bottom and a plunger entering from the top). The cylinder is heated by external bands and weight is placed on the plunger to force the polymer specimen through the orifice. The result is reported in grams per 10 minutes.

F

Fadometer: An apparatus for determining the resistance of materials to fading by exposing them to ultraviolet rays of approximately the same wavelength as those found in sunlight.

Failure analysis: The science of analyzing failures (product) employing a step-by-step method of elimination.

Falling weight impact tester: *See* Drop impact test.

Fatigue failure: The failure or rupture of a plastic article under repeated cyclic stress, at a point below the normal static breaking strength.

Fatigue limit: The stress below which a material can be stressed cyclically for an infinite number of times without failure.

Fatigue strength: The maximum cyclic stress a material can withstand for a given number of cycles before failure.

Flammability: Measure of the extent to which a material will support combustion.

Flexural modulus: Within the elastic limit, the ratio of the applied stress on a test specimen in flexure to the corresponding strain in the outermost fiber of the specimen. Flexural modulus is the measure of relative stiffness.

Flexural strength: The maximum stress in the outer fiber at the moment of crack or break.

Foamed plastics (cellular plastics): Plastic with numerous cells dispersed throughout its mass. Cells are formed by a blowing agent or by the reaction of the constituents.

G

Gel permeation chromatography (GPC): A newly developed column chromatography technique employing a series of columns containing closely packed rigid gel particles. The polymer to be analyzed is introduced at the top of the column and then is eluted with a solvent. The polymer molecules diffuse through the gel at rates that depend on their molecular size. As they emerge from the columns, they are detected by a differential refractometer coupled to a chart recorder, on which a molecular weight distribution curve is plotted.

Gel point: The stage at which liquid begins to gel, that is, exhibits pseudoelastic properties.

Gel time: Gel time is the interval of time between introduction of the catalyst and the formation of a gel.

Glossmeter: An instrument for measuring specular gloss at various angles.

H

Hardness: The resistance of plastic materials to compression and indentation. Brinnel hardness and shore hardness are major methods of testing this property.

Haze: The cloudy or turbid aspect of appearance of an otherwise transparent specimen caused by light scattered from within the specimen or from its surfaces.

Heat deflection temperature (HDT): Temperature at which a standard test bar deflects 0.010 in. under a stated load of either 66 or 264 psi.

Hooke's law: Stress is directly proportional to strain.

Hoop stress: The circumferential stress in a material of cylindrical form subjected to internal or external pressure.

Hue: The attribute of color perception by means of which an object is judged to be red, yellow, green, blue, purple, or intermediate between some adjacent pair of these.

Hygroscopic: Materials having the tendency to absorb moisture from air. Plastics, such as nylons and ABS, are hygroscopic and must be dried prior to molding.

Hysteresis: The cyclic noncoincidence of the elastic loading and the unloading curves under cyclic stressing. The area of the resulting elliptical hysteresis loop is equal to the heat generated in the system.

I

Impact strength: Energy required to fracture a specimen subjected to shock loading.

Impact test: A method of determining the behavior of material subjected to shock loading in bending or tension. The quantity usually measured is the energy absorbed in fracturing the specimen in a single blow.

Indentation hardness: Resistance of a material to surface penetration by an indentor. The hardness of a material as determined by the size of an indentation made by an indenting tool under a fixed load, or the load necessary to produce penetration of the indentor to a predetermined depth.

Index of refraction: The ratio of velocity of light in vacuum (or air) to its velocity in a transparent medium.

Infrared analysis: A technique frequently used for polymer identification. An infrared spectrometer directs infrared radiation through a film or layer of specimen and measures and records the relative amount of energy absorbed by the specimen as a function of wavelength or frequency of infrared radiation. The chart produced is compared with correlation charts for known substances to identify the specimen.

Inherent viscosity: In dilute solution viscosity measurements, inherent viscosity is the ratio of the natural logarithm of the relative viscosity to the concentration of the polymer in grams per 100 mL of solvent.

Intrinsic viscosity: In dilute solution viscosity measurements, intrinsic viscosity is the limit of the reduced and inherent viscosities as the concentration of the polymeric solute approaches zero and represents the capacity of the polymer to increase viscosity.

ISO: Abbreviation for the International Standards Organization.

Isochronous (equal time) stress–strain curve: A stress-strain curve obtained by plotting the stress versus corresponding strain at a specific time of loading pertinent to a particular application.

Izod impact test: A method for determining the behavior of materials subjected to shock loading. Specimen supported as a cantilever beam is struck by a weight at the end of a pendulum. Impact strength is determined from the amount of energy required to fracture the specimen. The specimen may be notched or unnotched.

K

K factor: A term sometimes used for thermal insulation value or coefficient of thermal conductivity. *See also* Thermal conductivity.

L

Luminous transmittance (light transmittance): The ratio of transmitted light to incident light. The value is generally reported in percentage of light transmitted.

M

Melt index test: Melt index test measures the rate of extrusion of a thermoplastic material through an orifice of specific length and diameter under prescribed conditions of temperature and pressure. Melt index value is reported in grams per 10 minutes for specific condition.

Metamerism: Metamerism is a phenomenon of change in the quality of color match of any pair of colors as illumination or observer or both are changed.

Modulus of elasticity (elastic modulus, Young's modulus): The ratio of stress to corresponding strain below the elastic limit of a material.

Molecular weight: The sum of the atomic weights of all atoms in a molecule. In high polymers, the molecular weight of individual molecules varies widely, therefore, they are expressed as weight average or number average molecular weight.

Molecular weight distribution: The relative amount of polymers of different molecular weights that comprise a given specimen of a polymer.

Monochromatic light source: A light source capable of producing light of only one wavelength.

Monomer: (*mono-mer; single-unit*) A monomer is a relatively simple compound that can react to form a polymer (multiunit) by combination with itself or with other similar molecules or compounds.

N

Necking: The localized reduction in cross section that may occur in a material under stress. Necking usually occurs in a test bar during a tensile test.

Newtonian behavior: A flow characteristic evidenced by viscosity that is independent of shear rate, that is, the shear stress is directly proportional to shear rare. Water and mineral oil are typical Newtonian liquids.

Non-Newtonian behavior: The behavior of liquid that does not satisfy the requirement for a Newtonian liquid as defined. The flow of molten polymers is generally non-Newtonian, producing lower viscosities at higher rates of shear.

Notch sensitivity: Measure of reduction in load-carrying ability caused by stress concentration in a specimen. Brittle plastics are more notch sensitive than ductile plastics.

O

Orientation: The alignment of the crystalline structure in polymeric materials so as to produce a highly uniform structure.

Oxygen index: The minimum concentration of oxygen expressed as a volume percent, in a mixture of oxygen and nitrogen that will just support flaming combustion of a material initially at room temperature under the specified conditions.

P

Peak exothermic temperature: The maximum temperature reached by reacting thermosetting plastic composition is called peak exothermic temperature.

Photoelasticity: An experimental technique for the measurement of stresses and strains in material objects by means of the phenomenon of mechanical birefringence.

Poisson's ratio: Ratio of lateral strain to axial strain in an axial loaded specimen. It is a constant that relates the modulus of rigidity to Young's modulus.

Polarized light: Polarized electromagnetic radiation whose frequency is in the optical region.

Polarizer: A medium or a device used to polarize the incoherent light.

Polymer (*poly-many; mer-unit*): A polymer is a high-molecular-weight organic compound whose structure can be represented by a repeated monomeric unit.

Polymerization: A chemical reaction in which the molecules of monomers are linked together to form polymers.

Proportional limit: The greatest stress that a material is capable of sustaining without deviation from proportionality of stress and strain (Hooke's law).

R

Refractive index: *See* Index of refraction.

Relative humidity: The ratio of the quantity of water vapor present in the atmosphere to the quantity that would saturate it at the existing temperature. It is also the ratio of the pressure of water vapor present to the pressure of saturated water vapor at the same temperature.

Relative viscosity: Ratio of kinematic viscosity of a specified solution of the polymer to the kinematic viscosity of the pure solvent.

Rheology: The science dealing with the study of material flow.

Rheometer: *See* Extrusion plastometer (rheometer).

Rockwell hardness: Index of indentation hardness measured by a steel ball indenter. *See also* Indentation hardness.

S

Secant modulus: The ratio of total stress to corresponding strain at any specific point on the stress–strain curve.

Shear strength: The maximum load required to shear a specimen in such a manner that the resulting pieces are completely clear of each other.

Shear stress: The stress developing in a polymer melt when the layers in a cross section are gliding along each other or along the wall of the channel (in laminar flow).

Shear rate: The overall velocity over the cross section of a channel with which molten or fluid layers are gliding along each other or along the wall in laminar flow.

Shore hardness: *See* Indentation hardness.

S–N diagram: Plot of stress (S) against the number of cycles (N) required to cause failure of similar specimens in a fatigue test.

SPE: Abbreviation for Society of Plastics Engineers.

Specific gravity: The ratio of the weight of the given volume of a material to that of an equal volume of water at a stated temperature.

Specimen: A piece or a portion of a sample used to conduct a test.

Spectrophotometer: An instrument that measures transmission or apparent reflectance of visible light as a function of wavelength, permitting accurate analysis of color or accurate comparison of luminous intensities of two sources of specific wavelengths.

Specular gloss: The relative luminous reflectance factor of a specimen at the specular direction.

SPI: Abbreviation for Society of Plastics Industry.

Spiral flow test: A method for determining the flow properties of a plastic material based on the distance it will flow under controlled conditions of pressure and temperature along the path of a spiral cavity using a controlled charge mass.

Strain: The change in length per unit of original length, usually expressed in percent.

Stress: The ratio of applied load to the original cross-sectional area expressed in pounds per square inch.

Stress concentration: The magnification of the level of applied stress in the region of a notch, crack, void, inclusion, or other stress risers.

Stress optical sensitivity: The ability of some materials to exhibit double refraction of light when placed under stress is referred to as stress-optical sensitivity.

Stress relaxation: The gradual decrease in stress with time under a constant deformation (strain).

Stress–strain diagram: Graph of stress as a function of strain. It is constructed from the data obtained in any mechanical test where a load is applied to a material and continuous measurements of stress and strain are made simultaneously.

Surging: Pressure rise in a pipeline caused by a sudden change in the rate of flow or stoppage of flow in the line. These changes of pressure cause elastic deformation of the pipe walls and changes in the density of fluid column.

T

Tensile impact energy: The energy required to break a plastic specimen in tension by a single swing of a calibrated pendulum.

Tensile strength: Ultimate strength of a material subjected to tensile loading.

Thermal conductivity: The ability of a material to conduct heat. The coefficient of thermal conductivity is expressed as the quantity of heat that passes through a unit cube of the substance in a given unit of time when the difference in temperature of the two faces is one degree.

Thermogravimetric analysis (TGA): A testing procedure in which changes in the weight of a specimen are recorded as the specimen is progressively heated.

Thermomechanical analysis (TMA): A thermal analysis technique consisting of measuring physical expansion or contraction of a material or changes in its modulus or viscosity as a function of temperature.

Thermoplastic: A class of plastic material that is capable of being repeatedly softened by heating and hardened by cooling. ABS, PVC, polystyrene, and polyethylene are thermoplastic materials.

Thermosetting plastics: A class of plastic materials that will undergo a chemical reaction by the action of heat, pressure, catalysts, and so on, leading to a relatively infusible, nonreversible state. Phenolics, epoxies, and alkyds are examples of typical thermosetting plastics.

Torsion: Stress caused by twisting a material.

Torsion pendulum: An equipment used for determining dynamic mechanical properties of plastics.

Toughness: The extent to which a material absorbs energy without fracture. The area under a stress–strain diagram is also a measure of toughness of a material.

Tristimulus colorimeter: The instrument for color measurement based on spectral tristimulus values. Such an instrument measures color in terms of three primary colors: red, green, and blue.

U

Ultrasonic testing: A nondestructive testing technique for detecting flaws in material and measuring thickness based on the use of ultrasonic frequencies.

Ultraviolet: The region of the electromagnetic spectrum between the violet end of visible light and the x-ray region, including wavelengths from 100 to 3900 Å. Photon of radiations in the UV area have sufficient energy to initiate some chemical reactions and to degrade some plastics.

V

Vicat softening point: The temperature at which a flat-ended needle of 1-mm^2 circular or square cross section will penetrate a thermoplastic specimen to a depth of 1 mm under a specified load using a uniform rate of temperature rise.

Viscometer: An instrument used for measuring the viscosity and flow properties of fluids.

Viscosity: A measure of resistance of flow due to internal friction when one layer of fluid is caused to move in relationship to another layer.

W

Water absorption: The amount of water absorbed by a plastic article when immersed in water for a stipulated period of time.

Weathering: A broad term encompassing exposure of plastics to solar or ultraviolet light, temperature, oxygen, humidity, snow, wind, pollution, and so on.

Weatherometer: An instrument used for studying the effect of weather on plastics in accelerated manner using artificial light sources and simulated weather conditions.

Y

Yellowness index: A measure of the tendency of plastics to turn yellow upon long-term exposure to light.

Yield point: Stress at which strain increases without accompanying increase in stress.

Yield strength: The stress at which a material exhibits a specified limiting deviation from the proportionality of stress to strain. Unless otherwise specified, this stress will be the stress at the yield point.

Young's modulus: The ratio of tensile stress to tensile strain below the proportional limit.

Z

Zahn viscosity cup: A small U-shaped cup suspended from a looped wire, with an orifice of any one of five sizes at the base. The entire cup is submerged in test sample and then withdrawn. The time in seconds from the moment the top of the cup emerges from the sample until the stream from the orifice first breaks is the measure of viscosity.

APPENDIX D

TRADE NAMES*

Trade Name	Product	Company
Acrylite	Acrylic resin and sheet	Cyro
Affinity	Polyolefin plastomers	Dow
Aim	Advanced styrenic resin	Dow
Alathon	HDPE	Occidental
Alcryn	Halogenated polyolefin	DuPont
Amodel	Polyphthalamide	Amoco
APEC	High heat PC	Bayer
Aristech	Acrylic resin	Aristech
Aurum	Polyimide resin	Mitsui Toatsu
Avimid K	Polyimide	DuPont
Barex	Acrylonitrile copolymer	BP Chemical
Bayblend	ABS/Polycarbonate	Bayer
Cadon	SMA and SMA/ABS 2	Bayer
Calibre	PC	Dow
Capron	Nylon 6	Allied Signal
Celanex	PBT, PBT/PET alloy	Ticona
Celcon	Acetal and acetal elastomer	Ticona
Centrex	AES/ASA	Bayer
Crystalor	Polymethylpentene	Phillips
Cycolac	ABS	G.E.
Cycoloy	ABS/PC alloy	G.E.
Cyrolite	Acrylic Resin	Cyro

*A complete all inclusive trade name directory can also be viewed on the web site www.plastics.com.

Handbook of Plastics Testing and Failure Analysis, Third Edition, by Vishu Shah
Copyright © 2007 by John Wiley & Sons, Inc.

Trade Name	Product	Company
Delrin	Acetal, acetal/elastomer alloy	DuPont
Denka	ABS	Showa (Japan)
Desmopan	TPU	Bayer
Dowlex	LDPE/LLDPE	Dow
Durethan	Nylon 6	Bayer
Dylark	SMA and SMA/PS alloy	Arco
Dylene	PS	Arco
Ektar	Polyester-PETG	Eastman
Engage	Polyolefin elastomers	Dow Elastomers
Estamid	Polyamide (CTPE)	Dow
Estane	Urethane (TPE)	B.F. Goodrich
Eval	EVOH	Eval
Fiberloc	PVC glass reinforced	Polyone
Flexomer	LLDP/EVA	Union Carbide
Fluon	TFE	ICI
Foraflon	PVDF	Atochem
Forar	HDPE	Amoco
Fortiflex	HDPE, MDPE	Solvay
Fortilene	Polypropylene	Solvay
Fortron	Polyphenylene sulfide	Ticona
Geloy	ASA and ASA/PVC alloys	G.E.
Geolast	PP/Nitrile (TPE)	Advanced Elastomer Systems
Geon	PVC	Polyone
Grilamid	Nylon 12	EMS American Grilon
GTX	PPE/Nylon	G.E.
Halar	ECTFE	Ausimont
Hostalen GUR	UHMWPE	Ticona
Hytrel	Polyetherester (TPE)	DuPont
Isoplast	Rigid PU	Dow
K-Resin	TPE	Phillips Chemical
Kadel	Polyketone	Amoco
Kamax	Acrylic	Rohn & Haas
Kel-F	CTFE	3M
Kevlar	Aramid fiber	DuPont
Kraton	TPE	KRATON Polymers
Kynar	PVDF	Atochem
Lexan	PC	G.E.
Lucite	Acrylic	DuPont
Lupiace	PPE	Mitsubishi
Lupital	Acetal resin	Mitsubishi
Luran	SAN	BASF
Lustran	ABS, SAN copolymer	Bayer
Lustrex	PS	Novacor
Magnum	ABS	Dow
Makroblend	PC blends	Bayer
Makrolon	PC	Bayer

Trade Name	Product	Company
Maranyl	PA 66	LNP
Marlex	HDPE	Phillips
Merlon	PC	Bayer
Mindel	Polysulfone blends	Amoco
Minlon	Mineral reinforced PA66	DuPont
Mylar	PET film	DuPont
NAS 10	Acrylic styrene copolymer	Novacor
Noryl GTX	PPO/PS alloy	G.E.
Novodur	ABS	Bayer
Paxon	HDPE	Paxon Polymers
Pellethane	TPU	Dow
Petrothene	PE	Quantum
Plexiglas	Acrylic sheet	AtoHaas
Polyman	ABS/PVC alloy	Schulman
Polystrol	PS	BASF
Prevail	ABS/TPU	Dow
Prevex	PPE/PS alloy	G.E.
Pro-Fax	PP	Himont
Pulse	PC/ABS alloy	Dow
Radel	Polyethersulfone	Amoco
Rilsan	PA 11, 12	Atochem
Royalite	ABS and ABS alloys	Uniroyal
Rynite	PET	DuPont
Ryton	PPS	Phillips
Saran	PVDC	Dow
Styron	PS	Dow
Supec	Polyphenylene sulfide	G.E.
Surlyn	Ionomer	DuPont
Teflon	FEP, PFA, TFE film	DuPont
Tefzel	ETFE	DuPont
TempRite	CPVC	B.F. Goodrich
Tenite	Cellulosic compounds	Eastman Chemical
Terblend	ASA/PC alloy	BASF
Terluran	ABS	BASF
Texin	TPU	Bayer
Torlon	Polyamideimide	Amoco
Toyolac	ABS	Toray
TPX	Polymethylpentene	Mitsui
Triax	ABS/PA	Bayer
Tyril	SAN Copolymer	Dow
Udel	Polysulfone	Amoco
Ultem	Polyetherimide	G.E.
Ultradur	Polyester-PBT	BASF
Ultraform	Acetal	BASF
Ultramid	PA	BASF
Ultrason	PES	BASF

Trade Name	Product	Company
Uvex	CAB	Eastman
Valox	PBT	G.E.
Vectra	Liquid crystal polymer	Ticona
Vespel	PA	DuPont
Vydyne	PA	Monsanto
Xenoy	PC/polyester alloy	G.E.
XT	Acrylic	Cyro
Xydar	Liquid crystal polymer	Amoco
Zylar	Acrylic	Novacor
Zytel	PA	DuPont

APPENDIX E

STANDARDS ORGANIZATIONS

American Gas Association (AGA), Arlington, Virginia
AGA is the association of the gas distribution industry. The association is mainly responsible for research, standardization, and information related to the production, distribution, and utilization of gas. www.aga.org

American National Standards Institute (ANSI), New York, New York
ANSI is a nonprofit organization consisting of members from commerce, industry, professionals, trade consumers, labor organizations, and government. ANSI in cooperation with these groups identifies the needs for standards and sets priorities for their completion, assigns development work, supplies a standards-writing organization with effective procedures and management services, and approves standards as American National Standards. www.ansi.org

American Society for Quality (ASQ), Milwaukee, Wisconsin
ASQC is a technical and professional organization responsible for developing and publishing standards related to quality control. www.asq.com

ASTM International, Philadelphia, Pennsylvania
ASTM International, originally known as American Society for Testing and Materials (ASTM), is a scientific and technical organization formed for "The development of standards on characteristics and performance of materials, products, systems and services, and the promotion of related knowledge." ASTM International is the world's largest source of Voluntary Consensus Standards. www.astm.org

Handbook of Plastics Testing and Failure Analysis, Third Edition, by Vishu Shah
Copyright © 2007 by John Wiley & Sons, Inc.

National Association of Metal Finishers, Washington, DC

The National Association of Metal Finishers (NAMF) is an association of owners and managers of metal finishing companies and their suppliers. Its purpose is to advance, protect and perpetuate the surface finishing industry, and to develop the highest standards of service, quality, and conduct. Today NAMF is comprised of more than 900 member companies worldwide. www.namf.org

American Society of Safety Engineers, Park Ridge, Illinois

This is a technical society interested in the advancement of the profession and professional development of its members. The organization develops standards for the profession and professional safety engineer and participates in standards policy bodies. www.asse.org

Association of Home Appliance Manufacturers, Chicago, Illinois

The association develops voluntary appliance performance standards and makes safety recommendations to Underwriters Laboratories and the American Gas Association, represents the industry in consumer and government relations, compiles statistics, sponsors certification programs, and provides consumer appliance information, educational materials, and technical aids. www.aham.org

BSI Group, London, UK

Founded in 1901, the BSI Group is a leading business services provider to organizations worldwide. The Group has over 4500 employees in 110 countries and provides independent certifications of management systems and products, commodity inspection services, product testing services, the development of private, national, and international standards, management systems training, and information on standards and international trade. The group is divided into four divisions: British Standards Institute, BSI Management Systems, BSI Product Services, and BSI Inspectorate. www.bsi-global.com

European Organization for Quality (EOQ)

The European Organization for Quality (EOQ) is an autonomous, non-profit-making association under Belgian law. The EOQ is the European interdisciplinary organization striving for effective improvement in the sphere of quality management as the coordinating body and catalyst of its Full Member Organizations (FMOs). The EOQ was established in 1956, and its present membership is comprised of 34 national European quality organizations, as well as institutions, companies, and individuals from all over the world. www.eoq.org

European Organization for Conformity Assessment (EOTC)

The European Organization for Conformity Assessment, an independent and non-profit-making European body, was established in April 1990 by the European Commission, the European Free Trade Association (EFTA), and the European Standards Bodies. The EOTC is run by a General Assembly (GA), with Board of Administrators (BoA), elected from the GA, responsible for developing policy and strategic planning, meeting four times per year. The EOTC members are national and European bodies representing those with a stake in conformity assessment. EOTC acts as a focal point for conformity assessment in Europe, but does not itself test or certify products and services. www.eotc.be

ETL SEMKO

The ETL SEMKO division of Intertek is a leader in testing, certifying, and inspecting products for global market access. The ETL Listed Mark is issued to products that have met the requirements of product safety standards in the United States and Canada. www.intertek-etlsemko.com

EUROLAB (A European Organization for Laboratory Testing)

Eurolab is the European Federation of National Associations of Measurement, Testing, and Analytical Laboratories. The function of EUROLAB is to provide an interface between the testing community and other concerned parties, to accelerate development and harmonization of test methods, to promote mutual acceptance of test results, and to provide expertise in the field of testing. www.eurolab.org

FM Global, Norwood, Massachusetts

FM Global is one of the world's largest commercial and industrial property insurance and risk management organizations specializing in property protection. FM Global research, an integral part of FM Global, focuses on four primary areas of loss phenomena and loss control, protection research, materials research, structural research, and risk and reliability. Many standards dealing with flammability, wind damage, fire sprinklers, and so on, have been developed. www.fmglobal.com

Industrial Safety Equipment Association, Arlington, Virginia

This association represents the manufacturers of personal protection equipment and machinery safeguard devices. www.safetyequip.org

International Association of Plumbing and Mechanical Officials (IAPMO), Los Angeles, California

IAPMO sponsors the uniform plumbing code (UPC) that is used in over 2500 jurisdictions in the United States and is a mandatory code for 10 states. www.iapmo.org

International Organization for Standardization

Formed in 1947, ISO is a worldwide federation of some 90 national standards bodies. ISO promotes the development of standardization and related activities to facilitate the international exchange of goods and services. It is also ISO's purpose to develop intellectual, scientific, technological, and economic cooperation among member countries. ISO is made up of technical committees, subcommittees, working groups, and *ad hoc* study groups. All these groups represent the manufacturers, suppliers, users, engineers, testing laboratories, public services, governments, consumer groups, and research organizations of each of the member countries. www.iso.org

Juvenile Products Manufacturer's Association, Moorestown, New Jersey

The association, which represents juvenile furniture manufacturers, is responsible for developing juvenile product safety performance standards. www.jpma.org

National Electrical Manufacturers Association (NEMA), Washington, D.C.

The association has over 200 separate standards publications for electrical apparatus and equipment. The NEMA membership includes over 500 major electrical manufacturing companies in the United States. www.nema.org

National Fire Protection Association (NFPA), Boston, Massachusetts
NFPA is involved in the standards-making field under which codes, standards, and recommended practices are developed as guides to engineering for reducing loss of life and property by fire. Standards are published yearly as national fire codes. www.nfpa.org

National Institute of Standards and Technology, Gaithersburg, Maryland
NIST is an agency of the U.S. Department of Commerce. NIST's mission is to develop and promote measurement, standards, and technology to enhance productivity, facilitate trade, and improve the quality of life. www.nist.gov

National Safety Council (NSC), Chicago, Illinois
National Safety Council's objective is to determine and evaluate methods and procedures that prevent accidents and minimize injury and economic loss resulting from accidents. The National Safety Council also provides leadership to expedite the adoption and the use of those methods and procedures that best serve the public interest. www.nsc.org

National Sanitation Foundation (NSF), Ann Arbor, Michigan
NSF has published standards and criteria under which testing and certification services are currently extended to over 1600 manufacturers who use the NSF seal on over 25,000 items of equipment or products. www.nsf.org

Polyurethane Manufacturer's Association, Chicago, Illinois
The organization is a private nonprofit trade association of companies involved in the manufacture of solid polyurethane thermosetting elastomers and related chemicals and equipment suppliers. www.pmahome.org

Quality Management Institute
QMI is the Canadian registrar of quality systems. It is part of the Canadian Standards Association (CSA). www.qmi.com

Society of Plastics Industry (SPI), New York, New York
SPI is a trade and technical society of over 1200 companies in all branches of the plastics industry interested in quality standards, research, uniform accounting, wage rate surveys, codes, public relations, informative labeling, safety, fire prevention, food packaging, and so on. www.plasticsindustry.org

Underwriters Laboratories (UL)
UL, an independent organization devoted to testing for public safety, was established to maintain and operate laboratories for examination and testing the safety of devices, systems, and related materials. www.ul.com

APPENDIX F

TRADE PUBLICATIONS

PLASTICS

Injection Molding Magazine, Canon Communications LLC, 11444 W. Olympic Blvd., Suite 900, Los Angeles, CA 90064. www.immnet.com

Modern Plastics, Canon Communications LLC, 11444 W. Olympic Blvd., Suite 900, Los Angeles, CA 90064. www.modplas.com

Plastics Engineering, Society of Plastics Engineers, 14 Fairfield Drive, Brookfield Center, CT 06804. www.4spe.org

Plastics Molding & Fabricating, Vance Publishing, 800 Liberty Drive, Libertyville, IL 60048. www.plasticsmachining.com

Plastics News, Crain Communications, Inc., 1725 Merriman Road, Akron, OH 44313-5283. www.plasticsnews.com

Plastics Technology, Gardner Publications, 6915 Valley Avenue, Cincinnati, OH 45244. www.plasticstechnology.com

Reinforced Plastics, Elsevier Science, PO Box 945, New York, NY 10159-0945. www.reinforcedplastics.com

Moldmaking Technology, Communications Technology, Inc., 301 S. Main Street, Suite 1 West, Doylestown, PA 18901. www.moldmakingtechnology.com

Additives & Compounding, Elsevier Science, PO Box 945, New York, NY 10159-0945. www.addcomp.com

Tool & Moldmaking Product News, Access Communications North American, Inc., 21361-B Pacific Coast Highway, Malibu, CA 90265. www.Tool-moldmaking.com

Plastics Machinery & Auxiliaries, Canon Communications LLC, 11444 W. Olympic Blvd., Suite 900, Los Angeles, CA 90064. www.pma-magazine.com

Handbook of Plastics Testing and Failure Analysis, Third Edition, by Vishu Shah
Copyright © 2007 by John Wiley & Sons, Inc.

534

OTHERS

Design News, Reed Elsevier, Inc., 29 West 34th Street, 8th Floor, New York, NY 10001. www.designnews.com

Medical Devices & Diagnostic Industry, Canon Communications, 11444 W. Olympic Blvd., Los Angeles, CA 90064. www.devicelink.com/mddi/

Quality, Quality Magazine, 1050 IL Route 83, Suite 200, Bensenville, IL 60106. www.qualitymag.com

Product Design and Development, Reed Elsevier, Inc., 301 Gibraltar Drive, Box 650, Morris Plains, NJ 07950-0650. www.pddnet.com

Macromolecules, American Chemical Society Member & Subscriber Services, PO Box 3337, Columbus, OH, 43210. http://pubs.acs.org/journals/mamobx

Applied Spectroscopy, Society of Applied Spectroscopy, 201B Broadway Street, Frederick, MD 21701-6501. www.s-a-s.org

Spectroscopy, Advanstar Communications, 485 Route One South, Building F, Iselin, NJ 08830. www.spectroscopymag.com

Polymer, Elsevier Science, P.O. Box 945, New York, NY 10159-0945. http://www.sciencedirect.com/science/journal/00323861

Journal of Analytical and Applied Pyrolysis, Elsevier Science, P.O. Box 945, New York, NY 10159-0945. www.sciencedirect.com/science/journal/01652370

Journal of Chromatography, Elsevier Science, P.O. Box 945, New York, NY 10159-0945. http://www.sciencedirect.com/science/journal/00219673

Journal of Polymer Science, John Wiley & Sons, 111 River Street, Hoboken, NJ 07030-5774. http://www3.interscience.wiley.com/cgi-bin/jhome/36444

Laboratory Equipment, Reed Elsevier, 301 Gibraltar Drive, Morris Plains, NJ 07950. www.labequipmag.com

*INDUSTRY PUBLICATIONS

Dupont Magazine, E.I. Dupont de Nemours and Co., Wilmington, Delaware 19898. www.dupont.com

Compounding Lines, RTP Company, 580 E. Front Street, Winona, MN 55987. www.rtpcompany.com

FLOWfront, Moldflow Corporation, 430 Boston Post Road, Wayland, MA 01778-9910. www.moldflow.com

*These periodicals are circulated without charge to qualified individuals. A complete all inclusive trade publications listing can also be viewed on the web site. www.plastics.com

APPENDIX G

INDEPENDENT TESTING LABORATORIES

American Research and Testing Inc.
14934 S. Figueroa Street
Gardena, CA 90248
(800) 538-1655 or (310) 538-9709
www.americanresearch.com

Ashland Specialty Chemical Company
5200 Blazer Parkway
Technical Center West
Dublin, OH 43017
(614) 790-3278 or (800) 545-8779
www.ashlandanalytical.com

Atlas Weathering Services Group
South Florida Test Service
17301 Okeechobee Road
Miami, Florida 33018
(305) 824-3900
www.atlaswsg.com

Battelle Memorial Institute
425 King Avenue
Columbus, OH 43201
(614) 424-5214
www.batelle.org/chembio/default.htm

Benz Materials Testing Instruments
73 Maplehurst Ave.
Providence, RI 02908
(401) 331-5650
www.benztesters.com

Bodycote Polymer Broutman
   Laboratory
1975 North Ruby Street
Melrose Park, IL 60160
(708) 236-5360
www.na.bodycote-mt.com

Handbook of Plastics Testing and Failure Analysis, Third Edition, by Vishu Shah
Copyright © 2007 by John Wiley & Sons, Inc.

Broutman, L.J., Associates Ltd.
3424 South State St.
Chicago, IL 60616
(312) 842-4100

Chemical Electro Physics Corp.
705 Yorklyn Road
Hockessin, DE 19707
203-239-1378
www.cep-corp.com

Chemir Analytical Services.
2672 Metro Boulevard
Maryland Heights, MI 63043
(314) 291-6620 or (800) 659-7659
www.chemir.com

Com-Ten Industries
6405 49th Street N.
Pinellas Park, FL 33781
(727) 520-1200
www.com-ten.com

CRT Labs Inc. (California Resin
    Testing)
1680 Main
Orange, CA 92867
(714) 283-2032
www.crtlabs.com

C.W. Brabender Instruments, Inc.
50 East Wesley Street
South Hackensack, NJ 07606
(201) 343-8425
www.cwbrabender.com

Datapoint Labs
95 Brown Road Suite 102
Ithaca, NY 14850
(888) DATA-4-CAE
www.datapointlabs.com

Delsen Testing Laboratories, Inc.
1024 Grand Central Avenue
Glendale, CA 91201
(818) 247-4106 or (888) 433-5736
www.delsen.com

Detroit Testing Laboratory, Inc.
7111 E. Eleven Mile Road
Warren, MI 48092-2709
(586) 754-9000
www.dtl-inc.com

Dow Corning Analytical Solutions
P.O. Box 994, Mail # C041C1
Midland, MI 48686-0994
(877) 322-8378
http://dowcorning.com/analytical

Dynisco Polymer Test
730 Hemlock Road
Morgantown, PA 19543
(508) 541-6206
www.dynisco.com

Engineered Polymers International,
    LLC (EPI)
2842 Progress Road
Madison, WI 53716
(608) 661-2800
www.engineeredpolymers.com

Gaynes Labs Incorporated
9708 Industrial Drive
Bridgeview, IL 600455
(708) 233-6655
www.nrinc.com/gayness

Ghesquire Plastic Testing, Inc.
20450 Harper Avenue
Harper Woods, MI 48225
(313) 885-3535

GretagMacbeth USA
617 Little Britain Road
New Windsor, NY 12553
(800) 622-2384 or (845) 565-7660
www.gretagmacbeth.com

GTI Testing Laboratories
1700 S. Mount Prospect Avenue
Des Plaines, IL 60018
(866) GTI-LABS
www.gastechnology.org/gtilabs

Harrop Industries, Inc.
3470 E. Fifth Avenue
Columbus, OH 43219
(614) 231-3621
www.harropusa.com

Hauser Laboratories
4750 Nautilus Court South unit A
Boulder, CO 80301
(720) 406-4800
www.hauserlabs.com

HunterLab
11491 Sunset Hills Road
Reston, VA 20190
(703) 471-6870
www.hunterlab.com

IMPACT Analytical
1910 West Saint Andrew Road
Midland, MI 48640
(989) 832-5555
www.impactanalytical.com

Intertek Caleb Brett
27611 La Paz Road Suit C
Laguna Niguel, CA 92677
(949) 448-4100
www.intertek-cb.com

Jordy Associates Inc.
4 Mill Street
Bellingham, MA 02019
(508) 966-1301
www.jordiassoc.com

Kars' Advanced Materials, Inc.
2528 W. Woodland Drive
Anaheim, CA 92801
(714) 527-7100
www.karslab.com

MATCO Associates, Inc.
4640 Campbell's Run Road
Pittsburgh, PA 15205
(800) 221-9090
www.matcoinc.com

Materials Engineering, Inc.
47W605 Indian Creek Trial
Virgil, IL 60151
(630) 3365-9060
www.material-engr.com

Measurement Technology Corp. (Measure
   Tech)
4396 Round Lake Road West
St. Paul, MN 55112-3923
(651) 623-7651
www.measuretech.com

Measurement Technology Inc.
4240 Lock Highland Parkway
Roswell, GA 30075
(770) 587-2222
www.mti-phoenix.com

Microbac Laboratories, Inc.
Hauser Division
4750 Nautilus Court South
Boulder, CO 80301
720.406.4805 (P)
www.microbac.com

Monarch Analytical Laboratories, Inc.
349 Tomahawk Drive
Maumee, OH 43537
(419) 897-9000
www.testing@monarchlab.com

MTS Systems Corp./ Materials Testing
 Division
14,000 Technology Drive
Eden Prairie, MN 55344
(800) 944-1687
www.mts.com

Neolytica
3606 W. Liberty Road
Ann Arbor, MI 48103
(800) 704-4034
www.neolytica.com

OCM Test Laboratories, Inc.
3883 East Eagle Drive
Anaheim, CA 92807
(714) 630-3003
www.ocmtestlabs.com

Orange County Material Test
 Laboratories Inc.
3883 E. Eagle Drive
Anaheim, CA 92807
(714) 630-3003 X 222
www.ocmtestlabs.com

Plastics Technology Laboratories, Inc.
50 Pearl Street
Pittsfield, MA 01201
(413) 499-0983
www.ptli.com

Polyhedron Laboratories, Inc.
10626 Kinghurst Street
Houston, TX 77099
(281) 879-8600
www.polyhedronlab.com

Polymer Solutions Inc.
1872 Pratt Drive #1375
Blacksburg, VA 24060-6363
(540) 961-4300 or (877) 961-4341
www.polymersolutions.com

Q-Panel Lab Products
800 Canterbury Road
Cleveland, OH 44145
(440) 835-8700
www.q-panel.com

RTS—Rexnord Technical
 Services
5101 West Beloit Road
Milwaukee, WI 533214
(414) 643-3067
www.rts-rexnord.com

Scientific Process and Research, Inc.
67 Veronica Avenue
Somerset, NJ 08873
(732) 846-3477
www.spar.com

Seal Laboratories
250 North Nash Street
El Segundo, CA 90245
(310) 322-2011
www.seallabs.com

SEM Lab, Inc.
20219 10th Place SE
Snohomish, Washington 98290
(425) 335-4400
www.sem-lab.com

SGS U.S. Testing Co., Inc.
291 Fairfield Avenue
Fairfield, NJ 07004
(973) 575-5252
www.ustesting.sgsna.com

Skeist Inc.
375 Route 10
Whippany, NJ 07891
(973) 515-2020
www.skeistinc.com

Southwest Research Institute
6220 Culebra Road P.O. Drawer
    28510
San Antonio, TX 78228
(210) 684-5111
www.swri.org

Springborn Smithers Laboratories
790 Main Street
Wareham, MA 02571-1075
(508) 295-2550
www.springbornsmithers.com

Tandex Test Labs, Inc.
15849 Business Ctr. Drive
Irwindale, CA 91706
(800) 729-8378 or (626) 962-7166
www.tandexlabs.com

The Madison Group
505 S. Rosa Road Suite 124
Madison, WI 53719
(608) 231-1907
www.madisongroup.com

Underwriters Laboratories
333 Pfingsten Road
Northbrook, IL 60062
(847) 272-8800
www.ul.com

University of Massachusetts Lowell
    Plastics Engineering Dept.
1 University Avenue
Lowell, MA 01854-2881
(978) 934-3420
www.uml.edu

VTEC Laboratories, Inc.
212 Manida Street
Bronx, NY 10474
(718) 542-8248 X 3
www.vteclabs.com

X-Rite, Inc.
3100 44th Street
S.W. Grandville, MI 49418
(616) 534-7663
www.x-rite.com

APPENDIX H

SPECIFICATIONS

The following is the index of material specifications. A complete copy of these specifications may be obtained by writing to the agency responsible for publishing the particular specification.

ABS
ASTM D 3011
FED L-P-1183

ACETAL
ASTM D4181
FED L-P-392
MIL-P-46137

ACRYLIC
ASTM D788
BS 3412
FED L-P-380
MIL-I-46058C
MIL-P-19735B

CELLULOSE ACETATE
ASTM D706
FED L-P-397
MIL-M-3165A

CELLULOSE ACETATE BUTYRATE
ASTM D707
L-P-349

CELLULOSE PROPIONATE
ASTM D1562
MIL-P-46074

CHLORINATED POLY(VINYL CHLORIDE)
ANSI/ASTM D1784-78

Handbook of Plastics Testing and Failure Analysis, Third Edition, by Vishu Shah
Copyright © 2007 by John Wiley & Sons, Inc.

DENSITY
ANSI/ASTM D1248-74
ANSI/ASTM D2581-73
BS 3412-1976
MIL-P-51431(EA)

EPOXY
ASTM D3013
MIL-P-46069
MIL-P-46892

MELAMINE
ASTM D709
MIL-M-14

MELAMINE–FORMALDEHYDE
ASTM D704-62(1975)
ISO2112-1977

NYLON
ASTM D4066
ASTM D789
ISO R1874
FED L-P-410
FED L-P-395
MIL-M19887
MIL-M-20693

PHENOL-FORMALDEHYDE
ASTM D700-75
MIL-P-47134(MI)
MIL-R-3745

PHENOLIC
AMS3823B
AMS3830
AMS3837
AMS3858A
ASTM D700
ISO800
MIL-M-14
FED L-P-1125

POLY(ARYL SULFONE ETHER)
MIL-P-46133A(MR)

POLY(METHYL METHACRYLATE)
ANSI/ASTM D788-78A

POLY(VINYL CHLORIDE)
ASTM D1755
ASTM D1784
ASTM D2287
ASTM F437
FED L-P-535
ISO1060
MIL-P-47136

POLYBUTYLENE
ASTM D2581

POLYCARBONATE
ASTM D3935
L-P-393A(2)
MIL-P-81390

POLYESTER
ASTM D4507
ASTM D3220
MIL-M24519
FED L-P-383

POLYETHYLENE
ASTM D1248
ASTM D2513
ASTM D2103
FED L-P-378
FED L-P-390

POLYPHENYLENE OXIDE
MIL-P-46131
ASTM D2874

POLYPHENYLENE SULFIDE
ASTM D4067
MIL-M24519

POLYPROPYLENE
ASTM D4101
ASTM D4104
FED L-P-394
MIL-P-46109

POLYSTYRENE
ASTM D3011
ASTM D703
ASTM D1892
ISO1622
FED L-P-396
MIL-S-676
MILP-19644C
MIL-C-26861

POLYSULFONE
MIL-P-46120A(MR)

POLYTETRAFLUOROETHYLENE
ASTM D1457
ASTM D1710
ASTM D3295
MIL-R-8791

POLYURETHANE
ASTM D3574
ASTM D412

PROPYLENE
ANSI/ASTM D2853-70(1)
ASTM D2146-77

SILICONE
MIL-M-14
ASTM D1418

STYRENE
ANSI/ASTM D1788-78A
ASTM D1892-78
ASTM D703-78
MIL-R-82483A(OS)

STYRENE-ACRYLONITRILE
ASTM D3011
ASTM D1431
FED L-P-399A

UREA-FORMALDEHYDE
ANSI/ASTM D705-62(19)
ISO2112-1977
MIL-HDBK-700(MR)

URETHANE
ASTM D2000-77A
MIL-HDBK-700(MR)

VINYL CHLORIDE
ANSI/ASTM D1784-78
ANSI/ASTM D2287-78
ANSI/ASTM D2474-78
ASTM D729-78
ISO2798-1974

Summary—ISO and ASTM Methods

ISO Method	Title	ASTM No.	Title
60–1977	Apparent density of material that can be poured from a specified funnel	D 1895–89(09)	Test methods for apparent density, bulk factor, and pourability of plastic materials
61–1976	Apparent density of molding material that cannot be poured from a specified funnel	D 1895–89(90)	Test methods for apparent density, bulk factor, and pourability of plastic materials
62–1980	Determination of water absorption	D 570–81(88)	Test for Water Absorption of Plastics
75–1987	Plastics and ebonite—Determination of temperature of deflection under load	D 648–82(88)	Test method of deflection temperature of plastics under flexural load
171–1980	Determination of bulk factor of molding materials	D 1895–89(90)	Test methods for apparent density, bulk factors, and pourability of plastics materials
178–1975	Determination of flexural properties of rigid plastics	D 790M–86	Test methods for flexural properties of plastics and electrical insulating material (metric)
179–1982	Plastics—Determination of Charpy impact strength of rigid materials	D 256–90b	Test methods for impact resistance of plastics and insulating materials
180–1982	Plastics—Determination of Izod impact strength of rigid materials	D 256–90b	Test method for impact resistance of plastics and insulating material
181–1981	Plastics—Determination of flammability characteristics of rigid plastics in the form of small specimens in contact with an incandescent rod	D 757–77	Test method for incandescence resistance of rigid plastics in a horizontal position
291–1977	Standard atmospheres for conditioning and testing	D 618–61(90)	Methods of conditioning
292–1967 (see 1133–1981)	Determination of the melt flow index of polyethylene and polyethylene compounds	D 1238–90b	Test method for flow rates of thermoplastics by extrusion plastometer
306–1987	Plastics—Thermoplastic materials—determination of the Vicat softening temperature	D 1525–87	Test method for vicat softening temperature of plastics
307–1984	Plastics—Polyamides—Determination of viscosity number	D 2857–87	Test method for dilute solution viscosity of polymers
458/1–1985	Plastics—Determination of stiffness in torsion of flexible materials (general method)	D 1043–87	Test method for stiffness properties of plastics as a function of temperature by means of a torsion test

489–1983	Plastics—Determination of the refractive index of transparent plastics	D 542–90	Index of refraction of transparent organic plastics
R 527–1966	Determination of tensile properties	D 638M–89	Test method for tensile properties of plastics (metric)
537–1989	Testing with the torsion pendulum	D 2236–81	Test method for dynamic mechanical properties of plastics by means of a torsional pendulum
846–1978	Behavior under the action of fungi and bacteria—Evaluation by visual exam or measurement of change in mass of physical properties.	G 21–90 and G 22–79(90)	Practice for determining resistance of synthetic polymer material to fungi. Practice for determining resistance of plastics to bacteria
868–1985	Plastics and ebonite—Determination of indentation hardness by means of a durometer (shore hardness)	D 2240–86	Test method for rubber property—durometer hardness
899–1981	Determination of tensile creep of plastics	D 2990–90	Test methods for tensile, compressive, and flexural creep and creep rupture of plastics
974–1980	Determination of the brittleness temperature by impact	D 746–79(87)	Test method for brittleness temperature of plastics and elastomers by impact
1133–1981	Determination of the melt flow rate of thermoplastics	D 1238–90b	Test method for flow rate of thermoplastics by extrusion plastometer
1183–1987	Methods of determining the density and relative density of non-cellular plastics	D 1505–90	Test method for density of plastics by the density—gradient technique
1628/1–1984	Guidelines for the standardization of methods for the determination of viscosity number of polymer in dilute solution. Part 1: General conditions	D 2857–87	Test method for dilute solution viscosity of polymers
2039/1–1987	Plastics—Determination of hardness. Part 1: Ball indentation method	D 785–89	Test method for Rockwell hardness of plastics and electrical insulating materials
3146–1985	Determination of melting temperature of semicrystalline polymers—Optical method	D 2117–82(88)	Test method for melting point of semicrystalline polymers
3268–1978	Glass reinforced materials—Determination of tensile properties	D 639–90 and D 3039–76(89)	Test method for tensile properties of plastics. Test method for tensile properties of fiber-resin composites
3582–1978	Cellular plastic and rubber mats—Lab assessment of horizontal burning char of small specimens subject to small flame	D 635–88	Test method for rate of burning and/or extent and time of burning of self-supporting plastics in a horizontal position

Summary—ISO and ASTM Methods—(Continued)

ISO Method	Title	ASTM No.	Title
4574–1978	PVC resins for general use—Determination of hot plasticizer absorption	D 3367–75 (1989)	Test method for plasticizer sorption of poly (vinyl chloride) resins under applied centrifugal force
4589–1984	Plastics—Determination of the oxygen index	D 2863–87	Method for measuring the minimum oxygen concentrate to support candle-like combustion of plastics (oxygen-index)
4607–1978	Methods of exposure to natural weathering	D 1435–85	Practice for outdoor weathering of plastics
4608–1984	Plastics—Homopolymer and copolymer resins (vinyl chloride) for general use—Determination of plasticizer absorption at room temperature	D 3367–75(90)	Test method for plasticizer sorption of poly (Vinyl chloride) resins under applied centrifugal force
4892–1981	Plastics—Methods of exposure to laboratory light sources	D 1499–84	Practice for operating light-and water-exposure apparatus (carbon-arc type) for exposure of plastics
		D 2565–89	Practice for operating xenon-arc type (water-cooled) light-exposure apparatus with and without water for exposure of plastics
6186–1980	Determination of pourability	D 1895–89(90)	Test methods for apparent density, bulk factor, and pourability of plastic materials
6239–1986	Plastics—Determination of tensile property by use of small specimens	D 638–90	Test method for tensile properties of plastics
6602–1985	Plastics—Determining of flexural creep by three-point loading	D 2990–90	Tensile, compressive and flexural creep and creep rupture of plastics
6603/1–1985	Plastics—Determination of multiaxial impact behavior of rigid plastics. Part 1: Falling dart method	D 3029–90	Impact resistance of rigid plastic sheeting of parts by means of a tup (falling weight)

Military Specifications

MIL-P-77	Cast Polyester or Diallylpthlalate sheet and rod
MIL-P-78A	Engraving Stock Rigid laminated sheet
MIL-P-79C	Thermoset Rod and Tube, Melamine and Phenolic. Glass, Cotton and Paper Reinforced
MIL-P-80	Acrylic Sheet, Anti-Electrostatic Coated
MIL-I-631	Electrical Insulation Tubing, Film, Sheet and Tape, Vinyl, Polyethylene and Polyester
MIL-I-742C	Fiberglas Thermal Insulation board
MIL-P-997C	Thermoset Silicone Resin Sheets, Glass Reinforced
MIL-Y-1140E	Fiber Glass Yarn, Cord Sleeving, Tape and Cloth
MIL-P-3054A	Polyethylene Special Material
MIL-P-3086	Non-Rigid Polyamide (Nylon) Resin
MIL-P-3115B	Thermoset Phenolic Sheet, Paper Reinforced
MIL-P-3158C	Insulation Tape and Cord Glass, Resin Filled
MIL-I-3190B	Insulation Sleeving, Flexible, Treated
MIL-I-3825A	Insulation Tape, Electrical, Self-Fusing
MIL-P-4640A	Polyethylene Film for ballon use
MIL-P-5425B	Acrylic Sheet, Heat Resistant
MIL-P-5431A	Phenolic, Graphite Filled Sheet, Rods, Tubes and Shapes
MIL-P-6264B	Vinyl Copolymers, Unplasticized unpigmented and unfilled
MIL-I-7444B	Insulation Sleeving, Flexible Electrical
MIL-I-7798A	Insulation Tape, Electrical, Pressure Sensitive
MIL-P-8059A	Thermoset Phenolic Resin Sheet and Tubes. Asbestos Paper and Cloth Reinforced
MIL-P-8184	Acrylic Plastic Sheet, Modified
MIL-P-8257	Polyester Base, Cast Transparent Sheet,Thermosetting
MIL-P-8587A	Cellulose-Acetate Sheet Colored, Transparent
MIL-P-8655A	Thermoset Phenolic Sheet, Postforming Cotton Reinforced
MIL-P-1394C	Laminated Plastic Sheet, Copper-Clad
MIL-P-9969	Polyurethane, Rigid, unicellular, Foam-In-Place for packaging
MIL-P-13436A	Filled Phenolic Sheet, Uncured
MIL-P-13491	Polystyrene Sheet, Rod and Tube
MIL-P-13949D	Copper-Clad, laminated Plastic Sheets (Paper Based and Glass Base)
MIL-P-14591B	Plastic Film, Non Rigid, Transparent
MIL-P-15305C	Thermoset Phenolic Sheet, Cotton Reinforced
MIL-P-15037E	Thermoset Melamine Resin Sheet, Glass Reinforced
MIL-P-15047B	Thermoset Phenolic Resin Sheet, Nylon Reinforced
MIL-I-15126F	Insulation Tape, Electrical, Pressure Sensitive and Thermoset Adhesive
MIL-P-16413	Methyl Methacrylate Molding Materials
MIL-P-16414	Cellulose Acetate Butyrate Molding Material
MIL-P-16416	Cellulose Acetate Molding Material
MIL-P-17091B	Polyamide (Nylon) Resin Rods, Sheets and Parts
MIL-P-17276	Cellulose Acetate Sheet

Military Specifications

MIL-P-17549C	Fibrous Glass Reinforced Plastic Laminates. For Marine Applications
MIL-P-18057A	Insulation Sleeving, Flexible Silicone Rubber Coated Glass
MIL-P-18080	Vinyl, Flexible, Transparent, Optical Quality
MIL-P-18177C	Thermoset Epoxy Sheet, Glass Reinforced
MIL-P-18324C	Thermoset Phenolic, Cotton Reinforced, Moisture Resistant
MIL-N-18352	Nylon Plastic, Flexible Molded or Extruded
MIL-I-18622A	Insulation Tape, Electrical, Pressure Sensitive Silicone Rubber Treated Glass
MIL-I-18746A	Insulation Tape, Glass Fabric TFE Coated
MIL-M-19098	Molding Plastics, Polyamide (Nylon), and Molded and Extruded Polyamide Plastic
MIL-I-19161A	Plastics Sheet, Teflon TFE and Glass Cloth Laminated
MIL-I-19166A	Insulation Tape Electrical, Pressure Sensitive, High Temperature Glass
MIL-P-19336C	Plastics Sheets, Polyethylene, Virgin and Borated, Neutron Shielding
MIL-P-19468A	Plastics Rods Molded and Extruded Teflon TFE
MIL-P-19735B	Molding, Acrylic, Colored and White, Heat Resistant, for Lighting Fixtures
MIL-P-19833B	Glass Filled Diallylpthlalate Resin
MIL-P-19904	Plastic Sheet ABS Copolymer, Rigid
MIL-M-20693A	Plastic Molding Material, Rigid, Polyamide
MIL-P-21094A	Cellulose Acetate, Optical Quality
MIL-P-21105C	Plastic Sheet, Acrylic, Utility Grade
MIL-P-21347B	Plastic Molding Material, Polystyrene, Glass Fiber Reinforced
MIL-M-21470	Polychlorotrefluoroethylene Resin for Molding
MIL-I-21557B	Insulation Sleeving, Electrical, Flexible Vinyl Treated Glass Fiber
MIL-P-21922A	Plastic Rods and Tubes Polyethylene
MIL-P-22035	Plastic Sheet, Polyethylene
MIL-P-22076A	Insulation Sleeving Electrical, Flexible Low Temperature
MIL-P-22096A	Plastic, Polyamide (Nylon) Flexible molding and extrusion Material
MIL-I-22129C	Insulation Sleeving, Electrical, Non-Rigid Teflon TFE Resin
MIL-P-22241A	Plastic Sheet and Film, Teflon TFE
MIL-P-22242	Cancelled—Refer to MIL-P-22241
MIL-P-22270	Plastic Film, Polyester, Polyethylene Coated (For I.D. Cards)
MIL-P-22296	Plastic Tubes and Tubing, Heavy Wall, Teflon TFE Resin
MIL-P-22324A	Thermoset Epoxy Resin Sheet, Paper Reinforced
MIL-T-22742	Insulation Tape, Electrical, Pressure Sensitive, Telfon TFE Resin
MIL-P-22748A	Plastic Material for Molding and Extrusion, High Density Polyethylene and Copolymers
MIL-I-23053A	Insulation Sleeving, Electrical, Flexible, Heat-Shrinkable
MIL-T-23142	Film Tape, Pressure Sensitive
MIL-P-23536	Plastic Sheets, Virgin and Borated Polyethylene
MIL-I-23594A	Insulation Tape, Electrical, High Temperature, Teflon, Pressure Sensitive

Military Specifications

MIL-I-24204	Nomex Film
MIL-P-24191	Plastic Sheet, Acrylic, Colored and White, Heat Resistant Shipboard Application
MIL-P-25374A	Plastic Sheet, Acrylic, Modified, Laminated
MIL-P-25395A	Heat-Resistant, Glass Fiber Base Polyester Resin,
MIL-P-25395A	Heat-Resistant, Glass Fiber Base Polyester Resin, Low Pressure Laminated Plastic
MIL-P-25421A	Glass Fiber Base—Epoxy Resin Low Pressure Laminated Plastic
MIL-P-25518A	Silicone Resin, Glass Fiber Base, Low-Pressure Laminated Plastic
MIL-P-25690A	Plastic Sheets and Parts, Modified Acrylic Base, Mono-lithic, Crack Propaqation Resistant—Covers Stretched Acrylic. .060″ Thru .675″ in Thickness
MIL-P-25770A	Thermoset Phenolic Resin Sheet, Asbestos Reinforced
MIL-P-26692	Plastic Tubes and Tubing, Polyethylene
MIL-P-27538	Plastic Sheet FEP Fluorocarbon unfilled, Copper-Clad
MIL-P-27730A	Tape, Anti-Seizing, Teflon TFE
MIL-P-40619	Plastic Material, Cellular, Polystyrene
MIL-P-43037	Thermoset Phenolic Resin Rod, Nylon Reinforced
MIL-T-43036	Tape, Pressure Sensitive, Filament Reinforced Plastic Film
MIL-P-43081	Plastic Low-Molecular Weight Polyethylene
MIL-P-46036	Chlorotrefluoroethlene Polymer—Sheets, Rods and Tubes (Plaskon)
MIL-P-46040A	Phenolic Sheet, Heat Resistant, Glass Fabric Reinforced
MIL-P-46041	Plastic Sheet, Flexible Vinyl
MIL-P-46060	Plastic Material Nylon
MIL-P-46112	Plastic Sheet and Strip, Polyimade H-Film
MIL-P-46115	Plastic Molding and Extrusion Material, Polyphenylene Oxide PPO
MIL-P-46120	Plastic Molding and Extrusion Material Polysulfone
MIL-P-46122	Plastic Molding Material, Polyvinylidene Fluoride— Kynar
MIL-P-46129	Plastic Molding and Extrusion Material, Polyphenylene Oxide, Modified—Noryl
MIL-P-46131	Polyphenylene Oxide, Modified, Glass Filled
MIL-P-52189	Thermoset Phenolic Resin Tube, Nylon Reinforced
MIL-P-55010	Plastic Sheet, Polyethylene Terephthalate
MIL-P-81390	Plastic Molding Material, Polycarbonate, Glass Fiber Reinforced
MIL-P-82540	Polyester Resin, Glass Fiber Base, Filament Wound Tube

Federal Specifications

L-P-315	Polyethylene Pipe
L-P-349A	Cellulose Acetate Butyrate, Molding and Extrusions
L-P-370	Copolymer of Vinyl and Vinylidene Chloride Plastic Film
L-P-375B	Vinyl Chloride Plastic Film, Flexible
L-P-377A	Polyethylene Terephthalate Plastic Film
L-P-378A	Polyethylene Plastic Film, Thin Gage
L-P-380	Methacrylate Molding Material
L-P-383	Glass Fiber Base, Low Pressure Polyester Resin Laminated

Military Specifications

L-P-395A	Polychlorotrifluoroethylene (KEL-F) Molding Material
L-P-386	Cellular Urethane, Flexible
L-P-387	Polyethylene Low and Medium Density-Molding Material
L-P-389A	FEP Fluorocarbon for Molding and Extrusion
L-P-390A	Low and Medium Density Polyethylene Molding Material
L-P-391A	Methacrylate Sheets, Rods, and Tubes, Cast
L-P-392A	Acetal Material, Injection Molding and Extrusion
L-P-393A	Polycarbonate Material, Molding and Extrusions
L-P-394A	Polypropylene Material for Injection Moldings and Extrusion
L-P-395A	Polyamide (Nylon) Molding Material, Glass Fiber Filled
L-P-396	Polystyrene Molding Material
L-P-397	Cellulose Acetate, Molding Material
L-P-398	Styrene Butadiene Molding Material
L-P-399	Styrene-Acrylonitrile Molding Material
L-P-401	Urea-Formaldehyde Molding Material
L-P-403	Polytetrafluoroethylene (Teflon TFE) Molding Material
L-P-410	Polyamide (Nylon), Rigid, Sheets, Rods, Tubes, and Molded Parts
L-P-501	Polyvinylidene Chloride (Saran) Molded
L-P-503	Polyvinyl Chloride Rod, Solid and Rigid
L-P-504B	Cellulose Acetate Sheet and Film
L-P-505B	Shatter-Resistant, Rigid, Reinforced, Translucent Corrugated Sheet, Polyester, Acrylic or Combination
L-P-506	Polystyrene, Biaxially Oriented Sheet and Film
L-P-507	Acrylic Sheet, Extruded
L-P-508	Laminated, Decorative and Non-Decorative Plastic Sheet
L-P-509A	Thermoset Sheet, Rod and Tube
L-P-510A	Polyvinyl-Chloride Sheet, Rigid, High Impact
L-P-511	Thermoset Phenolic Sheet, Cotton Reinforced, Post-Forming
L-P-512A	Polyethylene Sheet
L-P-513A	Thermoset Phenolic Sheet, Paper Reinforced
L-P-514A	Adhesive Coated, Paper-Backed, Plastic Sheet
L-P-516A	Thermosetting Resins Cast from Monomers Sheet and Rods
L-P-517A	Scribe-Coated Plastic Sheet
L-P-519B	Tracing, Glazed and Matte Finish Plastic Sheet
L-P-523	FEP Fluorocarbon Extruded Sheet and Film
L-P-524	Polyethylene, Laminated, Nylon Reinforced Sheet
L-P-527A	Styrene-Butadiene Sheet
L-P-528A	Cellulose Acetate and Polyester Sheet Adhesive Coated
L-P-535	Polyvinylchloride and PVC-Vinyl Acetate Copolymer Film Rigid
L-P-540	Polyvinylchloride Tube Heavy Wall Rigid
L-P-545	Polyethylene Tubing, Flexible
L-P-590	Polyethylene Sheets, Rods, Tubing
L-P-1036	Polyvinylchloride Heavy Wall
L-P-1040	Polyvinylfluoride Sheets and Strips
L-P-1125	Phenolic Resin Molding Material
L-P-1174	Chlorotrifluoroethylene Copolymer Extruded
L-P-1183	ABS Rigid Molding Material

ASTM/DOD CROSS-REFERENCE TABLES

The following ASTM/DOD corss-reference tables indicate those ASTM standards pertaining to plastics that have been adopted by the Department of Defense as replacements for military or federal specifications or standards.

TABLE 1. Specifications for Thermoplastic Materials

Specification[a]	ASTM Standard
Federal	
L-P-349C	D 707
L-P-385C (6 Oct 1988)	D 1430
L-P-389A (7 Jul 1987)	D 2116
L-P-392A (5 Oct 1988)	D 4181
L-P-393A (29 Jun 1988)	D 3935
L-P-394B (6 Oct 1988)	D 4101
L-P-395C (31 Dec 1985)	D 4066
L-P-396B (6 Oct 1988)	D 4549
L-P-397C	D 706
L-P-398B	D 4549
L-P-399B (6 Oct 1988)	D 4203
L-P-403C (5 Aug 1983)	D 1457
L-P-506A (7 Jul 1987)	D 1463
L-P-523D	D 3368
L-P-1035A (6 Oct 1988)	. . .
L-P-1041A	D 729
L-P-1183B	D 1788
Military	
MIL-A-19468A (9 Oct 1987)	D 1710
MIL-M-19887A(SH) (12 Feb 1985)	D 4066
MIL-M-20693B (26 Aug 1982)	D 4066
MIL-P-22096B (19 Aug 1982)	D 4066
MIL-I-22129C	D 3295
MIL-P-22241B (Type I)	D 3293
MIL-P-22241B (Type II)	D 3308
MIL-P-22241B (Type III)	D 3369
MIL-P-22985B (27 Aug 1981)	D 787
MIL-P-46074B(MR) (4 Jan 1982)	D 1562
MIL-P-46109C(MR) (15 Jan 1986)	D 4101
MIL-P-46160(MR) (26 May 1982)	D 3221
MIL-P-46161A(MR)	D 3220
MIL-P-46174(MR) (2 Aug 1984)	D 4067
MIL-P-46180(MR) (3 Aug 1981)	D 4066
MIL-P-46181(MR) (1 Nov 1985)	D 4066

[a]Date of cancellation is in parentheses. No date indicates that the federal specification is still valid.

TABLE 2. Test Methods for Plastics

Methods in FED-STD-406	ASTM Standard	Methods in FED-STD-406	ASTM Standard
1011	D 638	3041	D 637
1012	D 651	3051	D 523
1013	D 882		
1021	D 695	4011	D 495
1031	D 790	4021	D 150
1032	D 1502	4031	D 149
1041	D 732	4041	D 257
1042	D 3846	4042	D 257
1051	D 953	4052	D 257
1061	D 671	5011	D 792
1062	D 671	5012	D 792
1063	D 2990	5021	D 2734
1071	D 256	5031	D 617
1072 (CWOR)[a]		5041 (CWOR)	
1073 (CWOR)			
1074	D 3029	6011	D 756
1075 (CWOR)		6022	G 23
1081	D 785	6023	G 53
1082	D 2240	6024	D 1435
1083	D 2240	6031 (CWOR)	
1084	D 2240	6041 (Active)	
1091	D 1242	6051 (Active)	
1092	D 1044	6052 (CWOR)	
1093	D 673	6053	F 484
1101	D 621	6054 (CWOR)	
1111	D 229	6061 (CWOR)	
1121	D 1004	6062 (CWOR)	
1131	D 1893	6071	B 117, G 43
		6081	D 1203
2011	D 648	6091	G 21
2021	D 635		
2022	D 568	7011	D 543
2023	D 2863	7021	D 494
2031	D 696	7031	D 570
2032	D 696	7032	E 96
2041	D 569	7041	
2051	D 746	7051	D 793
		7061	C 613
3011	D 542	7071	D 4350
3022	D 1003	7081 (Active)	
3031	E 167		
3032	D 1494		

[a] CWOR, canceled without replacement.

Cross-Reference List of ASTM/ISO Plastics Standards

Similar ASTM Standard	Standard Topic	Similar ISO Standard
	Specimen Preparation	
D0955	Measuring Shrinkage from Mold Dimensions of Molded Thermoplastics	294-4
D3419	In-Line Screw-Injection Molding of Test Specimens from Thermosetting Compounds	10724
D3641	Injection Molding Test Specimens of Thermoplastic Molding and Extrusion Materials	294-1, 2, 3
D4703	Compression Molding Thermoplastic Materials into Test Specimens, Plaques, or Sheets	293
D524	Compression Molding Test Specimens of Thermosetting Molding Compounds	95
D6289	Measuring Shrinkage from Mold Dimensions of Molded Thermosetting Plastics	2577
	Mechanical Properties	
D0256	Determining the Pendulum Impact Resistance of Notched Specimens of Plastics	180
D0638	Tensile Properties of Plastics	527-1, 2
D0695	Compressive Properties of Rigid Plastics	604
D0785	Rockwell Hardness of Plastics and Electrical Insulating Materials	2039-2
D0790	Flexural Properties of Unreinforced and Reinforced Plastics and Electrical Insulating Materials	178
D0882	Tensile Properties of Thin Plastic Sheeting	527-3
D1043	Stiffness Properties of Plastics as a Function of Temperature by Means of a Torsion Test	458-1
D1044	Resistance of Transparent Plastics to Surface Abrasion	9352
D1708	Tensile Properties of Plastics by Use of Microtensile Specimens	6239
D1822	Tensile-Impact Energy to Break Plastics and Electrical Insulating Materials	8256
D1894	Static and Kinetic Coefficients of Friction of Plastic Film and Sheeting	6601
D1922	Propagating Tear Resistance of Plastic Film and Thin Sheeting by Pendulum Methods	6383-2
D1938	Tear Propagation Resistance of Plastic Film and Thin Sheeting by a Single Tear Method	6383-1
D2990	Tensile, Compressive, and Flexural Creep and Creep-Rupture of Plastics	899-1, 2
D3763	High-Speed Puncture Properties of Plastics Using Load and Displacement Sensors	6603-2
D4065	Determining and Reporting Dynamic Mechanical Properties of Plastics	6721-1
D4092	Dynamic Mechanical Measurements on Plastics	6721
D4440	Rheological Measurement of Polymer Melts Using Dynamic Mechanical Procedures	6721-10

Cross-Reference List of ASTM/ISO Plastics Standards—*(Contnined)*

Similar ASTM Standard	Standard Topic	Similar ISO Standard
D5023	Measuring the Dynamic Mechanical Properties of Plastics Using Three Point Bending	6721-3
D5026	Measuring the Dynamic Mechanical Properties of Plastics in Tension	6721-5
D5045	Plane-Strain Fracture Toughness and Strain Energy Release Rate of Plastic Materials	572
D5083	Tensile Properties of Reinforced Thermosetting Plastics Using Straight-Side Specimens	3268
D5279	Measuring the Dynamic Mechanical Properties of Plastics in Torsion	6721

Thermoplastic Materials

D1239	Resistance of Plastic Films to Extraction by Chemicals	6427
D1693	Environmental Stress-Cracking of Ethylene Plastics	4599
D1928	Preparation of Compression-Molded Polyethylene Test Sheets and Test Specimens	293 and 1872-2
D2581	Polybutylene (PB) Plastics Molding and Extrusion Materials	8986-1 and 8986-2
D2765	Determination of Gel Content and Swell Ratio of Crosslinked Ethylene Plastics	10147
D2951	Resistance of Types III and IV Polyethylene Plastics to Thermal Stress-Cracking	TR1837
D4020	Ultra-High-Molecular-Weight Polyethylene Molding and Extrusion Materials	11542-1&2
D4976	Polyethylene Plastics Molding and Extrusion Materials	1872-1&2
D0788	System for Poly(methyl Methacrylate) (PMMA) Molding and Extrusion Compounds	FDIS 8257-1 ISO 8257-2
D4802	Poly(methyl Methacrylate) Acrylic Plastic Sheet	ISO 7823-1 and -2
D3965	Rigid Acrylonitrile–Butadiene–Styrene (ABS) Compounds for Pipe and Fittings	7245
D4203	Styrene Acrylonitrile (SAN) Injection and Extrusion Materials	4894-1 4894-2
D4549	Polystyrene Molding and Extrusion Materials (PS)	1622-1&2 2897-1&2
D4634	Styrene–Maleic Anhydride Materials (S/MA)	10366-1 10366-2
D4673	Acrylonitrile–Butadiene–Styrene (ABS) Molding and Extrusion Materials	2580-1 2580-2
D3012	Thermal Oxidative Stability of Propylene Plastics Using a Biaxial Rotator	4577
D4101	Propylene Plastic Injection and Extrusion Materials	1873-1&2
D5857	Propylene Plastic Injection and Extrusion Materials for International Use	1873-1&2
D1203	Volatile Loss from Plastics Using Activated Carbon Methods	176
D1243	Dilute Solution Viscosity of Vinyl Chloride Polymers	DIS 1628-2

Cross-Reference List of ASTM/ISO Plastics Standards—*(Continued)*

Similar ASTM Standard	Standard Topic	Similar ISO Standard
D1705	Particle Size Analysis of Powdered Polymers and Copolymers of Vinyl Chloride	1624
D1755	Poly(Vinyl Chloride) Resins	1264 (pH only)
D1784	Rigid Poly(Vinyl Chloride) (PVC) Compounds and Chlorinated Poly(Vinyl Chloride) (CPVC) Compounds	1163-1 1163-2
D1823	Apparent Viscosity of Plastisols and Organosols at High Shear Rates by Extrusion Viscometer	3219
D1824	Apparent Viscosity of Plastisols and Organosols at Low Shear Rates by Brookfield Viscometer	2555
D2115	Oven Heat Stability of Poly(Vinyl Chloride) Compositions	305
D2287	Nonrigid Vinyl Chloride Polymer and Copolymer Molding and Extrusion Compounds	2898-1 2898-2
D3030	Volatile Matter (Including Water) of Vinyl Chloride Resins	1269
D3367	Plasticizer Sorption of Poly(Vinyl Chloride) Resins Under Applied Centrifugal Force	4608
D422	Thermal Stability of Poly(Vinyl Chloride) (PVC) Resin	305
D0789	Determination of Relative Viscosity, Melting Point, and Moisture Content of Polyamide (PA)	960 (water only)
D4066	Nylon Injection and Extrusion Materials	1874-1 1874-2
D1430	Polychlorotrifluoroethylene (PCTFE) Plastics	12086-1 12086-2
D1710	Polytetrafluoroethylene (PTFE) Basic Shapes, Rod and Heavy-Walled Tubing	13000-1 13000-2
D2116	FEP-Fluorocarbon Molding and Extrusion Materials	12086-1 12086-2
D3159	Modified ETFE-Fluoropolymer Molding and Extrusion Materials	12086-1 12086-2
D3275	E-CTFE-Fluoroplastic Molding, Extrusion, and Coating Materials	12086-1 12086-2
D3307	PFA-Fluorocarbon Molding and Extrusion Materials	12086-1 12086-2
D3308	PTFE Resin Skived Tape	13000-1 13000-2
D4441	Aqueous Dispepsions of Polytetrafluoroethylene	12086-1 12086-2
D4591	Determining Temperatures and Heats of Transitions of Fluoropolymers by Differential Scanning Calorimetry	12086-1 12086-2
D4894	Polytetrafluoroethylene (PTFE) Granular Molding and Ram Extrusion Materials	12086-1 12086-2
D4895	Polytetrafluoroethylene (PTFE) Resins Produced from Dispersion	12086-1 12086-2
D5575	System and Basis for Specification of Copolymers of Vinylidene Fluoride (VDF) with Other Fluorinated Monomers	12086-1 12086-2

Cross-Reference List of ASTM/ISO Plastics Standards—*(Continued)*

Similar ASTM Standard	Standard Topic	Similar ISO Standard
D5675	Fluoropolymer Micro Powders	12086-1 12086-2
D6040	SG to Test Methods for Unsintered PTFE Extruded Film of Sheet	12086-1 12086-2
D3935	Polycarbonate (PC) Unfilled and Reinforced Material	7391-1 7391-2
D4349	Polyphenylene Ether (PPE) Materials	DIS 15103-1 DIS 15103-2
D4181	Acetal (POM) Molding and Extrusion Materials	DIS 9988-1 DIS 9988-2
D4507	Thermoplastic Polyester (TPES) Materials	7792-1 7792-2
D4550	Thermoplastic Elastomer–Ether–Ester (TEEE)	14910-1 14910-2
D5927	Thermoplastic Polyester (TPES) Injection and Extrusion Materials Based on ISO Test Methods	7792-1 7792-2

Reinforced Thermosetting Plastics

D2584	Ignition Loss of Cured Reinforced Resins	1172
D2734	Void Content of Reinforced Plastics	7822
D3846	In-Plane Shear Strength of Reinforced Plastics	4585(?)
D3914	In-Plane Shear Strength of Pultruded Glass-Reinforced Plastic Rod	4585(?)

Film and Sheeting

D1709	Impact Resistance of Plastic Film by the Free-Falling Dart Method	7765-1, 2
D2578	Wetting Tension of Polyethylene and Polypropylene Films	8296
D2732	Unrestrained Linear Thermal Shrinkage of Plastic Film and Sheeting	11501
D3354	Blocking Load of Plastic Film by the Parallel Plate Method	11502
D4321	Package Yield of Plastic Film	4591

Cellular Plastics and Elastomers

D1056	Flexible Cellular Materials—Sponge or Expanded Rubber	6916-1
D1621	Compressive Properties of Rigid Cellular Plastics	844
D1622	Apparent Density of Rigid Cellular Plastics	845
D1623	Tensile and Tensile Adhesion Properties of Rigid Cellular Plastics	1926
D2126	Response of Rigid Cellular Plastics to Thermal and Humid Aging	2796
D2842	Water Absorption of Rigid Cellular Plastics	2896
D2856	Open-Cell Content of Rigid Cellular Plastics by the Air Pycnometer	4590
D3453	Flexible Cellular Materials—Urethane for Furniture and Automotive [CUSHIONING, Bedding, and Similar Applications]	5999

Cross-Reference List of ASTM/ISO Plastics Standards—*(Continued)*

Similar ASTM Standard	Standard Topic	Similar ISO Standard
D3574-A	Flexible Cellular Materials—Slab, Bonded, and Molded Urethane Foams	845
3574-B1	Flexible Cellular Materials—Slab, Bonded, and Molded Urethane Foams	2439
3574-B2	Flexible Cellular Materials—Slab, Bonded, and Molded Urethane Foams	2439
3574-C	Flexible Cellular Materials—Slab, Bonded, and Molded Urethane Foams	3386-1
3574-D	Flexible Cellular Materials—Slab, Bonded, and Molded Urethane Foams	1856
3574-E	Flexible Cellular Materials—Slab, Bonded, and Molded Urethane Foams	1798
3574-F	Flexible Cellular Materials—Slab, Bonded, and Molded Urethane Foams	8067
3574-G	Flexible Cellular Materials—Slab, Bonded, and Molded Urethane Foams	7231
3574-H	Flexible Cellular Materials—Slab, Bonded, and Molded Urethane Foams	8307
3574-13	Flexible Cellular Materials—Slab, Bonded, and Molded Urethane Foams	3385
3574-14	Flexible Cellular Materials—Slab, Bonded, and Molded Urethane Foams	None
3574-J	Flexible Cellular Materials—Slab, Bonded, and Molded Urethane Foams	2440
3574-K	Flexible Cellular Materials—Slab, Bonded, and Molded Urethane Foams	2440
D3575	Flexible Cellular Materials Made from Olefin Polymers	7214
D3676	Rubber Cellular Cushion Used for Carpet or Rug Underlay	5999
D3748	Evaluating High Density Rigid Cellular Thermoplastics	9054
D4274-A	Testing Polyurethane Polyol Raw Materials: Determination of Hydroxyl Numbers of Polyols	CD 14900
D4274-B	Testing Polyurethane Polyol Raw Materials: Determination of Hydroxyl Numbers of Polyols	CD 14900
D4274-C	Testing Polyurethane Polyol Raw Materials: Determination of Hydroxyl Numbers of Polyols	CD 14900
D4659	Polyurethane Raw Materials: Determination of Specific Gravity of Isocyanates	10349-11
D4660	Polyurethane Raw Materials: Determination of Isomer Content of Isocyanates	TR9372
D4661	Polyurethane Raw Materials: Determination of Total Chlorine in Isocyanates	4615
D4661	Polyurethane Raw Materials: Determination of Total Chlorine in Isocyanates	7725
D4669	Polyurethane Raw Materials: Determination of Specific Gravity of Polyols	10349–11
D4672	Polyurethane Raw Materials: Determination of Water Content of Polyols	CD 14897

Cross-Reference List of ASTM/ISO Plastics Standards—*(Continued)*

Similar ASTM Standard	Standard Topic	Similar ISO Standard
D4876	Polyurethane Raw Materials: Determination of Acidity of Crude or Modified Isocyanates	CD 14898
D4877	Polyurethane Raw Materials: Determination of APHA Color in Isocyanates	6271
D4878	Polyurethane Raw Materials: Determination of Viscosity of Polyols	3219
D4889	Polyurethane Raw Materials: Determination of Viscosity of Crude or Modified Isocyanates	3104
D4890-B	Polyurethane Raw Materials: Determination of Gardner and APHA Color of Polyols	6271
D5155	Polyurethane Raw Materials: Determination of the Isocyanate Content of Aromatic Isocyanates	CD 14896
D6099	PU RM: Acidity in Moderate to High Acidity Aromatic ISOS	CD 14898
	Thermal Properties	
D0635	Rate of Burning and/or Extent and Time of Burning of Self-Supporting Plastics in a Horizontal Position	1210
D0648	Deflection Temperature of Plastics Under Flexural Load	75-1, -2
D0746	Brittleness Temperature of Plastics and Elastomers by Impact	974
D1238	Flow Rates of Thermoplastics by Extrusion Plastometer	1133
D1525	Vicat Softening Temperature of Plastics	306
D1790	Brittleness Temperature of Plastic Sheeting by Impact	8570
D1929	Ignition Properties of Plastics	871
D2863	Measuring the Minimum Oxygen Concentration to Support Candle-Like Combustion of Plastics (Oxygen Index)	4589-2
D3417	Heats of Fusion and Crystallization of Polymers by Thermal Analysis	11357-3
D3801	Measuring the Comparative Extinguishing Characteristics of Solid Plastics in a Vertical Position	1210B
D3835	Determination of Properties of Polymeric Materials by Means of a Capillary Rheometer	11443
D4804	Determining the Flammability Characteristics of Non-Rigid Solid Plastics	9773
D4986	Horizontal Burning Characteristics of Cellular Polymeric Materials	9772
D5025	A Laboratory Burner Used for Small-Scale Burning Tests on Plastic Materials	IEC 695-11-3 and 10093
D5048	Measuring the Comparative Burning Characteristics and Resistance to Burn-Through of Solid Plastics Using 125-MM Flame	10351
D5207	Calibration of 20- and 125-mm Test Flames for Small-Scale Burning Tests on Plastic Materials	IEC 695-11-3, -4

Cross-Reference List of ASTM/ISO Plastics Standards—*(Contnined)*

Similar ASTM Standard	Standard Topic	Similar ISO Standard
	Optical Properties	
D0542	Index of Refraction of Transparent Organic Plastics	489
D1003	Haze and Luminous Transmittance of Transparent Plastics	14782/13468
	Analytical Methods	
D0494	Acetone Extraction of Phenolic Molded or Laminated Products	308/59
D1505	Density of Plastics by the Density-Gradient Technique	1183
D1601	Dilute Solution Viscosity of Ethylene Polymers	1191
D1603	Carbon Black in Olefin Plastics	6964
D1921	Particle Size (Sieve Analysis) of Plastic Materials	4610
D2124	Analysis of Components in Poly(Vinyl Chloride) Compounds Using an Infrared Spectrophotometric Technique	1265
D2857	Dilute Solution Viscosity of Polymers	307, 1628-1
D3749	Residual Vinyl Chloride Monomer in Poly(Vinyl Chloride) Homopolymer Resins by Gas Chromatographic Headspace Technique	6401
D4322	Residual Acrylonitrile Monomer in Styrene–Acrylonitrile Copolymers and Nitrile Rubber by Headspace Gas Chromatography	4581
D5630	Ash Content in Thermoplastics	3451-1, 6427
	DEGRADABLE Plastics	
D5209	Determining the Aerobic Biodegradation of Plastic Materials in the Presence of Municipal Sewer Sludge	14852
D5210	Determining the Anaerobic Biodegradation of Plastic Materials in the Presence of Municipal Sewer Sludge	CD 14853
D5271	Determining the Aerobic Biodegradation of Plastic Materials in an Activated-Sludge-Wastewater-Treatment System	14851
D5338	Determining Aerobic Biodegradation of Plastic Materials Under Controlled Composting Conditions	14855
D5511	Determining Anaerobic Biodegradation of Plastic Materials Under High-Solids Anaerobic-Digestion Conditions	CD 15985

APPENDIX I

CHARTS AND TABLES

Guide to the Effect of Chemical Environments on Plastics

PLASTIC MATERIAL	Aromatic Solvents		Aliphatic Solvents		Chlorinated Solvents		Weak Bases and Salts		Strong Bases		Strong Acids		Strong Oxidants		Esters and Ketones		24-hr Water Absorption
Temperature (°F)	77	200	77	200	77	200	77	200	77	200	77	200	77	200	77	200	% Change by Weight
Acetals*	1–4	2–4	1	2	1–2	4	1–3	2–5	1–5	2–5	5	5	5	5	1	2–3	0.22–0.25
Acrylics	5	5	2	3	5	5	1	3	2	5	4	4–5	5	5	5	5	0.2–0.4
Acrylonitrile–butadiene–styrenes (ABS)	4	5	2	3–5	3–5	5	1	2–4	1	2–4	1–4	5	1–5	5	3–5	5	0.1–0.4
Aramids (aromatic polyamide)	1	1	1	1	1	1	2	3	4	5	3	4	2	5	1	2	0.6
Cellulose acetates (CA)	2	3	2	3	3	4	2	3	3	5	3	5	3	5	5	5	2–7
Cellulose acetate butyrates (CAB)	4	5	1	3	3	4	2	4	3	5	3	5	3	5	5	5	0.9–2.0
Cellulose acetate propionates (CAP)	4	5	1	3	3	4	1	2	3	5	3	5	3	5	5	5	1.3–2.8
Diallyl phthalates (DAP, filled)	1–2	2–4	2	3	2	4	2	3	2	4	1–2	2–3	2	4	3–4	4–5	0.2–0.7
Epoxies	1	2	1	2	1–2	3–4	1	1–2	1	2	2–3	3–4	4	4–5	2	3–4	0.01–0.10
Ethylene copolymers (EVA) (ethylene–vinyl acetates)	5	5	5	5	5	5	1	2	1	5	1	5	1	5	2	5	0.05–0.13
Ethylene/tetrafluoroethylene copolymers (ETFE)	1	1	1	1	1	1	1	1	1	1	1	1	1	1	1	1	<0.03
Fluorinated ethylene propylenes (FEP)	1	1	1	1	1	1	1	1	1	1	1	1	1	1	1	1	<0.01
Perfluoroalkoxies (PFA)	1	1	1	1	1	1	1	1	1	1	1	1	1	1	1	1	<0.03
Polychlorotrifluoroethylenes (CTFE)	1	1	1	1	3	4	1	1	1	1	1	1	1	1	1	1	0.01–0.10

Fluorocarbons

Guide to the Effect of Chemical Environments on Plastics (Continued)

PLASTIC MATERIAL	Aromatic Solvents		Aliphatic Solvents		Chlorinated Solvents		Weak Bases and Salts		Strong Bases		Strong Acids		Strong Oxidants		Esters and Ketones		24-hr Water Absorption
Temperature (°F)	77	200	77	200	77	200	77	200	77	200	77	200	77	200	77	200	% Change by Weight
Polytetrafluoroethylenes (TFE)	1	1	1	1	1	1	1	1	1	1	1	1	1	1	1	1	0
Furans	1	1	1	1	1	1	2	2	2	2	1	1	5	5	1	1	0.01–0.20
Ionomers	2	4	1	4	4	4	1	4	1	4	2	4	1	5	1	4	0.1–1.4
Melamines (filled)	1	1	1	1	1	1	2	3	2	3	2	3	2	3	1	2	0.01–1.30
Nitriles (high barrier alloys of ABS or SAN)	1	4	1	2–4	1–4	2–5	1	2–4	1	2–4	2–5	5	3–5	5	1–5	5	0.2–0.5
Nylons	1	1	1	1	1	2	1	2	2	3	5	5	5	5	1	1	0.2–1.9
Phenolics (filled)	1	1	1	1	1	1	2	3	3	5	1	1	4	5	2	2	0.1–2.0
Polyallomers	2	4	2	4	4	5	1	1	1	1	1	3	1	4	1	3	<0.01
Polyamide-imides	1	1	1	1	2	3	1	1	3	4	2	3	2	3	1	1	0.22–0.28
Polyarylsulfones (PAS)	4	5	2	3	4	5	1	2	2	2	1	1	2	4	3	4	1.2–1.8
Polybutylenes (PB)	3	5	1	5	4	5	1	2	1	3	1	3	1	4	1	3	<0.01–0.3
Polycarbonates (PC)	5	5	1	1	5	5	1	5	5	5	1	1	1	1	5	5	0.15–0.35
Polyesters (thermoplastic)	2	5	1	3–5	3	5	1	3–4	2	5	3	4–5	2	3–5	2	3–4	0.06–0.09
Polyesters (thermoset-glass fiber filled)	1–3	3–5	2	3	2	4	2	3	3	5	2	3	2	4	3–4	4–5	0.01–2.50
Polyethylenes (LDPE-HDPE—low density to high density)	4	5	4	5	4	5	1	1	1	1	1–2	1–2	1–3	3–5	2	3	0.00–0.01

Material																	Water absorption (%)
Polyethylenes (UHMWPE—ultrahigh molecular weight)	3	4	3	4	3	4	1	1	1	1	1	1	1	1	1	4	<0.01
Polyimides	1	1	1	1	1	1	2	3	4	5	3	2	2	5	1	1	0.3–0.4
Polyphenylene oxides (PPO) (modified)	4	5	2	3	4	5	1	1	1	1	1	1	2	2	2	3	0.06–0.07
Polyphenylene sulfides (PPS)	1	2	1	1	1	2	1	1	1	1	1	1	1	2	1	1	<0.05
Polyphenylsulfones	4	4	1	1	5	5	1	1	1	1	1	1	1	1	3	4	0.5
Polypropylenes (PP)	2	4	2	4	2–3	4–5	1	1	1	1	1	2–3	2–3	4–5	2	4	0.01–0.03
Polystyrenes (PS)	4	5	4	5	5	5	1	5	1	5	4	5	5	5	4	5	0.03–0.60
Polysulfones	4	4	1	1	5	5	1	1	1	1	4	1	1	1	3	4	0.2–0.3
Polyurethanes (PUR)	3	4	2	3	4	5	2–3	3–4	2–3	3–4	2–3	4	3–4	4	4	5	0.02–1.50
Polyvinyl chlorides (PVC)	4	5	1	5	5	5	1	5	1	5	1	2	5	5	5	5	0.04–1.00
Polyvinyl chlorides—chlorinated (CPVC)	4	4	1	2	5	5	1	2	1	2	1	2	3	3	4	5	0.04–0.45
Polyvinylidene fluorides (PVDF)	1	1	1	1	1	1	1	1	1	2	1	1	2	2	3	5	0.04
Silicones	4	4	2	3	4	5	1	2	4	5	3	4	4	5	2	4	0.1–0.2
Styrene acrylonitriles (SAN)	4	5	3	4	3	5	1	3	1	3	1	3	3	4	4	5	0.20–0.35
Ureas (filled)	1	3	1	3	1	3	2	3	2	3	4	2	5	3	1	2	0.4–0.8
Vinyl esters (glass fiber filled)	1	3	1–2	2–4	1–2	4	1	3	1	3	1	2	2	3	3–4	4–5	0.01–2.50

RATING CODE: 1. no effect or inert, **2.** slight effect, **3.** mild effect, **4.** softening or swelling, **5.** severe degradation

MATERIALS
Some plastics incorporate a broad family of materials with widely differing properties in these environments. Special grades are also available with properties outside the ranges listed.

ENVIRONMENT
Chemical effects in plastics are defined by conditions, which include stress cracking, etching or chemical attack (ASTM D-543), staining (ASTM D-543), and swelling or solvating (ASTM D-2299) and swelling or solvating (ASTM D-543). These conditions are influenced by the time of exposure, temperature and concentrations or combination of these.
* Courtesy of *Plastics World* and Shell Chemical Co.

TEMPERATURE
In many cases the difference between the 77°F and 200°F ratings is simply a heat distortion effect with no degradation.

WATER ABSORPTION
In ordinary tap and salt water there is almost no chemical attack, however, ion-free water (such as distilled or deionized water) can show varying degrees of attack. Water absorption values have been included to quantify the most common effect of water.
*Widespread ratings exist since acetal copolymers are more inert than homopolymers in many "corrosive" environments.

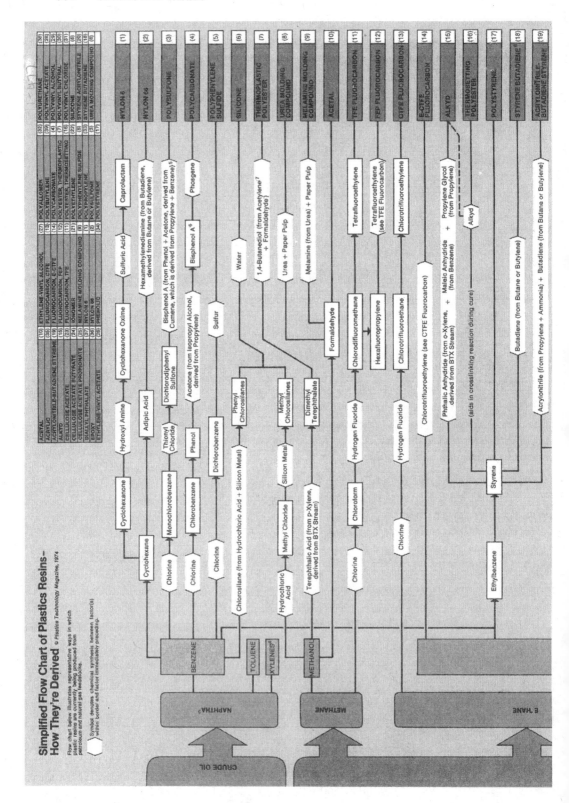

Simplified Flow Chart of Plastics Resins—
How They're Derived © Plastics Technology Magazine, 1974

Flow chart below illustrates representative ways in which plastic resins are currently being produced from petroleum and natural gas feedstocks.

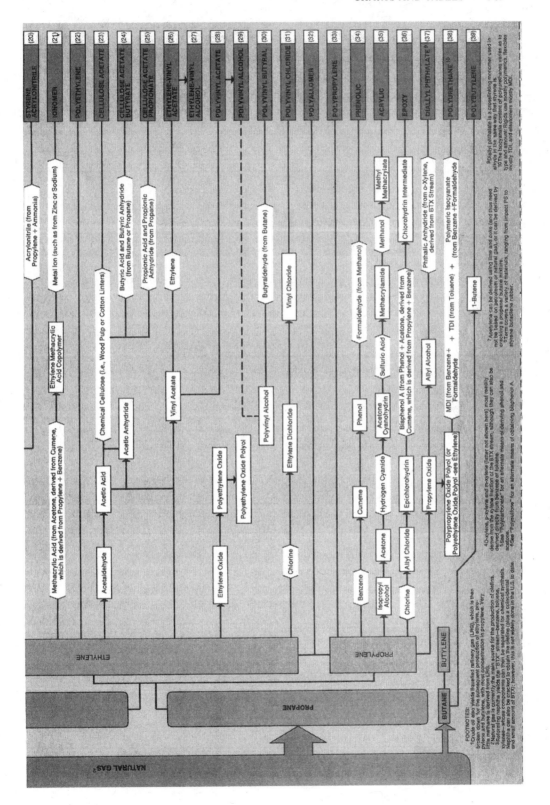

How to Convert Centigrade Temperature to Fahrenheit and Vice Versa

Centigrade °C = 5/9 (°F − 32) Fahrenheit °F = (9/5 × °C) + 32

°C		°F	°C		°F	°C		°F	°C		°F
−17.8	0	32	8.89	48	118.4	35.6	96	204.8	271	520	968
−17.2	1	33.8	9.44	49	120.2	36.1	97	206.6	277	530	986
−16.7	2	35.6	10.0	50	122.0	36.7	98	208.4	282	540	1004
−16.1	3	37.4	10.6	51	123.8	37.2	99	210.2	288	550	1022
−15.6	4	39.2	11.1	52	125.6	37.8	100	212.0	293	560	1040
−15.0	5	41.0	11.7	53	127.4	38	100	212	299	570	1058
−14.4	6	42.8	12.2	54	129.2	43	110	230	304	580	1076
−13.9	7	44.6	12.8	55	131.0	49	120	248	310	590	1094
−13.3	8	46.4	13.3	56	132.8	54	130	266	316	600	1112
−12.8	9	48.2	13.9	57	134.6	60	140	284	321	610	1130
−12.2	10	50.0	14.4	58	136.4	66	150	302	327	620	1148
−11.7	11	51.8	15.0	59	138.2	71	160	320	332	630	1166
−11.1	12	53.6	15.6	60	140.0	77	170	338	338	640	1184
−10.6	13	55.4	16.1	61	141.8	82	180	356	343	650	1202
−10.0	14	57.2	16.7	62	143.6	88	190	374	349	660	1220
−9.44	15	59.0	17.2	63	145.4	93	200	392	354	670	1238
−8.89	16	60.8	17.8	64	147.2	99	210	410	360	680	1256
−8.33	17	62.6	18.3	65	149.0	100	212	413	366	690	1274
−7.78	18	64.4	18.9	66	150.8	104	220	428	371	700	1292
−7.22	19	66.2	19.4	67	152.6	110	230	446	377	710	1310
−6.67	20	68.0	20.0	68	154.4	116	240	464	382	720	1328
−6.11	21	69.8	20.6	69	156.2	121	250	482	388	730	1346
−5.56	22	71.6	21.1	70	158.0	127	260	500	393	740	1364
−5.00	23	73.4	21.7	71	159.8	132	270	518	399	750	1382
−4.44	24	75.2	22.2	72	161.6	138	280	536	404	760	1400
−3.89	25	77.0	22.8	73	163.4	143	290	554	410	770	1418
−3.33	26	78.8	23.3	74	165.2	149	300	572	416	780	1436
−2.78	27	80.6	23.9	75	167.0	154	310	590	421	790	1454
−2.22	28	82.4	24.4	76	168.8	160	320	608	427	800	1472
−1.67	29	84.2	25.0	77	170.6	166	330	626	432	810	1490
−1.11	30	86.0	25.6	78	172.4	171	340	644	438	820	1508
−0.56	31	87.8	26.1	79	174.2	177	350	662	443	830	1526
−0	32	89.6	26.7	80	176.0	182	360	680	449	840	1544
0.56	33	91.4	27.2	81	177.8	188	370	698	454	850	1562
1.11	34	93.2	27.8	82	179.6	193	380	716	460	860	1580
1.67	35	95.0	28.3	83	181.4	199	390	734	466	870	1598
2.22	36	96.8	28.9	84	183.2	204	400	752	471	880	1616
2.78	37	98.6	29.4	85	185.0	210	410	770	477	890	1634
3.33	38	100.4	30.0	86	186.8	216	420	788	482	900	1652
3.89	39	102.2	30.6	87	188.6	221	430	806	488	910	1670
4.44	40	104.0	31.1	88	190.4	227	440	824	493	920	1688
5.00	41	105.8	31.7	89	192.2	232	450	842	499	930	1706
5.56	42	107.6	32.2	90	194.0	238	460	860	504	940	1724
6.11	43	109.4	32.8	91	195.8	243	470	878	510	950	1742
6.67	44	111.2	33.3	92	196.7	249	480	896	516	960	1760
7.22	45	113.0	33.9	93	199.4	254	490	914	521	970	1778
7.78	46	114.8	34.4	94	201.2	260	500	932	527	980	1796
8.33	47	116.6	35.0	95	203.0	266	510	950	532	990	1814

Decimal Equivalents of Fractions of One Inch

1/64	0.015	625	17/64	0.265	625	33/64	0.515	625	49/64	0.765	625
1/32	0.031	250	9/32	0.281	250	17/32	0.531	250	25/32	0.781	250
3/64	0.046	875	19/64	0.296	875	35/64	0.546	875	51/64	0.796	875
1/16	0.062	500	5/16	0.312	500	9/16	0.562	500	13/16	0.812	500
5/64	0.078	125	21/64	0.328	125	37/64	0.578	125	53/64	0.828	125
3/32	0.093	750	11/32	0.343	750	19/32	0.593	750	27/32	0.843	750
7/64	0.109	375	23/64	0.359	375	39/64	0.609	375	55/64	0.859	375
1/8	0.125	000	3/8	0.375	000	5/8	0.625	000	7/8	0.875	000
9/64	0.140	625	25/64	0.390	625	41/64	0.640	625	57/64	0.890	625
5/32	0.156	250	13/32	0.406	250	21/32	0.656	250	29/32	0.906	250
11/64	0.171	875	27/64	0.421	875	43/64	0.671	875	59/64	0.921	875
3/16	0.187	500	7/16	0.437	500	11/16	0.687	500	15/16	0.937	500
13/64	0.203	125	29/64	0.453	125	45/64	0.703	125	61/64	0.953	125
7/32	0.218	750	15/32	0.468	750	23/32	0.718	750	31/32	0.968	750
15/64	0.234	375	31/64	0.484	375	47/64	0.734	375	63/64	0.984	375
1/4	0.250	000	1/2	0.500	000	3/4	0.750	000	1	1.000	000

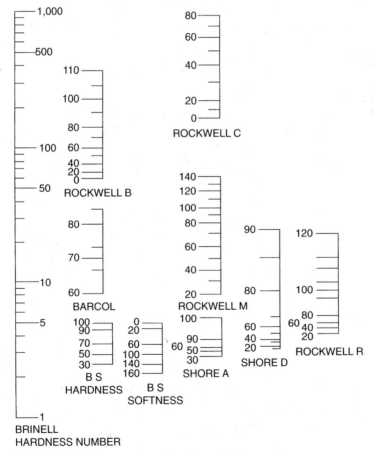

Approximate comparison of hardness scales.

COMPARISON BETWEEN SHORE AND ROCKWELL HARDNESS RANGES

Hardness can be considered in several ways. A Rockwell number is directly related to the resistance of a rigid material to penetration by a steel ball. Soft plastics and rubbers are tested for resistance to penetration by the shore hardness method. Here is a general comparison between the most common scales of hardness:

CONVENTIONAL ELASTOMERS									THERMOPLASTICS									THEROMO-SETS	
					THERMOPLASTIC ELASTOMERS														
20 A	30 A	40 A	50 A	60 A	70 A	80 A	90 A	100 A											
							40 D	50 D	60 D	70 D	80 D	90 D							
									50 R	60 R	70 R	80 R	90 R	100 R	110 R	120 R	130 R	140 R	150 R
SOFT																			HARD

Miscellaneous Measurement Conversions

Multiply This	By This	To Obtain This
acre	43,560	ft^2
	1.562×10^{-3}	sq mi
	160	sq rod
	0.4047	hectare
Angstrom unit	3.937×10^{-9}	in.
	1×10^{-4}	μ
	0.1	m$^\mu$
	1×10^{-8}	cm
	1×10^{-10}	m
are	3.954	sq rod
	100	m^2
	119.6	yd^2
	0.02471	acre
atmosphere	760	mm Hg at 0°C
	29.92	in Hg at 0°C
	406.79	in H$_2$O at 4°C
	33.899	ft H$_2$O at 4°C
	407.16	in H$_2$O at 15°C
	33.93	ft H$_2$O at 15°C
	1.0333	kg/cm^2
	10.333	kg/m^2
	1.01325×10^6	dyne/cm^2
	14.696	lb/in.^{2}
	2,116.32	lb/ft^2
	1.0133	bar
bar	0.9869	atm
	750	mm Hg (0°C)
	10,197	kg/m^2
	1×10^6	dyne/cm^2
	14.50	lb/in.^{2}
barrel	42	gal
Btu	777.98	ft-lb
	3.930×10^{-4}	hp-hr
	2.931×10^{-4}	kwh
	2.520×10^{-1}	cal (kg)
	107.6	kg-m
	1,055	joule
Btu/sec	1,055	watt
Btu/min	0.0236	hp
	17.58	watt
Btu/lb	0.55556	cal/gm
bushel	35.238	l
	0.3524	hectoliter
	4	peck
	1.2444	cu ft
calorie (gram)	3.968×10^{-3}	Btu
calorie/gm	1.8	Btu/lb
carat	0.2	g

Miscellaneous Measurement Conversions—*(Continued)*

Multiply This	By This	To Obtain This
cental	100	lb
centimeter	0.032808	ft
	0.393700	in.
	10,000	μ
	1×10^8	Å
	0.01	m
	10	mm
centimeter of Hg (0°C)	0.01316	atm
	0.1934	lb/in.^{2}
	27.845	lb/ft^2
	0.44604	ft H_2O at 4°C
	135.95	kg/m^2
	13,332	dyne/cm^2
centimeter^2	100	mm^2
	1×10^{-4}	m^2
	0.1550	in.^{2}
	0.00108	ft^2
	1.196×10^{-4}	yd^2
centimeter^3	0.061	in.^{3}
	3.531×10^{-5}	ft^3
centimeters^3	1.3079×10^{-6}	yd^3
	1×10^{-6}	in.^{3}
	0.99997	ml
	0.0338	oz (US fl)
	0.0351	oz (Brit. fl)
	10.567×10^{-4}	qt (US fl)
	8.7988×10^{-4}	qt (Brit. fl)
	2.6417×10^{-4}	gal (US)
	2.1997×10^{-4}	gal (Brit.)
centimeter/second	0.0328	fl/sec
	1.9685	ft/min
	0.06	m/min
	3.728×10^{-4}	mi/min
	0.02237	mi/hr
	0.03600	km/hr
centipoise	6.72×10^{-4}	lb/sec-ft
	3.60	kg/hr-m
chain (surveyors)	100	link
	66	ft
	20.117	m
cheval-vapeur	735.499	watt
	0.9863	hp
cord	128	ft^3
	3.625	stere
day	86,400	sec
	1,440	min
	0.143	wk
	0.0028	yr

Miscellaneous Measurement Conversions— *(Continued)*

Multiply This	By This	To Obtain This
decimeter	3.937	in.
	0.328	ft
dram (dry)	0.0625	oz (wt)
	1.7718	g
dram (fluid)	3.6967	cc
	0.125	oz (fl)
dyne	1.020×10^{-3}	g
	2.248×10^{-6}	lb
dyne/centimeter^2	1×10^{-6}	bar
	10.197×10^{-4}	g/cm^2
	1.4504×10^{-5}	lb/in.^{2}
ell	45	in.
erg	9.4805×10^{-10}	Btu
	1	dyne-cm
	7.376×10^{-8}	ft-lò
	2.388×10^{-11}	cal (kg)
fathom	6	ft
firkin	9	gal
foot	30.48	cm
	0.3048	m
foot H$_2$O (4°C)	0.0295	atm
	0.883	in Hg (0°C)
	2.419	cm Hg (0°C)
	0.4335	lb/in.^{2}
	62.427	lb/ft^2
	304.79	kg/m^2
foot^2	0.111	yd^2
	3.587×10^{-8}	mi^2
	2.296×10^{-5}	acre
	0.0929	m^2
	929	cm^2
foot^3	1,728	in.^{3}
	0.037	yd^3
	28,316	cc
	0.02832	m^3
	28.316	l
	7.4805	gal (US)
	6.2288	gal (Brit.)
foot^3 H$_2$O	62.42	lb (4°C)
	62.36	lb (15°C)
foot^3/min	0.4720	l/sec
	472	cc/sec
	448.831	gal/min
foot-pound	0.1383	kg-m
foot-pound/min	3.030×10^{-5}	hp
foot/min	0.01667	ft/sec
	0.0114	mph
	0.508	cm/sec

Miscellaneous Measurement Conversions—*(Continued)*

Multiply This	By This	To Obtain This
	0.005	m/sec
	0.3048	m/min
	0.0183	km/hr
foot/second	0.01136	mi/min
	0.6818	mph
	30.48	cm/sec
	18.288	m/min
	1.097	km/hr
furlong	660	ft
	220	yd
	40	rod
gallon (US)	0.8327	gal (Brit.)
	128	oz
	8	pt
	3785.4	cc
	3.785	l
	0.00379	m^3
	231	in.^{3}
	0.1337	ft^3
	0.00495	yd^3
	8.3378	lb H_2O (60°F)
gallon (British)	1.2009	gal (US)
	4,546	cc
	4.546	l
	277.419	in.^{3}
	0.16054	ft^3
	10	lb H_2O (60°F)
	160	oz (Brit fl)
gallon/minute (US)	8.0208	ft^3/hr
	0.06308	l/sec
gill (US)	4	oz
grain	0.00229	oz (avoir.)
	0.0648	g
gram	0.0353	oz (avoir.)
	0.0022	lb
	15.432	grain
gram/centimeter^2	9.6784×10^{-4}	atm
	0.7356	mm Hg (0°C)
	0.0289	in Hg (0°C)
	0.3284	ft H_2O (60°F)
	0.0142	lb/in.^{2}
	2.0482	lb/ft^2
	980.665	dyne/cm^2
	10	kg/m^2
gram/centimeter^3	62.428	lb/ft^3
	8.345	lb/gal (US)
	0.0361	lb/in.^{3}
	1	g/ml

Miscellaneous Measurement Conversions— *(Continued)*

Multiply This	By This	To Obtain This
gram/liter	1,000	ppm
	0.0624	lb/ft^3
gravity	32.174	ft/sec^2
	980.665	cm/sec^2
hand	4	in.
hectare	2.471	acre
	100	are
	107,640	ft^2
	10,000	m^2
hogshead	63	gal
horsepower	42.418	Btu/min
	33,000	ft-lb/min
horsepower	550	ft-lb/sec
	0.7457	kw
	10.688	cal (kg)/min
	1.014	cheval-vapeur
hour	3,600	sec
	0.04167	day
	0.0059	wk
hundredweight (cwt)	100	lb (short)
	112	lb (long)
inch	1,000	mil
	0.083	ft
	0.02778	yd
	2.54	cm
	0.0254	m
	2.54×10^8	Å
inch Hg (0°C)	13.595	in. H$_2$O
	1.133	ft H$_2$O
	0.0334	atm
	25.4	mm Hg
	34.5	g/cm^2
	345.3	kg/m^2
	33,864	dyne/cm^2
	0.4912	lb/in.^{2}
	70.73	lb/ft^2
inch H$_2$O (4°C)	0.00245	atm
	0.07355	in Hg
	25.399	kg/m^2
	5.2022	lb/ft^2
	0.0361	lb/in.^{2}
inch^2	6.4516	cm^2
	6.4516×10^{-4}	m^2
	0.0069	ft^2
	0.00077	yd^2
inch^3	16.387	cc
	5.7870×10^{-4}	ft^3
	1.6387×10^{-5}	m^3

Miscellaneous Measurement Conversions— *(Continued)*

Multiply This	By This	To Obtain This
	2.14335×10^5	yd^3
	0.003606	gal (Brit.)
	0.004329	gal (US)
	0.01639	l
	0.5541	oz (US fl)
	0.01488	qt (dry)
	0.01732	qt (US fl)
kilogram	35.274	oz (avoir)
	2.2046	lb
	9.842×10^{-4}	ton (long)
	0.0011	ton (short)
kilogram-meter	7.2330	ft-lb
kilogram/meter	0.67197	lb/ft
kilogram/meter^2	0.07356	mm Hg (0°C)
	0.00142	lb/in.^{2}
	9.6777×10^{-5}	atm
	0.20482	lb/ft^2
kilogram/meter^3	0.06243	lb/ft^3
	0.001	g/cc
kiloliter	35.317	ft^3
kilometer	3,280.8	ft
	0.53959	mi (naut)
	0.62137	mi
	1,093.6	yd
kilometer/hour	27.7778	cm/sec
	54.68	ft/min
	0.9113	ft/sec
	0.5396	knot
	16.667	m/min
	0.27778	m/sec
kilometer^2	1.076×10^7	ft^2
	1.196×10^6	yd^2
	0.386	mi^2
	1×10^6	m^2
	247.1	acre
kilowatt	56.884	Btu/min
	1.3410	hp
kilowatt-hour	3,413	Btu
	1.3410	hp-hr
knot	51.479	cm/sec
	6,080.2	ft/hr
	1.15155	mph
league	3	mi
light-year	9.4637×10^{12}	km
	5.88×10^{12}	mi
link	0.66	ft
liter	61.025	in.^{3}
	0.035	ft^3

Miscellaneous Measurement Conversions—*(Continued)*

Multiply This	By This	To Obtain This
	0.2642	gal (US)
	33.814	oz (US fl)
	1.05668	qt (US fl)
	0.8799	qt (Brit fl)
liter/second	15.8507	gal/min
meter	1×10^{10}	Å
	3.2808	ft
	39.370	in.
	5.3959×10^{-4}	mi (naut.)
	6.2137×10^{-4}	mi
	1.0936	yd
	1×10^{9}	μ
meter/minute	0.05468	ft/sec
	0.06	km/hr
	0.03728	mph
meter^2	0.01	are
	2.471×10^{-4}	acre
	10.7639	ft^2
	1,550	in.^{2}
	3.8610×10^{-7}	mi^2
	1.19598	yd^2
meter^3	35.314	ft^3
	61,023	in.^{3}
	1.3079	yd^3
	264.73	gal
	999.973	l
	1,056.7	qt (US fl)
micron	1×10^{4}	Å
	1×10^{-4}	cm
	3.937×10^{-5}	in.
mile (statute)	8	furlong
	63,360	in.
	1.60935	km
	1,609.35	m
	0.8684	mi (naut)
	320	rod
	1,760	yd
mile (nautical)	6,080.2	ft
	1.85325	km
	1.1516	mi (statute)
miles/hour	44.704	cm/sec
	88	ft/min
	0.8684	knot
	26.82	m/min
milligram	3.5274×10^{-5}	oz (avoir)
	5.6438×10^{-4}	dram
	2.2046×10^{-6}	lb

Miscellaneous Measurement Conversions—(Continued)

Multiply This	By This	To Obtain This
millimeter	0.03937	in.
	1,000	μ
millimeter Hg (0°C)	1.316×10^{-3}	atm
	1,333.22	dyne/cm^2
	1.3595	g/cm^2
	13.595	kg/m^2
	2.7845	lb/ft^2
millimicron	10	Å
	1×10^{-7}	cm
minute	6.9444×10^{-4}	day
	9.9206×10^{-5}	wk
month (mean calendar)	30.4202	day
	730.085	hr
month (mean calendar)	43,805	min
	2.6283×10^8	sec
ounce (avoirdupois)	16	dram
	437.5	grain
	0.91146	oz (troy)
	3.125×10^{-5}	ton (short)
	28.35	g
ounce (fluid)	29.5737	cc
	1.8047	in.^{3}
	8	dram
	0.25	gill
	0.029573	l
	0.03125	qt
ounce (British fl)	28.413	cc
pace	2.5	ft
parts per million	0.0584	grain/gal
peck	537.6	in.^{3}
	8.8096	l
	16	pint
pint (dry)	550.61	cc
	33.6003	in.^{3}
	0.5506	l
pint (fluid)	473.179	cc
	28.875	in.^{3}
	128	dram
	4	gill
	0.473168	l
pound	256	dram
	7,000	grain
	453.5924	g
	0.45359	kg
	1.2153	lb (troy)
pound/foot	1.48816	kg/m
pounds/foot^2	4.7252×10^{-4}	atm
	4.7880×10^{-4}	bar
	478.78	dyne/cm^2

Miscellaneous Measurement Conversions— *(Continued)*

Multiply This	By This	To Obtain This
	0.48824	g/cm^2
	4.8824	kg/m^2
	0.35913	mm Hg (0°C)
	6.9445×10^{-3}	$lb/in.^2$
pound/in.^{2}	0.068046	atm
	70.307	g/cm^2
	703.07	kg/m^2
	51.715	mm Hg (O°C)
pound/inch^3	27.68	g/cc
pound/foot^3	0.016018	g/cc
	16.018	kg/m^3
	5.787×10^{-4}	$lb/in.^3$
quart (dry)	0.03125	bushel
	1,101.23	cc
	0.03889	ft^3
	1.1012	l
quart (liquid)	946.358	cc
	57.749	in.^{3}
	0.03342	ft^3
	256	dram
	8	gill
	0.946333	l
quire	24	sheet
rod	0.25	chain
	16.5	feet
	0.025	furlong
	198	in.
	25	link
	5.029216	m
	3.125×10^{-3}	mi
rood	0.25	acre
second	0.01667	min
	2.7778×10^{-4}	hr
	1.1574×10^{-5}	day
slug	32.174	lb
	14.594	kg
span	9	in.
stere	1	m^3
	999.973	l
stone	14	lb
	6.3503	kg
ton	20	cwt
	907.1846	kg
	0.89286	ton (long)
ton (long)	22.4	cwt
	1,016.047	kg.
	1.12	ton (short)

Miscellaneous Measurement Conversions—*(Continued)*

Multiply This	By This	To Obtain This
ton (metric)	1,000	kg
	2,204.62	lb
watt	44.254	ft-lb/min
	$1.34.0 \times 10^{-3}$	hp
	3.41304	Btu/hr
week	168	hr
	10,080	min
	604,800	sec
yard	91.4402	cm
	5.68182×10^{-4}	mi
year	365.256	day
	8,766.144	hr

Conversions of Units Useful in Plastics Technology

Property	Typical USA Units	SI Units	Conversion Factor
Specific gravity			1.000 E+00
Specific volume	in.^{3}/lb$_m$	m^3/kg	3.613 E−05
Elongation	%	%	1.000 E+00
Strength[a]	lb$_f$/in.^{2}	MPa	6.895 E−03
Modulus[a]	lb$_f$/in.^{2}	MPa	6.895 E−03
Impact strength			
(a) Izod	ft·lb/in notch	J/m notch	5.423 E+01
(b) Charpy	ft·lb/in.^{2}	kJ/m^2	2.135 E+00
(c) FDI, DDI	ft·lb	J	1.356 E+00
(d) Gradner	ft·lb	J	1.113 E−01
(e)Tensile	ft·lb/in.^{2}	kJ/m^2	2.135 E+00
Vicat	°F	°C	5/9 (°F−32)
HDTUL	°F @ 264 PSI	°C @ 1.82 MPa	5/9 (°F−32)
	°F @ 66 PSI	°C @ 0.455 MPa	5/9 (°F−32)
Thermal conductivity	BTU in./h·ft^2°F	W/m/°C	1.441 E−01
Specific heat	BTU/lb$_m$/°F	J/kg/°C	4.184 E−01
Thermal expansion	in./in./°F	mm/mm/°C	1.796 E+00
Viscosity	poise	Pa·sec	1.000 E−01
Dynamic mechanical Properties	dynes/cm^2	MPa	1.000 E−07

[a] Strength and modulus can be tensile, flexural, shear, or compressive.

Conversion Table for the Material Scientist

Quantity	Unit	Symbol	Break down
Force	Newton	N	$1\,kg*m/(s*s)$
Force	Dyne	Dyne	$1\,g*cm/(s*s)$
Pressure	Pascal	Pa	$1\,N/(m*m)$
Energy	Joule	J	$1\,N*m$
Absolute viscosity	Pascal second	Pa*s	$1\,N*s/(m*m)$
Absolute viscosity	Poise	Poise	$1\,g/(cm*s)$

Other Unit Conversions:

Absolute viscosity = Density times kinematic viscosity
1 Tex = 1 gram mass per 1,000 meters
1 Denier = 1 gram mass per 9,000 meters
1 Radian = (57.30) degrees
1 Calorie = (4.187) joules
1 Joule = (10^7) erg
1 Joule = (0.000948) BTU
1 Newton = (10^5) dyne

Definitions:
1. English units are defined by ASTM as "US Customary."
2. The pound force (lbf) is the force required to support the standard pound body against gravity.
3. The kilogram force (kgf) is the force required to support the standard kilogram mass against gravity.
4. The kilogram mass (kgm) is equal to the mass of the international kilogram prototype.
5. The Newton is that force which will impart to a 1 kilogram mass an acceleration of 1 meter per sec and per second.
6. Force = mass * acceleration
7. Weight = mass * gravity
8. 1 kai = 10^3 pai; 1 Mai = 10^6 pai; 1 pai = 1 lbf/in.^{2}
9. The standard value of gravity is $9.807\,m/s^2$ or $32.17\,ft/s^2$.
10. Prefixes: Mega (M) 10^6, kilo (k) 10^3, milli (m) 10^{-3}, micro (μ) 10^{-6}.

Conversion Table for the Material Scientist

Example:

$$(5.04\,\text{in})\frac{25.40\text{ mm}}{1\text{ in}} = 128\text{ mm}$$

Example:

$$(9.68\text{ ft*lb})\frac{0.1383\text{ kgf*m}}{1\text{ ft*lb}} = 1.34\text{ kgf*m}$$

	ENGL = (?) SI	SI = (?) ENGL	SI = (?) MET	MET = (?) SI	ENGL = (?) MET	MET = (?) ENGL
Force	1 lbf = (4.448) N	1 N = (0.2248) lbf	1 N = (0.1020) kgf	1 kgf = (9.807) N	1 lbf = (0.4536) kgf	1 kgf = (2.205) lbf
Length	1 in. = (25.40) mm	1 mm = (.03937) in.	1 mm = (1) mm	1 mm = (1) mm	1 in. = (25.40) mm	1 mm = (.03937) in.
Stress	1 psi = (6895) Pa	$1\text{ Pa} = (1.450*10^{-4})\text{ psi}$	$1\text{ Pa} = (0.1020)\dfrac{\text{kgf}}{\text{m}^2}$	$1\dfrac{\text{kgf}}{\text{m}^2} = (9.807)\text{ Pa}$	$1\text{ psi} = (703.2)\dfrac{\text{kgf}}{\text{m}^2}$	$1\dfrac{\text{kgf}}{\text{m}^2} = (1.422*10^{-3})\text{ Ps}$
Speed	$1\dfrac{\text{in.}}{\text{min}} = (25.40)\dfrac{\text{mm}}{\text{min}}$	$1\dfrac{\text{mm}}{\text{min}} = (.03937)\dfrac{\text{in.}}{\text{min}}$	$1\dfrac{\text{mm}}{\text{min}} = (1)\dfrac{\text{mm}}{\text{min}}$	$1\dfrac{\text{mm}}{\text{min}} = (1)\dfrac{\text{mm}}{\text{min}}$	$1\dfrac{\text{in.}}{\text{min}} = (25.40)\dfrac{\text{mm}}{\text{min}}$	$1\dfrac{\text{mm}}{\text{min}} = (.03937)\dfrac{\text{in.}}{\text{min}}$
Energy	1 ft*lb = (1.356) J	1 J = (0.7376) ft*lb	1 J = (0.1020) kgf*m	1 kgf*m = (9.807) J	1 ft*lb = (0.1383) kgf*m	1 kgf*m = (7.234) ft*lb
Mass	1 lbm = (0.4536) kg	1 kg = (2.205) lbm	1 kg = (1) kg	1 kg = (1) kg	1 lbm = (0.4536) kg	1 kg = (2.205) lbm
Temperature	F = (K*1.8) − 459.7	K = (F + 459.7)/1.8	K = C + 273.15	C = K − 273.15	F = (C*1.8) + 32	C = (F − 32)/1.8
Absolute viscosity	$1\dfrac{\text{lbf*s}}{\text{ft}^2} = (47.87)\text{Pa*s}$	$1\text{Pa*s} = (.02089)\dfrac{\text{lbf*s}}{\text{ft}^2}$	1 Pa * s = (10) Poise	1 Poise = (.1) Pa * s	$1\dfrac{\text{lbf*s}}{\text{ft}^2} = (478.7)\text{poise}$	$1\text{ Poise} = (2.089*10^{-3})\dfrac{\text{lbf*s}}{\text{ft}^2}$
Modulus	1 psi = (6895) Pa	$1\text{ Pa} = (1.450*10^{-4})\text{ psi}$	$1\text{ Pa} = (0.1020)\dfrac{\text{kgf}}{\text{m}^2}$	$1\dfrac{\text{kgf}}{\text{m}^2} = (9.807)\text{Pa}$	$1\text{ psi} = (703.2)\dfrac{\text{kgf}}{\text{m}^2}$	$1\dfrac{\text{kgf}}{\text{m}^2} = (1.42*10^{-3})\text{ psi}$
Fracture toughness	$1\text{ksi}\sqrt{\text{in.}} = (1.099)\text{Mpa}\sqrt{\text{m}}$	$1\text{MPa}\sqrt{\text{m}} = (.9100)\text{ksi}\sqrt{\text{in.}}$	$1\text{MPa}\sqrt{\text{m}} = (3.226)\dfrac{\text{kgf}}{\text{mm}^{1.5}}$	$1\dfrac{\text{kgf}}{\text{mm}^{1.5}} = (.3100)\text{MPa}\sqrt{\text{m}}$	$1\text{ksi}\sqrt{\text{in.}} = (3.547)\dfrac{\text{kgf}}{\text{mm}^{1.5}}$	$1\dfrac{\text{kgf}}{\text{mm}^{1.5}} = (0.2819)\text{ksi}\sqrt{\text{in.}}$
Stiffness	$1\dfrac{\text{lbf}}{\text{in.}} = (.1751)\dfrac{\text{N}}{\text{mm}}$	$1\dfrac{\text{N}}{\text{mm}} = (5.710)\dfrac{\text{lbf}}{\text{in.}}$	$1\dfrac{\text{N}}{\text{mm}} = (.1020)\dfrac{\text{kgf}}{\text{mm}}$	$1\dfrac{\text{kgf}}{\text{mm}} = (9.807)\dfrac{\text{N}}{\text{mm}}$	$1\dfrac{\text{lbf}}{\text{in.}} = (.01786)\dfrac{\text{kgf}}{\text{mm}}$	$1\dfrac{\text{kgf}}{\text{mm}} = (56.01)\dfrac{\text{lbf}}{\text{in.}}$

Comparison of Materials

Tensile Yield Strength[a]10³ psi (MPa)		
Material ↓	High	Low
Cobalt & its alloys	290 (1999)	26 (179)
Low alloy hardening steels; wrought, quenched & tempered	288 (1986)	76 (524)
Stainless steels, standard martensitic grades; wrought, heat treated	275 (1896)	60 (414)
Rhenium	270 (1862)	—
Ultra-high strength steels; wrought, heat treated	270 (1862)	170 (1172)
Stainless steels, age hardenable; wrought, aged	237 (1634)	105 (724)
Nickel & its alloys	230 (1586)	10 (69)
Stainless steels, specialty grades; wrought, 60% cold worked	226 (1558)	102 (703)
Tungsten	220 (1517)	—
Molybdenum & its alloys	210 (1448)	82 (565)
Titanium & its alloys	191 (1317)	27 (186)
Carbon steels, wrought; normalized, quenched & tempered	188 (1296)	58 (400)
Low alloy carburizing steels; wrought, quenched & tempered	178 (1227)	62 (427)
Nickel-base superalloys	172 (1186)	40 (276)
Alloy steels, cast; quenched & tempered	170 (1172)	112 (772)
Stainless steels; cast	165 (1138)	31 (214)
Tantalum & its alloys	168 (1089)	48 (331)
Steel P/M parts; heat treated	154 (1062)	75 (517)
Ductile (nodular) irons, cast	150 (1034)	40 (276)
Copper casting alloys[b]	140 (965)	9 (62)
Stainless steels, standard austenitic grades; wrought, cold worked	140 (965)	75 (517)
Columbium & its alloys	135 (931)	35 (241)
Iron-base superalloys; cast, wrought	134 (924)	40 (276)
Cobalt base superalloys, wrought	116 (800)	35 (241)
Bronzes, wrought[b]	114 (786)	14 (97)
Heat treated low alloy constructional steels; wrought, mill heat treated	110 (758)	90 (621)
High copper alloys, wrought[b]	110 (758)	9 (62)
Stainless steels, standard martensitic grades; wrought, annealed	105 (724)	25 (172)
Cobalt base superalloys, cast	100 (689)	75 (517)
Heat treated carbon constructional steels; wrought, mill heat treated	100 (690)	42 (290)
Hafnium	96 (662)	32 (221)
Brasses, wrought[b]	92.5 (638)	10 (69)
Aluminum alloys, 7000 series	91 (627)	14 (97)
Alloy steels, cast; normalized & tempered	91 (627)	38 (262)
Copper-nickel-zincs, wrought[b]	90 (620)	18 (124)
Copper nickels, wrought[b]	85 (586)	13 (90)
Malleable irons, pearlitic grades; cast	80 (552)	45 (310)
High strength low alloy steels; wrought, as-rolled	80 (552)	42 (290)
Stainless steels, specialty grades; wrought, annealed	80 (552)	27 (186)
Stainless steels, standard ferritic grades; wrought, cold worked	80 (552)	45 (310)
Carbon steels, wrought; carburized, quenched & tempered	77 (531)	46 (317)
Carbon steel, cast; quenched & tempered	75 (517)	—
Stainless steel (410) P/M parts; heat treated	75 (517)	—
Steel P/M parts; as-sintered	75 (517)	30 (207)
Coppers, wrought[b]	72 (496)	10 (69)
Aluminum alloys, 2000 series	66 (455)	10 (69)

Material ↓	High	Low
Ductile (nodular) austenitic irons, cast	65 (448)	28 (193)
Zinc foundry alloys	64 (441)	30 (207)
Zinc alloys, wrought	61 (421)	23 (159)
Stainless steels, standard ferritic grades; wrought, annealed	60 (414)	35 (241)
Aluminum alloys, 5000 series	59 (407)	6 (41)
Aluminum alloys, 6000 series	59 (379)	7 (48)
Aluminum casting alloys	55 (379)	8 (55)
Carbon steels, cast; normalized & tempered	55 (379)	48 (331)
Stainless steels, standard austenitic grades; wrought, annealed	55 (379)	30 (207)
Stainless steel P/M parts, as-sintered	54 (372)	40 (276)
Rare earths	53 (365)	9.5 (66)
Zirconium & its alloys	53 (365)	15 (103)
Depleted uranium	50 (345)	35 (241)
Aluminum alloys, 4000 series	46 (317)	—
Thorium	45 (310)	26 (179)
Magnesium alloys, wrought	44 (303)	13 (90)
Silver	44 (303)	8 (55)
Carbon steels, cast; normalized	42 (290)	38 (262)
Beryllium & its alloys	40 (276)	5 (34)
Aluminum alloys, 3000 series	36 (248)	6 (41)
Carbon steel, cast; annealed	35 (241)	—
Malleable ferritic cast irons	35 (241)	32 (221)
Palladium	30 (207)	5 (34)
Gold	30 (207)	—
Magnesium alloys, cast	30 (207)	12 (83)
Polyimides, reinf	28 (193)	5 (34)
Platinum	27 (186)	2 (14)
Iron P/M parts; as-sintered	26 (179)	11 (76)
Aluminum alloys, 1000 series	24 (165)	4 (28)
Polyphenylene sulfide, 40% gl reinf	21 (145)	—
Polysulfone, 30–40% gl reinf	19 (131)	17 (117)
Acetal, copolymer, 25% gl reinf	18.5 (128)	—
Styrene acrylonitrile, 30% gl reinf	18 (124)	—
Phenylene oxide based resins, 20–30% gl reinf	17 (117)	14.5 (100)
Poly (amide-imide)	17 (117)	13.3 (92)
Polystyrene, 30% gl reinf	14 (97)	—
Zinc die-casting alloys	14 (96)	—
Polyimides, unreinf	13 (90)	7.5 (52)
Nylons, general purpose	12.6 (87)	7.1 (49)
Polyether sulfone	12.2 (84)	—
Polyphenylene sulfide, unreinf	11 (76)	—
Polysulfone, unreinf	10.2 (70)	—
Acetal, homopolymer, unreinf	10 (69)	—
Nylon, mineral reinf	10 (69)	9 (62)
Polypropylene, gl reinf	10 (69)	6 (41)
Polystyrene, general purpose	10 (69)	5.0 (34)
Phenylene oxide based resins, unreinf	9.6 (66)	7.8 (54)
Acetal, copolymer, unreinf	8.8 (61)	—

Material ↓	High	Low
ABS/Polycarbonate	8.0 (55)	—
Lead & its alloys[b]	8 (55)	1.6 (11)
Polyaryl sulfone	8 (55)	—
ABS/Polysulfone (polyaryl ether)	7.5 (52)	—
Acrylic/PVC	7.0 (48)	6.5 (45)
Tin & its alloys	6.6 (45)	1.3 (7)
ABS/PVC, rigid	6.0 (41)	—
Polystyrene, impact grades	6.0 (41)	2.8 (19)
Polypropylene, general purpose	5.2 (36)	4.8 (33)
ABS/Polyurethane	4.5 (31)	3.7 (26)
Polypropylene, high impact	4.3 (30)	2.8 (19)

[a] At 0.2% offset for metals, unless otherwise noted; tensile strength at yield or plastics, per ASTM D638. [b] At 0.5% offset.

Specific Strength,[a] 10^3 in. (10^3 m)		
Material ↓	High	Low
Graphite-epoxy, 56–60 v/o graphite	3645 (92)	98 (2.5)
Boron-epoxy, 55 v/o boron	2829 (72)	177 (4.5)
Polyesters, thermoset pultrusions	1429 (36)	345 (8.8)
Titanium & its alloys	1043 (26.5)	171 (4.3)
Stainless steels, standard martensitic grades; wrought, heat treated	982 (24.9)	214 (5.44)
Ultra-high strength steels; wrought, heat treated	954 (24.2)	601 (15.3)
Aluminum alloys, 7000 series	892 (22.6)	144 (3.6)
Cobalt & its alloys	878 (22.3)	89 (2.3)
Stainless steels, age hardenable; wrought, aged	842 (21.4)	373 (9.47)
Nickel & its alloys	687 (17.5)	36 (0.9)
Magnesium alloys, wrought	667 (17.0)	268 (6.8)
Carbon steels; wrought; normalized, quenched & tempered	665 (16.9)	205 (5.21)
Aluminum alloys, 2000 series	647 (16.4)	101 (2.6)
Vinylidene chloride copolymer, oriented	645 (16)	242 (6.1)
Aluminum alloys, 5000 series	602 (15.3)	62 (1.6)
Alloy steels, cast; quenched & tempered	600 (15.2)	396 (10.1)
Ductile (nodular) irons, cast	592 (15.0)	158 (4.01)
Aluminum alloys, 6000 series	561 (14.2)	72 (1.8)
Aluminum casting alloys	539 (13.7)	86 (2.2)
Beryilium & its alloys	533 (13.5)	75 (1.9)
Nylons, 30% gl reinf	521 (13)	396 (10)
Nickel-base superalloy	518 (13.1)	143 (3.6)
Titanium carbide-base cermets	515 (13)	—
Stainless steels, standard austenitic grades; wrought, cold worked	495 (12.6)	265 (6.73)
Aluminum alloys, 4000 series	474 (12.0)	—
Polyester, thermoplastic, PET, 45% & 30% glass reinf	459 (11.7)	288 (7.3)
Iron-base superalloys; cast, wrought	457 (11.6)	137 (3.48)
Magnesium alloys, cast	456 (11.6)	185 (4.7)

Material ↓	High	Low
Copper casting alloys	433 (11.0)	33 (0.84)
Molybdenum & its alloys	420 (10.7)	221 (5.6)
Polycarbonate, 40% & 20% gl reinf	418 (11)	327 (8.3)
Columbium & its alloys	394 (10.0)	115 (2.9)
Stainless steels, standard martensitic grades; wrought, annealed	375 (9.53)	89 (2.26)
Styrene acrylonitrile, 30% gl reinf	367 (9.3)	—
Aluminum alloys, 3000 series	363 (9.22)	61 (1.5)
Bronzes, wrought	355 (9.0)	54 (1.4)
Rhenium	355 (9.0)	—
Cobalt base superalloys	351 (8.9)	116 (2.9)
Plastic foams, rigid, integral skin, reinf	351 (8.9)	105 (2.7)
High copper alloys, wrought	341 (8.7)	30 (0.76)
Polyester, thermoplastic, PBT, 40% & 15% glass reinf	328 (8.3)	245 (6.2)
Alloy steels, cast; normalized & tempered	322 (8.18)	134 (3.40)
Acetal, copolymer; 25% gl reinf	319 (8.1)	—
Plastic foams; rigid, integral skin, unreinf	319 (8.1)	48 (1.2)
Tungsten	314 (7.9)	—
Styrene acrylonitrile, unreinf	312 (7.9)	247 (6.3)
Nylons, unreinf	307 (7.8)	173 (4.4)
Brasses, wrought	300 (7.6)	34 (0.86)
Malleable pearlitic cast irons	300 (7.62)	169 (4.29)
Polystyrene, 30% gl reinf	298 (7.6)	—
Stainless steels, standard ferritic grades; wrought, cold worked	291 (7.39)	164 (4.17)
Copper-nickel-zincs, wrought	285 (7.2)	57 (1.4)
Carbon steels, wrought; carburized, quenched & tempered	272 (6.91)	163 (4.14)
Polyester, thermoplastic, PBT/PET blind, 30% & 15% glass reinf	267 (6.8)	231 (5.9)
Carbon steel, cast; quenched & tempered	265 (6.73)	—
Polyaryl sulfone	265 (6.7)	—
Copper nickels, wrought	263 (6.7)	41 (1.0)
Polystyrene, general purpose	263 (6.7)	132 (3.4)
Tantalum & its alloys	263 (6.7)	5 (0.13)
Acrylics, mldgs	253 (6.4)	133 (3.4)
Boron carbide	247 (6.3)	—
Polyphenylene oxide based resins, unreinf	246 (6.2)	200 (5.1)
Aluminum alloys, 1000 series	245 (6.2)	41 (1.04)
Ductile (nodular) austenitic irons, cast	238 (6.05)	102 (2.59)
Tungsten carbide-base cermets	236 (6)	—
Acrylic, cast	235 (6.0)	141 (3.6)
Polyacrylate, unfilled	233 (5.9)	—
Silicone	224 (5.7)	—
Coppers, wrought	233 (5.7)	31 (0.8)
Zirconium & its alloys	221 (5.6)	64 (1.62)
Polycarbonate, unreinf	218 (5.5)	193 (4.9)
Stainless steels, standard ferritic grades; wrought, annealed	218 (5.54)	127 (3.23)
Alumina ceramic	214 (5.4)	—
Polyphenylene sulfide, unreinf	208 (5.3)	—

Material ↓	High	Low
Hafnium	204 (5.2)	68 (1.72)
Polyurethane plastics	198 (5.0)	106 (2.7)
Ethyl cellulose	195 (5.0)	73 (1.9)
Carbon steels, cast; normalized & tempered	194 (4.92)	170 (4.32)
Stainless steels, standard austenitic grades; wrought, annealed	194 (4.93)	106 (2.69)
Acetal, homopolymer, standard	192 (4.9)	—
ABS/Polycarbonate	191 (4.9)	—
Polyester, thermoplastic, PBT, 45% & 35% glass/ mineral reinf	187 (4.75)	170 (4.3)
Polyaryl ether	183 (4.6)	—
Fluorocarbon, ETFE & ECTFE; gl reinf	179 (4.5)	—
Cellulose acetate	174 (4.4)	65 (1.7)
Acetal, copolymer, standard	173 (4.4)	—
Silicon carbide	172 (4.4)	—
Polypropylene, general purpose	169 (4.3)	148 (3.8)
Chlorinated polyvinyl chloride	162 (4.1)	135 (3.4)
Polyester, thermoplastic, PBT, 30% & 10% mineral filled	161 (4.1)	145 (3.6)
Cellulose nitrate	160 (4.1)	140 (3.6)
PVC, PVC-acetate, rigid	160 (4.1)	110 (2.8)
Cellulose acetate butyrate	160 (4.1)	70 (1.8)
Polyethylene, high molecular weights	159 (4.0)	—
Polystyrene, impact grades	156 (4.0)	86 (2.2)
Acetal homopolymer; 20% gl reinf	152 (3.9)	—
Mullite	151 (3.8)	—
Cellulose acetate propionate	149 (3.8)	92 (2.3)
Rare earths	149 (3.8)	88 (2.2)
Carbon steels, cast; normalized	148 (3.8)	134 (3.40)
Silicon nitride	148 (3.8)	—
Magnesia	147 (3.7)	—
ABS/PVC	143 (3.6)	62 (1.6)
Beryllia	139 (3.5)	—
Acrylic/PVC	137 (3.5)	116 (2.9)
Natural rubber	136 (3.5)	—
Fluorocarbon, PVF_2	134 (3.4)	113 (2.9)
Malleable irons, ferritic grades; cast	134 (3.40)	123 (3.12)
Polymethylpentene	133 (3.4)	—
Synthetic isoprene	132 (3.4)	—
Vinylidene chloride copolymer, unoriented	129 (3.3)	65 (1.7)
Polyethylene, high density	128 (3.3)	84 (2.1)
Allyl diglycol carbonate	125 (3.2)	104 (2.6)
Urethane, polyether and polyester	125 (3.2)	—
Nitrile	124.7 (3.2)	—
Carbon steel, cast; annealed	124 (3.15)	—
Chromium carbide-base cermets	124 (3.1)	—
Propylene ethylene	121 (3.1)	—
Ionomer	118 (3.0)	—

Material ↓	High	Low
Silver	116 (3.0)	—
Thermoplastic elastomers	112.5 (2.9)	—
Boron nitride	112 (2.8)	—
Polypropylene, high impact	108 (2.7)	92 (2.3)
Thorium	107 (2.7)	61.9 (1.57)
Ethylene propylene, ethylene propylene diene	107 (2.7)	—
Silica	104 (2.6)	—
Ethyl butene	103 (2.6)	—
Styrene butadiene	102.9 (2.6)	—
Isobutylene–isoprene	100 (2.5)	—
Forsterite	95 (2.4)	—
Polysulfide	94 (2.4)	—
Chloroprene	89 (2.3)	—
Polybutadiene	88 (2.2)	—
Cordierite	87 (2.2)	—
Zirconia	86 (2.2)	—
Chlorinated polyethylene	83 (2.1)	—
Ethylene vinyl acetate	82 (2.1)	41 (1.0)
Fluorocarbon, PTFE	82 (2.1)	32 (0.8)
Polybutylenes, homopolymer	75 (1.9)	67 (1.7)
Polyethylene, low density	75 (1.9)	27 (0.7)
Fluorocarbon, ETFE & ECTFE; unreinf	74 (1.9)	65 (1.7)
Fluorocarbon, PTFCE	74 (1.9)	60 (1.5)
Depleted uranium	73 (1.85)	51.2 (1.3)
Polyethylene, medium density	71 (1.8)	59 (1.5)
Ethylene ethyl acrylate	71 (1.8)	47 (1.2)
Palladium	69.1 (1.75)	11.5 (0.292)
Polybutylenes, copolymer	68 (1.7)	28 (0.7)
Steatite	68 (1.7)	—
Polyacrylate	62.5 (1.6)	—
Epichlorohydrin	60 (1.5)	—
Chlorosulfonated polyethylene	58 (1.5)	—
Zirconia	58 (1.5)	—
Fluorocarbon, PFA	55 (1.4)	—
Propylene oxide	54.1 (1.4)	—
Fluorocarbon, FEP	45 (1.1)	32 (0.8)
Fluorocarbon	43 (1.1)	—
Gold	42 (1.07)	—
Platinum	34.8 (0.88)	2.58 (0.065)
Tin & its alloys	23 (0.6)	5 (0.13)
Lead & its alloys	20 (0.51)	4 (0.1)

[a] Strength-weight ratio determined by dividing tensile yield strength of metals and ultimate tensile strength of nonmetallics by density.

Modulus of Elasticity in Tension, 10^4 psi (10^{-4} MPa)

Material ↓	High	Low
Silicon carbide	95 (65.5)	13 (8.96)
Tungsten carbide-base cermets	94.8 (65)	61.6 (42.5)
Tungsten carbide	94 (64.8)	65 (44.8)
Osmium	80 (55.1)	—
Iridium	79 (54.5)	—
Titanium, zirconium, hafnium borides	73 (50.3)	71 (48.95)
Ruthenium	68 (46.9)	—
Rhenium	68 (46.9)	—
Boron carbide	65 (44.8)	42 (28.96)
Boron	64 (44.1)	—
Tungsten	59.0 (40.6)	—
Beryllia	58 (39.9)	39 (26.9)
Titanium carbide-base cermets	57 (39.3)	42 (28.96)
Rhodium	55 (37.9)	—
Titanium carbide	55 (37.9)	36 (24.8)
Molybdenum & its alloys	53 (36.5)	46 (31.7)
Tantalum carbide	53 (36.5)	—
Magnesia	50 (34.5)	35 (24.1)
Alumina ceramic	50 (34.5)	30 (20.7)
Columbium carbide	49 (33.8)	—
Beryllium carbide	46 (31.7)	30 (20.7)
Chromium	42 (28.9)	—
Beryllium & its alloys	42.0 (28.9)	27.0 (18.6)
Graphite-epoxy composites	40 (27.6)	20 (13.4)
Cobalt base superalloys	36.0 (24.8)	29.0 (19.9)
Zirconia	35 (24.1)	23 (15.8)
Nickel & its alloys	34.0 (23.4)	19.0 (13.1)
Cobalt & its alloys	33.6 (23.1)	30.0 (20.7)
Nickel-base superalloys	33.5 (23.1)	18.3 (12.6)
Iron-base superalloys; cast & wrought	31 (21.4)	28 (19.3)
Silicon nitride	31 (21.4)	9 (6.2)
Alloy steels; cast	30 (20.7)	29 (20)
Boron-epoxy composites	30 (20.7)	—
Carbon steels; cast	30 (20.7)	—
Carbon steels, carburizing grades; wrought	30 (20.7)	29 (20)
Carbon steels, hardening grades; wrought	30 (20.7)	29 (20)
Depleted uranium	30 (20.7)	20 (13.8)
Stainless steels, age hardenable; wrought	30 (20.7)	28 (19.3)
Stainless steels, specialty grades; wrought	30 (20.7)	27 (18.6)
Ultrahigh-strength steels; wrought	30 (20.7)	27 (18.6)
Stainless steels; cast	29 (20)	24 (16.5)
Stainless steels, standard austenitic grades; wrought	29 (20)	28 (19.3)
Stainless steels, standard ferritic grades; wrought	29 (20)	—
Stainless steels, standard martensitic grades; wrought	29 (20)	—
Boron–aluminum composites	28 (19.3)	—
Malleable irons, pearlitic grades; cast	28 (19.3)	26 (17.9)
Tantalum & its alloys	27.0 (18.6)	21.0 (14.4)
Ductile (nodular) irons; cast	25 (17.2)	22 (15.2)
Malleable ferritic cast irons,	25 (17.2)	—

Material ↓	High	Low
Platinum	25 (17.2)	—
Gray irons; cast	24 (16.5)	9.6 (6.62)
Copper nickels, wrought	22.0 (15.1)	18.0 (12.4)
Mullite	21 (14.5)	—
Zircon	21 (14.5)	—
Ductile (nodular) austenitic irons; cast	20 (13.8)	13 (8.96)
Hafnium	20 (13.8)	—
Copper casting alloys	19.3 (13.3)	11.0 (7.6)
Vanadium	19 (13.1)	18 (12.4)
High copper alloys, wrought	19.0 (13.1)	17.0 (11.7)
Coppers, wrought	18.7 (12.9)	17.0 (11.7)
Titanium & its alloys	18.5 (12.7)	11.0 (7.6)
Copper–nickel-zincs, wrought	18.0 (12.4)	18.0 (12.4)
Palladium	18.0 (12.4)	—
Brasses, wrought	18.0 (12.4)	15.0 (10.3)
Bronzes, wrought	17.5 (12.0)	16.0 (11.0)
Polycrystalline glass	17.3 (11.9)	12.5 (8.6)
Columbium & its alloys	16.0 (11.0)	11.5 (7.9)
Silicon	15.5 (10.7)	—
Zirconium & its alloys	14.0 (9.6)	13.8 (9.5)
Zinc alloys, wrought	14.0 (9.6)	6.2 (4.3)
Rare earths	12.2 (8.4)	2.2 (1.5)
Gold	12.0 (8.2)	—
Aluminum alloys, 4000 series	11.4 (7.9)	—
Silver	11.0 (7.6)	—
Boron nitride	11 (7.6)	7 (4.8)
Aluminum alloys, 2000 series	10.8 (7.4)	10.2 (7.0)
Silica	10.5 (7.24)	—
Aluminum alloys, 7000 series	10.4 (7.2)	10.3 (7.1)
Aluminum alloys, 5000 series	10.3 (7.1)	10.0 (6.9)
Thorium	10.3 (7.1)	—
Aluminum alloys, 1000 series	10.0 (6.9)	10.0 (6.9)
Aluminum alloys, 3000 series	10.0 (6.9)	10.0 (6.9)
Aluminum alloys, 6000 series	10.0 (6.9)	10.0 (6.9)
Thorium	10.0 (6.9)	—
Tin & its alloys	7.7 (5.3)	6.0 (4.1)
Cordierite	7 (4.8)	—
Magnesium alloys, wrought	6.5 (4.5)	6.0 (4.1)
Magnesium alloys, cast	6.5 (4.5)	6.5 (4.5)
Polyesters, thermoset, pultrusions, general purpose	6.0 (4.14)	2.3 (1.59)
Epoxy, gl laminates	5.8 (4.0)	3.3 (2.28)
Glass fiber-epoxy composites	5 (3.44)	—
Bismuth	4.6 (3.17)	—
Polyimides, gl reinf	4.5 (3.1)	—
Carbon graphite	4.0 (2.75)	0.6 (0.41)
Graphite, pyrolytic	4.0 (2.75)	—
Phenolics; reinf	3.3 (2.28)	0.35 (0.24)
Alkyds	2.9 (2.0)	1.9 (1.31)
Graphite, recrystallized	2.7–1.5 (1.86–1.03)	0.85–0.8 (0.59–0.55)

Material ↓	High	Low
Hickory (shag bark)	2.2 (1.51)	—
Locust (black)	2.1 (1.44)	—
Polyester, thermoplastic, PET, 45% & 30% glass reinf	2.1 (1.45)	1.3 (0.90)
Birch (yellow)	2.0 (1.37)	—
Douglas fir (coast type)	2.0 (1.37)	—
Lead & its alloys	2.0 (1.38)	—
Pine (long needle, ponderosa)	2.0 (1.37)	1.3 (0.89)
Polyesters, thermoset, reinf mldgs	2.0 (1.38)	1.2 (0.83)
Ash (white)	1.8 (1.24)	—
Graphite, gen pur	1.8 (1.24)	0.5 (0.34)
Maple (sugar)	1.8 (1.24)	—
Oak (red, white)	1.8 (1.24)	—
Styrene acrylonitrile; 30% gl reinf	1.8 (1.24)	—
Beech	1.7 (1.17)	—
Carbon & graphite, fibrous reinf	1.8 (1.24)	0.3 (0.2)
Graphite, premium	1.7 (1.17)	0.7 (0.48)
Walnut (black)	1.7 (1.17)	—
Polycarbonate, 40% gl reinf	1.7 (1.17)	0.86 (0.59)
Spruce (sitka)	1.6 (1.10)	—
Poplar (yellow)	1.6 (1.10)	—
Carbon, petroleum coke base	1.6 (1.10)	2.3 (1.58)
Indium	1.57 (1.08)	—
Basswood	1.5 (1.03)	—
Elm (rock)	1.5 (1.03)	—
Polysulfone, 30–40% gl reinf	1.5 (1.03)	1.1 (0.76)
Cypress (Southern bald)	1.4 (0.96)	—
Nylons; 30% gl reinf	1.4 (0.97)	1.0 (0.69)
Polyester, thermoplastic, PBT, 40% & 15% glass reinf	1.4 (0.97)	0.8 (0.55)
Cedar (Port Orford)	1.3 (0.89)	—
Cottonwood (black)	1.3 (0.89)	—
Phenylene oxide based resins; 20–30% gl reinf	1.3 (0.90)	0.93 (0.64)
Redwood (virgin)	1.3 (0.89)	—
Acetal, copolymer; 25% gl reinf	1.25 (0.86)	—
Carbon, anthracite coal base	1.2 (0.82)	0.6 (0.41)
Diallyl phthalates, reinf	1.2 (0.83)	0.6 (0.41)
Fir (balsam)	1.2 (0.83)	—
Hemlock (Eastern, Western)	1.2 (0.83)	1.5 (1.03)
Pine (Eastern white)	1.2 (0.83)	—
Polybutadienes	1.2 (0.83)	0.4 (0.28)
Polystyrene, 30% gl reinf	1.2 (0.83)	—
Polyphenylene sulfide, 40% gl reinf	1.12 (0.77)	—
Fluorocarbon, ETFE & ECTFE; gl reinf	1.1 (0.76)	—
Melamines, cellulose electrical	1.1 (0.76)	1.0 (0.69)
Cedar (Eastern red)	0.9 (0.62)	—
Polyimides, unreinf	0.70 (0.48)	0.45 (0.31)
Polyesters, thermoset, cast, rigid	0.65 (0.45)	0.15 (0.10)
Acetal, homopolymer; unreinf	0.52 (0.36)	—
Acrylics, cast, general purpose	0.50 (0.34)	0.35 (0.24)
Acrylics, mldgs	0.50 (0.34)	0.23 (0.16)

Material ↓	High	Low
Nylon, mineral reinf	0.5 (0.34)	—
Polystyrene, general purpose	0.50 (0.34)	0.46 (0.32)
Styrene acrylonitrile; unreinf	0.50 (0.34)	0.40 (0.28)
Nylons; general purpose	0.48 (0.33)	0.28 (0.19)
Polyphenylene sulfide, unreinf	0.48 (0.33)	—
Polystyrene, impact grades	0.47 (0.32)	0.15 (0.10)
Epoxies, cast	0.45 (0.31)	0.05 (0.03)
Polycarbonate, unreinf	0.45 (0.31)	0.34 (0.23)
Acrylonitrile butadiene styrene (ABS)	0.42 (0.29)	0.29 (0.20)
Acetal, copolymer; unreinf	0.41 (0.28)	—
Phenylene oxide based resins; unreinf	0.38 (0.26)	0.36 (0.25)
ABS/Polycarbonate	0.37 (0.26)	—
Acrylic/PVC	0.37 (0.26)	0.34 (0.23)
Polyaryl sulfone	0.37 (0.26)	—
Polysulfone; unreinf	0.36 (0.25)	—
Polyether sulfone	0.35 (0.24)	—
ABS/PVC, rigid	0.33 (0.23)	—
ABS/Polysulfone (Polyaryl ether)	0.32 (0.22)	—
Allyl diglycol carbonate	0.30 (0.21)	—
Fluorocarbon, PTFCE	0.30 (0.21)	0.19 (0.13)
Fluorocarbon, ETFE & ECTFE; unreinf	0.24 (0.17)	—
ABS/Polyurethane	0.22 (0.15)	0.16 (0.11)
Polypropylene, general purpose	0.22 (0.15)	0.16 (0.11)
Polymethylpentene	0.21 (0.14)	—
Fluorocarbon, PVF	0.2 (0.14)	0.17 (0.12)
Vinylidene chloride copolymer, oriented	0.20 (0.138)	—
Polypropylene, high impact	0.13 (0.09)	—
Polyethylene, high molecular weight	0.1 (0.069)	—
Fluorocarbon, FEP	0.07 (0.05)	0.05 (0.03)
Fluorocarbon, PTFE	0.07 (0.05)	0.04 (0.03)
Vinylidene chloride copolymer, unoriented	0.07 (0.048)	—
Polybutylene, homopolymer	0.036 (0.025)	0.034 (0.023)
Polybutylene, copolymer	0.034 (0.023)	0.012 (0.008)
Polyacrylate, unfilled	0.29 (0.20)	—
Polyethylenes, low density	0.027 (0.019)	0.020 (0.014)
PVC, PVC-acetate, non-rigid	0.003 (0.0021)	0.0004 (0.00027)

Specific Stiffness, 10^4 in. (10^5 m)

Material ↓	High	Low
Silicon carbide	819 (208)	—
Boron	792 (201)	—
Boron carbide	714 (181)	—
Graphite-epoxy composites	701 (178)	351 (89)
Beryllium & its alloys	560 (142)	403 (102)
Beryllia	537 (136)	—
Columbium & its alloys	466.4 (118.4)	377.0 (95.7)
Boron-epoxy composites	410 (104)	—
Magnesia	388 (99)	—
Beryllium carbide	357 (91)	—
Alumina ceramic	357 (91)	—

Material ↓	High	Low
Boron-aluminum composites	280 (71)	—
Silicon nitride	270 (69)	—
Titanium carbide-base cermets	219 (56)	—
Silicon	184 (47)	—
Glass, polycrystalline	184 (47)	—
Mullite	176 (45)	—
Columbium carbide	173 (44)	—
Gold	171.9 (43.6)	—
Tungsten carbide-base cermets	171 (43)	—
Chromium	162 (41.1)	—
Titanium, zirconium, hafnium borides	159 (40)	—
Ruthenium	154.2 (39.2)	—
Tungsten carbide	152 (39)	—
Zirconia	150 (38)	—
Boron nitride	138 (35)	—
Molybdenum & its alloys	124.3 (31.5)	106 (26.9)
Magnesium alloys; wrought	123.7 (31.5)	98.5 (25.1)
Rhodium	123 (31.2)	—
Aluminum alloys, 4000 series	117.5 (29.8)	—
Cobalt base superalloys	109.1 (27.7)	96.6 (24.5)
Silica	109 (28)	—
Stainless steels, age hardenable; wrought	107 (27.2)	99 (25.1)
Aluminum alloys, 7000 series	106.2 (27.0)	101.9 (25.8)
Alloy steels; cast	106 (26.9)	102 (25.9)
Carbon steels, carburizing grades; wrought	106 (26.9)	102 (25.9)
Carbon steels; cast	106 (26.9)	—
Carbon steels, hardening grades; wrought	106 (26.9)	103 (26.2)
Iron-base superalloys; cast, wrought	106 (26.9)	96 (24.4)
Ultra-high strength steels; wrought	106 (26.9)	95 (24.1)
Aluminum alloys, 2000 series	105.8 (26.8)	103.0 (26.2)
Aluminum alloys, 5000 series	105.2 (26.7)	105.1 (26.6)
Malleable irons, pearlitic grades;cast	105 (27.7)	98 (24.9)
Stainless steels, standard ferritic grades; wrought	105 (26.7)	—
Nickel-base superalloys	104.0 (26.4)	65.3 (16.6)
Stainless steels, standard martensitic grades; wrought	104 (26.4)	—
Aluminum alloys, 6000 series	103.1 (26.2)	102.0 (25.9)
Cobalt & its alloys	103.1 (26.2)	101.8 (25.8)
Aluminum alloys, 1000 series	102.0 (25.9)	—
Aluminum alloys, 3000 series	102.0 (25.9)	101.0 (25.6)
Nickel & its alloys	101.8 (25.8)	67.4 (17.1)
Titanium & its alloys	101.1 (25.7)	69.6 (17.7)
Magnesium alloys; cast	100 (25.4)	98.5 (25.1)
Ductile (nodular) irons; cast	98 (24.9)	87 (22.1)
Tantalum carbide	98 (24.9)	—
Osmium	97.6 (24.8)	—
Iridium	97.1 (24.7)	—
Zirconia	97 (24.7)	—
Malleable irons, ferritic grades; cast	96 (24.4)	—
Gray irons; cast	92 (23.4)	37 (9.4)
Rhenium	89.4 (31.6)	—

Material ↓	High	Low
Vanadium	85.9 (21.8)	81.4 (20.6)
Tungsten	84.3 (21.4)	—
Cordierite	78 (20)	—
Ductile (nodular) austenitic irons; cast	73 (18.5)	48 (12.2)
Glass fiber-epoxy composites	67 (17)	—
Bronzes, wrought	61.5 (15.6)	54.5 (13.8)
Copper casting alloys	59.7 (15.1)	40.8 (10.3)
High copper alloys; wrought	58.8 (14.9)	57.0 (14.4)
Zirconium & its alloys	58.7 (14.9)	58.3 (14.8)
Brasses, wrought	58.4 (14.8)	50.6 (12.8)
Coppers, wrought	57.8 (14.7)	52.9 (13.4)
Copper-nickel-zincs, wrought	57.3 (14.5)	56.9 (14.4)
Copper nickels, wrought	56.0 (14.2)	55.7 (14.1)
Lead & its alloys	48.7 (12.3)	—
Tantalum & its alloys	45.0 (11.4)	34.7 (8.8)
Depleted uranium	43.9 (11.1)	29.3 (7.44)
Hafnium	42.5 (10.8)	—
Palladium	41.4 (10.5)	—
Styrene acrylonitrile; 30% gl reinf	37 (9.4)	—
Polyester, thermoplastic, PET 45 & 30% glass reinf.	34.4 (8.7)	23.2 (5.9)
Rare earths	34.4 (8.7)	20.3 (5.15)
Polyesters, thermoset, reinf mldgs	34 (8.6)	20 (5.1)
Platinum	32.2 (8.17)	—
Polycarbonate, 40% & 20% gl reinf	31.0 (7.9)	17.6 (4.5)
Nylons; 30% gl reinf	29 (7.4)	21 (5.3)
Silver	29.0 (7.3)	—
Phenylene oxide based resins; 20–30% gl reinf	28 (7.1)	21 (5.3)
Polysulfone; 30–40% gl reinf	28 (7.1)	21 (5.3)
Tin & its alloys	26.6 (6.7)	22.9 (5.8)
Polystyrene; 30% gl reinf	26 (6.6)	—
Polyester, thermoplastic, PBT, 40% & 15% glass reinf	24.1 (6.1)	15.7 (4.0)
Thorium	24.4 (6.2)	—
Acetal, copolymer; 25% gl reinf	22 (5.6)	—
Polyphenylene sulfide; 40% gl reinf	19 (4.8)	—
Fluorocarbon, ETFE & ECTFE; gl reinf	16.4 (4.2)	
Polyimides, unreinf	13.5 (3.4)	8.7 (2.2)
Polystyrene, general purpose	13 (3.3)	12 (3.0)
Styrene acrylonitrile	13 (3.3)	10.4 (2.6)
Bismuth	12.9 (3.27)	—
Polystyrene, impact grades	12.2 (3.1)	3.9 (0.99)
Acrylics, mldgs	12.0 (3.1)	5.5 (1.4)
Acrylics, cast, general purpose	11.8 (3.0)	8.2 (2.1)
Polycarbonate, unreinf	10.2 (2.6)	7.7 (1.95)
Acrylonitrile butadiene styrene (ABS)	10.1 (2.6)	7.0 (1.8)
Acetal, homopolymer; unreinf	10 (2.5)	—
Polyphenylene sulfide; unreinf	10 (2.5)	—
Phenylene oxide based resins; unreinf	9.7 (2.5)	9.2 (2.3)
Nylon; mineral reinf	9.4 (2.4)	—
ABS/polycarbonate	8.6 (2.2)	—
Acetal, copolymer; unreinf	8.0 (2.0)	—

Material ↓	High	Low
Polysulfone; unreinf	8.0 (2.0)	—
ABS/PVC, rigid	7.8 (2.0)	—
Acrylic/PVC	7.8 (2.0)	7.2 (1.8)
Polyaryl sulfone	7.6 (1.9)	—
Polyether sulfone	7.1 (1.8)	—
Polypropylene, general purpose	6.8 (1.7)	4.9 (1.2)
Polyacrylate, unfilled	6.74 (1.7)	—
Allyl diglycol carbonate	6.3 (1.6)	—
Indium	5.94 (1.5)	—
ABS/polyurethane	5.7 (1.5)	4.2 (1.1)
Polypropylene, high impact	4 (1.0)	—
Fluorocarbon, ETFE & ECTFE; unreinf	3.9 (0.99)	—
Fluorocarbon, PTFCE	3.9 (0.99)	2.5 (0.64)
Fluorocarbon, PVF	3.1 (0.79)	2.7 (0.69)
Polyethylene, high molecular weight	2.9 (0.74)	—
Vinylidene chloride copolymer, unoriented	1.13 (0.29)	—
Polybutylene, homopolymer	1.09 (0.28)	1.03 (0.26)
Polybutylene, copolymer	1.05 (0.27)	0.37 (0.09)
Fluorocarbon, FEP	0.90 (0.23)	0.65 (0.17)
Fluorocarbon, PTFE	0.88 (0.22)	0.5 (0.13)
Polyethylene, low density	0.82 (0.21)	0.61 (0.15)

Hardness of Metals, Brinell

Material ↓	High	Low
White & alloy irons; cast	700	130
Osmium	670	300
Low alloy steels, wrought; normalized, quenchod & tempered	627	202
Stainless steels, wrought martensitics hardened & tempered	580	180
Rhenium	555	331
Molybdenum & its alloys	555	179
Nickel & its alloys	534	75
Stainless steels, cast	470	130
Tungsten	443	330
Low alloy steels, wrought; carburized, quenched & tempered	429	212
Copper casting alloys	415	35
Alloy steels, cast; quenched & tempered	401	262
Rhodium	401	100
Iridium	351	200
Gray irons; cast	350	140
Ruthenium	350	200
Nickel-base superalloys	341	302
Titanium & its alloys	331	—
Ductile (nodular) irons; cast	300	140
Hafnium	285	277
Malleable irons, pearlitic grades; cast	269	163
Stainless steels, standard martensitic grades, wrought; annealed	260	150
Ductile (nodular) austenitic cast irons	240	130
Tantalum & its alloys	237	—
Alloy steels, cast; normalized & tempered	217	137

Material ↓	High	Low
High strength low alloy steels, 42,000–65,000 psi yld str, wrought; as rolled	190	149
Depleted uranium	187	—
Zirconium & its alloys	179	112
Stainless steels, standard austenitic grades, wrought; annealed	170	143
Aluminum alloys, 7000 series	160	60
Malleable ferritic cast irons	156	110
Aluminum casting alloys	145	40
Aluminum alloys, 2000 series	135	45
Zinc foundry alloys	125	85
Aluminum alloys, 4000 series	120	—
Aluminum alloys, 6000 series	120	25
Palladium	118	40
Columbium & its alloys	114	—
Platinum	106	40
Aluminum alloys, 5000 series	105	28
Zinc die-casting alloys	91	82
Silver	90	26
Magnesium alloys, wrought	82	46
Magnesium alloys, cast	80	50
Aluminum alloys, 3000 series	77	28
Rare earths	77	17
Gold	66	25
Aluminum alloys, 1000 series	44	19
Tin & its alloys	29	5
Lead & its alloys	17	4.7
Indium	0.9	—

Hardness of Plastics, Rockwell R

Material ↓	High	Low
Melamine, cellulose electrical	125	115
Polyphenylene sulfide; unreinf	124	—
Polyphenylene sulfide; 40% gl reinf	123	—
Chlorinated polyvinyl chloride	122	117
Nylon, mineral reinf	121	119
Polyester, theroplastic, PBT; unreinf	120	117
Polyester, thermoplastic, PET, 45% & 30% glass reinf	120	—
Polyester, thermoplastic, PBT/PET blend, 30% and 15% gl reinf	120	119
Cellulose acetate	120	49
Nylons, general purpose	120	111
PVC, PVC-acetate; rigid	120	110
Polysulfone; unreinf	120	—
Phenylene oxide based resins; unreinf	119	115
Polyester, thermoplastic; PBT, 45% & 35% glass/mineral reinf	119	114
Polyester, thermoplastic, PBT; 30% gl reinf	119	118
Polyester, thermoplastic, PBT, 30% & 10% mineral filled	118	112
ABS/Polycarbonate	117	—
ABS/Polysulfone (polyaryl ether)	117	—
Acrylonitrile butadiene styrene (ABS)	115	75

Material ↓	High	Low
Fluorocarbon, PTFCE	115	110
Polypropylene; gl reinf	115	90
Cellulose acetate butyrate	114	23
Fluorocarbon, PVF	110	—
Cellulose acetate propionate	109	57
ABS/PVC	102	—
Polypropylene, general purpose	100	80
Fluorocarbon, ETFE & ETCFE, unreinf	95	—
ABS/Polyurethane	82	70
Fluorocarbon, PTFE	55	35

Hardness of Rubber & Elastomers, Shore A

Material ↓	High	Low
Thermoplastic elastomers	100	35
Urethane, polyether and polyester	100	80
Styrene butadiene	100	40
Natural rubber	100	30
Chlorinated polyethylene	95	40
Chloroprene	95	40
Chlorosulfonated polyethylene	95	50
Silicone	90	25
Polyacrylate	90	40
Nitrile	90	20
Fluorocarbon	90	65
Ethylene propylene, ethylene propylene diene	90	30
Epichlorohydrin	80	40
Isobutylene–isoprene	80	40
Polybutadiene	80	45
Polysulfide	80	20
Propylene oxide	80	40
Synthetic isoprene	80	40

Hardness of Ceramics, Mohs

Material ↓	High	Low
Mullite	18	14
Columbium carbide	10	9
Beryllium carbide	9+	—
Tantalum carbide	9+	—
Beryllia	9	—
Alumina	9	—
Titanium carbide	9	8
Zirconium carbide	9	8
Zirconia	8	7
Zircon	8	—
Steatite	7.5	—
Thoria	7	—
Magnesia	6	—
Porcelain enamel	6	$3\frac{1}{2}$
Calcia	6.0	3.3
Boron and aluminum nitrides	2	—

Maximum Service Temperature of Nonmetallics,[a] F (K)

Material ↓	High	Low
Thoria	4890[a] (2972)	—
Beryllia	4350[a] (2672)	—
Calcia	4350 (2672)	—
Magnesia	4350[a] (2672)	—
Zirconia	4350[a] (2672)	—
Silicon carbide	4200 (2589)	—
Boron carbide	4100 (2533)	—
Alumina ceramic	3540[a] (2222)	—
Mullite	3200 (2033)	—
Forsterite	1832 (1273)	—
Steatite	1832 (1273)	—
Mica, phlogopite	1800 (1255)	1400 (1033)
Mica, ceramoplastic	1500 (1116)	700 (644)
Porcelain enamel	1500 (1088)	700 (644)
Mica, natural muscovite	1110 (872)	—
Silicone coatings	1000 (811)	—
Tungsten carbide flame sprayed	1000 (811)	—
Mica, glass bonded	700 (644)	—
Polyimides	600 (589)	—
Silicone elastomer	600 (589)	—
Silicone plastics, reinf	600 (589)	—
Epoxies, cycloaliphatic, unreinf	550 (561)	480 (522)
Fluorocarbon, PTFE	550 (561)	—
Phenolics, reinf	550 (561)	220 (378)
Fluorocarbon coatings	550 (561)	—
Glass, soda lime (tempered)	550 (561)	—
Epoxies	500 (533)	250 (394)
Epoxy novolacs, unreinf	500 (533)	450 (505)
Fluorocarbon elastomers	500 (511)	—
Glass, borosilicate (tempered)	500 (533)	—
Polyaryl sulfone	500 (533)	—
Polyphenylene sulfide	500 (533)	—
Alkyds, reinf	450 (505)	300 (422)
Felt, TFE fluorocarbon	450 (505)	—
Melamines, reinf	430 (494)	170 (350)
Diallyl phthalates	400 (478)	300 (422)
Epoxy amine coatings	400 (477)	—
Fluorocarbon, FEP	400 (478)	—
Fluorocarbon, CTFE	390 (472)	350 (450)
Polyacrylate elastomer	375 (464)	—
Felt, nylon	350 (450)	—
Felt, polyester	350 (450)	—
Phenolic coatings	350 (450)	—
Polyesters, thermoplastic, PBT & PTMT	350 (450)	—
Polyesters, thermoset, reinf	350 (450)	300 (422)
Polysulfones	350 (450)	340 (444)
Ethylene propylene, ethylene propylene diene	325 (436)	—
Chlorosulfonated polyethylene elastomer	325 (436)	—

Material ↓	High	Low
Chloronated polyethylene elastomers	300 (422)	—
Epichlorohydrin elastomers	300 (422)	—
Fluorocarbon, ETFE & ECTFE	300 (422)	—
Fluorocarbon, PVF$_2$	300 (422)	—
Isobutylene–isoprene elastomers	300 (422)	—
Nitrile elastomers	300 (422)	—
Nylons	300 (422)	175 (353)
Polypropylene, general purpose	300 (422)	225 (380)
Polyurethane coatings	300 (422)	—
Propylene oxide elastomers	300 (422)	—
Thermoplastic elastomers	300 (422)	—
Chlorinated polyether	290 (416)	—
Polymethylpentene	275 (408)	—
Felt, polypropylene	250 (394)	—
Polyaryl ether	250 (394)	—
Polycarbonate	250 (394)	—
Polyetryene, high density	250 (394)	175 (353)
Polypropylene, high impact	250 (394)	200 (366)
Polysulfone elastomers	250 (394)	—
Polyurethane plastics, unreif	250 (394)	190 (361)
Urethane polyether and polyester	250 (394)	—
Chloroprene elastomers	240 (389)	—
Propylene ethylene	240 (389)	190 (361)
Acrylics	230 (383)	125 (325)
Acrylonitrile butadiene styrene	230 (383)	130 (328)
Chlorinated polyvinyl chloride	230 (383)	—
Felt, rayca viscose	225 (380)	—
Polybutylene	225 (380)	—
Cellulose acetate	220 (378)	140 (333)
Cellulose butyrate	220 (378)	140 (333)
Cellulose propionate	220 (378)	155 (341)
Phenylene oxide based resins	220 (378)	175 (353)
ABS/Polycarbonate	220 (378)	—
Acetal copolymer	212 (373)	—
Allyl diglycol carbonate	212 (373)	—
Polybutylene, low density	212 (373)	180 (355)
Alkyd coatings	200 (366)	—
Chloroprene coatings	200 (366)	—
Polyurethane elastomers	200 (366)	—
Polyester coatings	200 (366)	—
Styress acrylonitrile	200 (366)	140 (333)
Acetal bomopolymer	195 (364)	—
Ethylene ethyl acrylate	190 (361)	—
Polyurethane plastics	190 (361)	—
Ethyl cellulose	185 (358)	115 (319)
Acrylic coatings	180 (355)	—
Lonomer	180 (355)	160 (344)
Natural rubber	180 (355)	—
Nitrocellulose coatings	180 (355)	—
Styrene butadiene elastomers	180 (355)	—

Material ↓	High	Low
Synthetic isoprene	180 (355)	—
Polystyrenes	175 (353)	140 (333)
Polviryl chloride	175 (353)	140 (333)
Ureas cellulose reinf	170 (350)	—
Polyviayl coatings	150 (339)	—
Cellulose nitrate	140 (333)	—

[a] At zero stress.

Heat Deflection Temperature of Plastics,[a] F (R)

Material ↓	High	Low
Silicoae plastics	>900 (>755)	—
Polyimdes	680 (633)	582 (588)
Poly(amide–imide)	545 (558)	520 (544)
Epoxy cycloaliphatic diepexides, cast, rigid	525 (547)	300 (422)
Polyaryl sulfone	525 (547)	—
Polybutadienes	500 (533)	—
Nylons 30% gl reinf	495 (530)	420 (498)
Polyesters, thermoplastic, PET, 45% & 35% glass reinf	442 (501)	428 (493)
Epoxy tovolacs	425 (491)	422 (300)
Polypterylene sulfide; 40% gl reinf	425 (491)	—
Polyester thermoplastic, PBT, 40% & 15% glass reinf	405 (480)	375 (464)
Alkyds	>400(>478)	350 (450)
Epoxy, mineral gl reinf	400 (478)	340 (444)
Melamnes	400 (478)	265 (403)
Polyesters, thermoset, cast rigid	400 (478)	120 (322)
Polyesters, thermoset, reinf mldgs	400 (478)	375 (464)
Polyesters, thermoset, pultrusions	400 (478)	—
Polyester sulfone	400 (478)	—
Plastic foams, rigid, integral skin; gl reinf	390 (472)	162 (345)
Polyester, thermoplastic PBT 45% & 35% glass/mineral reinf	390 (472)	340 (444)
Polyester, thermoplastic, PBT, 30% & 10% mineral filled	385 (469)	150 (339)
Polyester, thermoplastic PBTFET blend, 30% & 15% gl reinf	380 (466)	320 (433)
Polysulfones	365 (458)	345 (447)
Vinylidene chloride copolymer	339 (150)	328 (130)
Polyacrylate, unfilled	320 (433)	—
Acetals	325 (436)	212 (373)
Phenylene oxide based resins, 20–30% gl reinf	310 (428)	282 (412)
ABS/Polysulfone	300 (422)	—
Nylon; mineral reinf	300 (422)	—
Polypropylene; gl reinf	300 (422)	250 (394)
Polycarbonates	295 (419)	260 (400)
Fluorocarbon, ETFE & ECTFE; gl reinf	285 (414)	—
Polyphenylene sulfide, unreinf	278 (410)	—
Plastic foams, rigid, integral skin; unreinf	>270(>405)	94 (307)
Polyphenylene oxide based resins; unreinf	265 (403)	212 (373)
Acrylonitrile butadiene styrene (ABS)	245 (391)	180 (355)

Material ↓	High	Low
Chlorinated polyvinyl chloride	234 (385)	202 (368)
Fluorocarbon, PVF$_2$	232 (384)	—
Epoxy, cast, rigid	230 (383)	—
ABS/Polycarbonate	220 (378)	—
Fluorocarbon, PTFE; ceramic reinf	220 (378)	170 (350)
Nylons, general purpose	220 (378)	155 (341)
Polystyrenes	220 (378)	210 (372)
Styrene acrylonitrile	220 (378)	210 (372)
ABS/Polyurethane	207 (370)	201 (367)
Cellulose acetate butyrate	196 (364)	118 (321)
Cellulose acetate	188 (360)	117 (320)
Acrylic/PVC	185 (358)	160 (344)
Fluorocarbon, PTFCE	178 (354)	151 (339)
Cellulose acetate propionate	173 (351)	129 (327)
Fluorocarbon, ETFE & ECTFE; unreinf	170 (350)	160 (344)
Epoxy, cast, flexible	155 (341)	90 (305)
Polybutylene, homopolymer	140 (333)	120 (322)
Polypropylene, general purpose	140 (333)	135 (330)
Polypropylene, high impact	140 (333)	120 (322)
Polyester, thermoplastic, PBT unreinf	130 (328)	123 (324)

^{a} At 264 psi (1.82 MPa) stress.

Specific Heat, Btu/lb·F (J/kg·K)		
Material ↓	High	Low
Polyester, thermoset, cast, rigid	0.55 (2301)	0.30 (1255)
Polyethylenes, high density	0.55 (2301)	0.46 (1925)
Polyethylenes, low & medium density	0.55 (2301)	0.53 (2218)
Nylons	0.5 (2092)	0.3 (1255)
Epoxies, cast, rigid	0.5 (2092)	0.4 (1673)
Polypropylene, high impact	0.48 (2008)	0.45 (1883)
Beryllium & its alloys	0.45 (1883)	—
Propylene, general purpose	0.45 (1883)	—
Cellulose acetate	0.42 (1757)	0.3 (1255)
Cellulose acetate butyrate	0.4 (1674)	0.3 (1255)
Phenolics, reinf	0.40 (1674)	0.28 (1172)
Cellulose acetate propionate	0.4 (1674)	0.3 (1255)
ABS/Polysulfone (polyaryl ether)	0.35 (1464)	—
Acetals	0.35 (1464)	—
Acrylics	0.35 (1464)	0.33 (1381)
Polyester, thermoset; reinf mldgs	0.35 (1464)	0.20 (837)
Polystyrene, general purpose	0.35 (1464)	0.30 (1255)
Polystyrene, impact grade	0.35 (1464)	0.30 (1255)
Wrought magnesium alloys	0.346 (1448)	0.245 (1025)
Silicon carbide, (0–2550°F; 255–1672 K)	0.34 (1422)	0.285 (1192)
Beryllium carbide (85–210°F; 303–372 K)	0.334 (1397)	—
Fluorocarbon, PVF$_2$	0.33 (1381)	—
Styrene acrylonitrile	0.33 (1381)	—

Material ↓	High	Low
Vinylidene chloride copolymer	0.32 (1339)	—
Polyimides, unreinf	0.31 (1297)	0.27 (1130)
Boron	0.307 (1284)	—
ABS/Polycarbonate	0.3 (1255)	—
Allyl diglycol carbonate	0.3 (1255)	—
Polycarbonate; unreinf	0.30 (1255)	—
Acrylic/PVC	0.29 (1213)	—
Fluorocarbon, FEP	0.28 (1172)	—
Polyesters, thermoset, pultrusions	0.28 (1172)	0.24 (1004)
Polyimides, gl reinf	0.27 (1130)	—
Magnesia (400°F; 477 K)	0.26 (1088)	—
Polyphenylene sulfide	0.26 (1088)	—
Polystyrene; 30% gl reinf	0.26 (1088)	—
Beryllia (200°F; 366 K)	0.25 (1046)	—
Fluorocarbon, PTFE	0.25 (1046)	—
Magnesium alloys, cast	0.245 (1025)	0.245 (1025)
Polysulfone	0.24 (1004)	—
Aluminum alloys, 6000 series	0.23 (962)	0.23 (962)
Aluminum alloys, 7000 series	0.23 (962)	0.23 (962)
Aluminum alloys, 5000 series	0.23 (962)	0.22 (920)
Aluminum alloys, 2000 series	0.23 (962)	0.22 (920)
Aluminum alloys, 3000 series	0.22 (920)	0.22 (920)
Boron carbide	0.22 (920)	—
Aluminum alloys, 1000 series	0.22 (920)	—
Fluorocarbon, PTFCE	0.22 (920)	—
Calcia	0.21 (879)	0.075 (314)
Nickel-base superalloys	0.20 (837)	0.106 (443)
Forsterite	0.2 (837)	—
Alumina ceramic	0.19 (795)	—
Polycrystalline glass	0.190 (795)	0.18 (753)
Boron and aluminum nitrides	0.17 (711)	—
Zirconia	0.16 (669)	—
Silica	0.16 (669)	—
Silicoa	0.1597 (668.2)	—
Titanium & its alloys	0.155 (648)	0.110 (460)
Silicon nitride	0.15 (627)	—
Titanium, zirconium, hafnium borides	0.15 (627)	0.05 (209)
Cobalt & its alloys	0.14 (585)	—
Nickel & its alloys	0.14 (585)	0.091 (381)
Stainless steel, cast	0.14 (586)	0.11 (460)
Ductile (nodular) irons; cast	0.13 (544)	—
Gray irons; cast	0.13 (544)	—
Titanium carbide	0.13 (544)	—
Zinc foundry alloys	0.125 (523)	0.104 (410)
Cobalt base superalloys	0.12 (502)	0.09 (376)
Iron-base superalloys; cast, wrought	0.12 (502)	0.10 (418)
Specialty stainless steels; wrought	0.12 (502)	0.10 (418)
Stainless steels, wrought austenitics	0.12 (502)	—
Vanadium	0.12 (502.1)	—
Zircon	0.12 (502)	—

Material ↓	High	Low
Indium	0.117 (489)	—
Manganese	0.114 (476.9)	—
Alloy steels, cast	0.11 (460)	0.10 (418)
Carbon steels, cast	0.11 (460)	0.10 (418)
Carbon steels, hardening grades; wrought	0.11 (460)	0.10 (418)
Stainless steels, standard and ferritic & martensitic grades wrought	0.11 (460)	—
Chrominum	0.1068 (446.8)	—
High copper alloys, wrought	0.10 (418)	0.09 (376)
Bronzes, wrought	0.10 (418)	0.09 (376)
Copper, casting alloys	0.10 (418)	0.09 (376)
Zinc die-casting alloys	0.10 (418)	0.10 (418)
Zinc, wrought alloys	0.096 (402)	0.094 (393)
Coppers, wrought	0.092 (385)	0.092 (385)
Brasses, wrought	0.09 (376)	0.09 (376)
Copper-nickel-zinc, wrought	0.09 (376)	0.09 (376)
Copper-nickel, wrought	0.09 (376)	0.09 (376)
Zirconium carbide (400°F; 477 K)	0.08 (334)	—
Columbium carbide (400°F; 477 K)	0.08 (334)	—
Zirconium & its alloys	0.0659 (276)	—
Molybdenum & its alloys	0.065 (272)	—
Columbium and its alloys	0.065 (272)	0.060 (251)
Rhodium	0.059 (247)	—
Palladium	0.058 (243)	—
Ruthenium	0.057 (238)	—
Silver	0.056 (234)	—
Tungsten carbide +15% cobalt flame sprayed coatings	0.056 (234)	—
Tin & its alloys	0.054 (226)	—
Tungsten carbide +9% cobalt flame sprayed coatings	0.048 (201)	—
Tungsten carbide (400°F; 477 K)	0.04 (167)	—
Tantalum carbide	0.04 (167)	—
Tantalum & its alloys	0.036 (151)	—
Rhenium	0.035 (146)	—
Hafnium	0.035 (146)	—
Lead & its alloys	0.032 (134)	0.031 (130)
Gold	0.031 (130)	—
Tungsten	0.034 (142)	—
Iridium	0.031 (130)	—
Platinum	0.031 (130)	—
Osmium	0.031 (130)	—
Thorium	0.03 (126)	—
Depleted uranium	0.03 (126)	—
Bismuth	0.0294 (123.0)	—
Thorium	0.0282 (117.9)	—

Thermal Conductivity, Btu·ft/hr·sq·ft·F (W/m·K)		
Material ↓	High	Low
Silver plating	244 (422)	—
Silver (212°F, 373 K)	242 (419)	—
Copper, casting alloys	226 (391)	16 (28)
Coppers, wrought	226 (391)	112 (194)
Copper plating	222 (384)	—
High copper alloys, wrought	218 (377)	62 (107)
Graphite, pyrolytic	215 (272.1)	108 (186.9)
Gold (212°F, 373 K)	172 (298)	—
Gold plating	169 (292)	—
Graphite recrystallized	140–90 (242.3-155.8)	80–39 (138.5–67.5)
Aluminum alloys, 1000 series	135 (234)	128 (222)
Brassses, wrought	135 (234)	15 (26)
Aluminum alloys, 6000 series	125 (216)	99 (171)
Beryllium & its alloys (212°F, 373 K)	123 (213)	87 (151)
Bronzes, wrought	120 (208)	20 (35)
Aluminum alloys, 5000 series	116 (201)	67.4 (117)
Aluminum alloys, 2000 series	111 (192)	82.5 (143)
Aluminum alloys, 3000 series	111 (192)	93.8 (162)
Boron nitride	105 (182)	—
Tungsten (212°F, 373 K)	96.6 (167)	—
Beryllia (2190°F; 1472 K)	95.2 (164)	—
Graphite, premium	95 (164.4)	65 (112.5)
Graphite, gen pure	94 (162.7)	65 (112.5)
Aluminum casting alloys	92.5 (160)	51 (88)
Molybdenum & its alloys (212°F, 393 K)	84.5 (146)	—
Magnesium alloys; wrought	79 (137)	29 (50)
Zinc foundry alloys	72.5 (125)	66.3 (115.1)
Aluminum alloys, 7000 series	70 (121)	70 (121)
Carbon-graphite	66 (114.2)	18 (31.2)
Zinc die-casting alloys (158–284°F, 343–413 K)	65.5 (113)	62.9 (109)
Zinc plating	64.2 (111.1)	—
Zinc, wrought alloys	62.2 (108)	60.5 (105)
Magnesium alloys, cast	58 (100)	24 (42)
Tungsten carbide	50.8 (88)	16.5 (28.5)
Tungsten carbide-base cermets	50.1 (86.7)	25.7 (44)
Rhodium (212°F, 393 K)	50 (87)	—
Carbon & graphite, fibrous reinf	50 (86.5)	2 (3.5)
Nickel & its alloys	49.6 (86)	5.67 (10)
Silicon	48.4 (83.7)	—
Rhenium (212°F, 373 K)	43.7 (76)	—
Platinum (212°F, 373 K)	42 (73)	—
Indium	41.1 (71.1)	—
Palladium (212°F, 373 K)	41 (71)	—
Tin & its alloys	37 (64)	—
Nickel plating	34.4 (59.5)	—
Iridium (212°F, 373 K)	34 (59)	—
Columbium & its alloys (212°F, 373 K)	31.5 (55)	—
Tantalum & its alloys (212°F, 373 K)	31.5 (55)	—

Material ↓	High	Low
Copper nickels, wrought	31 (54)	17 (29)
Gray irons; cast (212°F, 373 K)	30 (51.9)	25 (43.3)
Malleable irons; cast	29.5 (51.1)	—
Ultrahigh-strength steels	29 (50.2)	17 (29.4)
Carbon & alloy steels; cast (212°F, 373 K)	27 (46.7)	—
Carbon steels, carburizing & hardening grades; wrought (212°F, 373 K)	27 (46.7)	—
Nickel-base superalloys (1600°F, 1144 K)	27 (47)	5.25 (9)
Copper-nickel-zincs, wrought	26 (45)	17 (29)
Silicon carbide (2200°F, 1478 K)	25 (43)	9 (15)
Thorium (392°F; 473 K)	21.8 (37.7)	—
Stainless steels, standard martensitic grades; wrought (212°F, 373 K)	21.2 (36.7)	11.7 (20.2)
Rare earths (82°F, 301 K)	21.0 (36)	0.023 (0.040)
Lead plating	20.1 (34.8)	—
Alumina flame sprayed coating	20 (34.6)	19 (32.9)
Ductile (nodular) irons; cast (212°F, 373 K)	20 (34.6)	18 (31)
Lead & its alloys (212°F, 373 K)	19.6 (34)	16 (28)
Sillcon nitride	19 (33)	8 (14)
Cr_2O_2 flame sprayed coating	18 (31.1)	—
Vanadium (212°F; 373 K)	17.9 (31.0)	—
Depleted uranium (212°F, 373 K)	17.2 (30)	—
Cobalt & its alloys (1300°F, 978 K)	16.6 (29)	5.2 (9)
Boron carbide (2200°F, 1478 K)	16 (27.7)	—
Cobalt base superalloys, cast (1300°F, 978 K)	15.8 (27)	11.9 (20.6)
Stainless steels, standard ferritic grades; wrought (212°F; 373 K)	15.6 (27)	13.8 (23.9)
Stainless steels; cast (212°F; 373 K)	15.2 (26.3)	7.5 (13)
Iron-base superalloys; wrought & cast (1100–1200°F; 866–922 K)	15 (26)	10.5 (18.2)
Specialty stainless steels; wrought (212°F; 373 K)	14.4 (24.9)	6.5 (11.2)
Cobalt base superalloys, wrought (1300°F, 978 K)	13.1 (22.6)	11.0 (19.0)
Tantalum carbide	12.8 (22)	—
Beryllium carbide (68–795°F; 243–697 K)	12.1 (20.9)	—
Stainless steels, age hardenable; wrought (212°F; 373 K)	12.1 (20.9)	8.7 (15.1)
Zirconium carbide	11.9 (20.5)	—
Titanium carbide	11.9 (20.5)	9.9 (17)
Titanium & its alloys	11.5 (20)	3.9 (7)
Carbon, anthracite coal base	11 (19.0)	7 (12.1)
Zirconium & its alloys (212°F, 373 K)	9.6 (17)	8.1 (14)
Stainless steels, standard austenitic grades; wrought (212°F; 373 K)	9.4 (16.3)	8.2 (14.2)
Porcelain enamel	9.0 (15.5)	6.0 (3.46)
Columbium carbide	8.23 (14.2)	—
Zirconium oxide flame sprayed coating	8 (13.8)	7 (12.1)
Cadmium plating	5.3 (9.17)	—
Tungsten carbide + 9% cobalt flame sprayed coating	5.3 (9.2)	—
Tungsten carbide + 15% cobalt flame sprayed coating	5.3 (9.2)	—
Carbon, petroleum coke base	5 (8.7)	3 (5.2)
Bismuth	4.83 (8.37)	—

Material ↓	High	Low
Calcia (1830°F; 1272 K) 9% pososity	4.12 (7.13)	—
Zircon	3.61 (6.2)	2.88 (4.9)
Forsterite	2.40 (4.1)	1.94 (3.3)
Cordierite	2.40 (4.1)	0.97 (1.7)
Polycrystalline glass	1.95 (3.4)	1.13 (1.9)
Steatite	1.94 (3.3)	1.45 (2.5)
Wood comp board, hardboard (tempered)	1.50 (2.6)	1.10 (1.65)
Magnesia (2190°F; 1472 K; 22% porosity)	1.47 (2.5)	—
Mullite	1.47 (2.5)	1.38 (2.4)
Wood comp board, particleboard (medium density)	1.0 (0.40)	—
Epoxies, molded	0.87 (0.5)	0.1 (0.17)
Silica, 0.9% porosity	0.80 (1.4)	—
Wood comp board, medium density	0.60 (1.04)	0.50 (0.87)
Polyimides, unreinf	0.57 (0.99)	0.21 (0.36)
Thermoplastic polyester, PBT, 30% carbon reinf	0.542 (0.938)	—
Zirconia (2190°F; 1472 K), 28% porosity	0.53 (0.9)	—
Silicone plastics	0.50 (0.87)	0.075 (0.13)
Wood comp board, structural insulating	0.45 (0.78)	0.27 (0.47)
Polyesters, thermoset, pultrusions	0.42 (0.73)	0.33 (0.57)
Mica, phlogophite	0.4 (0.69)	0.3 (0.51)
Mica, natural muscovite	0.36 (0.62)	0.25 (0.43)
Felt, flurocarbon (fine)	0.35 (0.605)	—
Phenolics, molding grades	0.309 (0.535)	0.116 (0.201)
Epoxies, cast, rigid	0.3 (0.52)	0.1 (0.17)
Polyimide, gl reinf	0.299 (0.517)	—
Mica, ceramoplastic	0.29 (0.50)	—
Nylons, 30% gl reinf	0.29 (0.50)	0.1 (0.17)
Melamine, gl fiber	0.28 (0.48)	—
Mica, glass bonded	0.24 (0.42)	—
Felt, polyester	0.24 (0.42)	0.175 (0.302)
Felt, rayon viscose (medium)	0.23 (0.40)	0.19 (0.33)
Acrylonitrile butadiene styrene	0.20 (0.35)	0.08 (0.14)
Melamine, cellulose electrical	0.20 (0.35)	0.17 (0.29)
Cellulose acetate	0.19 (0.33)	0.10 (0.17)
Cellulose acetate butyrate	0.19 (0.33)	0.10 (0.17)
Cellulose acetate propionate	0.19 (0.33)	0.10 (0.17)
Polyethylenes, low-high density & high molecular weight	0.19 (0.33)	—
Felt, nylon (fine)	0.183 (0.31)	—
Polyester, thermoplastic, PET, 45% & 30% glass reinf	0.183 (0.31)	—
ABS/Polysulfone	0.173 (0.299)	—
Polyphenylene sulfide; unreinf	0.167 (0.289)	—
Acetal, copolymer	0.16 (0.28)	—
Polysulfone, unreinf	0.15 (0.26)	—
Fluorocarbon, PTFCE	0.145 (0.251)	—
Nitrile	0.143 (0.247)	—
Polypropylene, high impact	0.143 (0.247)	—
Styrene butadiene	0.143 (0.247)	—
Fluorocarbon, PTFE	0.14 (0.24)	—
Fluorocarbon, PVC_2	0.14 (0.24)	—

Material ↓	High	Low
Nylons, general purpose	0.14 (0.24)	0.01 (0.17)
Polyesters, thermoset, reinf high strength mldgs	0.14 (0.24)	0.11 (0.19)
Acetal, homopolymer, unreinf	0.13 (0.22)	—
Acrylics	0.13 (0.22)	0.12 (0.21)
Fluorocarbon	0.13 (0.22)	—
Silicone	0.13 (0.22)	—
Polycarbonate, 40% gl reinf	0.128 (0.222)	—
Polyester, thermoplastic, PBT, 45% & 35% glass/mineral reinf	0.128 (0.222)	0.100 (0.17)
Phenylene oxide based resins, unreinf	0.125 (0.216)	0.092 (0.159)
Polybutylene, homopolymer	0.125 (0.216)	—
Allyl diglycol carbonate	0.12 (0.21)	—
Fluorocarbon, FEP	0.12 (0.21)	—
Polycarbonate, unreinf	0.118 (0.204)	0.113 (0.196)
Polystyrene; 30% gl reinf	0.117 (0.202)	—
Polypropylene, general purpose	0.113 (0.196)	0.10 (0.17)
Chloroprene	0.112 (0.19)	—
Thermoplastic polyester, PBT; unreinf & 30% gl reinf	0.108 (0.187)	0.092 (0.159)
PVC, PVC-acetate	0.10 (0.17)	0.07 (0.12)
Phenylene oxide based resins, 20% & 30% gl fiber reinf	0.096 (0.166)	0.092 (0.159)
Polyaryl sulfone	0.092 (0.159)	—
Polystyrene, general purpose	0.090 (0.156)	0.058 (0.10)
Polystyrene, impact grades	0.090 (0.156)	0.024 (0.042)
Thermoplastic elastomers	0.087 (0.15)	—
Acrylic/PVC	0.084 (0.145)	—
Natural rubber	0.082 (0.14)	—
Synthetic isoprene	0.082 (0.14)	—
ABS/Polycarbonate	0.079 (0.137)	—
Plastic foams, rigid, no surface skin	0.077 (0.133)	0.009 (0.016)
Chlorosulfonated polyethylene	0.065 (0.112)	—
Isobutylene-isoprene	0.053 (0.09)	—
Vinylidene chloride copolymer	0.053 (0.092)	—
Felt, wool	0.03 (0.52)	—
Felt, polypropylene (fine)	0.0216 (0.0375)	—
Thoria (2190 F; 1472 K), 17% porosity	0.0 (0.0)	—

Coeffcient of Thermal Expansion,^{a} 10^{-4} in./in./F (10^{-4} m/m/K)

Material ↓	High	Low
Polyethylene, medium & high density	167 (301)	83 (149)
ABS/polyurethane	121 (218)	116 (209)
Fluorocarbon, PFA	111 (200)	67 (121)
Polyethylenes, low density	110 (198)	89 (160)
Fluorocarbon, FEP	105 (189)	83 (149)
Cellulose acetate	90 (162)	44 (79)
Cellulose acetate butyrate	90 (162)	60 (108)
Cellulose acetate propionate	90 (162)	60 (108)
Vinylidene chloride copolymer	88 (158)	—
Fluorocarbon, PVF$_2$	85 (153)	—
Acrylonitrile butadiene styrene (ABS)	72 (130)	16 (29)

Material ↓	High	Low
Polybutylene, homopolymer	71 (128)	—
Phenylene oxide based resins; unreinf	68 (38)	59 (33)
Plastic foams, rigid, integral skin type; unreinf[b]	67 (121)	50 (90)
Acrylics	60 (108)	30 (54)
Allyl diglycol carbonate	60 (108)	—
Polypropylene, high impact	59 (106)	40 (72)
Polypropylene, general purpose	58 (104)	38 (68)
Polyester, thermoset, cast, rigid	56 (101)	39 (70)
Polystyrene, impact grades	56 (101)	22 (40)
Fluorocarbon, PTFE	55 (99)	—
Diallyl phthalates	50 (90)	22 (40)
Epoxy, cast, flexible	50 (90)	30 (54)
Nylons, general purpose	50 (90)	45 (81)
Nylon; mineral reinf	50 (90)	27 (49)
Polyester, thermoplastic, PBT/PET blend, 30% & 15% gl reinf	50 (90)	25 (45)
Silicone plastics	50 (90)	25 (45)
Polyester, thermoplastic, PBT, 30% & 10% mineral filled	48 (86)	42 (76)
Polystyrene, general purpose	48 (86)	33 (59)
Acetal, copolymer; unreif	47 (85)	—
Acetals; 20–25% gl reinf	47 (85)	20 (36)
ABS/PVC, rigid	46 (83)	—
Acetals, homopolymer; unreinf	45 (81)	—
Acrylic/PVC	44 (79)	35 (63)
Polyacrylate, unfilled	40 (72)	35 (63)
Polyester, thermoplastic, PBT, unreinf	40 (72)	—
Fluorocarbon, PTFCE	39 (70)	—
Chlorinated polyvinyl chloride	38 (68)	—
Polycarbonate; unreinf	38 (68)	18 (32)
Polyesters, thermoplastic, PBT, 45% & 35% glass/mineral reinf	38 (68)	28 (50)
ABS/polycarbonate	37 (67)	—
Styrene acryonitrile; unreinf	37 (67)	36 (65)
Polyester, thermoplastic, PBT, 40% & 15% glass reinf	35 (63)	28 (50)
Epoxy, cast, rigid	33 (59)	—
Polysulfone; unreinf	31 (56)	—
Alkyds	30 (54)	10 (18)
Polyphenylene sulfide; unreinf	30 (54)	—
Melamine, cellulose electrical	28 (50)	11 (20)
Polyimides; unreinf	28 (50)	25 (45)
Polyarylsulfone	26 (47)	—
Nylons; 30% gl reinf	25 (45)	12 (22)
Poly(amide-imide), high impact	24 (43)	—
Phenolics, mldg grades, general purpose	23 (41)	12 (22)
Polyimide; 15% graphite reinf	23 (41)	—
Polyphenylene sulfide	22 (40)	—
Magnesium alloys, wrought (68–750°F, 293–672 K)	21.8 (39.2)	14.0 (25.2)
Epoxy; mineral gl reinf	20 (36)	10 (18)
Fluorocarbon, PTFE, ceramic reinf	20 (36)	17 (31)
Phenylene oxide based resins; 20–30% gl reinf	20 (36)	14 (25)

Material ↓	High	Low
Polyester, thermoplastic, PET, 45% & 30% glass reinf	20 (36)	13 (23.4)
Zinc glloys wrought (68–212 F, 293–373 K)	19.3 (34.7)	6.0 (10.8)
Poly(amide-imede), high modulus	19 (34)	—
Polyester, thermoset; gl reinf high strength mlds	19 (34)	13 (23)
Phenolics, impact mldg grades	18 (32)	6.7 (12)
Polystyrene; 30% gl reinf	18 (32)	—
Fluorocarbon, ETFE & ECTFE; gl reinf	17 (31)	—
Lead & its alloys (68–212°F, 293–373 K)	16.3 (29.3)	16.0 (28.8)
Polysulfone; 30–40% gl reinf	16 (29)	12 (22)
Styrene acrylomitrile; 30% gl reinf	16 (29)	—
Zinc foundry alloys	15.5 (26.9)	12.9 (23.2)
Zinc die-casting alloys (68–212°F, 293–373 K)	15.2 (27.4)	–
Phenolics, heat resistant mldg grades	15 (27)	7.8 (14)
Fluorocarbon, ETFE & ECTFE; unreinf	14 (25)	—
Magnesium alloys, cast	14 (25)	—
Aluminum casting alloys (68–212°F, 293–373 K)	13.7 (24.7)	9.0 (16.2)
Aluminum alloys, 5000 series (68–212°F, 293–373 K)	13.4 (24.1)	13.1 (23.6)
Aluminum alloys, 3000 series (68–212°F, 293–373 K)	13.3 (23.9)	12.9 (23.2)
Aluminum alloys, 2000 series (68–212°F, 293–373 K)	13.2 (23.8)	12.4 (22.3)
Aluminum alloys, 1000 series (68–212°F, 293–373 K)	13.2 (23.8)	13.1 (23.6)
Aluminum alloys, 6000 series (68–212°F, 293–373 K)	13.1 (23.6)	12.9 (23.2)
Aluminum alloys, 7000 series (68–212°F, 293–373 K)	13.1 (23.6)	13.0 (23.4)
Tin & its alloys (32–212°F, 273–373 K)	13.0 (23.4)	—
Manganese (68 F, 293 K)	12.2 (22.0)	—
Copper casting alloys (68–572 F, 293–573 K)	12.0 (21.6)	9.0 (16.2)
Brasses, wrought (68–572°F, 293–573 K)	11.8 (21.2)	10.0 (18.0)
Bronzes, wrought (68–572°F, 293–573 K)	11.8 (21.2)	8.3 (14.9)
Silver (32–212°F, 273–373 K)	10.9 (19.6)	—
Aluminum alloys, 4000 series (68–212°F, 293–373 K)	10.8 (19.4)	—
Iron-base superalloys; wrought & cast (70–1500°F, 294–1089 K)	10.7 (19.3)	9.4 (16.9)
White & alloy cast irons (70°F, 294 K)	10.7 (19.3)	4.5 (8.1)
Specialty stainless steels; wrought (70–212°F, 294–373 K)	10.5 (18.9)	4.8 (8.6)
Ductile (nodular) austenitic irons; cast (70–400°F, 294–478 K)	10.4 (18.7)	7.0 (12.6)
Stainless steels; cast (70–1000°F, 294–811 K)	10.4 (18.7)	6.4 (11.5)
Stainless steels, standard, austenitic grades; wrought (32–212°F, 273–373 K)	10.4 (18.7)	8.3 (14.9)
Cobalt & its alloys (70–1800°F, 294–1255 K)	9.9 (17.8)	6.8 (12.2)
Nickel-base superalloys (70–200°F, 294–366 K)	9.89 (17.8)	5.92 (10.6)
Cobalt base-superalloys; cast (70–1800°F, 294–1255 K)	9.8 (17.6)	8.7 (15.7)
High copper alloys; wrought (68–572°F, 293–573 K)	9.8 (17.6)	9.0 (16.2)
Coppers; wrought (68–572°F, 293–573 K)	9.8 (17.6)	9.3 (16.7)
Nickel & its alloys	9.6 (17.2)	6.2 (11.2)
Copper nickets; wrought (68–572°F, 293–573 K)	9.5 (17.1)	9.0 (16.2)
Carbon-graphite	9.4 (16.9)	1.0 (1.8)
Cobalt base superalloys; wrought (70–1800°F, 294–1255 K)	9.4 (16.9)	9.0 (16.2)
Copper–nickel–zincs; wrought (68–572°F, 293–573 K)	9.3 (16.7)	9.0 (16.2)

Material ↓	High	Low
Polycarbonate; 40% gl reinf	9.3 (16.7)	—
Beryllium & its alloys (70°F, 294 K)	9.0 (16.2)	6.4 (11.5)
Carbon steels, carburizing grades; wrought (70–1200°F, 294–922 K)	8.4 (15.1)	—
Alloy steels; cast (70–1200°F, 294–922 K)	8.3 (14.9)	8.0 (14.4)
Carbon steels; cast (70–1200°F, 294–922 K)	8.3 (14.9)	—
Carbon steels, hardening grades; wrought (70–1200°F, 294–922 K)	8.3 (14.9)	7.5 (13.5)
Melamine; gl reinf	8.2 (14.8)	—
Stainless steels, age hardenable; wrought (70–212°F, 294–373 K)	8.2 (14.8)	5.3 (9.5)
Gold (68°F, 293 K)	7.9 (14.2)	—
Magnesia (68–2550°F, 293–1672 K)	7.78 (14)	—
Depleted uranium (70°F, 294 K)	7.7 (13.9)	—
Malleable irons, pearlitic grades; cast (68–212°F, 293–373 K)	7.5 (13.5)	—
Titanium carbide-base cermets (68–1200°F, 293–922 K)	7.5 (13.5)	4.3 (7.7)
Bismuth	7.39 (13.3)	—
Calcia (70–1000°F, 294–811 K)	7.0 (12.6)	—
Thorium (70°F, 294 K)	6.94 (12.5)	—
Gray irons; cast (32–212°F, 273–373 K)	6.8 (12.2)	6.0 (10.8)
Carbon & Graphite, fibrous reinf	6.7 (12.1)	1.8 (3.2)
Ductile (nodular) irons; cast (70–400°F, 294–478 K)	6.6 (11.9)	—
Stainless steels, standard ferritic grades; wrought (32–212°F, 273–373 K)	6.6 (11.9)	5.2 (9.4)
Palladium (68°F, 293 K)	6.5 (11.7)	—
Stainless steels, standard martensitic grades; wrought (32–212°F, 273–373 K)	6.2 (11.2)	5.5 (9.9)
Titanium & its alloys (68–1000°F, 293–811 K)	6.0 (10.8)	4.5 (8.1)
Chromium carbide-base cermets (68–576°F, 293–575 K)	6.0 (10.8)	—
Malleable irons, ferritic grades; cast (68–212°F, 293–373 K)	5.9 (10.6)	—
Beryllium carbide (77–1472°F, 298–1073 K)	5.8 (10.4)	—
Beryllia (68–2550°F, 293–1672 K)	5.28 (9.5)	—
Thoria (68–2550°F, 293–1672 K)	5.28 (9.5)	—
Ruthenium (68°F, 293 K)	5.1 (9.2)	—
Polyester, thermoplastic, PBT; 30% carbon reinf	5 (9)	—
Polyester, thermoset, pultrusions	5 (9)	3 (5.4)
Vanadium (70°F, 294 K)	4.96 (8.94)	—
Platinum (68°F, 293 K)	4.9 (8.8)	—
Titanium, zirconium, hafnium borides (70–4000°F, 294–2478 K)	4.8 (8.6)	4.2 (7.5)
Forsterite (68–212°F, 293–373 K)	4.72 (8.5)	—
Rhodium (68°F, 293 K)	4.7 (8.49)	—
Glass, soda lime	4.7 (8.46)	—
Boron (68–1380°F, 293–1022 K)	4.61 (8.3)	—
Tantalum carbide (77–1472°F, 298–1073 K)	4.6 (8.3)	—
Graphite, recrystallized (70°F, 294 K)	4.5 (8.1)	0.4 (0.09)

Material ↓	High	Low
Alumina ceramic (77–1830°F, 298–1272 K)	4.3 (7.74)	—
Boron nitride (70–1800°F, 294–1255 K)	4.17 (7.5)	—
Titanium carbide (77–1472°F, 298–1073 K)	4.1 (7.4)	3.7 (6.6)
Tungsten carbide	4.1 (7.4)	2.5 (4.5)
Steatite (68–212°F, 293–373 K)	3.99 (7.2)	3.33 (5.9)
Tungsten carbide-base cermets (68–1200°F, 293–922 K)	3.9 (7)	2.5 (4.5)
Columbium & its alloys; wrought (70°F, 294 K)	3.82 (6.9)	3.80 (6.8)
Iridium	3.8 (6.8)	—
Zirconium carbide (77–1472°F, 298–1073 K)	3.7 (6.6)	—
Rhenium (70°F, 294 K)	3.7 (6.7)	—
Osmium	3.6 (6.5)	—
Zirconium & its alloys (212°F, 373 K)	3.6 (6.5)	3.1 (5.6)
Chromium (68°F, 293 K)	3.4 (6.2)	—
Hafnium (70°F, 294 K)	3.4 (6.2)	—
Glass, polycrystalline (77–570°F, 298–572 K)	3.2 (5.7)	0.2 (0.36)
Zirconia (68–2190°F, 293–1472 K)[c]	3.1 (5.6)	—
Mullite (68–212°F, 293–373 K)	3.0 (5.4)	2.7 (4.8)
Molybdenum & its alloys (70°F, 294 K)	2.7 (4.9)	—
Silicon	2.6 (4.67)	—
Tungsten (70°F, 294 K)	2.5 (4.5)	—
Silicon carbide (0–2550°F, 255–1672 K)	2.4 (4.3)	2.17 (3.9)
Graphite, pyrolitic (70°F, 294 K)	2.2 (4.0)	1.1 (2.0)
Cordierite (68–212°F, 293–373 K)	2.08 (3.7)	—
Zircon (68–212°F, 293–373 K)	1.84 (3.3)	1.31 (2.3)
Glass, borosilicate	1.83 (3.2)	—
Boron carbide (0–2550°F, 255–1672 K)	1.73 (3.1)	—
Carbon, anthracite coal base (70°F, 294 K)	1.5 (2.7)	1.3 (2.3)
Carbon, petroleum coke base (70°F, 294 K)	1.5 (2.7)	1.3 (2.3)
Silicon nitride (70–1800°F, 294–1255 K)	1.37 (2.4)	—
Graphite, premium (70°F, 294 K)	0.5 (0.9)	0.963 (0.113)
Glass, 96% silica	0.44 (0.79)	—
Glass, fused silica (quartz)	0.31 (0.56)	—
Silica (68–2280°F, 293–1522 K)	0.28 (0.5)	—
Mica, phlogopite	0.27 (0.49)	0.144 (0.26)
Mica, natural muscovite	0.18 (0.32)	—
Mica, ceramoplastic	0.067 (0.12)	0.062 (0.11)
Mica, glass bonded	0.058 (0.10)	—
Graphite, general purpose	0.064 (0.115)	0.055 (0.099)

[a] For −22 to 86°F (243–303 K) for plastics; as indicated for other materials. [b] Polyethylene, polypropylene and polystyrene types. [c] Depends on degree of stabilization.

Dielectric Strength of Nonmetallics,[a] v/mil (10^3 v/m)

Material ↓	High	Low
Fluorocarbon, PFA	2000 (78.7)	—
Mica (step by step, 1/8 in.)	2000 (78.7)	1000 (39.37)
Chlorinated polyvinyl chloride	1500 (59.1)	1220 (48.0)
PVC, PVC-acetate, rigid	1400 (55.1)	725 (28.5)
Ionomer	1000 (39.4)	—
Polyesters, mermoplastic, PBT & PTMT, unreinf	750 (29.5)	540 (21.2)
Polyester, thermoplastic, PBT; 30% gl reinf	750 (29.5)	—
Polymethylpentenes	700 (27.6)	—
Acrylic/PVC	670 (26.37)	430 (16.9)
Polyallomer	650 (25.6)	500 (19.7)
Polypropylene, general purpose	650 (25.6)	—
Polypropylene, high impact	650 (25.6)	450 (17.7)
Polystyrene, impact grades	650 (25.6)	300 (11.8)
ABS/PVC, rigid	600 (23.6)	—
Cellulose acetate	600 (23.6)	250 (9.8)
Fluorocarbon, PTFCE	600 (23.6)	530 (20.9)
Fluorocarbon, FEP	600 (23.6)	500 (19.7)
Polybutadienes	600 (23.6)	400 (15.7)
Polyphenylene sulfide, unreinf	595 (23.4)	—
Acetals, 20–25% gl reinf	580 (22.8)	500 (19.7)
Polyimides, unreinf	560 (22.0)	310 (12.2)
Ethylene ethyl acrylate	550 (21.7)	—
Acrylics	530 (20.9)	400 (15.7)
Ethylene vinyl acetate	525 (20.7)	—
Styrene acrylonitrile; 30% gl reinf	515 (20.3)	—
Polystyrene, general purpose	>500 (>19.7)	—
Acetals, unreinf	500 (19.7)	—
Fluorocarbon, PTFE	500 (19.7)	400 (15.7)
Mica, glass bonded	500 (19.7)	—
Pherrylene oxide based resins, unreinf	500 (19.7)	400 (15.7)
Porcalain enamel	500 (19.7)	400 (15.7)
Styrene acrylonitrile, unreinf	500 (19.7)	400 (15.7)
Fluorocarbon, ETFE & ECTFE; unreinf	490 (19.3)	—
Polyphenylene sulfide	490 (19.3)	—
Polyethylenes, low-high density & high molecular weight	480 (18.9)	—
Polysulfone, 30% & 40% gl reinf	480 (18.9)	—
Polypropylene; gl reinf	475 (18.7)	317 (12.5)
Nylons, general purpose	470 (18.5)	385 (15.2)
Cellulose acetate propionate	450 (17.7)	300 (11.8)
Diallyl phthalates; reinf	450 (17.7)	350 (13.8)
Nylons; 30% gl reinf	450 (17.7)	400 (15.7)
Polycarbonate, unreinf	450 (17.7)	380 (15.0)
Polycarbonate, 40% gl reinf	450 (17.7)	—
Poly (amide–imide)	440 (17.3)	430 (16.9)
ABS/Polysulfone (polyaryl ether)	430 (16.9)	—
Phenolics	425 (16.7)	200 (7.8)
Polysulfone, unreinf	425 (16.7)	—
Acrylonitrile butadiene styrene	415 (16.3)	300 (11.8)

Material ↓	High	Low
Cellulose acetate butyrate	400 (15.7)	250 (9.8)
Ceramoplastic	400 (15.7)	270 (10.6)
Melamines, cellulose electrical	400 (15.7)	350 (13.8)
Phenolics, heat resistant midg grades	400 (15.7)	210 (8.3)
Phenolics, impact midg grades	400 (15.7)	300 (11.8)
Polystyrene, 30% gl reinf	396 (15.6)	—
Alkyds	350 (13.8)	290 (11.4)
Polyaryl sulfone	350 (13.8)	—
Polycrystalline glass	350 (13.78)	250 (9.84)
Phenolics, general purpose midg grades	340 (13.4)	234 (9.2)
Alumina ceramic	300 (11.81)	200 (7.87)
Mullite	300 (11.81)	—
Allyl diglycol carbonate	290 (11.4)	—
Steatite	280 (11.02)	145 (5.71)
Fluorocarbon, PVF_2	260 (10.2)	—
Zircon	250 (9.84)	—
Forsterite	250 (9.84)	—
Cordierite	230 (9.06)	140 (5.51)

[a] Short term.

Dielectric Constant of Nonmetallics[a]		
Material ↓	High	Low
Alumina ceramic	10.0	8.0
Zircon	10.0	8.0
Silicon nitride	9.4	—
Mica	8.7-5.4	—
Melamines	7.9	5.2
Fluorocarbon, PVF_2	7.5	—
Polycrystalline glass	7.13, 10^3 cps (10^3 Hz)	5.62
Cellulose acetate	7.0	3.2
Mullite	7.0	6.5
Phenolics	7.0	4.0
Ceramoplastic	6.9	6.8
Alkyds	6.8	4.5
Mica, glass bonded	6.7	—
Forsterite	6.5	6.2
Steatite	6.3, 6×10^1 cps (6×10^1 Hz)	5.9
Cordierite	6.23	4.02
Cellulose acetate butyrate	6.2	3.2
Polyesters, thermoset, cast, flexible	6.1	3.7
Nylons; 30% gl reinf	5.4	3.5
Epoxies	5.2	2.78
Epoxy anvolacs	5.1	4.3
Boron nitride	4.8	4.1
Polyesters, thermoset, general purpose reinf mldgs	4.75	4.55

Material ↓	High	Low
Glass, borosilicate	4.6	—
Polyesters, thermoset, cast, rigid	4.4	2.8
Silicone plastics	4.3	3.4
Acetal, homopolymer, 20% gl reinf	4.0	—
Polystyrene, impact grades	4.0	2.4
Acetal, copolymer, 25% gl reinf	3.9	—
Poly (amide–imide)	3.9	3.8
Polyimides; unreinf	3.9	3.5
Polyphenylene sulfide; 40% gl reinf	3.9	—
Allyl diglycol carbonate	3.8	3.5
Glass, 96% silica	3.8	—
Glass, fused silica (quartz)	3.8	—
Nylons, general purpose	3.8	3.5
Acetals, unreinf	3.7	—
Cellulose acetate propionate	3.7	3.4
Polyaryl sullone	3.7	—
Polyester, thermoplastic, PBT, 30% gl reinf	3.7	—
ABS/Polycarbonate	3.6	3.2
Polyether sulfone	3.5	—
Polycarbonate; 40% gl reinf	3.48	—
Acrylic/PVC	3.44	3.06
Polysulfone, 30% & 40% gl reinf	3.4	—
Styrene acrylonitrile, 30% gl reinf	3.4	—
Polybutadienes	3.3	—
Acrylonitrile butadiene styrene (ABS)	3.2	2.4
Polyesters, thermoplastic, PBT & PTMT, unreinf	3.2	3.1
Polyphenylene sulfide, unreinf	3.2	—
Polypropylenes	3.2	2.0
ABS/Polysulfone (polyaryl ether)	3.10	—
Polycarbonate; unreinf	3.1	3.0
Polysulfone; unreinf	3.0	—
Styrene acrylonitrile	3.0	2.6
Polystyrene; 30% gl reinf	3.0	—
Phenylene oxide based resins; 20% & 30% gl reinf	2.9	—
Acrylics	2.9	2.5
Phenylene oxide based resins, unreinf	2.7	2.6
Polystyrene, general purpose	2.7	2.5
Fluorocarbon, ETFE & ECTFE; unreinf	2.5	—
Flurocarbon, PTFCE	2.4	2.3
Polybutylene, copolymer	2.25	2.18
Polybutylene, homopolymer	2.25	—
Fluorocarbon, FEP	2.1	—
Fluorocarbon, PFA	2.1	—
Fluorocarbon, PTFCE	2.1	—
Polymethylpentenes	2.1	—
Polyethylene, cellular foam	1.84	1.05
Plastic foams, rigid, no surface skin	1.1	2.0

[a] At 10^6 cyc/sec (10^6 Hz) unless otherwise indicated.

Plastics by Performance Ability

Performance	Best	Good	Fair	Risk
Heat	PEI	Nylon	Acrylic	PVC
	PPS	PC	Acetal	PVC/ABS
	Polyester	PPO/PS	PC/polyester	Polyethylene
	(GR)	PPO/nylon	ABS	
	LCP	Polystyrene	POLYPROPYLENE	
	GR Grades		ASA	
			ASA/PVC	
Chemical	LCP	PC/polyester	Acrylic	None
	Nylon		PC	
	Acetal	PVC	PPO/PS	
	Polyester		ABS	
	PEI		Polystyrene	
	PPS		ASA	
	PPO/NYLON		ASA/PVC	
	Polyethylene			
	Polypropylene			
Dimensional	LCP	Acrylic	Nylon	Polyethylene
	PC	PC	Acetal	
	PEI	PPO/PS	PPO/nylon	Polypropylene
	GR grades	PC/polyester	Polyester	
	PVC	ABS		
		ASA		
		ASA/PVC		
		PPS		
		ABS/PVC		
		Polystyrene		
Strength	LCP	Acetal	PC/polyester	Polyethylene
	PEI	Acrylic	ASA/PVC	
	PPO/nylon	Nylon	ASA	Polypropylene
	Polyester	PPO/PS	PVC	
	(GR)	ABS		
	GR grades	ABS/PVC		
	PPS	Polystyrene		
Stiffness	LCP	Acetal	PPO/nylon	Polyethylene
	Polyester	Acrylic	PC/polyester	
	(GR)	Nylon	ASA/PVC	Polypropylene
	GR grades	PEI	ASA	
	PPS	PVC		
		PC		
		PPO/PS		
		ABS		
		ABS/PVC		
		Polystyrene		
Impact	PC	PPO/PS	Nylon	Acrylic
	PC/polyester	ABS	Acetal	GR
		ASA	PPO/nylon	Grades
		ASA/PVC	PEI	ABS/PVC
		PVC	ABS (FR)	PPS
			HIPS	Polyethylene
			Polypropylene	GPPS

Agency Approved Plastic Materials

The following table lists many plastic materials and corresponding approvals by or listings with regulatory agencies or organizations

Material Name	Color	FDA	USDA	NSF	3A Dairy	AG Canada	USP Class IV
ABS	Natural	✓	✓	✗	✓	✗	✗
Acetal copolymer	Natural	✓	✓	✓	✓	✓	✗
Celazole PBI & nbsp; polybenzimidazole	Black	✗	✗	✗	✗	✗	✗
CPVC	Gray	✗	✗	✓	✗	✗	✗
Duratron polyimide	Varies	✗	✗	✗	✗	✗	✗
Delrin acetal copolymer	Natural	✓	✓	✓	✓	✓	✗
Delrin AF acetal/PTFE blend	Brown	✗	✗	✓	✗	✗	✗
Ertalyte PET-P	Natural	✓	✓	✗	✓	✓	✗
Ertalyte PET-P	Black	✓	✓	✗	✗	✗	✗
Fluorosint 207 mica-filled PTFE	Natural	✓	✓	✗	✗	✗	✗
Fluorosint 500 mica-filled PTFE	Natural	✗	✗	✗	✗	✗	✗
Halar ECTFE ethylene-chlorotrifluoro-ethylene	Natural	✓	✗	✗	✗	✗	✗
Hydex 4101 PBT-polyester	Natural	✓	✓	✓	✓	✗	✗
Hydex 4101 PBT-polyester	Black	✓	✓	✗	✗	✗	✗
Hydex 4101L lubricated PBT-polyester	Natural	✓	✓	✗	✓	✗	✗
Hydlar ZF Kevlar reinforced nylon	Natural	✓	✓	✗	✓	✗	✗
Kynar PVDF polyvinylidene fluoride	Natural	✓	✓	✓	✓	✗	✓
MC901 heat-stabilized cast nylon 6	Blue	✗	✗	✗	✗	✗	✗
MC907 cast nylon 6	Natural	✓	✓	✗	✗	✗	✗
Noryl modified PPO	Black	✓	✓	✗	✗	✗	✗
Nylon extruded type 6/6	Natural	✓	✓	✓	✓	✗	✗
Nylatron (all) lubricant-filled nylon 6, 6/6	Gray, Blue	✗	✗	✗	✗	✗	✗
Nyloil-FG food grade oil-filled cast nylon 6	Naural	✓	✓	✗	✓	✗	✗
PEEK	Natural	✓	✗	✗	✗	✗	✓
Polycarbonate standard grades	Natural	✗	✗	✗	✗	✗	✗
Polycarbonate food grade only	Natural	✓	✗	✓	✗	✓	✓
Polyethylene LDPE—low density	Natural	✓	✗	✗	✗	✗	✗

The following table lists many plastic materials and corresponding approvals by or listings with regulatory agencies or organizations—(Continued)

Material Name	Color	FDA	USDA	NSF	3A Dairy	AG Canada	USP Class IV
Polyethylene HDPE—hi density	Natural	✓	✓	✗	✗	✗	✗
Polyethylene HDPE—pipe grade	Black	✗	✗	✓	✗	✗	✗
Polyethylene UHMW—ultrahigh mol wt	Natural	✓	✗	✗	✗	✗	✗
Polypropylene homopolymer	Natural	✓	✓	✗	✗	✗	✗
Polysulfone standard grades	Natural	✗	✗	✗	✗	✗	✗
Polysulfone food grade only	Natural	✓	✓	✓	✓	✗	✓
PVC type I	Gray	✗	✗	✗	✗	✗	✗
Rade® A polyethersulfone	Natural	✓	✗	✓	✗	✗	✗
Rade® R polyarylethersulfone	Natural	✗	✗	✗	✗	✗	✓
Rulon® 641 filled PTFE	White	✓	✗	✗	✗	✗	✗
Techtron PPS polyphenylene sulfide	Natural	✗	✗	✗	✗	✗	✗
PTFE polytetrafluoroethylene	White	✓	✗	✗	✗	✗	✗
FEP tetrafluorethylene-perfluorpropylene	White	✓	✗	✗	✗	✗	✗
PFA perfluoroalkoxy	White	✓	✗	✗	✗	✗	✗
Torlon PAI all grades	Varies	✗	✗	✗	✗	✗	✗
Ultem® 1000 polyetherimide	Natural, black	✓	✗	✗	✗	✗	✓
Vespel PI all grades	Varies	✗	✗	✗	✗	✗	✗
ANY & ALL Fiber-Reinforced Plastics including glass, carbon, etc.	—	✗	✗	✗	✗	✗	✗

Note: All information in this table was provided by respective stock shapes manufacturers and is subject to change without notice. Users of this information should verify agency approvals prior to specifying any plastic material.

DATABASES FOR PLASTICS MATERIALS

This section of Appendix I, which listed properties of most common plastics, elastomers, and rubbers in earlier editions, has been replaced with information about the four major electronic databases for plastic materials.

IDES: The Plastics Web™

IDES is a plastic materials information company with a vertical search engine for the plastics industry. IDES's mission is to provide tools for plastics professionals to quickly access technical plastic materials information and data sheets.

As a first step to fulfilling that mission, IDES developed a new approach to cataloging plastic materials information, which later translated to an online three-dimensional database, architected to hold and display highly technical and complex raw materials information. IDES is now widely recognized as the world's largest vertical search engine for plastic materials information and data sheets.

Prospector Plastics Search Engine
A *free* search engine accessing over 60,000 plastic material data sheets from more than 500 global resin suppliers. Data are updated every two weeks: www.ides.com/register

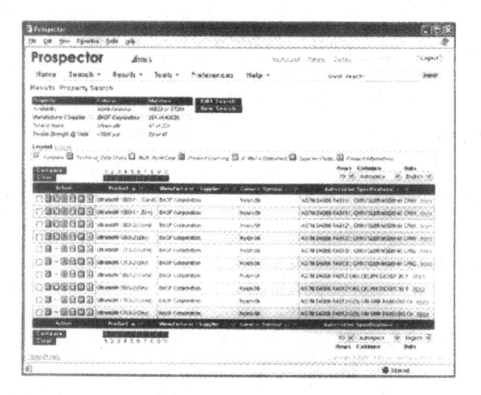

MatWeb: **Data Sheets for over 56,000 Metals, Plastics, Ceramics, and Composites**

The heart of MatWeb is a searchable database of material data sheets, including property information on thermoplastic and thermoset polymers such as ABS, nylon, polycarbonate, polyester, polyethylene and polypropylene. MatWeb is freely available and does not require registration.

The MatWeb database was started by a group of engineers who were excited about having an in-house database with material properties so that they could avoid searching through volumes of technical articles and books. This small group of engineers has grown to the current company, Automation Creations, Inc.
http://www.matweb.com/

PlasticsTechnology^{TM}: Material Database

In three steps, the Plastics Technology Materials Selection Database utility delivers a materials list based on your unique criteria.

In Step 1, apply specific "filters" to select the Plastics Materials that match your requirements. In Step 2, designate what "properties" to display with your tailored materials list. Step 3 builds your list.

http://www.plasticstechnology.com/
http://www.Plaspec.com/

CAMPUS

CAMPUS (Computer-Aided Material Preselection by Uniform Standards) has become the most successful and widely used materials database for plastics. Today more than 50 plastics producers worldwide are providing technical data on their products using CAMPUS. More than 200,000 copies have been distributed worldwide. This remarkable success is based on the unique concept, which combines uniform global protocols for acquisition and representation of data with user-friendly software. CAMPUS is the first and only database system which provides truly comparable data on products from different suppliers. The CAMPUS Windows program and material data are distributed for **free** from participating manufacturers. You can obtain the CAMPUS data by visiting the CAMPUS website to download CAMPUS.

www.campusplastics.com

M-Base Engineering + Software

M-Base is the leading international supplier of:

— material databases,
— material information systems,
— product information systems and
— material related simulation software.

http://www.m-base.de

APPENDIX J

PLASTICS/POLYMER EDUCATIONAL PROGRAMS IN THE UNITED STATES AND CANADA

University of Massachusetts, Lowell

B.S.:	Plastics Engineering
M.S.:	Plastics Engineering
D. Engr.:	Doctorate of Engineering

The Plastics Engineering Department was founded in 1954 on the recommendation of a study team from industry and academia. The B.S. program graduated the first class in 1958, and this program has been ABET accredited continuously since 1977. Subsequent graduate programs instituted are: M.S. in Plastics Engineering (1968), Ph.D. in Chemistry with an option in Polymer Science/Plastics Engineering (1980), Doctor of Engineering in plastics manufacturing, involving processes, materials, and design. The 15 faculty members have interdisciplinary backgrounds–the majority have industrial experience.

The primary mission of the Plastics Engineering Department at University of Massachusetts Lowell is to provide educational and research programs designed to prepare students for satisfying careers and to support and enhance the development of a sustainable and vigorous industrial sector.

The primary objective of the plastics engineering program is to provide a well-rounded basic level position from which to pursue a lifetime career of plastics engineering. The program is designed to prepare the students for professional engineering careers in the polymers industries, of which plastics is the largest; the others include manmade fibers, elastomers, coatings, and adhesives. The curriculum provides a sound foundation in basic science and engineering disciplines plus a detailed study of plastics materials, properties, processing, and design. Graduates are well qualified to pursue advanced studies leading to careers in research, development, production, or teaching. Owing to the uniqueness of

Handbook of Plastics Testing and Failure Analysis, Third Edition, by Vishu Shah
Copyright © 2007 by John Wiley & Sons, Inc.

the plastics engineering program at UML, there is no peer group of comparable departments at other American universities.

The Plastics Engineering Department excels, as evidenced by various national rankings of distinction. The talents of the faculty of the Plastics Engineering Department are world-renowned. The faculty are very active in professional societies, holding offices at both local and national levels. Three faculty members serve on editorial boards or recognized journals. Three faculty members are Fellows of SPE. In a typical year, faculty members publish 30–50 papers. Several faculty members have been honored with "best paper" awards in recent years. The Plastics Engineering Department runs the most successful industrial seminar program in the country. The Plastics Engineering Department has a 70-member industrial advisory group, which meets semiannually to advise on program development.

The Plastics Engineering Department has gradually accumulated five laboratories of small-scale industrial processing equipment—injection and compression molders, extruders, blow molders, thermoformers, and compounding equipment—and four laboratories of test equipment—rheology, mechanical, thermal and electrical properties, permeability, and accelerated aging. In addition, the polymer science group in the Chemistry Department has excellent instrumentation for polymer characterization. The Institute for Plastics Innovation has developed a high-bay processing lab, with injection molding and extrusion capabilities, a testing lab, a computer lab, and other small labs, conference rooms, workshop, and office space for both faculty and students.

The itinerary for the students enrolled in the plastics engineering program includes the basics of plastics engineering, engineering mechanics, plastics design, engineering science, plastics materials, plastics mold engineering, process control, product design, polymer structure and plastics properties, and physical labs. The students take advantage of state-of-the-art laboratories for computer design, characterization, processing and manufacturing, biodegradable research, and plastics innovation. Since its inception, the plastics Engineering program has produced more than 4000 graduates who are well received and well-compensated by the global plastics industry.

Ferris State University

A.A.S.:	Plastics Technology
B.S.:	Plastics Engineering Technology

The plastics program at Ferris State University began in 1969 when the university decided to include plastics technology in its growing Associate of Applied Science (A.A.S.) programs. The continued demand for and success of this program led to yet another highly regarded program aimed toward a Bachalor of Science (B.S.) degree in Plastics Engineering Technology in 1982. The focus of the curriculum is on manufacturing rather than research and development, allowing the program to achieve its objective of graduating students with hands-on manufacturing experience.

Students in the Plastics engineering Technology program learn how to set up, operate, and troubleshoot sophisticated plastics processing equipment. They also analyze the results of laboratory tests on plastics materials to determine their properties and working characteristics. Instruction is also provided in time and motion studies, cost estimating, quality control, plant layout, and using computers for solving manufacturing problems.

The required summer industrial internship between junior and senior years enables students to gain valuable experience and apply their knowledge to work in the plastics industry. A new Plastics engineering technology building is located on campus and includes laboratories for material testing, quality control, fabricating, painting and decorating, composites, and processing. The processing laboratories include injection molding, extrusion, blow molding, and thermoforming and mold repair shop.

Shawnee State University

A.A.S.:	Plastics Engineering Technology
B.S.:	Plastics Engineering Technology

The aim of the Plastics engineering technology program is to prepare the student to become a valuable and integral part of the plastics field. The Associate degree program deals with basic courses in properties of polymeric materials, injection molding, extrusion, blow molding thermosetting processes, and plastics manufacturing in general. Graduates of the A.A.S. degree program have the option of applying their two years directly into the bachelor's program and continuing for two more years.

The B.S. degree program emphasizes plastics processing operations and includes significant components in the areas of materials, mold design, and production methods. Advance processing, plant layout and material handling, testing of plastics, composites, plastic part design, mold design, and analysis are examples of the type of subjects covered under this program.

Eastern Michigan University

B.S.:   Plastics Technology

The plastics technology program at EMU is designed so that graduates have a broad understanding of the plastics industry, polymeric materials, plastics processes, plastics product design, and techniques for finishing and decorating. The program includes a core of fundamental courses designed to provide the students with both theoretical and practical knowledge.

Some of the required courses include plastics materials, plastics processing, plastics mold design and construction, product design, fabricating, and decorating. The students of this program may also gain on-the-job experience through cooperative opportunities with companies in the area.

Western Washington University

B.S.:   Plastics Engineering Technology

Students enrolled in Western Washington University Plastics Engineering Technology Program study the structure and characteristics of polymers, manufacturing methods, product design fundamentals, and modern processing methods for composites and polymeric materials. The PET curriculum prepares engineering technologists who understand and apply established scientific and engineering knowledge to support engineering activities in manufacturing environments.

The plastics Engineering Technology program is less theoretical than a typical engineering program and is more applications or "hands-on" oriented. The plastics processing laboratory consists of a microprocessor-controlled injection molding machine, extruder with profile, film and sheet manufacturing capabilities, extrusion blow molder, thermoformer, filament winder, and compression molder. The material testing laboratory is equipped with all types of testing equipment including a rheometer and a weatherometer. Solids modeling, FEA software, and mold-flow software, along with a 20-station CAD lab complement other computing facilities. A fully equipped tool room with soft-tooling facility helps the students learn about mold and die making.

Pittsburgh State University

B.S.E.T.: Plastics Engineering Technology

The Plastics Engineering Technology program emphasizes detailed technical knowledge as well as real world applications. The subjects range from general technology and processing to part and mold design. Classes also include thorough reviews of problem-solving techniques for both production and managerial issues.

The students have the choice of either a manufacturing option, which involves statistical process control and processing techniques, or a design option, which emphasizes plastics part design. The students also have the opportunity to participate in an industrial cooperative program to acquire "hands-on" experience by working for a local plastics company. The facilities for plastics engineering technology include laboratories that house more than 42 pieces of sophisticated production and testing equipment.

Some other institutions that offer plastics programs are also listed in the following table:

Institution	2-Yr	4-Yr	Grad	Degree
Ahuntsic College Montreal, QUE Canada	✓			D.E.C. in Science (Techniques of Plastics Materials)
Berkshire Community College Pittsfield, MA 01201				A.A.S. Mold-Making
California State Univ. at Long Beach Long Beach, CA 90840		✓	✓	B.A., M.A. Polymer Processing and Design and Teaching Credential
California State Univ. at Pomona Pomona, CA 91768		✓		Plastics Engineering Certificate Program
California State University at Chico Chico, CA 95929				B.S. Industrial Technology (Manufacturing Plastics Option)
California State University at Fresno Fresno, CA 93740		✓		B.S. Industrial Technology (Plastics Option)
Carnegie-Mellon University Pittsburgh, PA 15213			✓	M.S., Ph.D. Chemical Engineering (Polymer Concentration)
Case Western Reserve University Cleveland, OH 44106		✓	✓	B.S. Engineering (Polymer Science Major) M.S., Ph.D. Macromolecular Science
Central Carolina Technical College Sanford, NC 27330	✓			Diploma in Tool and Die
Cerritos College Norwalk, CA 90650	✓			A.A.S. Plastics Mfg. Technology
Cincinnati Technical College Cincinnati, OH 45223	✓			A.A.S. Plastics Engineering Technology
College de la region de l'amiante Quebec G6G 1N1 Canada	✓			A.A.S. Plastics Processing Technology
College of DuPage Glen Ellyn, IL 60137	✓			A.A.S. Plastics Technology
Community College of Rhode Island Warwick, RI 02886	✓			A.A.S. Plastics Process Technology
Cumberland County College Vineland, NJ 08360-0517	✓			A.A.S. Plastics Technology
Des Moines Area Community College Ankeny, IA 50021	✓			A.A.S. Applied Science (Tool and Die)
Drexel University Philadelphia, PA 19104		✓	✓	B.S. Materials Engineering (Polymer Specialization)
Eastern Michigan University Ypsilanti, MI 48197		✓	✓	B.S. Plastics Technology M.S. Industrial Technology
Elgin Community College Elgin, IL 60123	✓			Vocational Specialist Certificate Thermoplastics Injection Molding
El Paso Community College El Paso, TX 79998				480-Hour Certificate Program
Ferris State University Big Rapids, MI 49307	✓	✓		A.A.S., B.S. Plastics Engineering and Technology
Fitchburg State College Fitchburg, MA 01420	✓	✓		Certificate Plastics Technology (Nypro Institute) B.S. Industrial Technology
Florida Atlantic University Boca Raton, FL 33431	✓			Chemistry Polymers

Institution		Program
Georgia Institute of Technology Atlanta, GA 30332	✓	M.S., Ph.D. Chemical and Textile Engineering (Polymer Emphasis)
Grand Rapids Junior College Grand Rapids, MI 49503	✓	A.A.S. Plastics Technology
Hennepin Technical College Brooklyn Park, MN 55445	✓	A.A.S. Plastics Technology
Illinois Institute of Technology Chicago, IL 60616	✓	B.S., M.S., Ph.D Chemical Engineering (Polymer Option)
Indiana Vocational Technical College Indianapolis, IN 46206	✓	A.A.S. Plastics Manufacturing Technical Certificate
Indiana Vocational Technical College South Bend, IN 46619	✓	A.A.S. Plastics Manufacturing
Kalamazoo Valley Community College Kalamazoo, MI 49009	✓	A.A.S. Plastics Technology
Lakeshore Technical College Cleveland, WI 53015	✓	A.A.S. Applied Science (Plastics Concentration) Certificate Plastics Technology A.A.S. Plastics Technology
Laney College Oakland, CA 94607		
Lehigh University Bethlehem, PA 18015	✓	M.S., Ph.D. in Polymer Science and Engineering
Lorain County Community College Elyria, OH 44035-1691	✓	AAS Manufacturing Engineering Technology (Plastics Processing Option)
Los Angeles Trade and Technical College Los Angeles, CA 90015	✓	A.A.S. Plastics Technology
McMaster University Hamilton Ontario L9H 4M6 Canada	✓	M.Eng., Ph.D. Chemical Engineering (Plastics Concentration)
Michigan Technological University Houghton, MI 49931	✓	B.S., M.S., Ph.D. Chemistry and Chemical Engineering (Polymer Option)
Milwaukee Area Technical College Milwaukee, WI 53233	✓	Plastic Technology and 1-Year Diploma Industrial Plastics
Morehead State University Morehead, KY 40351	✓	B.S. Industrial Technology (Plastics Option)
Morrisville State College Morrisville, NY 13508	✓	A.A.S. Plastics Technology
North Carolina State University Raleigh, NC 27695-7907	✓	B.S. Materials Science and Engineering (Polymer Specialty Option)
North Dakota State University Fargo, ND 58105	✓	M.S., Ph.D. Chemistry (Polymer Specialty Option)
Northeast Wisconsin Technical Inst. Green Bay, WI 54303	✓	A.A.S. Mechanical Design (Mold-making)
Northern Alberta Inst. of Technology Edmonton, Alberta T5G 2R1 Canada	✓	A.A.S. Plastics Engineering Technology
Northern Illinois University Dekalb, IL 60115-2854	✓	B.S. emphasis in plastics
Penn State-Erie, The Behrend College Erie, PA 16563-0203	✓	A.A.S. Plastics Technology B.S. Plastics Engineering Technology M.S. Plastics Manufacturing
Pennsylvania State University University Park, PA 16802	✓	B.S., M.S., Ph.D. Polymer Sciences Materials Sciences

Institution	2-Yr	4-Yr	Grad	Degree
Pittsburg State University Pittsburg, KS 66762		✓		B.S.E.T. Plastics Engineering Technology
Polytechnic University Brooklyn, NY 11201		✓	✓	M.S., Ph.D. Polymer Science and Engineering
Rutgers University New Brunswick, NJ 08903			✓	M.S., Ph.D. Materials Science (Plastics Concentration)
San Diego State University San Diego, CA 92182-0269		✓		B.S. Industrial Technology
San Francisco State University San Francisco, CA 94132		✓	✓	B.S. Plastic Product Design and Mfg. M.A.
Shawnee State University Portsmouth, OH 45662	✓	✓		A.S., B.S. Plastics Engineering Technology
Stevens Institute of Technology Hoboken, NJ 07030		✓	✓	B.S., M.S., Ph.D. Chemistry (Polymer Concentration)
Sturgeon Creek Reg. Sec. School				3-year vocational program (high school)
Winnipeg Manitoba Canada R3J 1A5				Certificate of Proficiency (Plastics Technology) upon graduation
Syracuse University Syracuse, NY 13210		✓	✓	B.S. Chemistry (Polymer Option) M.S., Ph.D. Environmental Science (Polymer Concentration)
Trenton State College Hillwood Lakes CN 4700 Trenton, NJ 08650		✓		Offer a Few Elective Plastic Courses
University of Akron Akron, OH 44325-0301			✓	M.S., Ph.D. Polymer Engineering
University of Connecticut Storrs, CT 06268			✓	M.S., Ph.D. Materials Science (Plastics Concentration)
University of Detroit Detroit, MI		✓	✓	B.S., M.E., Ph.D. Chemical Engineering and Polymer Engineering
University of Florida Gainsville, FL		✓	✓	B.S., M.S., Ph.D. Polymer Science and Engineering
University of Illinois Urbana, IL 61801		✓	✓	B.S., M.S., Ph.D. Chemical Engineering (Polymer Concentration)
University of Louisville Louisville, KY 40292			✓	M.S., M. Eng., Ph.D. Chemical and Environmental Engineering (Plastics Concentration)
University of Lowell Lowell, MA 01854		✓	✓	B.S., M.S. Plastics Engineering, B.S. Industrial Engineering Technology, Ph.D. Plastics Engineering Option
University of Maine at Orono Orono, ME 04469		✓	✓	B.S., M.S., Ph.D. Chemical Engineering (Polymer Option)
University of Massachusetts Amherst, MA 01003			✓	M.S., Ph.D. Polymer Science and Engineering
University of Michigan Ann Arbor, MI 48106			✓	M.S., Ph.D. Macromolecular Science and Engineering
University of Minnesota Minneapolis, MN 55414		✓	✓	B.S., M.S., Ph.D. Chemical Engineering and Materials Science (Polymer Concentration)

Institution	Program	
University of Southern California Los Angeles, CA 90089-1211	B.S., M.A., Ph.D. Chemical Engineering (Plastics Concentration)	✓
University of Southern Mississippi Hattiesburg, MS 39406	B.S., M.S., Ph.D. Polymer Science	✓
University of Tennesse Knoxville, TN 37996-2200	M.S., Ph.D. Polymer Engineering B.S. with specialization in Polymer Engineering	✓ ✓
University of Toledo Toledo, OH 43606	AAS Plastics Technology B.S., M.S. Chemical Engineering (Polymer Option)	✓
University of Utah Salt Lake City, UT 84112	B.S., M.S., M.E., Ph.D. Materials Science and Engineering (Polymer Specialization)	✓
University of Wisconsin Madison, WI 53715	M.S., Ph.D. Chemical Engineering (Plastics Concentration)	✓
University of Wisconsin-Platteville Platteville, WI 53818	B.S. Industrial Studies with minor in Plastics Processing Technology/M.S.	✓
Virginia Polytechnic Institute Blacksburg, VA 24061	B.S. Chemical or Materials Engineering (Polymer Concentration)	✓
Washington State University Pullman, WA 99164-2920	M.S., Ph.D. Chemical Engineering Materials Science (Polymer Concentration)	✓
Washington University St. Louis, MO 63130	B.S. in Chemical Engineering M.S., D.Sc. Materials Science and Engineering (Plastics, Composites Concentration)	✓
Waterbury State Technical College Waterbury, CT 06708	Certificate Plastics Technology Graduate Certificate Program in Polymer Engineering	✓
Wayne State University Detroit, MI 48202		
Western Michigan University Kalamazoo, MI 49001	B.S. Mfg. Eng. Tech., B.S. Eng Graphics (Plastics Option)	✓
Western Washington University Bellingham, WA 98225	B.S. Plastics Engineering Technology	✓

Figure 15-5. New Application Checklist. (Courtesy of Bayer Corporation.)

New Application Checklist

This checklist includes critical considerations for new part development.
Its use will help provide a more rapid and more accurate recommendation.

Name _____   Date _____

Customer _____   Part _____

_____   _____

Project timing _____

Driving force _____

Current product _____

Its performance _____

Comments _____

Part Function — *What is the part supposed to do?* _____

_____

Appearance

Clear

☐ water clear

☐ very clear

☐ generally clear,   maximum haze level: _____

☐ transparent color, maximum haze level: _____

Comments: _____

Opaque

☐ high gloss

☐ medium gloss

☐ low gloss

   ☐ from the plastic   ☐ from paint   ☐ from the mold

Comments: _____

Colors desired: _____

   ☐ from the plastic   ☐ from paint   ☐ from both

Criticality of color match: _____ %

☐ daylight   ☐ tungsten light   ☐ fluorescent light   ☐ all (no metamerism allowed)

Comments: _____

Critical appearance areas — *please attach sketch*

	None	Invisible	Minor	OK
gate blemishes	☐	☐	☐	☐
sink marks	☐	☐	☐	☐
weld lines	☐	☐	☐	☐

Comments: _____

Critical structural areas — *please attach sketch*

Comments: _____

_____

Handbook of Plastics Testing and Failure Analysis, Third Edition, by Vishu Shah
Copyright © 2007 by John Wiley & Sons, Inc.

Required physical characteristics — *please attach sketch*

	not too important	from plastic	from design	from both
Rigidity	☐	☐	☐	☐
Strength (load bearing)	☐	☐	☐	☐
Heat resistance	☐	☐	☐	☐
Creep resistance	☐	☐	☐	☐
Impact resistance	☐	☐	☐	☐
Chemical resistance	☐	☐	☐	☐
Electrical properties	☐	☐	☐	☐

Details:

    applied load/stress   ☐ static load   ☐ pressure   ☐ cyclic

    amount   normal _____   min. _____   max. _____

    duration   normal _____   min. _____   max. _____

        frequency (if cyclic)   normal _____   min. _____   max. _____

    operating temperature   normal _____   min. _____   max. _____

    operating lifetime   normal _____   min. _____   max. _____

    Comments: _____

    impact resistance

        room temp.   acceptable _____   min. _____

        low temp., _____ °C/°F   acceptable _____   min. _____

    Comments: _____

    dimensional tolerances

        deflection (under stress)   acceptable _____   max. _____

        expansion (thermal)   acceptable _____   max. _____

        shrinkage (mold)   acceptable _____   max. _____

        creep   acceptable _____   max. _____

    Comments: _____

    electrical properties

        dielectric constant   acceptable _____   min. _____

        dissipation factor   acceptable _____   max. _____

        volume resistivity   acceptable _____   min. _____

        dielectric strength   acceptable _____   min. _____

    Comments: _____

    chemical resistance   (List chemicals, frequency & duration of exposure, part stress/strain level, and type of resistance required.)

_____

_____

permanence	not too important	from plastic	from paint, etc.
color stability, indoor	☐	☐	☐
color stability, outdoor	☐	☐	☐
property retention, outdoor	☐	☐	☐

    Comments: _____

Required physical characteristics — *continued*

    miscellaneous

        Rockwell hardness   target _____   min. _____   max. _____

        Others: _____

Regulatory Approvals Required?

- ☐ Underwriters Laboratory, Inc.
  - ☐ U.L. 94   rating _____   thickness _____
  - ☐ R.T.I.   electrical _____ °C   mechanical _____ °C   with impact _____ °C
- ☐ National Sanitation Foundation   type _____
- ☐ Federal Specifications (Mil. Specs.)   type _____
- ☐ Canadian Standards Administration   type _____
- ☐ Food and Drug Association   type _____
- ☐ U. S. Pharmacopeia   type _____
- ☐ Automotive Specifications   type _____
- ☐ Other: type _____

Comments: _____
_____
_____

Process

- ☐ Extrusion
  - ☐ profile extrusion
  - ☐ sheet extrusion — monolayer
  - ☐ sheet extrusion — co-extruded
  - ☐ thermoforming
  - ☐ extrusion/blow molding

  Comments: _____
  _____
  _____

- ☐ Injection Molding

  Comments: _____
  _____
  _____

- ☐ Secondary Operations
  - ☐ decorating
    - ☐ painting
    - ☐ plating
    - ☐ hot stamping
    - ☐ laminating
  - ☐ assembly
    - ☐ gluing
    - ☐ sonic welding
    - ☐ vibrational welding
    - ☐ mechanical assembly

  Comments (What is attached to what, difference in types of plastic, etc.?)
  _____
  _____

Customer Part Testing Requirements

_____
_____
_____
_____

Final Comments

_____
_____
_____
_____
_____
_____

INDEX

Handbook of Plastics Testing and Failure Analysis, Third Edition, by Vishu Shah
Copyright © 2007 by John Wiley & Sons, Inc.

CUSTOMER NOTE: IF THIS BOOK IS ACCOMPANIED BY SOFTWARE, PLEASE READ THE FOLLOWING BEFORE OPENING THE PACKAGE.

This software contains files to help you utilize the models described in the accompanying book. By opening the package, you are agreeing to be bound by the following agreement:

This software product is protected by copyright and all rights are reserved by the author and John Wiley & Sons, Inc. You are licensed to use this software on a single computer. Copying the software to another medium or format for use on a single computer does not violate the U.S. Copyright Law. Copying the software for any other purpose is a violation of the U.S. Copyright Law.

This software product is sold as is without warranty of any kind, either express or implied, including but not limited to the implied warranty of merchantability and fitness for a particular purpose. Neither Wiley nor its dealers or distributors assumes any liability of any alleged or actual damages arising from the use of or the inability to use this software. (Some states do not allow the exclusion of implied warranties, so the exclusion may not apply to you.)

WILEY